Man and Animal Relationship in Early Farming Communities of Western India, with Special Reference to Inamgaon

SOUTH ASIAN ARCHAEOLOGY SERIES

EDITED BY ALOK K. KANUNGO **No. 6**

Man and Animal Relationship in Early Farming Communities of Western India, with Special Reference to Inamgaon

Seema J. Pawankar

BAR International Series 1639
2007

Published in 2016 by
BAR Publishing, Oxford

BAR International Series 1639

South Asian Archaeology Series No. 6
Series Editor: Alok K. Kanungo

Man and Animal Relationship in Early Farming Communities of Western India, with Special Reference to Inamgaon

ISBN 978 1 4073 0062 7

BAR Publishing is the trading name of British Archaeological Reports (Oxford) Ltd. British Archaeological Reports was first incorporated in 1974 to publish the BAR Series, International and British. In 1992 Hadrian Books Ltd became part of the BAR group. This volume was originally published by John and Erica Hedges Ltd. in conjunction with British Archaeological Reports (Oxford) Ltd / Hadrian Books Ltd, the Series principal publisher, in 2007. This present volume is published by BAR Publishing, 2016.

Printed in England

BAR
PUBLISHING

BAR titles are available from:

BAR Publishing
122 Banbury Rd, Oxford, OX2 7BP, UK
EMAIL info@barpublishing.com
PHONE +44 (0)1865 310431
FAX +44 (0)1865 316916
www.barpublishing.com

Contents

List of Figures

Chapter II

Chapter III

Chapter IV

Chapter V

List of Tables

Chapter IV

Chapter V

Appendix A

List of Plates

Acknowledgement

During the preparation of this thesis many persons have helped me in various ways. I am glad to record my indebtedness to them.

I am greatly indebted to Dr. P.K. Thomas, my supervisor for his guidance, encouragement and untiring help. He has really taught me A to Z of Archaeozoology for which I will always be very grateful to him.

I wish to thank Prof. M. K. Dhavalikar, former Director of Deccan College for allowing me to study the faunal material from Inamgaon. I owe my sincere thanks to Prof. V. N. Misra and the Deccan College authorities for initially giving me a scholarship and necessary institutional help. Prof. Misra was always kind and considerate.

I am grateful to the Indian Council of Historical Research for awarding me a research fellowship for the completion of this thesis.

I am thankful to S. B. Ota, who showed me the right way to start my career in Archaeology. Dr. Sushma Deo and Dr. P. P. Joglekar for the computer help and providing useful suggestions. I also thank Ms. Kathleen Ryan of the University of Pennsylvania, USA, for valuable suggestions and comments.

I wish to thank Arati, Chandraratne and S. M. Atre for their support and various kinds of help, especially Y. Matsushima for helping me drawing and photography. I am grateful to all my dear friends especially, Bilal and Sukanya Sharma who were always with me in my hour of greatest need. I also thank to Dattoba for serving me hot cups of tea during my working hours.

I am thankful to the other staff members of the Department of Archaeology, Library and office of the Deccan College for the co-operation they offered me.

My special thanks to Dr. M. M. Hoque for his constant support and encouragement. I am dedicating this research work to my parents Mr. Jaydeo J. Pawankar and Mrs. Maya J. Pawankar with deepest sense of gratitude who are responsible to brought me up this stage.

Seema J. Pawankar

Introduction

1.1 Significance of Archaeozoological studies

Animals have played an important role in the subsistence of human societies right from the emergence of man on our planet some five million years ago to almost the present times. Until about 10,000 years ago all human societies everywhere in the world lived by hunting and gathering and then slowly the societies learned to domesticate plants and animals. Some groups in the Near East, Mesoamerica and probably in South East Asia learned to cultivate plants and animals around 10,000 years ago. These developments were of far reaching consequences in human history. Groups which learned agriculture and animal husbandry were able to settle down permanently in villages. The villages slowly developed into towns and cities. And around 5000 years ago the first civilizations developed in the valleys of Nile, Tigris-Euphrates and the Indus. The knowledge of agriculture spread rapidly among the hunting gathering communities through out the world. But even today there are societies which continue to live wholly or partially by hunting and gathering.

Since the earliest times man has depended on animals, including fish, birds, insects, etc. for his food. For hunter-gatherer societies animals constituted a major source of food. Man began his hunting career with preying on small land mammals and by the beginning of Acheulian stage was able to hunt or trap large mammals. In the Upper Palaeolithic period and particularly during the Mesolithic period trapping, snaring and catching of birds and fish assumed considerable importance. From the beginning of the Neolithic period the role of hunting began to decline, but man was now able to kill domestic animals for food. From this stage onwards man probably also began using animal milk as source of food. Our knowledge of the economic practices and food habits of the prehistoric populations is largely based on the surviving animal remains.

After man learnt to domesticate the animals, their role further increased in the economy. Now besides supplying food they also became source of energy. Ox, horse, buffalo, camel, etc. have been used in a variety of operations like ploughing, running irrigation devices and for thrashing harvested crops. These and other animals were used as pack animals as well as for riding. Several of them were used for drawing carts and chariots as also for running oil presses.

The relationship between man and animals transcended the purely utilitarian level even during the hunting stage. Man felt close emotional affinity to the animals which formed the source of his food and shared his environment. That explains the fact that a number of goat heads were buried around the skeleton of a Neanderthal man at Teshik Tash in Uzbekistan. At Buzahome a Neolithic site in Kashmir, Dog/wolf heads were buried with a human dead body. The Chalcolithic site of Ramapuram in Andhra Pradesh has yielded partial and complete burials of goat in Human graves. The Megalithic stone circles in the Vidarbha region of Maharashtra, belonging to the Iron Age period, have brought to light partial burials of horses in the form of skulls and lower extremities of limb bones. Pets or utilitarian animal to whom an individual was closely attached were buried with the person in his grave.

Therefore, the study of animal remains or the science of Archaeozoology is an integral part of Archaeology, dealing with a major part of the cultural history of the ancient human populations. However, a few decades back there was very little awareness among the archaeologists about the full scientific potential of the faunal assemblages from archaeological sites. In this process valuable data based on fauna were permanently lost. It is a welcome sign that in recent years archaeologists everywhere in the world are giving much emphasis on the faunal studies.

Study of these dry, dirty and dusty bones can tell us more about man-animal relationship and several aspects of the past cultures. Meticulously collected faunal assemblage from an archaeological site can throw light on animal domestication, food economy of ancient population, past environment, season of occupation of site, religious and social aspects, cultural contacts, activity area within site, secondary uses of animals, bone tool technology, techniques of butchery and hunting, palaeopathology, cooking practices, the distribution and migration of animals in the past and taphonomical aspects of the bones recovered.

1.2 Domestication of Animals

Domestication of animals is a highly developed man-animal relationship which has forced man to settle down and have a sedentary way of life. If one looks at the diversity of wild animal species exploited during the hunting-gathering and later periods, the number of domesticated animals is very few. Not all the wild animals were domesticated, only animals with certain characteristics like docile nature and sociability were found domesticated. This particularly shows the knowledge of the early human populations about the animal behaviour.

The process of domestication may have started much earlier than the 'Neolithic Revolution'. Child (1928) himself has stated that it is not a catastrophic change. The prelude to Neolithic Revolution must have been much longer than what is initially thought to be. Now more and more archaeological evidences are surfacing to support this. Domestication may have started in different times at different places, where the wild animals were available. The idea travels faster than the goods.

A domestic animal is defined as one whose breeding is largely controlled by humans. This is a very general definition of a domestic animal. Bokonyi (1989) defines the essence of domestication "as the capture and taming by man of animal species with particular behavioural characteristics, their removal from their natural living area and breeding community and their maintenance under controlled breeding conditions for mutual benefits". Domestication is also defined as a sort of symbiosis. Symbiosis is a biological process – the relationship between two living organisms for their mutual benefit. Ducos (1989) argues that man and animal are not partners. He believes that domestication is not a natural state; it exists, because human wished it and not animal. Man domesticated animals for food purposes and in later period he realized their importance for agriculture, for traction and for carrying people and goods. Man also started exploiting animals for their secondary products like milk, wool, dung, etc.

According to Ducos (1989), domestication can be said to exist when living animals are integrated as objects into the socio-economic organization of the human group, in the sense that, while living, those animals are objects for ownership, inheritance, exchange, trader, etc.

Bokonyi (1989) has given his opinion on Ducos's definition of domestication. He says that if domestication means integration of living animals as objects into socio-economic groups, one necessarily has to change the living conditions of the animals by isolating them, corralling them, etc. Inheritance, exchange, trade etc. are consequences of domestication not components.

Clutton-Brock's (1989) definition of domestication is very close to Bokonyi's definition. According to her a domesticated animal is one that has been bred in captivity for purposes of economic profit to a human community that maintains complete mastery over its breeding, organization of territory and food supply.

Zeuner (1963) thinks that social interaction or keeping pet animals by women and children may have led to the process of domestication. He also calls this as 'mothering instinct'. Actually this idea was put forth a century ago by Francis Galton (1883). This may be true for the domestication of dog, but to a lesser extent in animals like pigs, cattle, etc.

Binford (1968) and Flannery (1969) suggest that the stress situation caused by demographic pressure may have been a stimulus for the process of domestication. Due to demographic pressure and decline in resources, people were forced to move out of the 'optimal zones' into the 'marginal zones', where situation similar to the former was created by acquiring plants and animals from the 'optimal zones'. Probably in due course of time this led to the domestication of plants and animals.

During the period between 14000 and 8000 years BP., the culture, the attitudes and the behavior of people changed. Man became a farmer more than a hunter (Reed 1984). An increase of human population was the main factor, which stimulated people to cultivate crops (Cohen 1977). Due to continuous animal hunting and rise of population, the number of wild animals decreased, which could have compelled man to domesticate animals like sheep and goat (Khan 1975).

Soon after the Pleistocene period, we find evidence of dogs in southwestern Asia. According to Reed (1984), people while hunting might have taken the living young of killed animals back to camp. Such primitive pet wolf pups might have been the beginning of the domestication of wolves into dogs. According to some scholars, wolves also realized the advantages of hunting together with man, and joined him voluntarily to bring the domestic dog into being. Dog is the oldest domestic animal to have been used as food. Even today the dog is eaten in wide areas of the world in Asia and among the aborigines of South America (Herre 1969).

The domestication of cattle began in the early phase of agricultural development. In the early phase of agriculture, people started settling down by the river side on flood plains. According to Zeuner's (1963) 'crop robbers' hypothesis, during the early Neolithic period the domestication of animals like cattle, buffalo, elephant, rabbit and goose was a necessary step to control and prevent animal from stealing man's crops.

Issac's (1962) 'Religious' hypothesis of animal domestication; suggests that the motive for animal domestication was religious. He links it between the shape of animal horns and the shape of ancient deity, the moon. Probably the theory was connected with ox sacrifice in the agricultural societies as a sort of religious practice. It is believed that cattle domestication started much later in the sequence of domesticated animals, and was probably during the agricultural phase. Animals such as dog, sheep and goat were domesticated much earlier than cattle in the pre-agricultural phase. Thus, cattle sacrifice may not have been a stimulus for the origin of domestic animals or the process of domestication.

Domestication is a complex interaction between man and animal. Its consequences are influenced by society, economy, ideology, environment, way of life, etc. (Bokonyi 1989). The result of domestication is the domesticated animal. Therefore, domestication is both a cultural and biological process.

The beginning of cultivation and domestication of plants and animals is clearer in south western Asia than elsewhere (Reed 1984). The transition from food gathering to food producing took place in the southern alluvial plain of Mesopotamia, the low hills at the foot of the Zagros and Taurus mountains, the valleys of those mountains, the high plateaux of Turkey and the coastal area and the low hills of the eastern Mediterranean (Clason 1979).

Approximately 10,000 years ago in the Near East man domesticated sheep and goat. The domestication of goat completely changed the mode of life of man. Goat is a browser and plays an important role in the maintenance of pastures and arable fields in deforested areas (Khan 1975). When agriculture was properly established the importance of goat in the fertile regions was lost. This species was pushed to the semi-arid region to feed on thorny shrub vegetation or to the mountain ranges.

Wool producing races of sheep were developed in the temperate zones and they also provided meat and milk. Breeds are adapted to tropical zones, where man is not interested in wool; there they bred for meat and milk. Sheep were economically of greater importance for the hot semi-arid region than goat (Khan 1975).

Morphologically domestic animals differ from their wild ancestors in a number of ways, e.g. general body proportions, shapes of the horn, colour of the body, hair and fleece. A few of the characteristics are reflected in the skeleton. Some of the animals have become smaller in size like cattle, sheep, goat, pig, etc., while some have gained in size like horse, chicken, etc. It is interesting to note that some remained the same, like camel. Recent experiments show that the morphological changes occur after about thirty generations. In smaller animals like dog, sheep, goat, etc. each generation is 2-3years, hence the changes can be observed after 60-90 years whereas, in larger animals like cattle, buffalo, horse, etc. the change is visible in period of 120-150 years.

In India the earliest evidence for domesticated animals is reported from the Mesolithic sites, Bagor in the district of Bhilwara in Rajasthan and Adamgarh, in the district of Hoshangabad in the Central India. The date of the Mesolithic cultural phase is 6245 ± 200 B.P. at Bagor and at Adamgarh dates are 2765 ± 105 B.P., TF 116 and 7240 ± 125 B.P., TF 120 (Agrawal & Kusumagar 1968). The dating of the Mesolithic culture of Adamgarh is not clear. Two enigmatic C14 dates from the site have been determined and are based on uncharred bone and on shell. In both cases the probability of contamination of the samples is considered high (Thomas 1975), and dating is far from secure. Cattle, buffalo, sheep, goat, pig and ass are reported from Adamgarh. At Bagor sheep and goats were predominant followed by cattle. The Mesolithic culture represents a transition from hunting and gathering to food producing way of life. This is considered to be a gradual process. Evidence is lacking for local domestication of animals at both the sites. Hence, the process of domestication may have been brought to the settlement in an already domesticated state.

1.3. Chalcolithic Cultures of western India

The Chalcolithic culture in western India flourished in the Tapi, the Godavari, and the Krishna river valleys. The best black cotton soil occurs in this area which may have attracted the early farmers to settle down in this region.

The concentration of the Chalcolithic sites is found mainly in western Maharashtra. The beginning of settled life in Maharashtra can chronologically be placed into four cultural phases: the Savalda (2200 B.C.-1800 B.C.), Late Harappan (1800 B.C. - 1600B.C.), Malwa (1600 B.C. – 1400 B.C.) and Jorwe (1400 B.C. – 700 B.C.) based on terracotta wares from the Chalcolithic sites.

The number of Chalcolithic sites include Prakashe, Savalda, Bahurupa, Bahal, Kaothe, Tekwada in Tapi valley; Nasik, Jorwe, Daimabad, Nevasa and Apegaon in Godavari valley and Chandoli, Walki, Inamgaon and Songaon in Bhima valley. Bhima basin is in fact, a tributary basin of the Krishna but its length and drainage area in the Deccan marks it out as a distinct relief region (Deshpande 1948).

In Western Maharashtra, out of the total fourteen Chalcolithic sites excavated only Inamgaon, Kaothe and Walki excavation reports have detailed study of faunal remains. The reports of Nevasa, Apegaon, Daimabad and Nasik have given only the list of animal species represented, the rest of the4 excavation reports do not mention anything about faunal remains, except at times a small description of the animal motifs on terracotta wares.

1.4. The objective

The objective of this study is to find out the man-animal relationship in the ancient past, especially among the early farmers of western India. The main focus of this work is the study of faunal remains retrieved from the Chalcolithic site, Inamgaon, district Pune in Maharashtra, at present housed in the Archaeozoology laboratory of the Deccan College, Pune. The entire material has been analysed both qualitatively and quantitatively by using some improved archaeozoological techniques. The results of this study will be compared with the other Chalcolithic sites of Maharashtra from where faunal remains have been already studied such as Kaothe, Daimabad, Walki, Nasik, Nevasa, Apegaon, Tuljapur garhi and Adam. Probably this could throw some more light on the mode of exploitation of both domestic and wild animals in relation to the functional aspects of the Chalcolithic culture.

The Sites, Their Ecological Setting and Cultural Sequence

2.1. Chalcolithic sites

Over 200 Chalcolithic sites have been reported from Maharashtra, a majority of which belonged to the Early Jorwe period followed by the Malwa and very few of the Late Jorwe period. Of the total fourteen excavated Chalcolithic sites, in western Maharashtra only Inamgaon, Kaothe, Walki, Tuljapur Garhi and Adam (Fig. 2.1) have produced reports of detailed faunal studies. Nevasa, Apegaon, Daimabad and Nasik have given only brief reports do not mention anything about the faunal remains except for some small description of stray animal motifs depicted on the pottery. Published data from the above sites are also used for comparative purposes.

Fig.2.1. Map showing Chalcolithic sites referred in the Text.

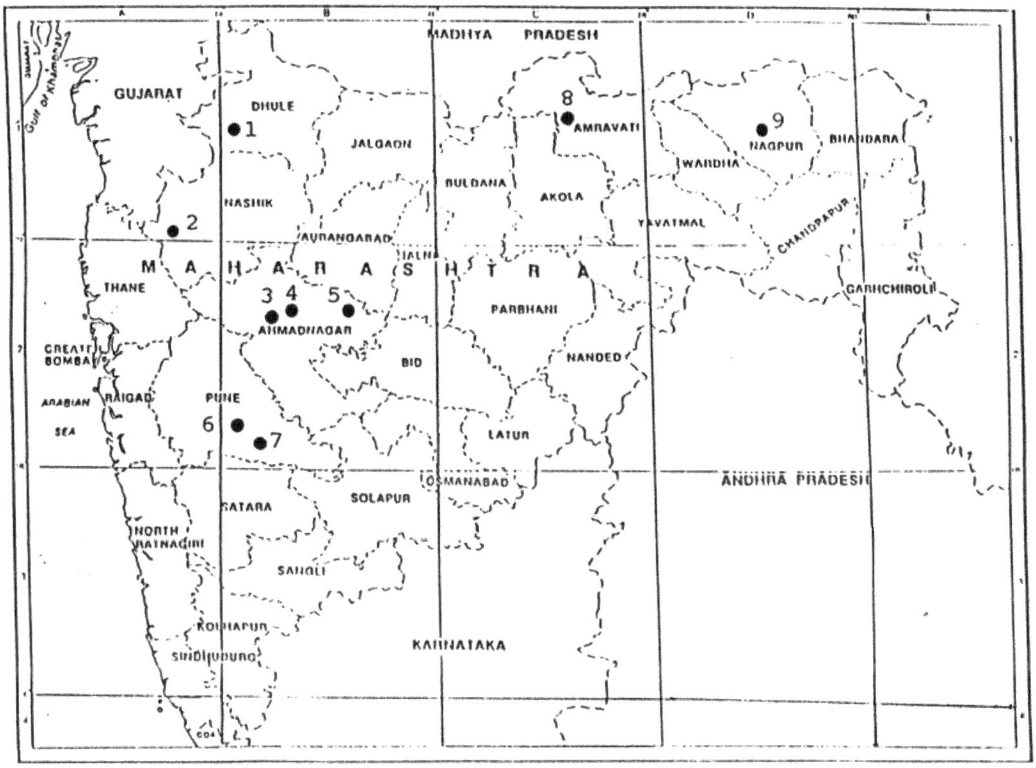

1. Kaothe, 2. Nasik, 3. Daimabad, 4. Nevasa, 5. Apegaon, 6. Walki, 7. Inamgaon, 8. Tuljapur Garhi and 9. Adam.

Inamgaon

The Chalcolithic site at Inamgaon (N lat. 18°36'; E long. 74°32') in Pune district is situated on the right bank of the river Ghod, a tributary of the Bhima. The total inhabited area roughly measures 550x430 m. The maximum length of the mound north-south is 215 m and the breadth east-west is 175 m. The total cultural deposit is 4.18 m thick. Three cultural phases were noticed and they are as follows:

Malwa Phase 1600 B.C. – 1400 B.C.
Early Jorwe Phase 1400 B.C. – 1000 B.C.
Late Jorwe Phase 1000 B.C. – 700 B.C.

This is one of the most extensive Chalcolithic sites in Maharashtra. This site was excavated by Late Prof. H. D. Sankalia, Prof. M. K. Dhavalikar and Dr. Z. D. Ansari from the Department of Archaeology, Deccan College Post Graduate and Research Institute, Pune for about twelve field seasons between 1968-1984 (Dhavalikar, *et.al.* 1988a). The Chalcolithic settlement at Inamgaon is situated on an alluvial fill terrace about 14 m above the bed level of the river Ghod. Geologically this area is covered by

Cretaceous-Eocene Deccan Trap basalts. Except the older alluvium of the Holocene age there are no later formations. This area falls in typical semi-arid zone with the mean annual rainfall 500 mm. The rainfall is received from the southwest monsoons during the four months from mid June to mid October. The soils around this region are black in colour. Black colour of the soil is due to the presence of clay-humus colloidal complex formed by bio-inorganic processes.

The vegetation is of scrub type. The xerophytic vegetation includes typical species like Bor (*Zizyphus jujube*), wild Babhul (*Acacia Arabica*), Khair (*Acacia catechu*), Chinch (*Tamarindus indica*), etc.

The food economy of the early farming communities at Inamgaon was based on subsistence agriculture, stock-raising, hunting and fishing. The plant remains from the site reveal that the inhabitants had knowledge of double cropping (Kajale 1988). Large rectangular structures are identified as granaries in the Malwa and the Early Jorwe phases. In the Malwa and Early Jorwe, the structures are rectangular in shape and are circular in the Late Jorwe phase (Dhavalikar 1988).

A major part of the faunal remains have been studied by Thomas (1988). The faunal assemblage comprises mammals, reptiles, birds, fishes and mollusks. Over 35 animal species are identified at Inamgaon. The domestic animals identified are cattle, buffalo, sheep, goat, dog, pig, horse and ass and the wild animal species are sambar, chital, blackbuck, four horned antelope, hare and mongoose, etc.

Walki

The ancient site Walki (N lat.18° 35'; E long. 74° 18') is located about 32 km west of Inamgaon, in the Pune district. It is situated on the right bank of the river Bhima near the confluence of river Mula. This was excavated by Prof. M. K. Dhavalikar for two field seasons between 1985 and 1987. The site was probably occupied around 1200 B.C. (Dhavalikar *et. al.* 1990b).

The site is located in the semi-arid zone of north Deccan where the annual rainfall is around 500 mm and most of it occurs during the monsoon months from June to October. The remaining eight months are characterised by extremely dry weather.

In the earlier part of this century, there was thick scrub forest, and small games were plentifully available and also small herds of deer and wild boar. But the increasing human population has been destroying the forest cover with the result that the landscape today is totally bare and almost all wild animals became totally extinct with the depletion of the forest cover.

The important feature of this Jorwe culture site is its circular huts with flimsy floors and had no solid walls which indicate their temporary nature. The settlement was occupied only during the winter season. The bone tool made of the shoulder bone of cattle had two perforations and was triangle in shape and the 35 cm long antler fragment which was hollow. These two tools are very significant and unique findings. Probably these tools were associated with agricultural operations.

More than 50,000 fragments of bones have been collected from this excavation. The 26 species identified include mammals, birds, reptiles and fishes. Cattle was the most predominant animal in the food economy and the second most important animal was blackbuck which was excessively hunted by the inhabitants of Walki (Joglekar 1992; Joglekar and Thomas 1995).

Kaothe

The Chalcolithic site, Kaothe (N lat.21° 00'; E long. 74° 18') is situated on the left bank of the river Kan, a tributary of the river Panjhara which is a major tributary of the river Tapi. This site was excavated under the leadership of Prof. M.K. Dhavalikar. Kaothe is located at a distance of half a kilometer west of the present village at an elevation of 415 m. The river Kan has developed a shallow meander and the Chalcolithic settlement is confined to the inner periphery of the meander on a 7 to 8 m high alluvial flat surface. The settlement is roughly oblong in shape and spread over an extensive area of about 30 hector.

Kaothe is a single culture site, cultural debris is very thin therefore, trenches were dug in three localities viz. KTE I located in the eastern side of the site, KTE II was in the central part, whereas KTE III was located to the northwest of KTE II. The single radiocarbon date available for the Kaothe is 1920 ±90 B.C. (uncalibradted). This provides a time bracket of 2000 – 1800 B.C. for Kaothe cultural phase (Dhavalikar, *et.al.* 1990 a).

The region around the ancient site is covered by Cretaceous – Eocene Deccan trap basalts and the common rock type is mainly compact and vesicular varieties of basalt.

This region can broadly be classified as hot semi-arid monsoonic. It receives an average rainfall of about 1200 mm. This area is less affected by droughts and scarcity conditions as compared to similar semi-arid region of Deccan upland drained by the rivers Godavari, Bhima and Krishna. Soils are pedocal type black soils or vertisols and are medium textured with clay content varying from 35 to 50 %. The black colour of the soil is mainly due to the presence of clay-humus colloidal complex formed by bio-inorganic process.

The natural vegetation around Kaothe consists of thorn and scrub type forest represented by the xerophytic plant species, includes Bor (*Zyzyphus jujuba*), Babhul (*Acacia arabica* wild), Chinch (*Tamrindus indica*), etc., and the wild animals of this region are fox, jackal, wolf and hare.

Thomas and Joglekar (1990) have studied the faunal remains and identified 38 species of animals which include mammals, reptiles, birds, fishes, crustaceans and molluscs. Domestic animals reported from here include cattle, buffalo, sheep, goat, pig and dog, while wild animals are nilgai, blackbuck, four-horned antelope, chital, sambar, hog deer, barking deer, wild pig, fox, civet, bear, langur, porcupine, mongoose and rat.

As compared to other some of the archaeological sites, the density of faunal assemblage at Kaothe is more considering the thickness of the deposit as well as the area excavated. Probably it reveals the fact that the inhabitants of Kaothe were more depended on animal food rather than agricultural products in their subsistence (Thomas and Joglekar 1990).

Pit dwelling is an important feature of this culture. The evidence from Palaeobotany, Zooarchaeology and Ethnographic parallels indicates a seasonal occupation at the site. These Savalda people may have occupied site only during monsoon. Their economy was based more on stock-raising, subsistence agriculture and hunting-fishing.

Nasik

Nasik (N lat. 20º 00'; E long. 73º 51') is an ancient town situated on both bank of the river Godavari. This site was excavated by H.D. Sankalia and S.B. Deo in 1950-51. The mound is nearly 30 m high, 243 m in length and 122 m in breadth. The habitation deposit reveals four cultural periods. Radiocarbon dates are not available hence; time period has been assigned with the help of pottery wares, coins and other artifacts. Chronology of the cultural periods is as follows:

Period I Chalcolithic or Early Bronze Age 500 B.C. -1000 B.C. to 1500 B.C. or earlier
Period II
A. Early Historic 200 B.C. - 300 B.C. to 400 B.C.
B. Later phase of Early Historic 50 A. D. to 200 A.D.
Period III Roman contact
Period IV
A. Early Muslim
B. Mughal 1400 A.D. to 1875 A.D.
C. Maratha

George (1955) has studied the animal remains from Nasik and identified domestic animals such as cattle, buffalo, sheep, goat, pig and dog. Wild animals include nilgai, four-horned antelope, chital, sambar and langur. All these animals are common in the Chalcolithic culture of Western India.

Daimabad

At present Daimabad (N lat. 19º 31'; E long.74º 42') is a deserted village situated on the left bank of the river Pravara, a tributary of the Godavari and was excavated by S.A. Sali (1986). The site measured about 1000 m in length and 500 m in breadth. The thickness of the occupational deposit is approximately 10 m. Climatically, This site is situated in semi-arid zone and the average rainfall is 600 mm per annum. The vegetation around Daimabad is of dry deciduous type. The wild animals like jackal, fox, wild hog, blackbuck, chinkara and hare are seen occasionally.

The occupational deposit consists of five cultural periods and is as follows:
Period I is Savalda culture, represented by 20-30 cm thick deposit of blackish brown colour, resting upon the black soil; dated 2200 B.C. – 2000 B.C.
Period II is the Late Harappan culture, dated 2000 B.C. – 1800 B.C.
Period III Daimabad culture, named after buff and cream Daimabad ware and dated 1800 B.C. – 1600 B.C.
Period IV Malwa culture, 1600 B.C. – 1400 B.C.
Period V Jorwe culture, 1400 B.C. – 1000B.C.

This site was occupied by the inhabitants from the last quarter of the third millennium B.C. to the end of the second millennium B.C. Subsistence economy was depended on agriculture and stock raising. By looking at the botanical evidences, people of Daimabad had the knowledge of double cropping; winter (*Kharif*) and summer (*Rabi*) crops. Chariot, elephant, rhinoceros and buffalo made of Bronze and terra-cotta animal figurines unearthed at the site point out the aesthetic and artistic qualities of the population at Daimabad.

The animal remains from Daimabad depict the presence of both wild and domestic animals (Badam 1986), which include cattle, buffalo, sheep, goat, pig, horse, ass, dog, cat, elephant as domestic and nilgai, blackbuck, four-horned antelope, chital, sambar and barasingha as wild. In addition to this avian and aquatic resources were also exploited.

Nevasa

The town Nevasa (N lat. 19° 34'; E long. 74° 54') is located on both the banks of the river Pravara, a major tributary of Godavari. Nevasa Budruk is situated on the northern bank of the river while Nevasa Khurd is on the southern bank of the river.

About half a mile to the west of the Nevasa Khurd, there are three mounds in a series, cut up by several gullies. The first mound on the southern bank is traditionally called Ladmod. It measures approximately 27 m north-south and 116 m east-west. Its western and southern slopes are very much distorted by earth-grabbers, whereas the northern edge is virtually a cliff overhanging the river. This mound is higher than the other three and rises about 21 m over the level of the Pravara.

The material from excavation and exploration in and around Nevasa has been grouped into six cultural periods and are as follows:
Period I Lower Palaeolithic (Middle Pleistocene) c. 1, 50, 000 years.
Period II Middle Palaeolithic (Upper Pleistocene) c. 25, 000 years.
Period III Chalcolithic c. 1, 500B.C. – 1000 B.C.
Period IV Early Historic c. 150 B.C. – 50 B.C.
Period V Late Early Historic (Indo-Roman) c. 50 B.C. – 200 A.D.
Period VI Medieval (Muslim-Maratha) c. 1, 400 A.D. – 1, 700 A.D.

Geologically, The river Pravara and the Godavari belong to a mature river system. At Nevasa, Godavari is 11 km away towards north, flowing almost parallel to the Pravara and meeting at Pravara sangam about 11 km to the west of Nevasa. The extensive field survey shows that these rivers flowed in the Middle Pleistocene times in a very wide valley and deposited large coarser gravels and fine silt.

A weathered deposit had been found in between the Chalcolithic period and the Early Historic period. This weathered deposit consisted of blackish soil of a thickness varying between 26-45 cm. This layer indicates the long time gap between the occupations of period III and period IV, due to abandonment of the site after about 1000 B.C.

The faunal assemblage from the site has been studied by Eapen (1960). A major part of the faunal remains consist bones of domestic animals such as cattle and sheep/goat followed by the bones of barasingha. However, in the Indo-Roman period cattle bones were predominant over others followed by sheep/goat. It is interesting to note that pig bones were absent in the collection.

Clason (1979) had restudied the faunal material from Nevasa. From the Chalcolithic period she had identified domestic animals such as cattle, buffalo, sheep, goat, horse, donkey, dog and cat. Wild animals include nilgai, four-horned antelope, blackbuck, sambar and chital. Remains of domestic fowl were found in the Chalcolithic as well as in Muslim-Maratha period. Maxilla of a young elephant was collected from the Indo-Roman period.

Apegaon

Apegaon (N lat. 19° 36' 30"; E long. 75° 29' 30"), situated on the left bank of river the Godavari in district Aurangabad. This site was excavated under the direction of S.B. Deo in 1976. The habitation mound locally called as Ramatirth measures about 100 m in length and 70 m in breadth. The total deposit was about two meters. Three cultural phases have been found and are assigned on the basis of ceramics (Deo, Dhavalikar and Ansari 1979).

Period I, Ramatirth culture 16th cent. B. C. Pottery ware from this cultural phase was called Ramatirth ware after the name of the mound. Ramatirth ware is medium thick in fabric with carded surface and with dull brown slip and geometric paintings in dull black. The average thickness of this cultural deposit was of 0.5 m. The excavators of Daimabad hold that the

Ramatirth ware is identical with the Savalda ware, whose stratigraphic position is much earlier. Houses of this cultural period were circular in shape with lime floors. There was no evidence of flora and fauna from this cultural phase.

Period II, Malwa-Jorwe culture 15th cent. B. C.
Period III, Late Jorwe culture 15th to 10th cent. B. C.

The subsistence economy of the inhabitants of Apegaon was based on agriculture and stock-raising. A variety of grains are found like pea, green gram, lentil, grass pea, wheat, barley, horse gram, black gram and Indian jujube which show that the people at Apegaon carried out farming activities both in the winter and summer.

Faunal remains from this site include the bones of domestic animals like cattle, buffalo, sheep and goat. There was no evidence of horse, dog and pig bones. Cattle were the predominant animal in the food economy of the inhabitants. Apart from domestic animals, wild animals such as chital, sambar?, blackbuck, nilgai, mongoose, etc. have also been identified (Badam 1979).

Tuljapur Garhi

Tuljapur Garhi (N lat.21° 10'; E long. 77° 35'), a Chalcolithic site, in district Amaravati is situated on the right bank of the river Purna, a tributary of the river Tapi. The concentration of the Chalcolithic sites is generally seen in western Maharashtra. This site is one of the first Chalcolithic settlements excavated in the Vidarbha region of Maharashtra. The site was excavated under the supervision of Shri. B.P. Bopardikar, Nagpur Excavation Branch of the Archaeological Survey of India in 1984-85. The length of the mound measures 210 m north-south and breadth 90 m east-west. The thickness of the cultural deposit is 1.5 m. The entire cultural deposit consists of Jorwe and Malwa cultural debris. There is one radio carbon date for the Malwa phase and three for the Jorwe phase. They are as follows:

Malwa phase 2870 ± 100 years B.P.
Jorwe phase 2410 ± 100 years B.P.
3310 ± 90 years B.P. and
340 ± 90 years B.P.

The food economy of the early farmers of the site was based on agriculture, stock raising and hunting. Fifteen plant species are identified. The botanical remains reveal that the inhabitants had cultivated both the winter and summer crops (Kajale 1988).

Thomas (1992) has studied the faunal remains of the site. Altogether fourteen animal species are identified including mammals, reptiles, birds and molluscs. Domestic animals like cattle, goat, dog and ass and wild animals such as four-horned antelope, chital, hog deer, wild pig and mongoose are identified. Cattle were predominant in both the cultural phases.

Adam

The site Adam (N lat. 21° 00'; E long. 79° 27'), is situated on the left bank of the Waghor of the Wainganga drainage system. Adam is located about 60 km southeast of Nagpur. The site measures approximately 8000 m east-west and 500 m. north-south and the height of the mound is 8 m from the surrounding plains. The site was excavated by Shri. Amarendra Nath, of the Nagpur Excavation Branch, Archaeological Survey of India (1988-89). Excavation reveals five fold cultural sequences and is as follows:
Period I, 50 cm thick and dated between 3rd and 2nd millennium B.C.
Period II, Vidarbha Chalcolithic dated roughly to the 1st quarter of the 2nd millennium B.C.
Period III has been assigned between circa 1000 B.C. and 500 B.C. This period is characterized by the introduction of Iron.
Period IV, Pre-Mauryan and Mauryan dated 500 B.C. and 150 B.C.
Period V, Attributed to the Bhadras, Mitras, Satavahans and Maratha rulers on the basis of numismatic and epigraphical evidences.

The faunal remains of the site have been studied by Thomas (in press). The fauna comprises mammals, reptiles, birds, fish and molluscs. The domestic animals represented are cattle, buffalo, sheep, goat, dog and pig. Among wild animals gaur, wild buffalo, wolf, jackal, wild pig, chital, sambar, barasingha, nilgai, blackbuck, etc. are identified.

2.2. General Ecological Setting of the Region

The development of a culture is governed by the ecological factors of the region. So, any cultural study will be incomplete without the mention of the ecology of the region.

2.2a. Geography

The area under study consists of a part of western India, mainly the western region of Maharashtra and Gujarat. Western Maharashtra region occupies the Sahyadri hill ranges, Deccan Plateau and Tapi Basin.

In Western India, the extensive basaltic sheet, which has a thickness of more than 2,000 m resulted from the consolidation of the lava during the Late Cretaceous and Early Eocene. The westward margin of the plateau in the relicts of the original plateau is preserved and the ancient surfaces of erosion are known as the Sahyadri. Its peak and plateaus are often covered with lateritic red soil and localized patches of the dense forest. The Sahyadri is a narrow crest zone and its width is 15-25 km.

The Tapi basin in the north is an asymmetrical valley with a gentle southern slope and an abrupt rise to the north in the Satpura hills. The General height of the basin is 200-300 m above sea level in contrast to 400-600 m of the plateau drained by the Godavari and the Krishna rivers (Singh 1971).

The Deccan trap occupies the largest area in western India. It extends almost from the southern flanks of the Narmada valley to the banks of the Krishna. It also occupies central and southern Kathiawar.

Gujarat region consisting of major and minor peninsulas, gulfs, islands, rann creeks, marshes, hills, plateaux and the coastal strip is covered by alluvium and miliolites (Deshpande 1948, 1971).

2.2b. Geology

Geologically in western Maharashtra, there is only one formation i.e. Deccan traps of the Cretaceous-Eocene age. The Deccan traps are represented by varieties of basalts such as compact, mygdoloidal, porphysitic, brecciated and tachylitic types. The exposed rocks do not show any folding and faulting which indicate the absence of a major tectonic activity in this area since early tertiary times. Veins of Chalcedony are common in basalts exposed in this area (Rajaguru 1988).

In Kutch is found the highest Gondwana groups, resting upon the marine Jurassic rocks and capped by the lower Cretaceous beds. Lithological marks in the form of plateaus, Umia sandstones and shale, Wadhwan sandstones, Narha Dark shale and flags overlain by sandstones, Dhrangadhra freestone and Ahmednagar sandstone of Idar indicate their upper Gondwana sequence (Deshpande 1971).

2.2c. Soils

The best black soils occur in the region of Tapi, lower Godavari and Krishna valley. May be this is the reason that early farmers had settled in these valleys. Black-cotton soils have resulted from the weathering of the Trap and its transport over long distance in the valleys. Black colour of the soil is mainly due to the presence of clay humus colloidal complex which is formed by bio-inorganic processes. Montmorillonite is the main clay mineral. These soils are often referred to as self ploughing soils. Mature, deep dark brown soils occur in alluvial flats and immature with no profile development. Shallow, light brown soils are found on hill slopes and rocky pediments. In narrow Sahyadrian zone occur lateritic soils. The soils are poor in organic matter with pH varying between 5.5- 6.5. The lateritic soils are often covered with forest (Singh 1971, Rajaguru 1988 and Pappu 1988).

Six types of soils occur in Gujarat; deep black soils or regur, medium black, coastal alluvium, alluvium-sandy loam, sandy alluvium and desert soil. There are two basic soil regions; plans and peninsulas. In south Gujarat the plain exhibits deep black soil and sandy loam soil in central Gujarat. The alluvial sandy soils occur in the western part of Mahsana Banaskantha. The northern part of Jamnagar, Rajkot and Surendranagar districts are covered with alluvium sandy soils derived from sandstones, cherts and conglomerates. Tracts of black cotton soil running from west to east on the basaltic surface occur in central Kathiawar. The Saurashtra coast, north Cambay and western coastal strip of Gujarat are clad with coastal alluvium and miliolites (Deshpande 1971).

2.2d. Drainage

Western Maharashtra region is drained by Tapi, Godavari and Krishna rivers. Except the Tapi basin which is drained to the Arabian Sea, the rest of the region is drained to the Bay of Bengal. More than half of the area is covered by the Godavari basin and one fifth of the area by the Tapi and remaining by the Krishna. The plateau of Maharashtra is drained by the Godavari in the north and the Krishna and its tributaries in the south.

In Gujarat, numerous streams abound from Kori creek in the west. They carve out their valleys before uniting with the Gulf of Kutch. North flowing streams unite with Gulf of Kutch and the little Rann, while south flowing streams carve out their valleys in the Surat sea coast, west flowing streams like Bhadar and Ojat merge in the Arabian Sea. The eastward drainage line like the Shatrunji and others discharge their water in the Gulf of Cambay. The Sabarmati and Mahi rivers form drainage net with their feeders. The Narmada and Tapi from east to west align along the rift valleys and unite with Gulf of Cambay. The Mahi, the Narmada and the Tapi form the main arteries of the Gujarat Plain. The Banas, the Sarswati, the Vishvamitri, the Dhadhar, the Kim, the Purna, the Ambika, the Auranga, the Damanganga, the Shatrunji, the Bhadar, the Aji are either merge in the Rann of Kutch or fall into the Gulf of Cambay and the Arabian Sea (Deshpande (1971).

2.2e. Climate

The region of western India is subtropical monsoonic and rainfall is mostly from the southwest monsoon (Rajaguru 1973). In Maharashtra the maximum temperature in winter varies from 30° C and 35°C. In summer the western part shows moderate heating, because of an effect of altitude and nearness to sea. The rain starts from June to October. During summer in Maharashtra and Gujarat the maximum temperature varies from 36.7° C to 43.3°C. In winter the minimum temperature is between 2°C and 18.3°C. from November to February (Deshpande 1971).

The distribution of rainfall is mainly governed by the physical features of the region. Coastal Maharashtra, Western Ghats and South Gujarat receive more than 2000 mm rainfall. These regions are covered by pedalfar type soils and thick deciduous and semi evergreen forests. On the other hand the plateau of Maharashtra, north Gujarat, Saurashtra and Kutch receive rainfall less than 1000 mm and these regions are covered by pedocal type soils and thin deciduous and thorny scrub forest (Rajaguru 1973).

2.2f. Flora and Fauna

A major part of this area falls in the semi-arid zone with low rainfall and the region is covered with xerophytic type thin deciduous and dry thorny and scrub forest type of vegetation like Bor (*Zizyphus jujuba*), Babhul (*Acacia Arabica* wild), Chinch (*Tamarindus indica*), etc.

The forests of the Western Ghats and of the Satpura range have left little wild life. The areas of Vidarbha have the best forests and have plentiful wild life. Sambar, chital, barking deer, nilgai, blackbuck, four-horned antelope, chinkara, wolf, jackal, fox, leopard, tiger, jungle cat, civets, bears, wild pig, gaur, wild buffalo, etc. are some of the mammals represented in this region.

The gazetteer of India (1965) mentions about the disappearance of crocodile, cheetah, two horned rhinoceros, Indian great bustard and deer species from the forests of the Western Ghats and Satpura range. Tiger was plentiful, particularly in Western Ghats and in the Mahabaleshwar plateau; however, in recent years they have become rare. Nilgai, blackbuck, chinkara, chowsingha and sambar were also plentiful in this area have become rare. Blackbuck once seen in thousands in the Deccan has now declined to the verge of extinction, are only found in a few pockets in the state. This may be due to extension of cultivation and reduction of forest areas and the grass lands.

Methodology

The faunal remains of Inamgaon comprise the material collected for 12 field seasons between 1968 -1984. A part of the faunal material collected between 1968 -1971 was studied by A.T. Clason (1979). Almost the same specimens were restudied by Badam (1977). Later on the entire faunal material has been taken by Thomas (1988) for a detailed study. The whole faunal assemblage consisting of 66847 bone fragments housed in the Deccan College was available to me for detailed analysis. The bones were clean and some what better preserved. All the identified fragments were sorted out and recorded, giving trench, layer and accession numbers. Now these bones were ready for the identification to species or generic levels, whatever was possible.

3.1. Species Identification

The first and most important step in faunal analysis is the identification of fragments of bones to the species they belong to. In many cases an accurate identification to the species level was possible with the help of the reference skeletons of various domestic and wild animals, housed in the Archaeozoology laboratory of the Deccan College.

Large sized animals like cattle, buffalo, horse, nilgai and sambar are almost of the same size range. With the help of certain morphological characters, it is easy to separate the bones of Bovidae, Cervidae and Equidae. When the species are closely related like cattle and buffalo, the identification of bones is problematic, because of their morphological similarities. Thomas (1977) has studied the osteological differences between cattle and buffalo. This was helpful in the analysis of faunal assemblage at Inamgaon. Whenever it was not possible to identify the bone fragments to species level, they have been assigned to general category called *Bos/Bubalus* group; certain fragments where it was impossible to tell whether they are cattle or buffalo or other large bovid are identified as large bovid.

The group of medium sized animals has several species of the same size range, such as sheep, goat, blackbuck, chinkara and small deer species. There are very minute morphological differences which require a careful study. Boessneck (1963) and Prummel and Frisch (1986) have studied the morphological difference in sheep and goat. Also the bones of sheep, goat and blackbuck are quite similar in appearance. Dr. Thomas has developed criteria to identify these animals (Thomas personal communication) and with the help of reference collection, it was easy to differentiate between sheep, goat and blackbuck. The author has found some more morphological differences among blackbuck, sheep and goat and they are as follows:

Mandible

1) Shape of the mandible is similar in blackbuck and goat and it is distinct in sheep. 2) In blackbuck the portion between the mandibular symphysis and the second premolar is flat, thin and long and in sheep and goat it is rounded, very thick and short. 3) The mental foramen is away from the second premolar and nearer to the symphysis in blackbuck and is in the centre of the symphysis and closer to the second premolar, in sheep and goat. 4) Mandible is thin in blackbuck and thick in sheep and goat. 5) Shape of the coronoid process of mandible is distinct in all the three species (Fig. 3.1).

Fig. 3.1. Mandible: Difference between **A.** *Antilope cervicapra*, **B.** *Capra hircus* and **C.** *Ovis aries*

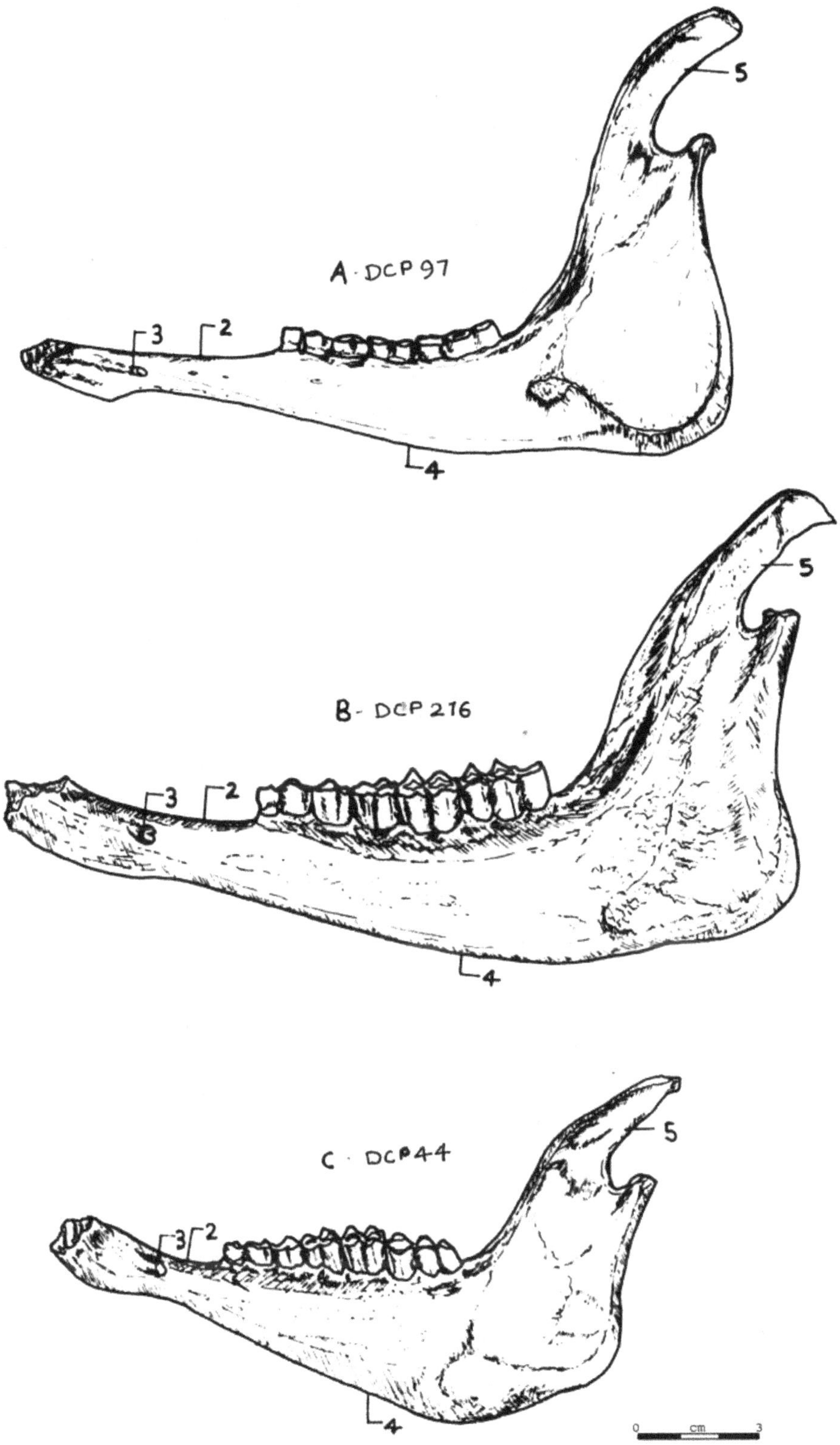

Scapula

1) Overall the shape of the scapula is long, narrow and elongated in blackbuck. In sheep and goat the scapula is broader. 2) The spine goes straight upwards in blackbuck and it is a little turned towards the anterior border of the scapula in sheep and goat. 3) In blackbuck the neck region is long and slender. It is long and broad in goat and thick and short in sheep. 4) The distance between the glenoid cavity and the spine is less in blackbuck than in sheep and goat. 5) On the posterior side of the neck region, there is a depression in blackbuck while it is flat in sheep and goat. 6) The glenoid cavity is deep in blackbuck and shallow in sheep and goat. 7) The tuber scapula is thin and curved in blackbuck and thick and broad in sheep and goat. 8) The anterior and posterior edges are sharp in blackbuck and are blunt in sheep and goat. 9) In blackbuck, from the medial side, the spine is not visible but in sheep and goat, it is visible. 10) In the neck region, the anterior and posterior borders are straight in blackbuck and goat. The anterior border is curved in sheep. 11) On the posterior border of the neck region, there is very prominent muscle striation in blackbuck; this is less prominent in goat and is absent in sheep (Fig. 3.2 a and b).

Fig.3.2a. Scapula: Difference between **A.** *Antilope cervicapra*, **B.** *Capra hircus* and **C.** *Ovis aries*

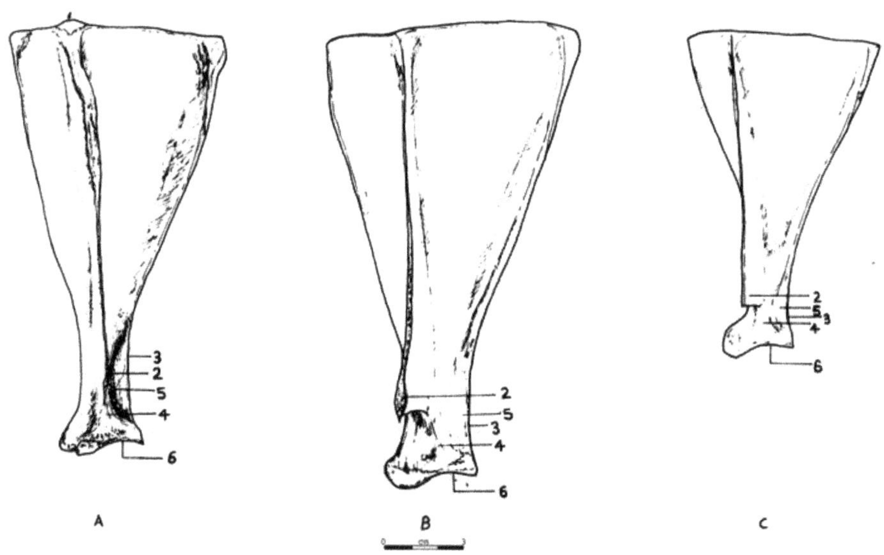

Fig. 3.2b. Scapula: Difference between **A.** *Antilope cervicapra*, **B.** *Capra hircus* and **C.** *Ovis aries*

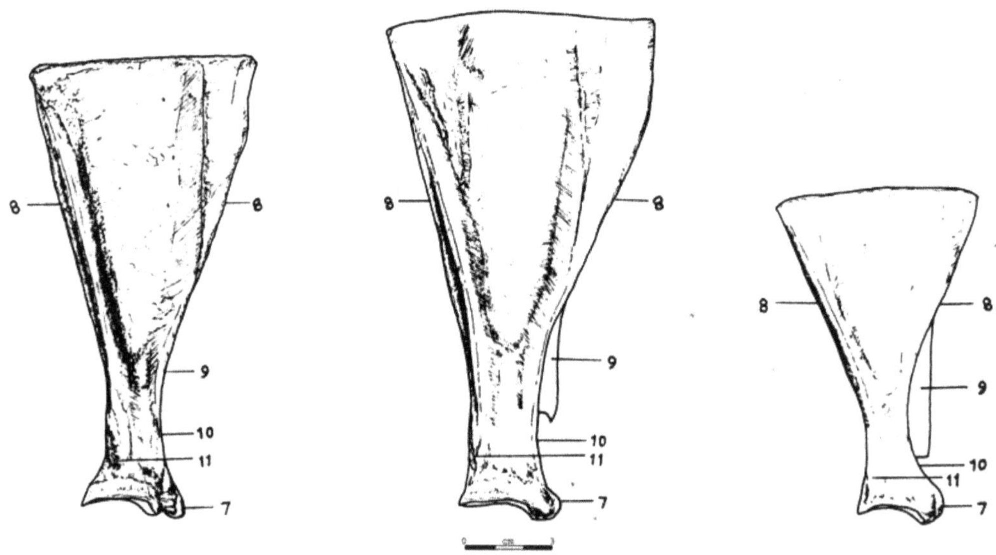

Humerus

1) The shaft of the humerus is short in length in blackbuck, slender in sheep and thick and long in goat. 2) The head is more elongated in blackbuck and flat in sheep and goat. 3) The tuberculum majus cranial is small and broad, it does not project upward from the shaft and the angle with the articular surface is less in blackbuck. In sheep and goat tuberculum majus cranial projects upward from the shaft itself and is very long and broad. The angle with the articular surface is greater than in blackbuck. 4) The neck region is more curved in blackbuck than in sheep and goat. 5) The deltoid ridge is sharper in blackbuck than in the sheep and goat. 6) The lateral and medial epicondyles are in the same plain in blackbuck. In sheep and goat the medial epicondyle is higher up than the lateral epicondyle. 7) The olecranon fossa is wide open in sheep and blackbuck and is slightly closed in goat. 8) The shaft above the olecranon fossa, towards the proximal side is raised up in blackbuck, flat in goat and slightly raised up in sheep. 9) In blackbuck, the lateral and medial epicondyles terminate where the lateral and medial condyles end. In sheep and goat both the epicondyles terminates where the olecranon ends. 10) The condyles merge with coronoid fossa in blackbuck and in sheep and goat there is a distance between the condyles and the shaft which make a deep and wider coronoid fossa. 11) The shaft is more curved near the distal end in blackbuck than in the sheep and goat. 12) The groove on the medial condyle is deeper in blackbuck than in the sheep and goat. 13) The medial epicondyle is extending straight distally beyond the medial condyle in blackbuck. In sheep and goat it is extending to the level of the medial condyle. 14) Metrically the thickness and the width of the distal end vary in all the three species. 15) There is a depression on tuberculum majus on cranio-lateral side in blackbuck and it is absent in sheep and goat (Fig. 3.3 a, b and c).

Radius and Ulna

1) The shaft is broad in blackbuck, slender in sheep and broader than blackbuck in goat. 2) The lateral tuberosity is well developed in blackbuck and poorly developed in goat and it is absent in sheep. 3) In blackbuck, on the shaft near the proximal lateral articular facet, there is a small depression which is absent in sheep and goat. 4) The lateral articular facet is projected upward in sheep and goat. It is very prominent in goat and absent in blackbuck. 5) Blackbuck has two very prominent ridges on the distal shaft which are absent in sheep and goat. 6) The shaft is curved on the lateral side in blackbuck. In sheep it is slightly curved but not in goat. 7) The medial edge is straight proximally with no lip formation in blackbuck. In sheep and goat the medial edge of the proximal articular surface has a distinct lip formation. 8) In blackbuck, the ulnar notch is very small and shallow and is 'V' shaped. In sheep and goat the ulnar notch is deep and 'U'shaped. 9) Anteriorly and posteriorly the outline of the proximal articular surface is distinct in all three species. 10) The intermediate facet of the distal epiphysis terminates towards the medial side of the shaft in blackbuck and in sheep and goat it ends towards the middle. 11) In blackbuck the intermediate and medial facets are shallow and in sheep and goat the facets are deep. 12) The interosseus space is very near the proximal end in blackbuck and in sheep and goat it is away from the proximal end. 13) The bone is long from processus anconeus to olecranon in blackbuck and short in sheep and goat. 14) The processus anconeus is projected in blackbuck and in sheep and goat there is a depression. 15) The lateral facet is extended laterally and the medial facet is projected in sheep and goat and not in blackbuck. 16) In blackbuck the olecranon is slender and in sheep and goat it is broad and projected upward. 17) On the medial side the olecranon has a depression in blackbuck and in sheep and goat it is flat. 18) The dorsal ridge of the ulna is sharp in blackbuck and rounded in sheep and goat. 19) The shape of the semilunar notch is different in blackbuck than in sheep and goat (Fig 3.4 a, b and c).

Metacarpal

1) The metacarpal is long and slender in blackbuck; short and slender in sheep and short and broad in goat. 2) On the anterior side of the shaft, the nutrient foramen is near to the distal end in blackbuck and in sheep and goat it is away from the distal end. 3) The distal trochleas are straight and the sagital ridge is sharp in blackbuck. In goat, both the trochleas are at an angle and in sheep, they are straight. 4) The distance between the distal trochleas and the shaft is short. The trochleas look very closely attached to the shaft in blackbuck. This distance is greater in sheep and goat and there is a small shallow pit between shaft and trochleas. 5) In blackbuck the inter trochlear faces are parallel and the space between the two trochleas is less as compared to sheep and goat. 6) The lateral facet at the proximal end is more angled in sheep and goat than blackbuck. 7) On the lateral side of the trochlea, the groove is deep in blackbuck and shallow in sheep and goat. 8) In blackbuck, the posterior side of the shaft, the edges are

sharp and there is a depression and muscle striation marks on the middle of the shaft. In sheep and goat, posteriorly the shaft is flat with no muscle striation (Fig. 3.5).

Fig. 3.3a. Humerus: Difference between **A.** *Antilope cervicapra*, **B.** *Capra hircus*

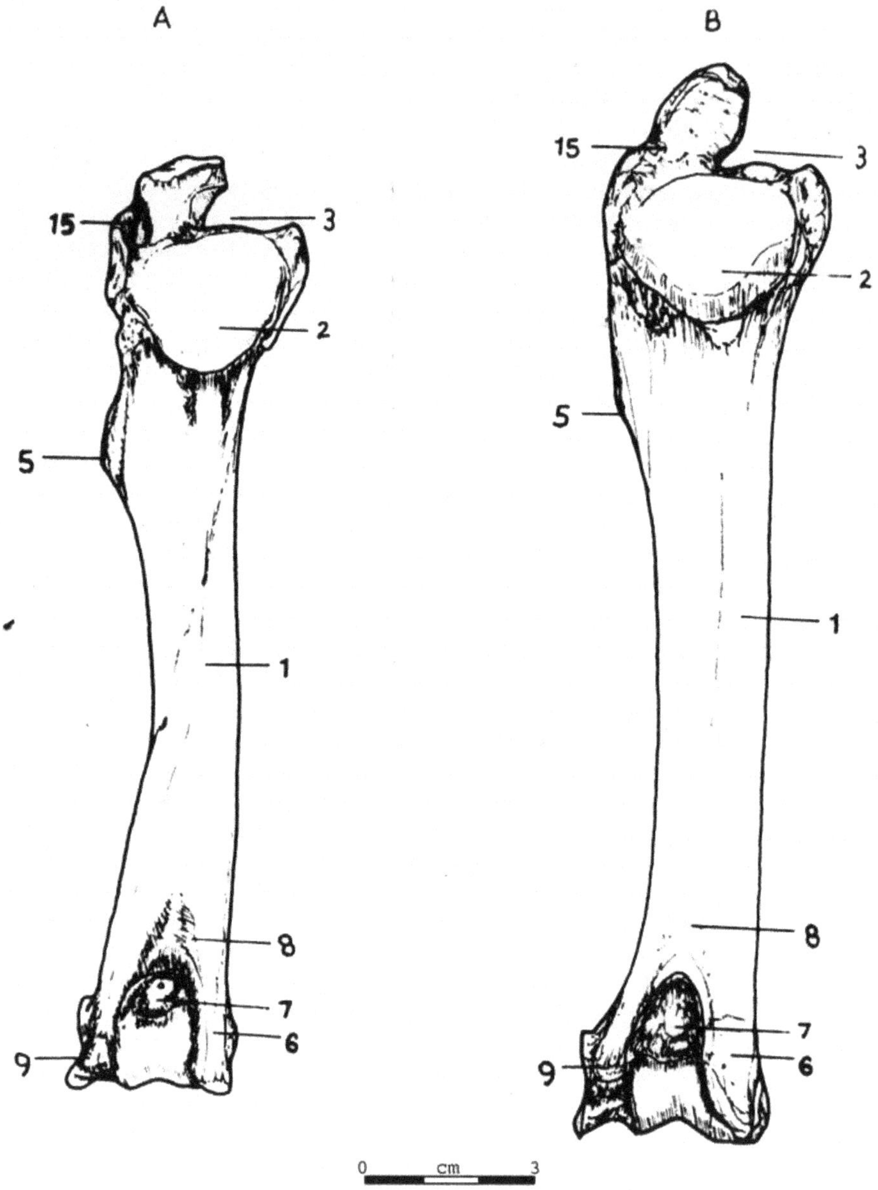

Fig. 3.3b. Humerus: Difference between **A.** *Antilope cervicapra*, **B.** *Capra hircus*

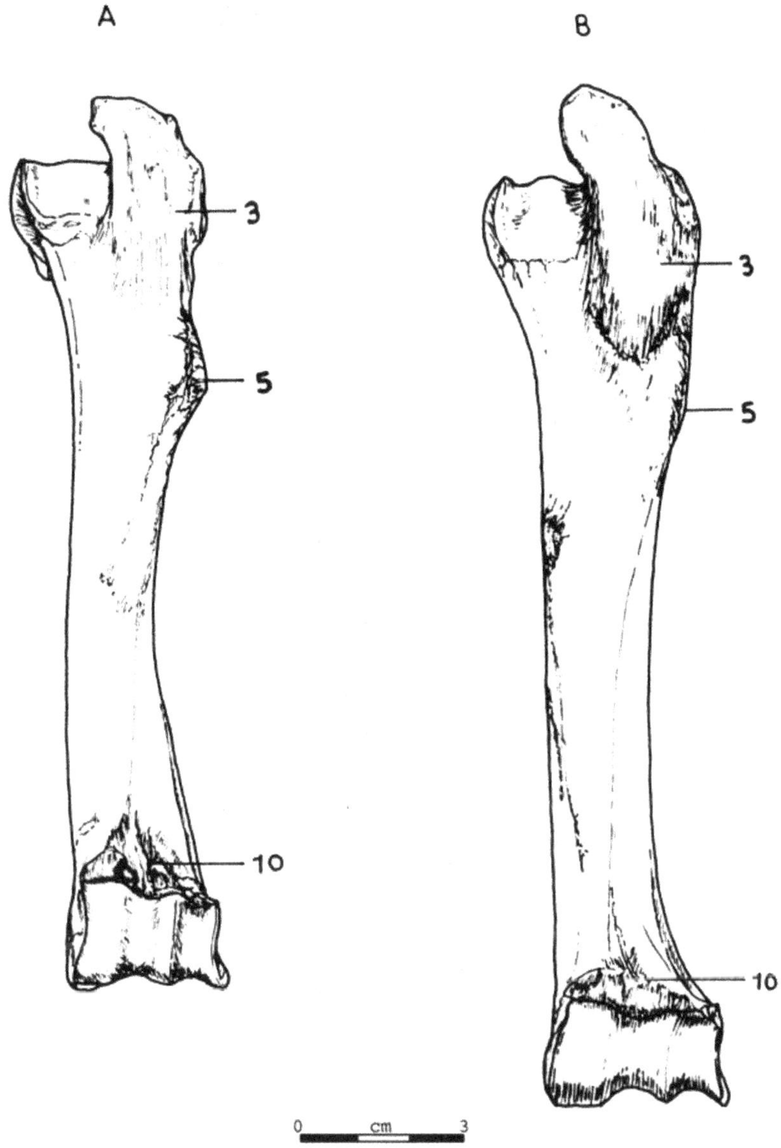

Fig. 3.3c. Humerus: Difference between **A.** *Antilope cervicapra*, **B.** *Capra hircus*

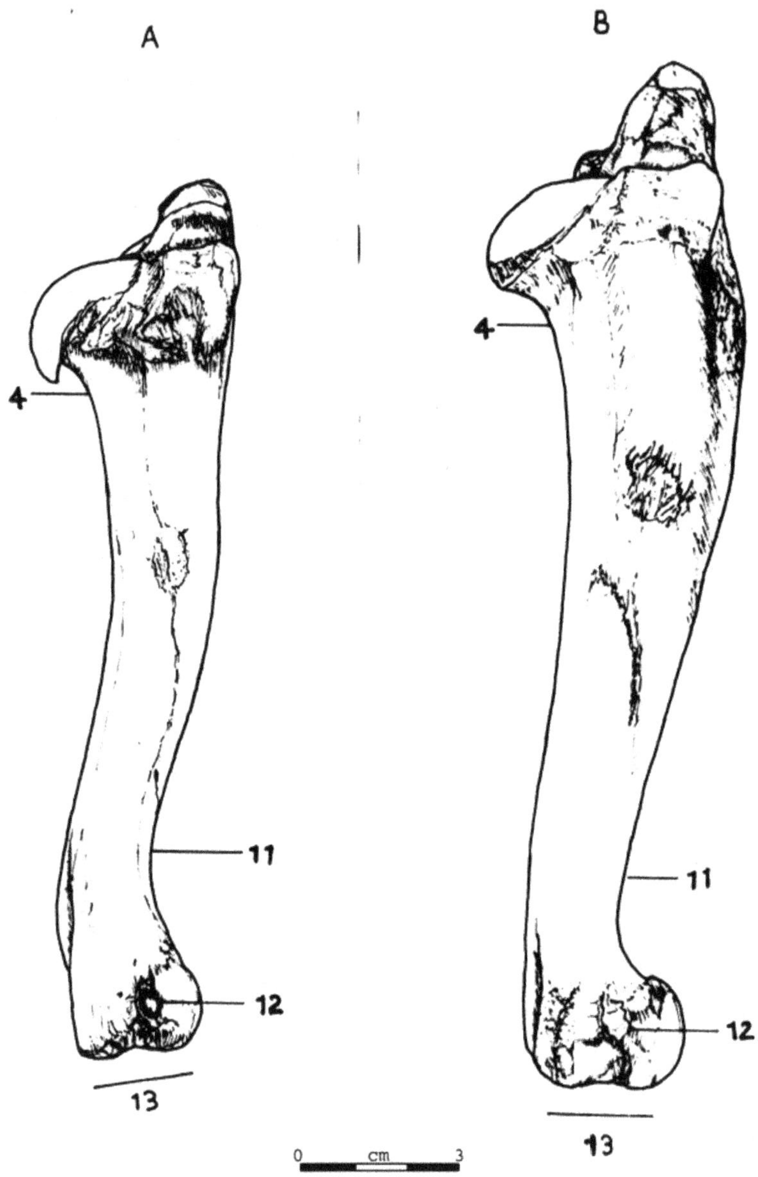

Fig. 3.4a. Radius: Difference between **A.** *Antilope cervicapra*, **B.** *Capra hircus*

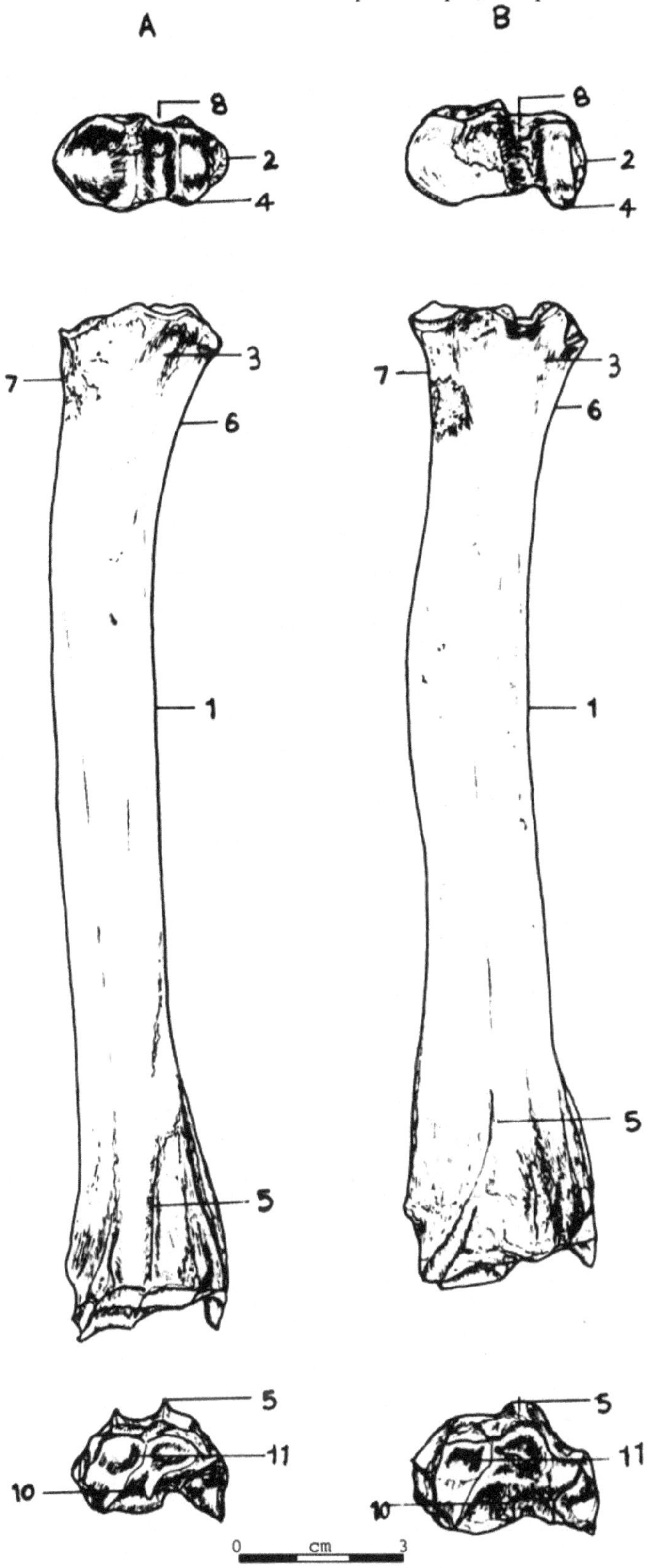

Fig. 3.4b. Radius and Ulna: Difference between **A.** *Antilope cervicapra*, B. *Capra hircus*

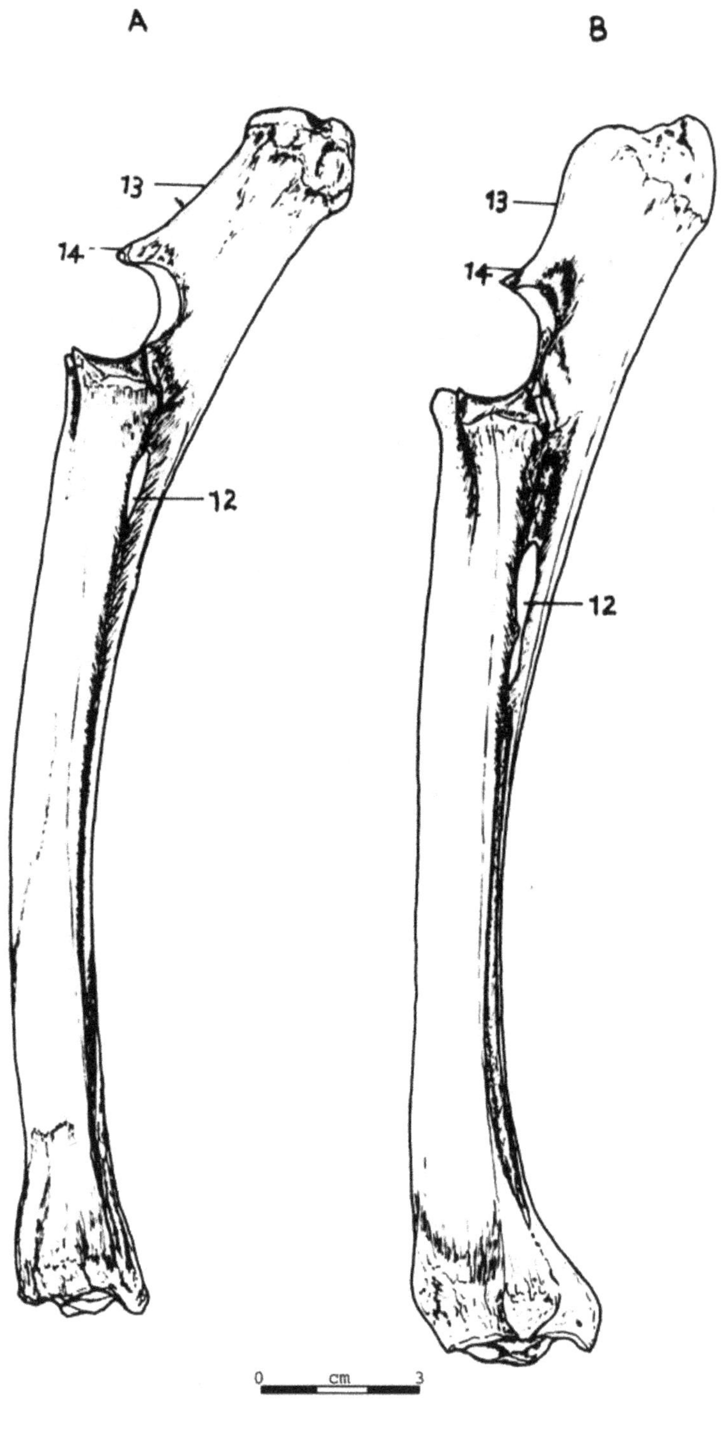

Fig. 3.4c. Ulna: Difference between **A.** *Antilope cervicapra*, **B.** *Capra hircus*

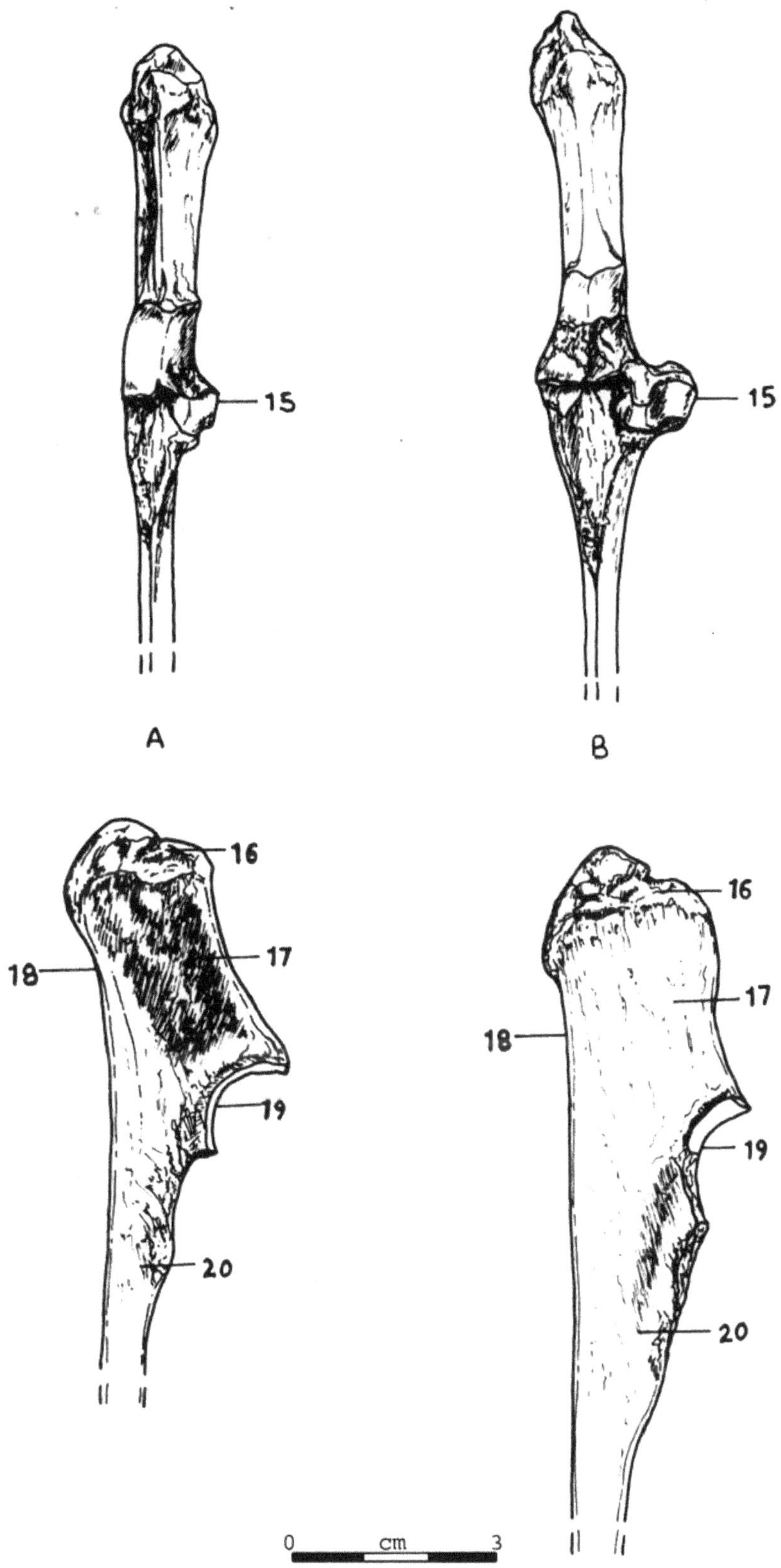

Fig. 3.5. Metacarpal: Difference between **A.** *Antilope cervicapra*, **B.** *Capra hircus* and **C.** *Ovis aries*

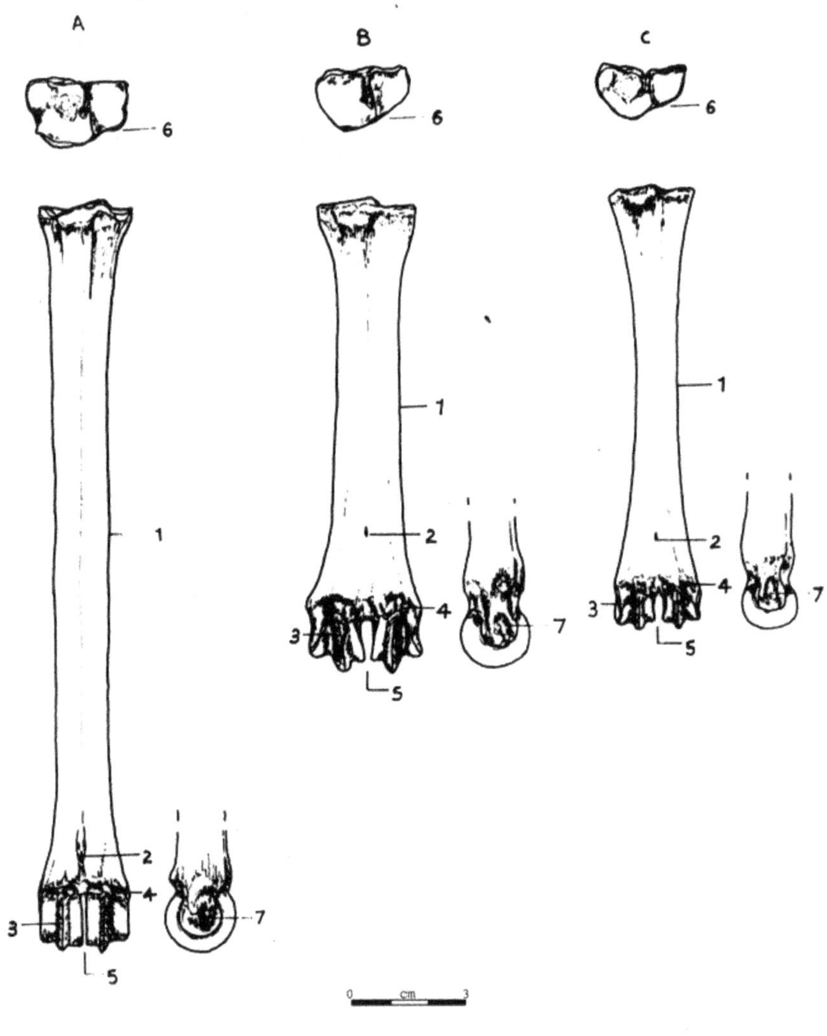

Pelvis

1) In blackbuck the iliatic wing is broad. The medial and lateral sides of the ilium are curved. In sheep and goat the iliatic wings are not broad and neither side is curved. 2) The iliatic shaft is short in blackbuck and long in both sheep and goat. 3) In blackbuck, on the iliatic shaft, just above the acetabulum on the lateral side, a very prominent groove is present. This groove is present on the iliatic shaft, above the acetabulum on the medial side and is prominent in goat but very less developed in sheep. 4) In blackbuck the ischiatic spine is very broad as compared to sheep and goat and muscle striations are present on it. Muscle striations re absent in sheep and goat. 5) The pubis is rounded and prominent muscle striations are seen in blackbuck. In sheep the pubis is round without muscle striations and in goat the pubis is round near the acetabulum and then it is flat and muscle striations are on it. 6) In blackbuck the acetabulum is wide open and the distance between the facet of the pubis and the facet of the ischium is greater. In sheep and goat the acetabulum is not wide open. It is bit closed and circular in shape and the distance between the two facets of pubis and ischium is less. 7) The obturatory foramen is circular in shape in blackbuck and oblong in sheep and goat. 8) The lip of the acetabulum in the ischial region is thick and sloping down in blackbuck and in sheep and goat the lip is thin and projecting upwards. 9) The tuber ischii is well pronounced in blackbuck and less pronounced in

sheep and goat. 10) In blackbuck the ischium is short and broad and the lateral side is more curved. In sheep and goat the ischium is long and narrow and the lateral side is more straight. 11) The pelvis symphysis is away is away from the obturatory foramen in blackbuck and further away in goat but in sheep it is closer. 12) The shape of ischium plate is different in all three species. 13) The ischium blade is angular in blackbuck and circular in sheep and goat (Fig 3.6).

Fig. 3.6. Pelvis: Difference between A. *Antilope cervicapra*, B. *Capra hircus* and C. *Ovis aries*

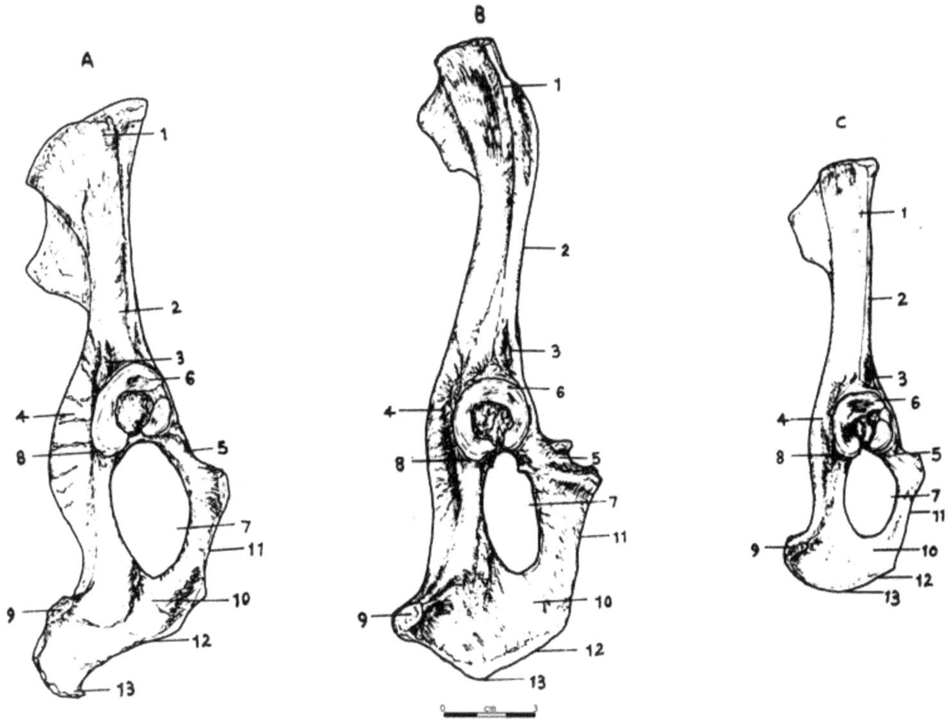

Femur

1) The femur head is perpendicular to the trochanter fossa in blackbuck, this portion is deep and in angle of 45°- 60°. In goat the head is almost in angle to trochanter fossa and in sheep this angle is wide. 2) The femur head is elongated in blackbuck and in sheep and goat it is rounded. 3) and 4) In blackbuck the curvature in the neck region and of the major trochanter is greater than in sheep and goat. 5) At the distal end, condyles are angled in blackbuck and are parallel in sheep and goat. 6) The minor trochanter is projected outwards and is very prominent in blackbuck and in sheep and goat it is less projected. 7) In blackbuck, proximally, the anterior side the shaft is flat and in sheep and goat it is bulbular. 8) The proximal end is not broad in sheep and goat and is broad in blackbuck. 9) On the anterior side, the nutrient foramen is on the middle of the shaft in blackbuck and is on the lateral side in sheep and goat. 10) In blackbuck the distance between the two epicondyles is much greater than in sheep and goat. 11) The medial epicondyle is higher up than the lateral epicondyle in blackbuck and in sheep and goat both the epicondyles are on the same level. 12) The medial epicondyle goes straight to the shaft and the lateral epicondyle turns towards the lateral side and slightly angled in blackbuck. In sheep and goat both the epicondyles are parallel and turned towards the lateral side (Fig. 3.7 a and b).

Fig. 3.7a. Femur: Difference between A. *Antilope cervicapra*, B. *Capra hircus*

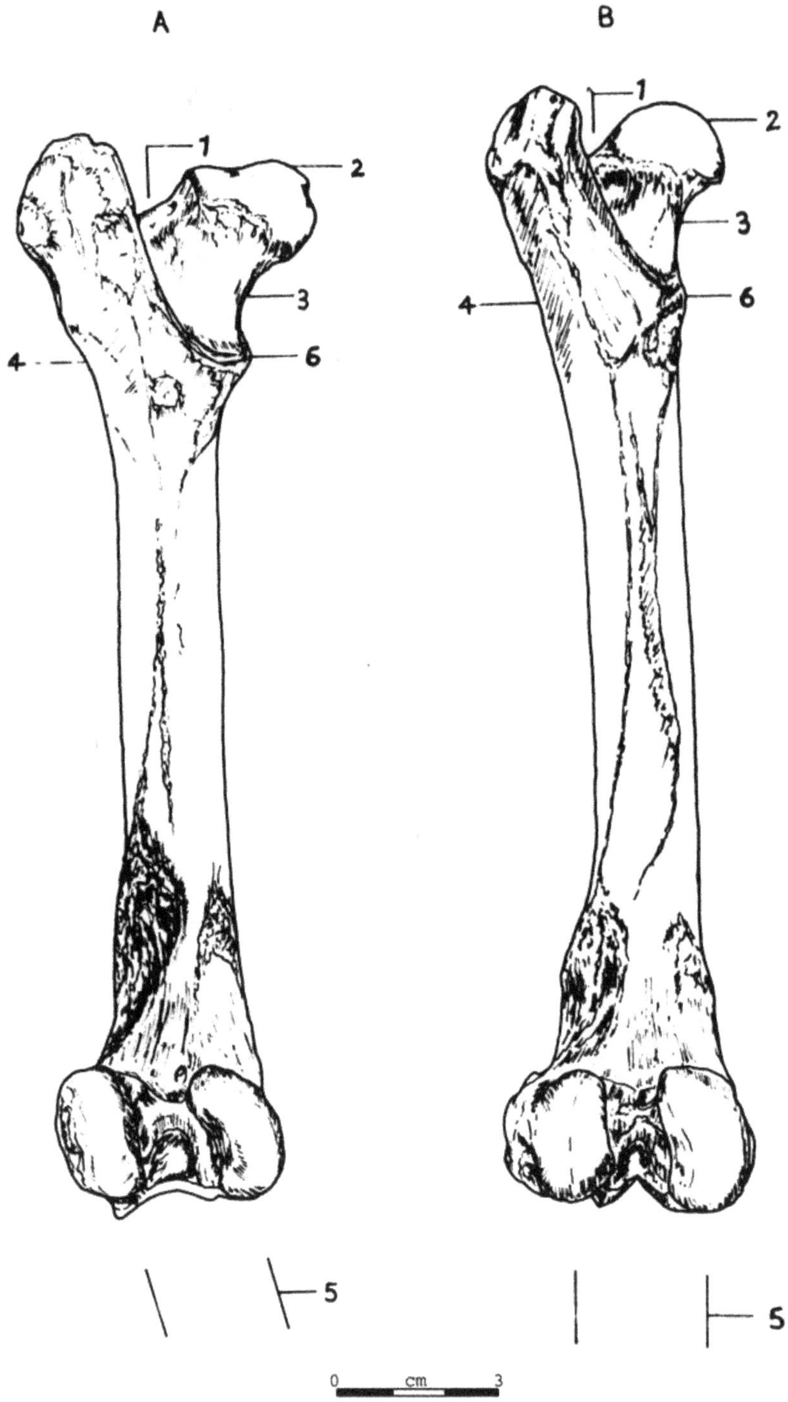

Fig. 3.7b. Femur: Difference between A. *Antilope cervicapra*, B. *Capra hircus*

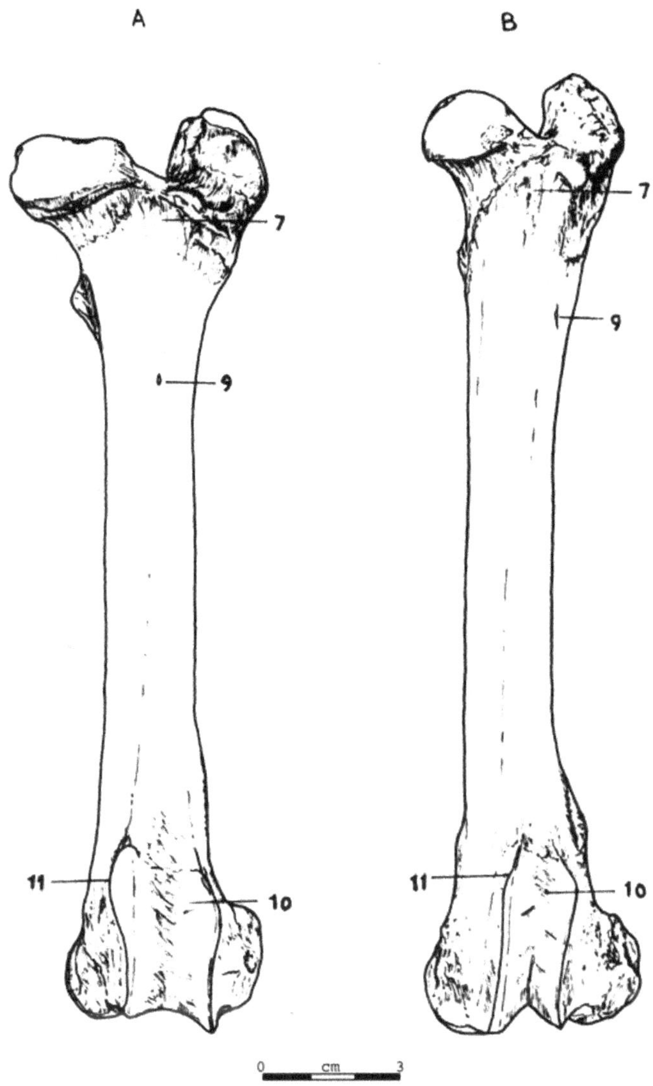

Tibia

1) At the proximal, the tuberosity is long and narrow in blackbuck and broad and short in sheep and goat. 2) The lateral side curvature is lesser in blackbuck than in sheep and goat. 3) In blackbuck, the tibial crest has turned more to the lateral side and in sheep and goat; it is only slightly curved to the lateral side. 4) On the anterior side, near the distal end, there is a small protuberance which is parallel with the lateral ridge in blackbuck and is angled in sheep and goat. 5) Medially, on the distal end, the nutrient foramen is present in blackbuck and is absent in goat and present in sheep laterally. 6) Marked portion in the figure is narrow in blackbuck and broad in sheep and goat. 7) On the proximal epiphysis the nutrient foramen is very deep in blackbuck and in goat it is shallow. In sheep there are many nutrient foramens, very small in size. 8) The groove between the tuberosity and the lateral articular facet is wide in blackbuck and in sheep and goat it is deep and 'U' shaped. 9) The marked portion is less prominent in blackbuck and more prominent in sheep and goat. 10) In blackbuck the space between the two articular facets is greater than in sheep and goat. 11) On the posterior side, the curve of the lateral articular facet is deep in blackbuck and shallow in sheep and goat. 12) In blackbuck the spine is very prominent but not in sheep and goat. 13) On the posterior side of the shaft, near

the proximal part, muscle striations are very prominent in blackbuck and are not prominent in sheep and goat. 14) In blackbuck the lateral and the medial grooves of the distal epiphysis are deep and shallow in sheep and goat. 15) The two small facets for the lateral malleolus are separated by a small groove which is not very deep and both the facets are smaller in size in blackbuck. This groove is deep but these facets are not separated by this groove and are a little bigger in size in sheep and goat. 16) The posterior edge of the distal articular surface has a very deep curve in blackbuck and in sheep and goat the curve is less prominent.17) The distal epiphysis is broad and square in shape in sheep and goat and is not broad in blackbuck. 18) On the medial side, at the distal end, the groove is absent in blackbuck and is present prominently in sheep and goat (Fig.3.8 a, b and c).

Fig. 3.8a. Tibia: Difference between A. *Antilope cervicapra*, B. *Capra hircus*

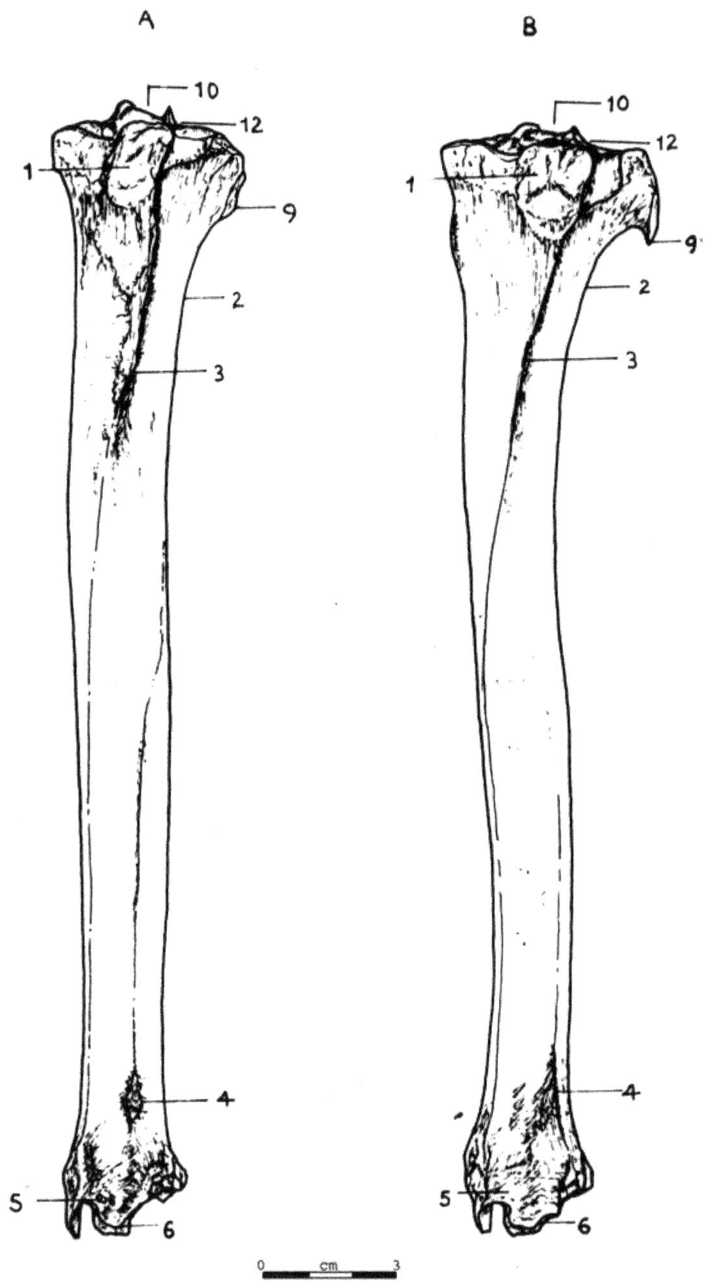

Fig. 3.8b. Tibia: Difference between A. *Antilope cervicapra*, B. *Capra hircus*

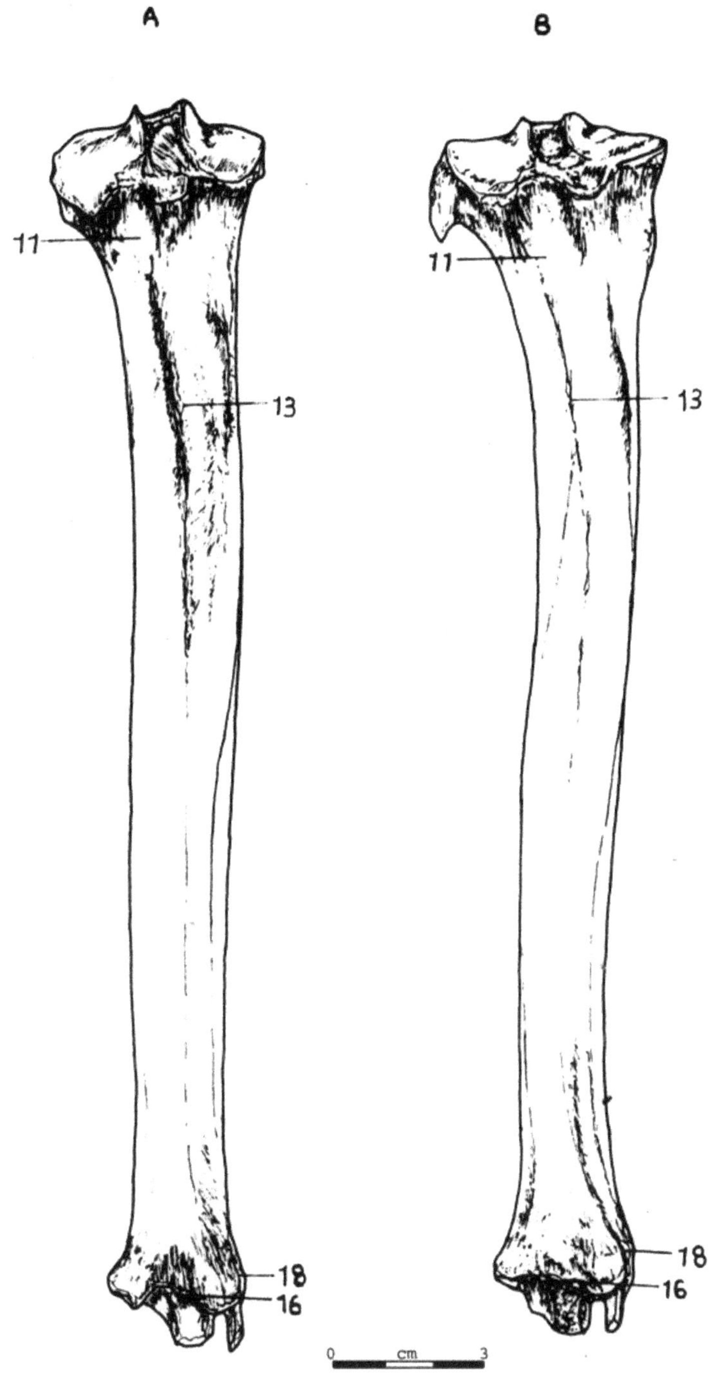

Fig. 3.8c. Tibia: Difference between A. *Antilope cervicapra*, B. *Capra hircus*

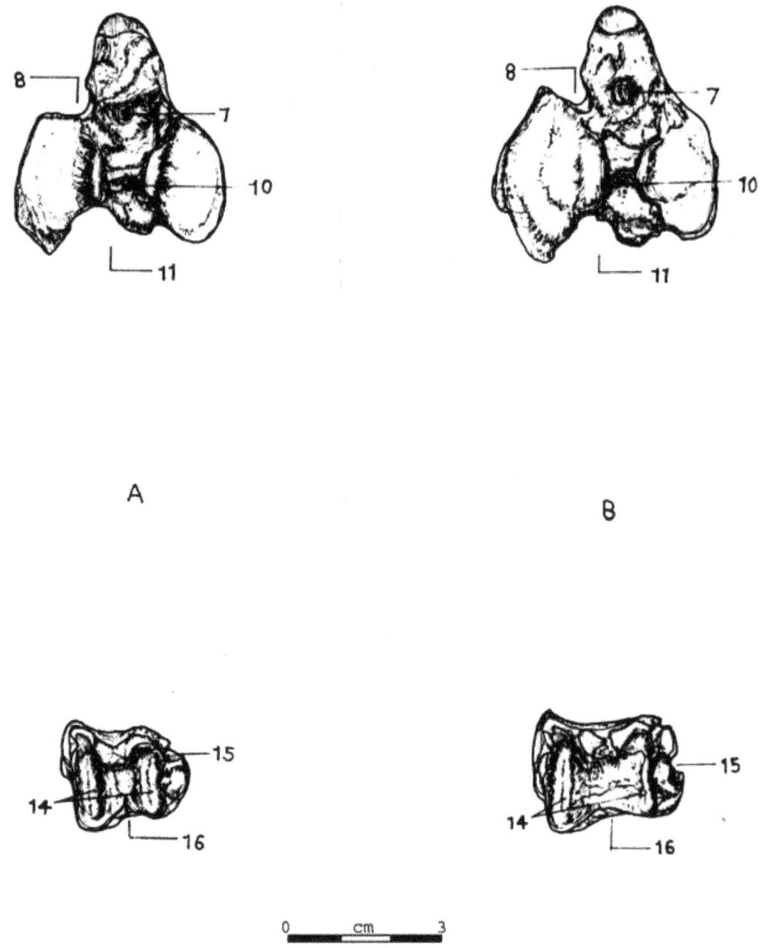

Metatarsal

1) The metatarsal is long and slender in blackbuck, short and slender in sheep and short and broad in goat. 2) The shaft is flat on the lateral and medial side but not at the distal end in blackbuck. It is not so in sheep and goat. 3) On the posterior, the edges of the lateral and the medial sides of the shaft are sharp and form a groove on the middle of the shaft in blackbuck. In sheep and goat, posteriorly the shaft is flat. 4) In blackbuck the nutrient foramen is very small on the distal part of the shaft anteriorly but absent posteriorly and it is present on the proximal part of the shaft and absent on anterior side. It is prominent on the distal part of the shaft on the anterior and posterior side in sheep and goat. The nutrient foramen is near to the distal end in blackbuck and in sheep and goat it is away from the distal end. 5) The facet is higher up and round in blackbuck and in sheep and goat it is lower and long. 6) The distance between the distal trochleas and the shaft is short. The trochleas look very closely attached to the shaft in blackbuck. This distance is greater in sheep and goat and there is a small shallow pit between shaft and trochleas. 7) The distal trochleas are straight and the saggital ridge is sharp in blackbuck. In goat, both the trochleas are at an angle and in sheep, they are straight. 8) In blackbuck the inter trochlear faces are parallel and the space between the two trochleas is less as compared to sheep and goat. 9) On the lateral side of the trochlea, the groove is deep in blackbuck and shallow in sheep and goat (Fig. 3.9).

Fig. 3.9. Metatarsal: Difference between **A.** *Antilope cervicapra*, **B.** *Capra hircus* and **C.** *Ovis aries*

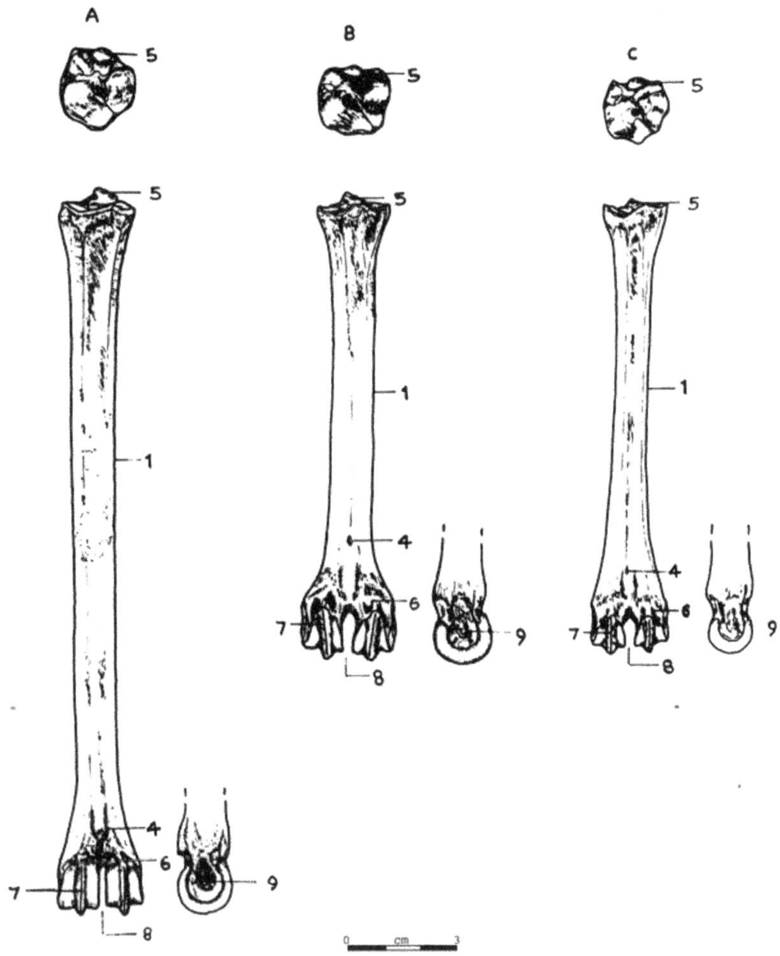

Astragalus

1) The anterior surface of the astragalus of blackbuck is round and bulbular and in sheep and goat it has a slight depression. 2) In blackbuck, on the anterior side of the medial side, near the distal end, the facet is poorly developed; it is well developed in sheep and goat. 3) Proximally, in between the lateral and medial sides, the depression is more deep in blackbuck and shallow in sheep and goat. 4) And distally the depression is deep in sheep and goat and shallow in blackbuck. 5) The proximal lateral condyle is blunt and rounded in blackbuck and pointed in sheep and goat. 6) The marked portion is shallow and closed in blackbuck and deep and opened in sheep and goat. 7) and 8) Posteriorly, in between the lateral and medial condyles, the groove is deep in blackbuck and shallow in sheep and goat. 9) At the posterior, on the medial side, the marked portion in the figure is less projected in blackbuck and more projected in goat than in sheep.10) On the posterior side, the distal medial condyle is angled in blackbuck and straight in sheep and goat. 11) Posteriorly on the distal condyle the groove is shallow in blackbuck and deep in sheep and goat. 12) Facets on the lateral side are different in blackbuck than in sheep and goat. 13) The facet at the distal end is elongated in blackbuck and rounded in sheep and goat. 14) On the anterior side, proximally, the edge of medial side is rounded in blackbuck and is projected out in sheep and goat.15) At the distal end of the medial side, the round facet is deep in blackbuck and shallow in sheep and goat (Fig. 3.10).

Fig. 3.10. Astragalus: Difference between **A.** *Antilope cervicapra*, **B.** *Capra hircus*

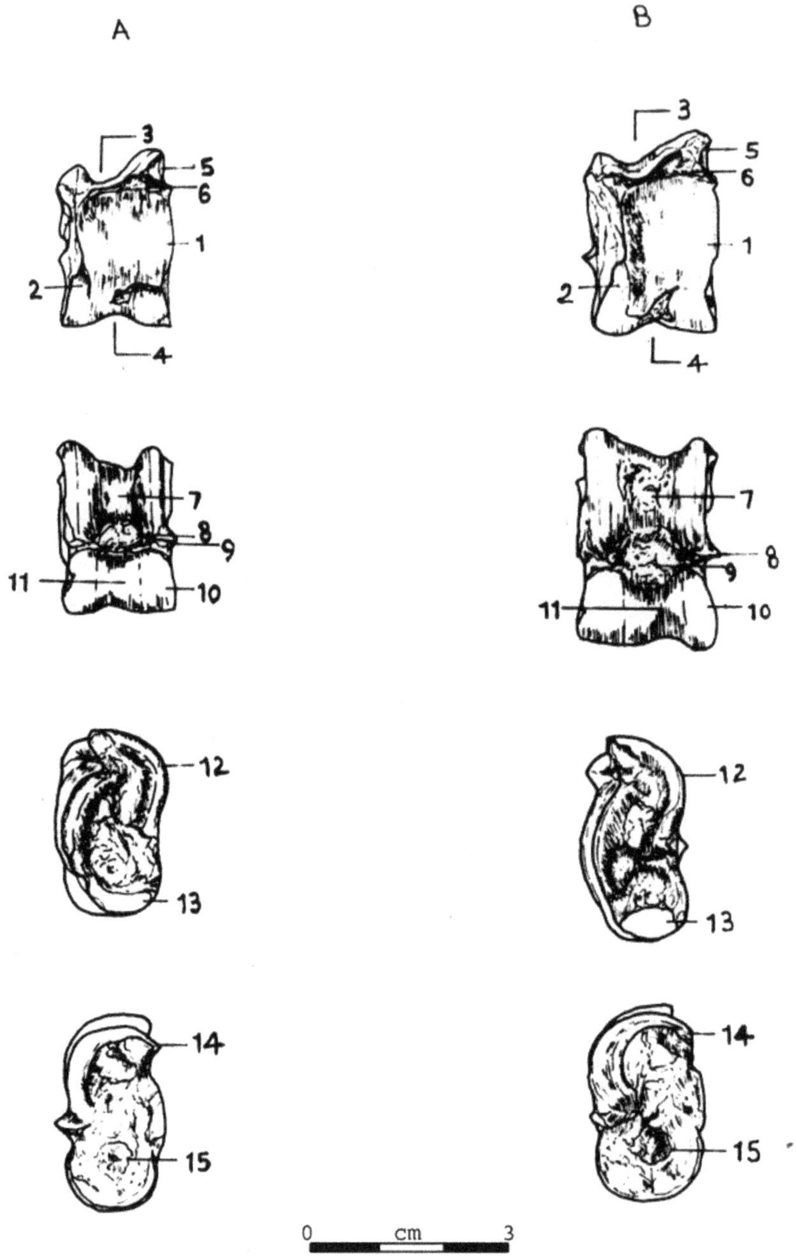

Calcaneum

1) In blackbuck corpus calcanei has a convexity and is straight in sheep and goat. 2) The tuber calcanei is pointed in blackbuck and in sheep and goat it is rounded. 3) In blackbuck the sustentaculum tali projects within the body and it projects outside the body in sheep and goat. 4) The medial protuberance is prominent and shape is different in blackbuck than the sheep and goat. 5) The shape of the lateral articular facet for the centrotarsal is less curved in blackbuck and is more curved in sheep and goat. 6) From the lateral side the sustentaculum tali is not visible in blackbuck and are visible in sheep and goat. 7) On the anterior side of the corpus calcanei, a sharp ridge is formed in blackbuck and it is rounded with no ridge in sheep and goat. 8) On the distal side proximal tuberosity has a marked depression in blackbuck and has a shallow depression in sheep and goat (Fig.3.11).

Fig. 3.11. Calcaneum: Difference between **A.** *Antilope cervicapra*, **B.** *Capra hircus*

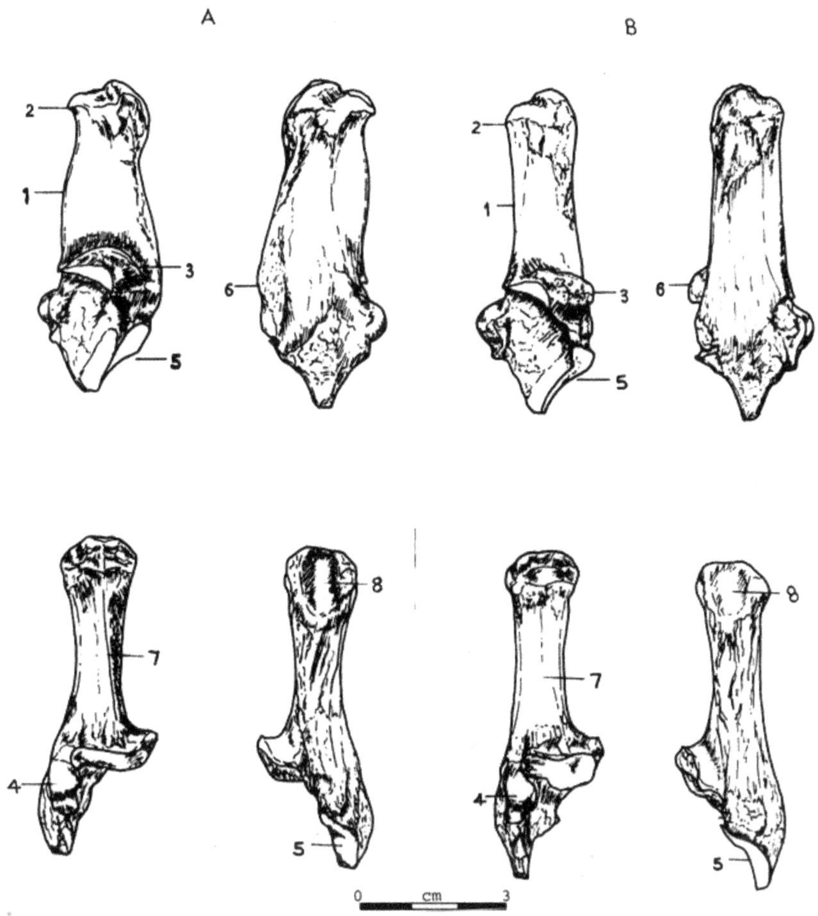

Centrotarsal

1) On the lateral side, the facet for the calcaneum is going straight up in blackbuck and in sheep and goat it is curved to the medial side. 2) The nutrient foramen is on the lateral articular facet in blackbuck and is in between the lateral articular facet and the facet for the calcaneum in sheep and goat. 3) The distance between the facets for the calcaneum and the astragalus is more in blackbuck and less in sheep and goat. 4) On the dorsal side the shape of the medial articular facet is different. In blackbuck the lateral facet is higher up and semicircular in shape and the medial facet is square in shape. In sheep and goat both the facets are elongated. 5) The groove in between the cranial and the caudal articular facets on the lateral side is shallow in blackbuck and in sheep and goat groove is deep (Fig. 3.12).

First Phalanx

1) The proximal articular facets are elongated and compressed on the lateral and medial sides and the medial articular facet is little projected anteriorly. In sheep and goat, it is more square in shape. 2) In blackbuck, the shaft is curved, slender and compressed from both sides and in sheep and goat the shaft is straight and flat on the cranial side and rounded on both sides. 3) The proximal articular facets are longer in blackbuck than in sheep and goat. 4) The groove on the lateral side of the distal end is shallow in blackbuck and deep in sheep and goat. 5) The lateral proximal articular facet is higher up in goat than in sheep and blackbuck. 6) Medially, at the distal end, the small groove is very prominent and deep in blackbuck and is poorly developed in sheep and goat. 7) In blackbuck, on the cranial side, the groove between the lateral and medial articular facets is triangular in shape and in sheep and goat it is square in shape. 8) The distal articular facets are not flaring out and

the groove between the facets is shallow in blackbuck and the facets are flaring in sheep and goat and the groove is deep. 9) On the posterior side, at the proximal end, the lateral and medial sides are closer in blackbuck and in sheep and goat there is deep 'U' shaped groove on the posterior side and the lateral and medial sides are broader. 10) Distal epiphysis is slender in blackbuck and broad in sheep and goat. 11) Posteriorly the shaft is concave in blackbuck and flat in sheep and goat (Fig. 3.13).

Fig. 3.12. Centrotarsal: Difference between **A.** *Antilope cervicapra*, **B.** *Capra hircus*

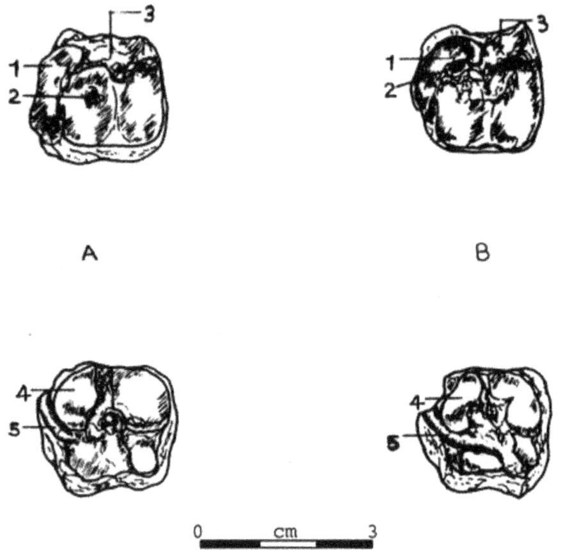

Fig. 3.13. First Phalanx: Difference between **A.** *Antilope cervicapra*, **B.** *Capra hircus*

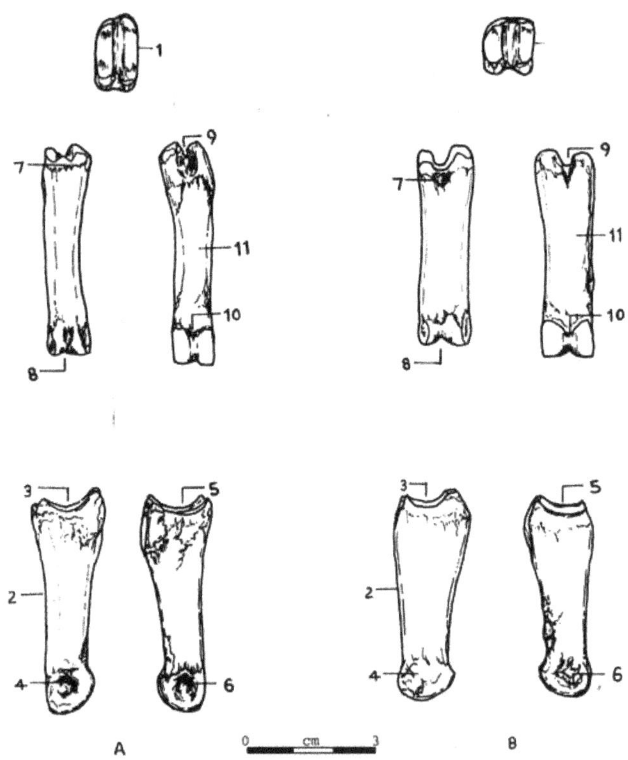

Second Phalanx

1) In blackbuck it is small in size and slender and in sheep and goat it is bigger in size, thick and broad. 2) The cranial side of the proximal end is projected upward in blackbuck and is flat in sheep and goat. 3) The lateral and medial distal articular facets are on the same plane in blackbuck and in sheep and goat the medial facet is projecting downwards. 4) In blackbuck the posterior line of the lateral side of the proximal epiphysis there is a projection and in sheep and goat it is straight with no projection. 5) the groove between the lateral and the medial distal articular facets is deep and both the facets terminate posteriorly almost at the same level in blackbuck and in sheep and goat the groove is shallow and posteriorly the lateral facet is longer than the medial one. 6) The proximal articular surface is elongated and narrow in blackbuck. In sheep and goat the proximal articular facets are wider and short and the proximal epiphysis is square in shape. 7) On the medial side at the distal end groove is deep in blackbuck and shallow in sheep and goat (Fig. 3.14).

Fig. 3.14. Second Phalanx: Difference between **A.** *Antilope cervicapra*, **B.** *Capra hircus*

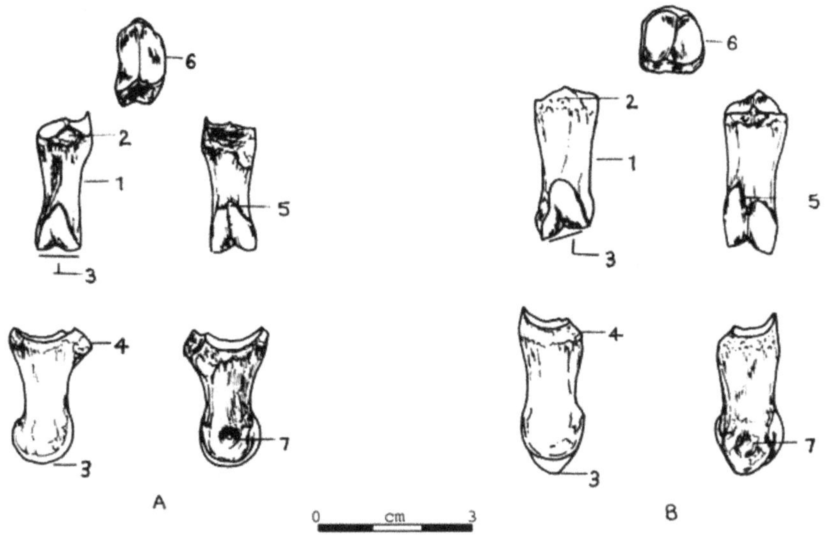

Third Phalanx

1) Overall the shape is triangular in blackbuck and semicircular in sheep and goat. 2) The proximal articular surface is semicircular in shape and deep in blackbuck and in sheep and goat it is right angled. 3) The posterior outline is straight in blackbuck and curved in sheep and goat. 4) Both the articular facets are parallel in blackbuck and not parallel in sheep and goat. 5) In blackbuck, on the cranial side of the proximal articular surface, there are two nutrient foramens and only one is present in sheep and goat. 6) A ridge is present on the medial side and it starts from the middle of the body in blackbuck and in sheep and goat it is at the base. 7) The anterior side is thick in blackbuck and thin in sheep and goat (Fig. 3.15).

Fig. 3.15. Third Phalanx: Difference between **A.** *Antilope cervicapra*, **B.** *Capra hircus*

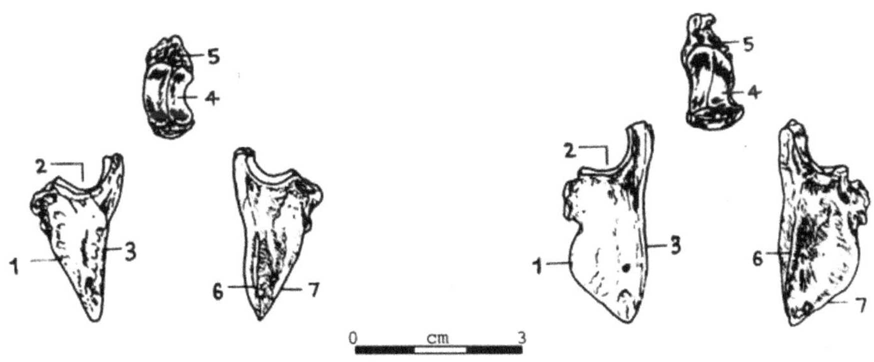

Generally the identification of wild and domestic pig is difficult and is based on the size and the comparative study of modern reference collection.

The fragments of ribs, sternum, sesamoids, carpals and small tarsals were not identified to the species level. These fragments were classified into broad groups like large bovids and small ruminants. The vertebrae of bovids, equids and carnivores were possible to distinguish and were put into broad groups like large bovids, small ruminants, etc. But very fragmented vertebrae collection such as centrum, neural spine and transverse process were grouped together as a large bovid size or small ruminant size.

Remains of small animals like rodents, lagomorphs, birds, fish and reptiles were identified with the help of reference skeletons. Otherwise they have been assigned in groups like birds, reptiles, fish and so on.

Molluscan shells were identified using the characters described by Tonapi (1980) and with the help from Dr. Arati Deshpande. Unidentified fragments have been classified according to their size as large, medium and small, they were counted and weighed.

3.2 Sex and Age Profiles of Subfossil Species

The determination of sex and age of both domestic and wild animals killed or hunted at archaeological sites can throw valuable light on many aspects of the cultural history and behaviour of ancient population. It is generally found that in a majority of the Indian protohistoric sites a large number of sub-adult male cattle have been killed between the age of 12 to 2 years, when the meat is tender and tasty. This selection is only possible when there is a stock of animals. Rather, this further substantiates the domestic character of the animal in addition to the qualitative markers. The thorough knowledge of the animal husbandry practices and the economic utility of animals can also be inferred from such studies. Another important aspect is the season of occupation of such sites which are not permanent settlements.

In the last few decades many scholars all over the world contributed a lot in the field of archaeozoology especially on the methodology of determining the age and sex of animals at death (Cornwall 1956; Silver 1963; Chaplin 1971; Uerpmann 1973; Grant 1975, 1978, 1982; Grigson 1982; Wilson 1982; Deniz and Payne 1982; Klein and Cruze-Uribe1984).

Sex determination

Grouping of animals into males, females and castrates can give information regarding animal husbandry practices related to the functional aspect of the animal in the general economy of a culture. In a majority of animals especially among the mammalian species some skeletal elements differ in morphology between males and females. The skulls and pelvic bones in a majority of mammals differ in males and females. In equids and suids the canine teeth are very prominent in males whereas in females they are absent or sometimes rudimentary. The female skulls are devoid of horns and antlers in certain bovids and cervids. Also when present they are different in size and shape in females. Likewise, the baculum is the characteristic feature of the male carnivores. Moreover, the pelvic bone in the females of mammalian species is shaped differently because of the child bearing character (Taber 1956; Boessneck 1969; Edwards, Marchinton and smith 1982; Klein and Cruz-Uribe 1984).

However, the criteria given above for sex determination could not be implemented to the faunal assemblage of Inamgaon due to very fragmentary nature of the bones. Hence for this sample sex determination of animals was mainly based on osteo-metric analyses.

Age determination

Aging of animals slaughtered at archaeological sites can provide information about the dietary preference of man, stock-raising and the season of occupation of site. Different methods are employed in the age determination of animals and they are as follows 1) tooth eruption and replacement, 2) tooth wear, 3) epiphysial fusion, 4) closure of cranial sutures and 5) development of antlers (Chaplin 1971; Moris 1972, 1978; Grigson 1982; Grant 1982).

Tooth wear is one of the oldest techniques for age determination. This is particularly applicable to herbivorous animals. Esref Deniz and Sebastian Payne (1982) are the main contributors to this methodology. They have worked on Anatolian sheep and goat. The sequence of eruption, replacement and dental wear patterns are the basis of their method which are easy to apply on mandibular dentition.

From the above group only two methods are used to estimate the age at death of animals at Inamgaon. First is the based on jaw bones with teeth and second on the state of fusion of epiphyses of certain bones. Accuracy of age determination is possible when the jaw bones are well preserved with dentition. Likewise, when epiphysial fusion has already taken place in bones further inference on age is rather difficult.

3.3 Quantitative Methods

Like identification of bones, the application of quantitative methods to archaeofauna is a basic necessity in the archaeological studies. This can throw light on the taxonomic abundance of various species and species preference.

1. Number of Identified Specimens (NISP)

To understand the relative importance of species in the culture, NISP is the basic counting unit. This has been used to quantify the abundance of taxa within a given faunal assemblage. There are many measures which can be derived NISP, like Minimum Number of Individuals, the size of population and weight of animals (Chaplin 1971; Grayson 1973, 1979, 1984; Gauthier 1984; During 1985).

NISP is the simple count of identifiable specimens of each species represented in the collection. It has both advantages and disadvantages. The advantages of the NISP are that, it is easy to calculate and there is no need for further numerical manipulation. NISP values are additive in nature. New NISP values of renewed excavations at a site can be added to the old one (Klein and Cruz-Uribe 1984).

It has some disadvantages also. The skeletons of some species have more parts than the skeletons of others and it ignores this fact. For example, primates have more number of bones like clavicles, supplementary metapodials and phalanges which are not found in other animals like bovids, cervids, etc.

NISP is very sensitive to bone fragmentation. Fragmentation does not affect equally to all the animal species and all bones. Highly fragmented assemblage will have higher NISP for all species. However, fragmentation does not affect equally to all species and all bones.

2. Minimum Number of Individuals (MNI)

It helps to determine the relative importance of various animals in the economy over a period of time. This can also be used as evidence for cultural changes or for changes in the local population (Krantz 1968). Minimum Number of Individual has been used in Paleontology for several decades before (Casteel 1979).

White (1953) was the first to introduce MNI in archaeozoology. There is simple way of calculating the MNI. This is done by classifying the bones to proximal or distal and right or left side. For example, if there are 9 distal right and 7 distal left humerii, the MNI will be 9. This is not the exact number as they might have come from more than 9 animals. Therefore, each of the bone is matched with the other (Bokonyi 1970; Chaplin 1970). According to Bokonyi (1970), other than matching the bones, one should also consider the state of fusion of the bone and the tooth wearing pattern in the case of dentition. This will not give the actual number of individual but, it does come closer. The MNI can never be larger than NISP (Klein and Cruz-Uribe 1984).

Techniques for defining MNI have become more refined over the years. According to Krantz (1968) formula, the numbers of pairs of a particular bone, unpaired of right side and unpaired of left side of the same are counted. The formula is as follows:

$$N = \frac{R^2 + L^2}{2P}$$

N = Number of animals in original population
R – Total number of right side bone
L = Total number of left side bone
P = Number of pairs established.

This method yields a higher total which is closer to the actual number of individual originally involved.

Chaplin (1971) suggests another way of counting MNI. He has proposed a formula as follows:

$$GMT = \frac{C^t}{2 + D^t}$$

GMT stands for the Grand Minimum Total
C^t stands for the total number of comparable paired elements

D^t stands for the total number of dissimilar elements

Fieller and Turner (1982) have suggested a formula which is modification of Krantz (1968) formula.

N = (LR) / P

Where N is the estimate of the original pairs and P is number of matched pairs of bone.

Perkins (1973) has also suggested a method to calculate the MNI. Before counting the MNI; consider the time period taken to form fossil assemblage, Preservation of fossil assemblage and recovery of fossil assemblage. According to his procedure 1) count the number of skeletal parts like distal humerus, proximal radius, etc. which are least affected by cultural and factors of preservation (A), 2) sum the number of bones or fragments of bones established in step 1 (B), 3) to determine the abundance of each species (f) divide the total number of bones or bone fragments established in step 2 by the number of skeletal parts in step 1 i.e. f = B/A.

While calculating the MNI of Inamgaon faunal assemblage, besides matching the bone, state of fusion, tooth wear pattern and the measurements have been taken into consideration.

MNI gives an idea about the actual number of animals killed, but it has some flaws. MNI values are not additive in nature. It involves complicated calculations which are time consuming. Species represented by only a few bones may have the same MNI as one represented by many bones. So, Gejvall (1969) and Perkins (1973) suggest that MNI should not be calculated for small sample size, there should be a minimum sample size.

3. Weight Method

Animals played an important role in prehistoric man's subsistence. Weight of all the bones is related to the meat weight of that animal, therefore, Kubasiewicz (1956) calculated the meat weight from bone weight of different species by using empirical values for their relationship. In this method bone weight is multiplied by a factor that is presumed to reflect the ratio between the bone weight and meat weight in live animals.

Cook and Treganza (1950) have proposed a formula for determining the meat weight. In this method, it was assumed that in mammals and birds, the bone weight represents 6% of the weight of an animal and recovered bones represents 50% loss of the bone originally present. The formula is as follows:

CT = 2 [16.67(x)]

Where, CT = the meat weight of the animal in an assemblage
X = the bone weight of animal from the assemblage

According to Kubasiewicz (1956) for pigs the weight of bone is between 5.6 and 8.3 %. To determine the meat weight of pigs from bone weight, the equation is as follows:

K1 = 12.05 (x)

K1 is the lower limit of predicted meat weight and x is the bone weight.

Or K2 = 14.28 (x)

Where K2 is the mean predicted meat weight

Or K3 = 17.86 (x)

Where K3 is the upper limit of predicted meat weight

He presented constants for pig, sheep and cattle.

Reed (1963) has also proposed formula as follows:

R = 0.7 [13.33 (x)]

Pitts and Bullard's (1968) studies indicated the bone weight to meat weight ratio in an average 10.77 + 2.05 %. According to Reynolds and Karlotki (1977), the mean value for the bone weight among mammals is 8.74 % of total body weight.

Casteel (1979) has given an equation by using McMeekan's (1940) data for pigs is

Y = aX<u>b</u>

Where, Y is the predicted meat weight
X is the bone weight
a and b are constants to be determined empirically
As bone weight exceeds from 100 gms to 100,000,000 gms, this above equation does not show accuracy.

Generally bone weight is never accurate (Prummel (1983). Even after washing and cleaning, some earth always remains on bones particularly in marrow hallows of the long bones which creates an error in bone weight.

Another method to calculate the meat weight is the modification of MNI method. First White (1953) has suggested and employed this method. In this method, the MNI for a species is multiplied by the consumable meat weight of that species (Klein and Cruz-Uribe 1984; Thomas 1988).

This method also has some drawbacks like if MNI of one cattle and ten blackbucks, where blackbucks are significant than the cattle but meat weight shows less significance of ten blackbucks. Usually it is assumed that all members of a species have the same meat weight without taking sex and age into consideration when actually this is not the case.

NISP, MNI and meat weight are the basic quantitative methods in archaeozoology. All these calculations depend on the faunal collection and sample size from the archaeological sites.

4. Standardization of data

Many times the area excavated indifferent parts may not be the same. It depends on the interest of the excavator, which cultural phase/s or layer/s is important to him. This can led to further serious anomalies in the faunal report. The excavation at Inamgaon, concentrated more on the Late Jorwe phase, since this phase of the Chalcolithic culture was not represented in many sites. The volume of the excavated soil from the Late Jorwe phase is 28195m^3, Early Jorwe phase 1611 m^3 and Malwa phase 977 m^3. Thus naturally the cultural materials were more from the Late Jorwe phase. Therefore, an attempt has been made to standardize the results of the faunal studies at one level with the assumption that the distribution of archaeofauna in the site is uniform for the excavated and the unexcavated areas, which may not be always true. However this criterion may be useful for comparing the three cultural phases at Inamgaon. Of the 16 layers at Inamgaon the area and the volume of the excavated deposit in layer 2 was maximum. Assuming that all the layers were excavated in the same pattern as in layer 2, a different picture emerges as shown in figures (no. 5.7, 5.8 and 5.17). Thus, the over representation of faunal remains of the Late Jorwe phase could be nullified by standardizing the faunal results (Pawankar and Thomas 1997).

5. Level of Identification

Always it may not be possible to identify all the bone fragments to species or even to genus levels. Fragments without any specific diagnostic character cannot be identified. Thus, to find out the level of identification the faunal remains from Inamgaon were classified into three groups and they are as follows:

1. NISP first level: Fragments which could be identified to the species level.
2. NISP second level: Fragments which could be identified into broad groups such as belong to Genus, large, medium and small sized mammals, birds, reptiles and fishes.
3. Unidentifiable: Fragments which could not be identified to any taxonomic unit or skeletal part.

The level of identification (Figs. 5.1-5.3) shows a high degree of fragmentation of bones in the Early Jorwe phase. The fragmentation of bones depends on many factors such as the taphonomy, the pattern of butchery, carcass utilization, cooking, the secondary uses of bones, excavation, transport, etc.

6. Measurability Index

The measurability index is the percentage of total measurable bone fragments in NISP level first, which includes complete and partially complete bones. The index (Fig. 5.4) calculated for the bones from Inamgaon shows that a large number of bones were measurable in the Late Jorwe phase and the degree of fragmentation is more in the Early Jorwe and Malwa phases which confirm the observations made from the Level of identification.

7. Horizontal distribution

Various activities such as butchering, cooking, storing, refuse dumping, manufacture of implements, etc. at a site take place in different areas. The horizontal or spatial distribution of faunal assemblage at a site can reveal some of these activity areas. For the Horizontal distribution the complete skeleton of animal is classified into three groups

A: High meat value bones which include scapula, humerus, radius, ulna, pelvis, femur, tibia and the vertebral column.

B: Less meat value bones include skull, mandible, rib, metacarpal and metatarsal.

C: No meat value bones like phalanges, carpals, tarsals and sesamoids.

The faunal assemblage at Inamgaon is fragmentary in nature. However, generally a set pattern is visualized in the breakage pattern of bones. The long bones, scapula, metacarpals and metatarsals are found broken in two parts, the proximal and distal ends. Similarly, pelvis is found in four fragments and mandible in three parts. Also almost 40 fragments of vertebrae, 24 fragments of ribs, 15 fragments of skulls are usually found. However, non-meat-bearing parts like phalanges, carpals, tarsals and sesamoids are generally devoid of cut marks and found complete. It is interesting to find what would be the normal distribution of bones if the fragmentation is as mentioned above taking into consideration some of the taphonomical factors. At Inamgaon the bones do not exhibit rolling marks or transportation and much predator or scavenging activities. Here an attempt has been made to achieve a normal distribution pattern based on the archaeofaunal fragmentation. Hence, the high meat value bones (A) and the less meat value bones (B) are added together and are divided by no-meat value bones (C) The expected value is 1.96. The expected value shows the normal distribution of bones which includes meat-bearing and non-meat-bearing parts depending on the anatomical skeletal parts. This expected value is calculated for the every trench in different layers and classified into three groups which show the normal distribution, high meat-bearing parts and the no-meat-bearing parts. Taking into consideration the loss of bones and taphonomical factors the normal distribution value has been fixed between the range of 1.5-2.5, more that 2.5 is considered high meat-bearing parts and less than 1.5 as a value for no-meat-bearing parts.

8. Skeletal representation

To understand the importance and the use of different parts of the skeletal parts in different animals, the skeletal elements were classified according to their anatomical positions, and a frequency for each of the elements has been worked out. This has been done for some of the major species of domestic and wild animals which contributed to the subsistence system at Inamgaon.

3.4. Metrical Analysis

Like quantitative methods, metrical analysis is also an important tool in archaeozoological studies. This helps in understanding various aspects such as:

1. Differentiation of animal species and individual variation
2. Size changes in wild animals through time and space
3. Determination of sex and sexual dimorphism
4. Differentiation of domestic and wild animals
5. Intra site comparisons of faunal remains

For the first time Durest (1926) used the measurements to interpret the archaeozoological data. Von den Driesch (1976) has produced a guideline to measure faunal remains from archaeological sites. This system of bone measurements is now used all over the world as a standard method.

Only complete and partially complete bones are measured using digital caliper. The whole data is transferred into the micro computer in Dbase III plus programme for further analysis.

Sexing of the blackbuck has been done by metrical analysis. The greatest length of the first phalanges, second phalanges and astragali versus the distal breadth of these bones are used to separate them as males and females. The first and second phalanges and astragali of known male (DCP 97) and female (DCP 83) are measured. The first and second phalanges of fore and hind limbs are measured separately. This known data is plotted with the data from Inamgaon. This has shown groups of males and females (Fig. 5.27-5.29) distinctly. The graphs of second phalanges and astragali reveal the two separate groups. For example, the greatest length of the female second phalanx is less than 23 cm and it is greater than 23 cm in male. The same procedure has been applied for the first phalanges and the astragali of blackbuck. The graph of the first phalanges shows that the measurements of the fore limb of male and the hind limb of female are overlapping between the 41.5 cm and 49.5 cm.

With the help of measurements of metacarpal and astragalus, the sex ratio for the cattle has been calculated. The measurements of the certain bones have been used to find out the stature or withers height of cattle. The estimation of the shoulder height in archaeozoological research helps to identify the size of an individual animal and also to assess the range of variation within a population. Kieswalter (1888) has introduced the height estimate to the sub-fossil assemblages. The height of the animal is based on the length of the bones (Lauwerier 1988). Different scholars have worked out equations to calculate the shoulder height for the domestic animals like cattle (Boessneck 1956; Zalkin 1960; Fock 1966; Matolcsi 1970; Bergstrom and Wijnaarden-Bakker 1983, 1988; Joglekar 1993) goat (Schramm 1967) sheep (Haak 1965; Teichert 1978), pig (Teichert 1990) dog (Harcourt 1974) and horse (Kieswalter 1988). Zalkin (1970) has found ut an equation to calculate the shoulder height of cattle and is as follows:

Height in cm = 18.30 x (medial length of the astragalus.

This formula is applied on the measurements of the astragali of cattle from Inamgaon.

3.5 Past and present distribution

Environmental interpretation is one of the major aspects of archaeozoological research. Generally the wild animals associated with the prehistoric cultures are often studied in detail for their habitats and past environmental reconstructions are made from such evidences. Wild animals need specific ecological settings. Four horned antelope, blackbuck, nilgai, chital and sambar have been found in a majority of the Chalcolithic sites in western Maharashtra. These are typical animals found in scrub forest and plains covered with grass areas. If one looks at the present day distribution of these animals many of them are not found in the same areas. It shows the changes in the past and present environment. The present day distribution of these animals has been studied from the literature published from the Forest Department of Maharashtra. The evidences of these animals from various archaeological sites are plotted in the present day distribution maps of wild animals to understand the environmental changes caused during the last 2000 years or more. However, it should mention here that the excavations were limited and confined to only a few districts of Maharashtra. Thus, an overall picture may not e available; even then the changes could be visualized.

Man and Animal in the Chalcolithic Culture of Maharashtra

Animals constituted a major source of food in the subsistence economy of the Chalcolithic population at Inamgaon. The fauna represented at an archaeological site mainly depends on the ecological setting of the region, the socio-economy and the exploitation pattern. Generally for the hunter-gatherer societies, wild animals played an important role in the food economy. But with the introduction of agriculture and animal domestication the subsistence system changed to a greater extent, now depending more on agricultural products and domestic animals than the wild animals. Inamgaon is such an early farming settlement in the Bhima valley which revealed a wide spectrum of diverse fauna consisting of mammals, reptiles, birds, fish, molluscs and crustaceans. A majority of these animals were utilized for food purpose.

More than 50 species of animals have been identified from the faunal assemblage of Inamgaon and they are as follows:

Mammals

Domestic animals
Bos indicus (cattle)
Bubalus bubalis (buffalo)
Capra hircus (goat)
Ovis aries (sheep)
Sus domesticus (domestic pig)
Canis familiaris (dog)
Equus caballus (horse)
Equus asisnus (ass)
Elephas maximus (elephant)

Wild animals
Canis lupus (wolf)
Canis aureus (jackal)
Felis chaus (jungle cat)
Panthera pardus (leopard)
Panthera tigris (tiger)
Sus scrofa (wild pig)
Bos gaurus (gaur)
Antilope cervicapra (blackbuck)
Gazella bennetti (chinkara)
Tetracerus quadricornis (four-horned antelope)
Boselaphus tragocamelus (nilgai)
Axis axis (chital)
Axis porcinus (hog deer)
Cervus unicolor (sambar)
*Muntiacus mu*ntjak (barking deer)
Hystrix indica (porcupine)
Herpestes edwardsi (mongoose)
Lepus nigricollis (hare)
Rattus rattus (rat)

Birds
Gallus domesticus
Gallus gallus
Anser anser
Pavo cristatus
Ardeotis nigriceps

Reptiles
Trionyx gangeticus
Kachuga sp.
Chitra indica
Crocodylus palustris
Varanus monitor
Snake

Fish
Cat fish
Labeo rohita
Ritha ritha
Wallago wallago

Molluscs
Lamellidens marginalis
Pyrreyssia corrugata
Turbo coronatus
Digoniostoma pulchella
Viviparus bengalensis
Pila globosa
Melania striatella tuberculata
Turbinella pyrum
Paphia gallus
Turritella sp.
Meretrix sp.
Marcia recens

Crustacea
Crabs

***Bos indicus* (cattle)**

The skeletal remain of cattle are present in all the 16 layers. A layer-wise distribution of identified specimens of this species and their weight are given in Table 4.11-4.16. Apparently, the degree of their representation varies in different stratigraphical units. The maximum percentage representation of this animal is noticed in layer 7, which accounted for 41.91% in the total collection of the bones from this layer (Table 4.7). The percentage of the identified fragments in Late Jorwe phase is 17.43%, in Early Jorwe phase 30.88% and in Malwa phase is 22.72%, suggests maximum utilization of this species in the Early Jorwe phase (Table 4.17) which was economically the most prosperous phase at Inamgaon.

Table 4.1. The Number of Identified Specimens in Layer 1.

Species	NISP	NISP %	Weight (gms)	Weight %
Bos indicus	257	13.02	6075	25.21
Bos/Bubalus	375	18.00	4973	20.68
Bubalus bubalis	17	0.86	485	2.01
Capra hircus	10	0.50	141	0.58
Ovis aries	1	0.05	5	0.02
Capra hircus/Ovis aries	12	0.61	135	0.56
Canis familiaris	10	0.51	75	0.31
Equus caballus	3	0.15	140	0.58
Axis axis	7	0.35	60	0.25
Axis porcinus	1	0.05	5	0.02
Cervus unicolor	7	0.35	150	0.62
Muntiacus muntjak	1	0.05	10	0.04
Deer sp.	1	0.05	15	0.06
Boselaphus tragocamelus	34	1.72	725	3.01
Tetracerus qudricornis	2	0.10	35	0.15
Antilope cervicapra	1025	51.92	9496	39.41
Gazella bennetti	10	0.51	135	0.56
Small Ruminant	116	5.88	607	2.52
Sus scrofa	4	0.20	65	0.27
Canis lupus	3	0.15	15	0.06
Felis chaus	2	0.10	10	0.04
Rattus rattus	7	0.35	14	0.06
Reptile	2	0.10	2	0.01
Trionyx gangeticus	10	0.51	90	0.37
Kachuga sp.	3	0.15	30	0.12
Turtle	1	0.05	5	0.02
Snake	1	0.05	1	0.004
Fish	4	0.20	16	0.07
Lamellidens sp.	27	1.37	202	0.84
Parreyssia corrugate	40	2.03	376	1.56
Turbo coronatus	1	0.05	5	0.02
Total	**1974**	**100%**	**24098**	**100%**

Table 4.2. The Number of Identified Specimens in Layer 2.

Species	NISP	NISP%	Weight (gms.)	Weight %
Bos indicus	777	14.21	17545	29.51
Bos/Bubalus	1332	24.37	13696	23.04
Bubalus bubalis	37	0.68	1160	1.95
Capra hircus	25	0.46	193	0.32
Ovis aries	9	0.16	101	0.17
Caprahircus/Ovis aries	26	0.47	211	0.35
Sus domesticus	20	0.37	231	0.39
Canis familiaris	33	0.60	270	0.46
Equus caballus	4	0.07	70	0.12
Equus asinus	1	0.02	10	0.02
Axis axis	8	0.15	115	0.19
Axis sp.	1	0.02	5	0.01
Cervus unicolor	5	0.09	120	0.20
Medium sized Ruminant	1	0.02	15	0.03
Deer sp.	3	0.05	15	0.03
Boselaphus tragocamelus	82	1.50	1460	2.46
Tetracerus quadricornis	11	0.20	52	0.09
Antilope cervicapra	2377	43.49	19856	33.40
Gazella bennetti	10	0.18	62	0.10
Small ruminant	313	5.73	1327	2.23
Sus scrofa	8	0.15	135	0.23
Canis lupus	11	0.20	81	0.14
Canis aureus	4	0.07	25	0.04
Canis sp.	7	0.13	29	0.05
Felis chaus	5	0.09	17	0.03
Panthera tigris	1	0.02	20	0.03
Panthera pardus	2	0.04	10	0.02
Herpestes edwardsi	1	0.02	5	0.01
Rattus rattus	1	0.02	1	0.001
Lepus nigricollis	1	0.02	5	0.01
Bird sp.	10	0.18	30	0.05
Gallus domesticus	1	0.02	1	0.001
Pavo cristatus	4	0.07	8	0.01
Anser anser	1	0.02	5	0.01
Reptile	3	0.05	20	0.03
Trianyx gangeticus	47	0.86	405	0.68
Kachuga sp.	7	0.13	60	0.10
Turtle	5	0.09	9	0.02
Fish	32	0.58	92	0.15
Lamellidens sp.	72	1.32	523	0.88
Digoniostoma pulchella	2	0.04	2	0.003
Parreyssia corrugata	164	3.00	1431	2.41
Viviparus bengalensis	1	0.02	5	0.01
Turbinella pyrum	1	0.02	20	0.03
Total	**5466**	**100 %**	**59453**	**100%**

Table 4.3. The Number of Identified Specimens in Layer 3.

Species	NISP	NISP%	Weight (gms.)	Weight %
Bos indicus	945	21.41	19013	38.17
Bos/Bubalus	959	21.73	10539	21.16
Bubalus bubalis	55	1.24	1785	3.58
Capra hircus	16	0.36	145	0.29
Ovis aries	2	0.05	15	0.03
Capra hircus/Ovis aries	22	0.50	164	0.33
Sus domesticus	7	0.16	55	0.11
Canis familiaris	19	0.43	155	0.31
Equus caballus	4	0.09	60	0.12
Bos gaurus	1	0.02	65	0.13
Axis axis	10	0.23	95	0.19
Cervus unicolor	9	0.20	310	0.62
Deer sp.	2	0.05	30	0.06
Boselaphus tragocamelus	98	2.22	1675	3.36
Tetracerus quadricornis	14	0.32	80	0.16
Antilope cervicapra	1820	41.24	13240	26.58
Gazella bennetti	6	0.13	27	0.06
Small ruminant	243	5.51	941	1.89
Sus scrofa	4	0.09	35	0.07
Canis lupus	6	0.13	60	0.12
Canis aureus	2	0.05	10	0.02
Canis sp.	2	0.05	6	0.01
Felis chaus	2	0.05	6	0.01
Herpestes edwardsi	5	0.11	25	0.05
Rattus rattus	2	0.05	2	0.004
Elephas maximus	3	0.07	205	0.41
Lepus nigricollis	4	0.09	4	0.01
Gallus domesticus	1	0.02	1	0.002
Gallus gallus	5	0.11	17	0.04
Reptile	5	0.11	22	0.05
Trionyx gangeticus	8	0.18	70	0.14
Chitra indica	1	0.02	10	0.02
Turtle	2	0.05	10	0.02
Varanus monitor	1	0.02	1	0.002
Crocodilia sp.	2	0.05	15	0.03
Fish	1	0.02	5	0.01
Lamellidens sp.	61	1.38	493	0.99
Digoniostoma pulchella	2	0.05	10	0.02
Corbicula sp.	1	0.02	10	0.02
Pila globosa	1	0.02	1	0.002
Parreyssia corrugata	60	1.36	406	0.81
Total	**4413**	**100%**	**49818**	**100%**

Table 4.4. The Number of Identified Specimens in Layer 4.

Species	NISP	NISP %	Weight (gms)	Weight %
Bos indicus	520	18.21	8943	32.48
Bos/Bubalus	865	30.30	8508	30.90
Bubalus bubalis	18	0.63	425	1.54
Capra hircus	2	0.07	6	0.02
Capra hircus/Ovis aries	21	0.73	122	0.44
Sus domesticus	5	0.17	35	0.13
Canis familiaris	14	0.49	90	0.33
Equus caballus	1	0.04	60	0.21
Bos gaurus	1	0.04	50	0.18
Axis axis	13	0.45	170	0.62
Cervus unicolor	9	0.31	70	0.26
Boselaphus tragocamelus	80	2.80	1231	4.47
Tetracerus qudricornis	2	0.07	10	0.04
*Antilope cervica*pra	1034	36.22	6667	24.21
Gazella bennetti	3	0.10	3	0.01
Small Ruminant	196	6.86	624	2.27
Canis aureus	1	0.04	5	0.02
Felis chaus	1	0.04	1	0.003
Rattus rattus	1	0.04	1	0.003
Lepus nigricollis	1	0.04	5	0.02
Bird sp.	1	0.04	5	0.02
Trionyx gangeticus	1	0.04	20	0.07
Kachuga sp.	1	0.04	5	0.02
Lamellidens sp.	25	0.87	186	0.67
Digoniostoma pulchella	1	0.04	1	0.003
Parreyssia corrugate	35	1.22	275	0.10
Viviparus bengalensis	3	0.10	15	0.06
Total	**2855**	**100%**	**27533**	**100%**

Table 4.5. The Number of Identified Specimens in Layer 5.

Species	NISP	NISP %	Weight (gms)	Weight %
Bos indicus	153	32.35	2755	46.22
Bos/Bubalus	145	30.66	1895	31.79
Bubalus bubalis	2	0.42	25	0.42
Ovis aries	1	0.21	10	0.17
Capra hircus/Ovis aries	7	1.48	40	0.67
Canis familiaris	5	1.06	25	0.42
Deer sp.	1	0.21	25	0.42
Boselaphus tragocamelus	9	1.90	110	1.85
Tetracerus qudricornis	1	0.21	5	0.08
*Antilope cervica*pra	75	15.86	537	9.01
Small Ruminant	49	10.36	209	3.51
Canis lupus	1	0.21	5	0.08
Trionyx gangeticus	1	0.21	5	0.08
Lamellidens sp.	15	3.17	230	3.86
Parreyssia corrugate	8	1.69	85	1.42
Total	**473**	**100%**	**5961**	**100%**

Table 4.6. The Number of Identified Specimens in Layer 6.

Species	NISP	NISP%	Weight (gms.)	Weight%
Bos indicus	205	24.46	3891	40.53
Bos/Bubalus	442	52.74	4481	46.68
Bubalus bubalis	6	0.71	75	0.78
Capra hircus	2	0.24	6	0.06
Ovis aries	1	0.12	5	0.05
Capra hircus/Ovis aries	11	1.31	66	0.69
Sus domesticus	2	0.24	10	0.10
Canis familiaris	4	0.48	30	0.31
Equus caballus	1	0.12	25	0.26
Boselaphus tragocamelus	3	0.36	25	0.26
Tetracerus quadricornis	2	0.24	15	0.16
Antilope cervicapra	91	10.86	584	6.07
Small ruminant	44	5.25	210	2.19
Sus scrofa	2	0.24	30	0.31
Canis lupus	2	0.24	6	0.06
Canis sp.	1	0.12	5	0.05
Felis chaus	1	0.12	5	0.05
Rattus rattus	1	0.12	1	0.01
Lamellidens sp.	13	1.55	110	1.15
Digoniostoma pulchella	1	0.12	5	0.05
Parreyssia corrugata	3	0.36	15	0.16
Total	**838**	**100 %**	**9600**	**100%**

Table 4.7. The Number of Identified Specimens in Layer 7.

Species	NISP	NISP%	Weight (gms.)	Weight %
Bos indicus	368	41.91	5731	54.54
Bos/Bubalus	346	39.41	3528	33.58
Bubalus bubalis	13	1.48	320	3.05
Capra hircus	2	0.23	15	0.14
Ovis aries	1	0.11	5	0.05
Capra hircus/Ovis aries	1	0.11	5	0.05
Canis familiaris	3	0.35	11	0.10
Bos gaurus	1	0.11	10	0.09
Axis axis	7	0.80	65	0.62
Cervus unicolor	2	0.23	40	0.38
Boselaphus tragocamelus	2	0.23	20	0.19
Tetracerus quadricornis	1	0.11	1	0.01
Antilope cervicapra	56	6.38	381	3.63
Small ruminant	39	4.45	150	1.43
Canis sp.	1	0.11	5	0.05
Felis chaus	1	0.11	5	0.05
Rattus rattus	13	1.48	13	0.12
Trionyx gangeticus	2	0.23	30	0.29
Lamellidens sp.	7	0.80	65	0.62
Digoniostoma pulchella	1	0.11	1	0.01
Parreyssia corrugata	10	1.14	56	0.53
Turbinella pyrum	1	0.11	50	0.47
Total	**878**	**100 %**	**10507**	**100 %**

Table 4.8. The Number of Identified Specimens in Layer 8.

Species	NISP	NISP%	Weight (gms.)	Weight%
Bos indicus	955	36.56	18016	49.07
Bos/Bubalus	1340	51.30	15216	41.44
Bubalus bubalis	59	2.26	1470	4.00
Capra hircus	7	0.27	45	0.12
Ovis aries	1	0.04	5	0.01
Capra hircus/Ovis aries	21	0.80	130	0.36
Sus domesticus	5	0.19	46	0.13
Canis familiaris	12	0.46	66	0.18
Equus caballus	1	0.04	20	0.05
Axis axis	3	0.11	40	0.11
Cervus unicolor	1	0.04	20	0.06
Muntiacus muntjak	1	0.04	10	0.03
Boselaphus tragocamelus	14	0.53	250	0.68
Tetracerus quadricornis	2	0.08	15	0.04
Antilope cervicapra	82	3.14	586	1.60
Gazella bennetti	5	0.19	25	0.07
Small ruminant	51	1,95	312	0.85
Sus scrofa	5	0.19	60	0.16
Canis lupus	1	0.04	15	0.04
Canis aureus	2	0.08	15	0.04
Rattus rattus	1	0.04	1	0.002
Gallus domesticus	1	0.04	5	0.01
Reptile	1	0.04	5	0.01
Trionyx gangeticus	4	0.15	20	0.06
Lamellidens sp.	8	0.31	60	0.16
Parreyssia corrugata	21	0.80	200	0.55
Paphia gallus	1	0.04	5	0.01
Turbo coronatus	2	0.08	25	0.07
Meretrix sp.	4	0.15	30	0.08
Marcia recens	1	0.04	5	0.01
Total	**2612**	**100 %**	**36718**	**100%**

Table 4.9. The Number of Identified Specimens in Layer 9.

Species	NISP	NISP%	Weight (gms.)	Weight%
Bos indicus	315	28.69	6045	50.86
Bos/Bubalus	549	50.00	4640	39.04
Bubalus bubalis	12	1.10	185	1.56
Capra hircus	3	0.27	25	0.21
Capra hircus/Ovis aries	13	1.18	65	0.55
Sus domesticus	1	0.09	55	0.46
Canis familiaris	7	0.64	35	0.30
Axis axis	3	0.27	35	0.30
Boselaphus tragocamelus	5	0.46	100	0.84
Antilope cervicapra	18	1.64	100	0.84
Small ruminant	163	14.85	540	4.54
Canis lupus	1	0.09	5	0.04
Canis sp.	1	0.09	5	0.04
Rattus rattus	1	0.09	1	0.008
Fish	1	0.09	5	0.04
Lamellidens sp.	2	0.18	10	0.08
Parreyssia corrugata	3	0.27	35	0.30
Total	**1098**	**100 %**	**11886**	**100%**

Table 4.10. The Number of Identified Specimens in Layer 10.

Species	NISP	NISP%	Weight (gms.)	Weight %
Bos indicus	89	15.03	1130	33.08
Bos/Bubalus	406	68.58	1770	51.81
Bubalus bubalis	7	1.18	120	3.51
Capra hircus	1	0.17	5	0.15
Capra hircus/Ovis aries	29	4.90	156	4.57
Canis familiaris	1	0.17	5	0.15
Boselaphus tragocamelus	1	0.17	10	0.29
Antilope cervicapra	4	0.68	25	0.73
Small ruminant	44	7.43	140	4.10
Turtle	1	0.17	5	0.15
Lamellidens sp.	7	1.18	35	1.02
Parreyssia corrugata	2	0.34	15	0.44
Total	**592**	**100 %**	**3416**	**100%**

Table 4.11. The Number of Identified Specimens in Layer 11.

Species	NISP	NISP%	Weight (gms.)	Weight%
Bos indicus	227	23.10	3935	34.79
Bos/Bubalus	503	51.17	5727	50.63
Bubalus bubalis	13	1.32	365	3.23
Capra hircus	7	0.71	32	0.28
Ovis aries	4	0.41	12	0.11
Capra hircus/Ovis aries	26	2.65	136	1.20
Sus domesticus	2	0.20	15	0.13
Canis familiaris	1	0.10	5	0.04
Axis axis	5	0.51	85	0.75
Deer sp.	1	0.10	10	0.09
Boselaphus tragocamelus	4	0.41	70	0.62
Tetracerus quadricornis	2	0.20	10	0.09
Antilope cervicapra	56	5.70	373	3.30
Small ruminant	114	11.60	421	3.72
Panthera pardus	2	0.20	10	0.09
Herpestes edwardsi	1	0.10	5	0.04
Rattus rattus	1	0.10	1	0.01
Fish	2	0.20	15	0.13
Lamellidens sp.	6	0.61	40	0.35
Parreyssia corrugata	6	0.61	45	0.40
Total	**983**	**100 %**	**11312**	**100%**

Table 4.12. The Number of Identified Specimens in Layer 12.

Species	NISP	NISP%	Weight (gms.)	Weight%
Bos indicus	417	28.66	7561	44.61
Bos/Bubalus	674	46.32	7336	43.29
Bubalus bubalis	12	0.82	345	2.03
Capra hircus	3	0.21	15	0.09
Ovis aries	2	0.14	6	0.03
Capra hircus/Ovis aries	24	1.65	133	0.78
Sus domesticus	6	0.41	40	0.24
Canis familiaris	6	0.41	31	0.16
Axis axis	3	0.21	20	0.12
Boselaphus tragocamelus	7	0.48	70	0.41
Tetracerus quadricornis	3	0.21	30	0.18

Antilope cervicapra	87	5.98	570	3.36
Gazella bennetti	1	0.07	5	0.03
Small ruminant	179	12.30	615	3.63
Panthera pardus	3	0.21	15	0.09
Herpestes edwardsi	1	0.07	1	0.01
Rattus rattus	8	0.55	8	0.05
Lepus nigricollis	1	0.07	1	0.01
Trionyx gangeticus	4	0.27	30	0.18
Lamellidens sp.	8	0.55	70	0.41
Digoniostoma pulchella	1	0.07	5	0.03
Parreyssia corrugata	4	0.27	35	0.21
Viviparus bengalensis	1	0.07	5	0.03
Total	**1455**	**100 %**	**16947**	**100%**

Table 4.13. The Number of Iidentified Specimens in Layer 13.

Species	NISP	NISP%	Weight (gms.)	Weight%
Bos indicus	115	22.29	2193	37.13
Bos/Bubalus	238	46.13	2691	45.56
Bubalus bubalis	2	0.39	55	0.93
Capra hircus	3	0.58	15	0.26
Capra hircus/Ovis aries	15	2.91	95	1.61
Sus domesticus	1	0.19	5	0.08
Canis familiaris	1	0.19	10	0.17
Axis axis	1	0.19	10	0.17
Cervus unicolor	1	0.19	10	0.17
Boselaphus tragocamelus	8	1.55	175	2.96
Antilope cervicapra	43	8.33	306	5.18
Gazella bennetti	2	0.39	10	0.17
Small ruminant	75	14.54	312	5.28
Sus scrofa	1	0.19	10	0.17
Canis sp.	5	0.97	5	0.08
Rattus rattus	5	0.97	5	0.08
Total	**516**	**100 %**	**5907**	**100%**

Table 4.14. The Number of Identified Specimens in Layer 14.

Species	NISP	NISP%	Weight (gms.)	Weight %
Bos indicus	276	38.55	5476	50.71
Bos/Bubalus	320	44.69	3920	36.30
Bubalus bubalis	8	1.12	200	1.85
Capra hircus	4	0.56	30	0.28
Capra hircus/Ovis aries	13	1.81	105	0.97
Sus domesticus	5	0.70	60	0.55
Canis familiaris	5	0.70	35	0.32
Bos gaurus	1	0.14	10	0.09
Axis axis	4	0.56	40	0.37
Cervus unicolor	1	0.14	20	0.19
Boselaphus tragocamelus	23	3.21	425	3.94
Antilope cervicapra	29	4.05	178	1.65
Gazella bennetti	1	0.14	5	0.05
Small ruminant	18	2.51	101	0.94
Canis lupus	2	0.28	20	0.19
Rattus rattus	2	0.28	2	0.02
Elephas maximus	1	0.14	50	0.46
Trionyx gangeticus	1	0.14	10	0.09
Kachuga sp.	1	0.14	5	0.05

Lamellidens sp.	1	0.14	106	0.98
Total	**716**	**100 %**	**10798**	**100%**

Table 4.15. The Number of Identified Specimens in Layer 15.

Species	NISP	NISP%	Weight (gms.)	Weight %
Bos indicus	176	23.66	3925	36.82
Bos/Bubalus	356	47.85	4620	43.34
Bubalus bubalis	11	1.48	325	3.05
Capra hircus	2	0.27	10	0.09
Ovis aries	1	0.13	5	0.05
Capra hircus/Ovis aries	11	1.48	65	0.61
Sus domesticus	7	0.94	75	0.70
Canis familiaris	5	0.67	30	0.28
Axis axis	2	0.27	45	0.42
Cervus unicolor	1	0.13	25	0.24
Muntiacus muntjak	5	0.67	25	0.24
Boselaphus tragocamelus	36	4.84	650	6.10
Tetracerus quadricornis	7	0.94	36	0.34
Antilope cervicapra	54	7.26	397	3.73
Small ruminant	43	5.78	211	1.98
Reptile	2	0.27	10	0.09
Lamellidens sp.	19	2.55	140	1.31
Digoniostoma pulchella	2	0.27	10	0.09
Parreyssia corrugata	4	0.54	55	0.52
Total	**744**	**100 %**	**10659**	**100%**

Table 4.16. The Number of Iidentified Specimens in Layer 16.

Species	NISP	NISP%	Weight (gms.)	Weight %
Bos indicus	73	37.63	1666	50.78
Bos/Bubalus	76	39.18	1170	35.66
Bubalus bubalis	2	1.04	80	2.44
Sus domesticus	1	0.51	5	0.15
Canis familiaris	2	1.04	15	0.46
Axis axis	1	0.51	15	0.46
Cervus unicolor	1	0.51	10	0.30
Boselaphus tragocamelus	5	2.58	100	3.05
Tetracerus quadricornis	1	0.51	5	0.15
Antilope cervicapra	17	8.76	113	3.45
Small ruminant	7	3.61	36	1.10
Canis lupus	3	1.55	25	0.76
Canis sp.	1	0.51	1	0.03
Trionyx gangeticus	1	0.51	10	0.30
Lamellidens sp.	3	1.55	30	0.91
Total	**194**	**100 %**	**3281**	**100%**

Bones belonging to the appendicular and axial skeletons of this animal are identified (Pl.I A&B). However, the appendicular parts are numerous compared to the axial (Table. 4.18). On the proximal and distal ends of long bones like humerus, radius,ulna, femur and tibia cut marks are noticed (Pl. II A). Cut marks are also found on the scapula and pelvis. It is interesting to note that the glenoid cavity was usually cut off from the main body of the scapula as it is done even today by the butchers. Similarly, in pelvis cut marks are observed around the region of acetabulum, in the part of illium and ischium, but not on the pubis. The metacarpals and metatarsals have cut marks on the shaft near the proximal and the distal articular ends. Phalanges, carpal and tarsal bones are well preserved and complete (Pl. II B). At times on the distal end of some first phalanges, longitudinal cut marks are noticed (Pl. III.A).

Plate I A. ***Bos indicus*****:** 1. Meatcarpal distal, 2. Metacarpal proximal, 3. Calcaneum, 4. Centrotarsal and 5. Astragalus.

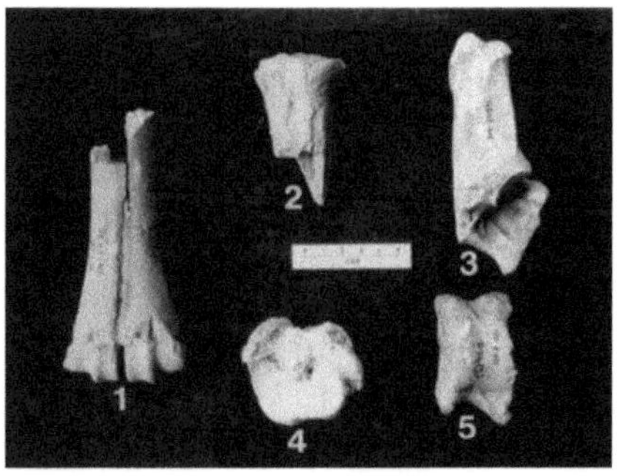

Plate I B. ***Bos indicus*****:** Mandibles

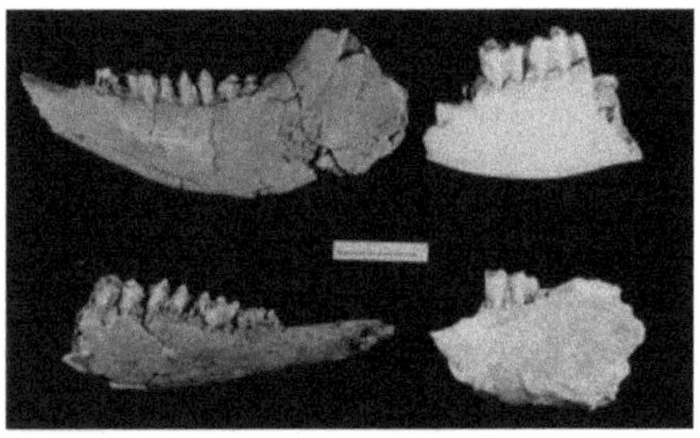

Plate II A. ***Bos indicus*****:** Cut marks on bones. 1st row- Humerus and 2nd row- Radius.

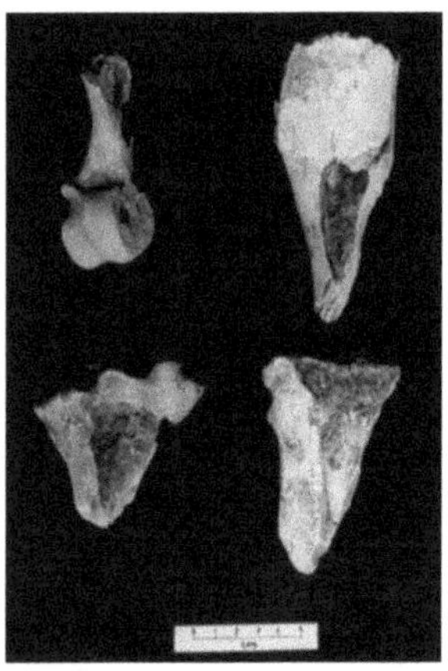

Plate II B. ***Bos indicus*****:** 1st row- First phalanges, 2nd row- Second phalanges and 3rd row- Third phalanges

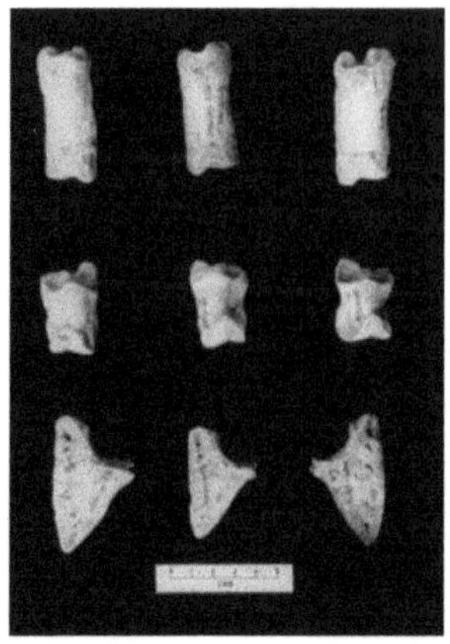

Plate III A. ***Bos indicus*****:** Cut marks on phalanges.

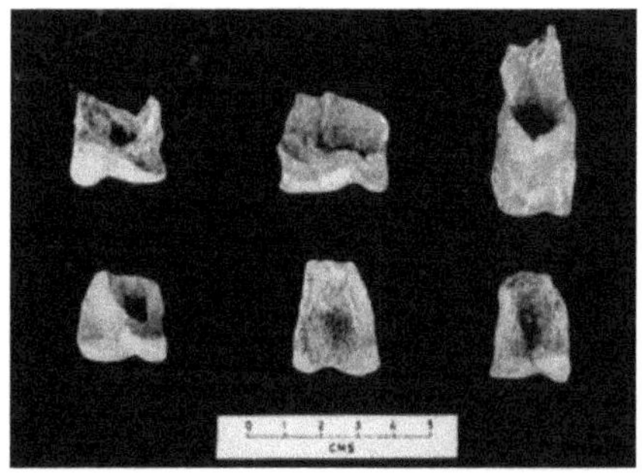

Table 4.17. Frequency of Identified specimens in the three cultural phases at Inamgaon

Species	Late Jorwe	%	Early Jorwe	%	Malwa	%	Total	%
Bos indicus	2652	17.43	2159	30.88	1057	29.13	5868	22.72
Bubalus bubalis	129	00.87	110	01.58	35	00.96	274	01.06
Bos/Bubalus	3676	24.18	3576	51.15	1664	45.85	8916	34.53
Capra hircus	53	00.35	22	00.32	12	00.33	87	00.34
Ovis aries	13	00.09	7	00.10	3	00.08	23	00.09
Capra/Ovis	88	00.58	101	01.45	63	01.74	252	00.98
Sus domesticus	32	00.21	10	00.14	20	00.55	62	00.24
Canis familiaris	81	00.53	28	00.40	19	00.52	128	00.49
Equus caballus	12	00.08	2	00.03	--	--	14	00.05
Equus asinus	1	00.01	--	--	--	--	1	00.01
Elephas maximus	3	00.02	--	--	1	00.03	4	00.01
Canis lupus	21	00.14	4	00.06	5	00.13	30	00.12
Cani aureus	7	00.05	2	00.03	--	--	9	00.03
Canis sp.	9	00.06	3	00.04	6	00.17	18	00.07
Felis chaus	10	00.06	2	00.03	--	--	12	00.05
Panthera pardus	2	00.01	2	00.03	3	00.08	7	00.03
Panthera tigris	1	00.01	--	--	--	--	1	00.01
Sus srofa	16	00.11	7	00.10	1	00.03	24	00.09
Bos gaurus	2	00.01	1	00.01	1	00.03	4	00.01
Antilope cervicapra	6331	41.64	307	04.39	230	06.34	6868	26.60
Gazella bennetti	29	00.19	5	00.07	4	00.11	38	00.15
Tetracerus quadricornis	30	00.20	7	00.10	11	00.30	48	00.19
Boselaphus tragocamelus	303	01.99	29	00.42	79	02.18	411	01.59
Small Ruminants	917	06.03	455	06.51	322	08.87	1694	06.56
Axis axis	39	00.26	18	00.26	11	00.30	68	00.26
Axis porcinus	1	00.01	--	--	--	--	1	00.01
Cervus unicolor	30	00.20	3	00.04	4	00.11	37	00.14
Muntiacus muntjak	1	00.01	1	00.01	5	00.14	7	00.03
Deer sp.	8	00.05	1	00.01	--	--	9	00.03
Herpestes edwardsi	6	00.04	1	00.01	1	00.03	8	00.03
Lepus nigricollis	6	00.04	--	--	1	00.03	7	00.03
Hystrix indica	--	--	--	--	2	00.06	2	00.01
Rattus rattus	11	00.07	17	00.24	15	00.41	43	00.17
Reptile	101	00.66	8	00.11	9	00.25	118	00.46
Birds	23	00.15	1	00.01	--	--	24	00.09
Fish	37	00.24	3	00.04	--	--	40	00.15
Molluscs	520	03.42	99	01.42	43	01.18	662	02.56
Crabs	2	00.01	--	--	2	00.06	4	00.01

Generally on the centrum portion of the vertebrae cut marks are observed. From atlas to caudal, the fragments of all the vertebrae are found. Mandibles are cut at the ascending ramus. Presently this part of the mandible is used in the preparation of soup (Thomas 1988). Cranial fragments with maxillary teeth and horn cores are also identified. The fragments of ribs are innumerable. Bifid dorsal spines of thoracic vertebrae the characteristic feature of the humped cattle are identified at Inamgaon. Charred cattle bones are recovered from all the cultural phases.

Bones are also quantified to find out the taxonomic abundance of each individual species represented in different layers (Table 4.19- 4.31). The maximum representation of this species is found in layer 8 by 45 animals. The Minimum Number of cattle killed are 107 (17.83%), 97 (32.44%) and 83 (32.30%) in the Late Jorwe, Early Jorwe and the Malwa phases, respectively. The bone fragments are also collected from the floors of all the structures in different layers. The settlement structures of layer 7 have yielded the maximum number of cattle bones. It appears that people depended more on beef in the subsistence economy at Inamgaon. Compared to other smaller species of animals exploited for subsistence, the bovids have contributed maximum quantity of meat because of the stocky nature of these animals. The weight of the meat possibly consumed from different animals also has been worked out for individual species in different layers (Table 4.32). It appears that cattle were predominant in the food economy in all the three cultural phases. The maximum amount of meat of this animal was consumed in layer 8 (9,000 kg). The meat weight shows that in the Late Jorwe phase 21,400 kg, Early Jorwe phase 19,400 kg and in the Malwa 16,600 kg meat has been consumed by occupants at Inamgaon.

Table 4.18. Bone distribution of *Bos indicus* in different layers at Inamgaon

Bones	**01**	**02**	**03**	**04**	**05**	**06**	**07**	**08**	**09**	**10**	**11**	**12**	**13**	**14**	**15**	**16**
HC/AN	10	07	12	--	02	--	--	--	02	--	--	--	--	--	--	--
CF	28	40	29	28	--	03	17	25	04	--	03	05	--	02	01	01
MN	18	49	42	11	08	08	11	27	11	01	02	12	03	04	09	01
SC	03	21	22	08	02	06	07	21	06	--	--	05	--	05	--	02
HU	03	21	20	09	02	10	04	29	14	--	04	11	02	05	10	02
RA	06	27	24	18	07	08	09	44	22	01	13	08	03	13	12	04
UL	03	15	20	11	01	02	05	07	05	01	03	04	01	06	05	03
MC	06	25	22	11	03	02	12	34	16	02	05	13	07	05	05	04
C/T	15	29	43	23	10	12	30	57	15	04	08	18	03	07	05	03
MP	07	28	35	19	06	09	06	51	11	03	15	24	04	09	07	03
PV	04	23	27	20	05	07	10	37	17	03	07	08	02	06	02	01
FE	10	23	13	06	03	05	06	35	11	01	--	06	02	07	01	02
TI	08	28	23	10	03	06	08	18	06	01	04	09	02	08	07	02
PA	--	04	04	01	01	03	06	10	02	--	01	05	--	03	01	02
MT	04	16	19	18	03	03	10	23	07	02	08	11	01	09	09	--
AS	16	24	35	19	03	05	14	54	17	03	13	23	11	20	09	05
CA	09	23	39	23	07	09	13	43	12	02	07	21	04	11	17	03
CT	05	13	18	10	06	02	03	23	09	02	07	11	03	08	05	05
H1	26	75	70	73	12	26	35	101	17	14	31	60	19	34	19	07
H2	14	56	77	41	19	18	22	73	19	09	18	42	17	34	16	05
H3	10	36	44	35	07	17	21	45	20	04	16	30	15	21	13	04
VB	22	96	134	42	24	21	55	109	44	06	23	24	05	17	--	05
RI	13	29	80	39	07	--	34	33	18	25	04	22	--	05	03	--
UJT	13	38	50	19	09	06	11	17	04	--	15	18	04	10	10	06
LJT	07	22	40	11	04	07	12	35	07	03	18	25	07	25	10	03
OTHERS	--	06	08	03	01	08	07	04	01	--	02	02	--	02	--	--
TOTAL	257	777	945	520	153	205	368	955	315	89	227	417	115	276	176	73

[Abbreviations for the table of this chapter : HC/AN – Horn Core /Antler, CF – Cranial Fragments, MN –Mandible, SC – Scapula, HU – Humerus, RA – Radius, Ul – Ulna, MC – Metacarpal, C/T – Carpal/Tarsal, MP – Metapodia, PV – Pelvis, FE – Femur, TI – Tibia, PA – Patella, MT – Metatarsal, AS – Astragalus, CA – Calcaneum, CT – Centrotarsal, H1 - 1st Phalanx, H2 – 2nd Phalanx, H3 – 3rd Phalanx, VB – vertebrae, RI – Rib, UJT – Upper Jaw Teeth, LJT - Lower Jaw Teeth, Others – Sesmoid, Hyoid, sterum, etc.]

The presence of the cattle bones and their abundance in all the Chalcolithic sites show the importance of this animal in the general economy of the early farming communities (Table 4.33). The manifold use of cattle may have been the reason for their abundance.

Table 4.19. Minimum Number of Individual animals (mammals) in different layers at Inamgaon.

Species	**01**	**02**	**03**	**04**	**05**	**06**	**07**	**08**	**09**	**10**	**11**	**12**	**13**	**14**	**15**	**16**
Bos indicus	12	25	30	32	08	09	20	45	11	04	08	31	08	19	13	11
Bubalus bubalis	01	04	04	07	01	02	03	08	02	02	03	06	01	02	04	02
Bos/Bubalus	03	11	14	13	03	06	06	34	03	03	05	20	04	06	03	05
Capra hircus	03	03	03	02	--	01	01	02	01	01	01	01	02	02	02	--
Ovis aries	01	02	01	--	01	01	01	01	--	--	02	01	--	--	01	--
Capra/Ovis	04	06	04	07	02	02	01	08	02	02	03	05	01	02	02	--
Sus domesticus	--	04	02	02	--	01	--	03	01	--	01	03	01	02	02	01
Canis familiaris	01	04	04	02	02	01	02	01	01	01	01	04	01	01	02	01
Equus caballus	01	01	02	01	--	01	--	01	--	--	--	--	--	--	--	--
Equus asinus	--	01	--	--	--	--	--	--	--	--	--	--	--	--	--	--
Elephas maximus	--	--	--	--	--	--	--	--	--	--	--	--	--	01	--	--
Canis lupus	01	02	02	--	01	01	--	01	01	--	--	--	--	01	--	02
Cani aureus	--	01	01	01	--	--	--	01	--	--	--	--	--	--	--	--
Canis sp.	--	01	01	--	--	01	01	--	01	--	--	--	01	--	--	01
Felis chaus	02	01	--	01	--	01	01	--	--	--	--	--	--	--	--	--
Panthera pardus	--	01	--	--	--	--	--	--	--	--	01	01	--	--	--	--
Panthera tigris	--	01	--	--	--	--	--	--	--	--	--	--	--	--	--	--
Sus srofa	02	03	01	--	--	01	--	01	--	--	--	--	01	--	--	--
Bos gaurus	--	--	01	01	--	--	01	--	--	--	--	--	--	01	--	--
Antilope cervicapra	36	89	65	59	07	06	11	16	02	01	04	16	06	04	04	08
Gazella bennetti	01	02	01	01	--	--	--	02	--	--	--	02	01	01	--	--
Tetracerus quadricornis	01	01	02	01	01	01	01	01	--	--	01	01	--	--	03	01
Boselaphus tragocamelus	05	08	12	11	02	02	01	04	02	01	01	04	01	03	03	03
Axis axis	03	04	02	02	--	01	02	01	01	--	01	01	01	02	01	01
Axis porcinus	01	--	--	--	--	--	--	--	--	--	--	--	--	--	--	--
Cervus unicolor	01	01	02	01	--	--	01	01	--	--	--	--	01	01	01	01
Muntiacus muntjak	01	--	--	--	--	--	--	01	--	--	--	--	--	--	02	--
Deer sp.	01	01	02	--	01	--	--	--	--	--	--	--	--	--	--	--
Herpestes edwardsi	--	01	03	--	--	--	--	--	--	--	01	--	--	--	--	--
Lepus nigricollis	--	01	02	01	--	--	--	--	--	--	--	01	--	--	--	--
Hystrix indica	--	--	--	--	--	--	--	--	--	--	--	--	--	--	01	--

Table 4.20. Minimum Number of Individual animals in **House No 132** of **Layer 1** at Inamgaon.

Species	132
Bos indicus	01
Capra hircus	01
Antilope cervicapra	06
Boselaphus tragocamelus	02
Axis axis	01

Table 4.21. Minimum Number of Individual animals in **Houses of Layer 2** at Inamgaon.

Species	23	108	118	125	126	129	130
Bos indicus	01	02	01	01	01	01	02
Bubalus bubalis	--	--	--	01	--	--	--
Bos/Bubalus	--	01	--	01	--	--	01
Capra/Ovis	--	--	01	--	--	01	--
Canis familiaris	--	--	01	--	--	--	--
Antilope cervicapra	01	01	01	01	01	03	02
Boselaphus tragocamelus	--	--	--	--	--	02	--
Axis axis	--	--	--	--	01	--	--
Lepus nigricollis	--	--	--	--	--	01	--

Table 4.22. Minimum Number of Individual animals in **Houses of Layer 3** at Inamgaon.

Species	102	103	106	107	126	127	128
Bos indicus	01	01	02	02	01	01	01
Bos/Bubalus	--	01	02	02	--	01	--
Capra/Ovis	--	--	--	01	--	--	--
Canis familiaris	--	--	01	--	--	--	--
Equus caballus	--	--	--	--	--	01	--
Antilope cervicapra	03	01	03	02	01	01	01
Boselaphus tragocamelus	01	--	01	01	--	--	--

Table 4.23. Minimum Number of Individual animals in **Houses of layer 4** at Inamgaon

Species	104	105	109	110	111	112	113	114	115	116	117	119	121	122	124
Bos indicus	02	02	--	03	02	01	01	01	03	01	01	01	--	01	01
Bubalus bubalis	--	--	--	01	--	--	--	--	01	--	--	02	--	01	--
Bos/Bubalus	01	--	--	--	02	--	--	--	01	--	--	02	--	02	--
Capra hircus	--	--	--	01	--	--	--	--	--	--	--	--	--	--	--
Capra/Ovis	01	--	--	01	01	--	01	--	--	--	--	--	--	--	--
Antilope cervicapra	03	03	--	04	03	02	01	01	04	01	01	02	02	01	01
Boselaphus tragocamelus	01	01	01	01	--	--	--	--	01	--	01	--	--	--	--
Axis axis	--	--	--	--	01	--	--	--	--	--	--	--	--	--	--

Table 4.24. Minimum Number of Individual animal in **House no. 49** of **layer 6** at Inamgaon

Species	49
Bos indicus	01
Bos/Bubalus	01
Antilope cervicapra	01

Table 4.25. Minimum Number of Individual animals in **Houses of layer 7** at Inamgaon

Species	39	42	51	51A	52	53	54	57	58	59
Bos indicus	02	01	02	01	01	02	02	01	--	--
Bubalus bubalis	--	--	01	--	--	--	--	--	--	--
Bos/Bubalus	--	--	02	--	--	--	--	--	--	--

Bos gaurus	--	--	01	--	--	--	--	--	--	--
Antilope cervicapra	--	--	02	--	01	01	01	01	--	--
Axis axis	--	--	01	--	--	--	--	01	--	--

Table 4.26. Minimum Number of Individual animals in **Houses of layer 4** at Inamgaon

Species	**61**	**62**	**63**	**64**	**65**	**60&67**	**68**	**69**	**70**	**71**	**72**	**73**
Bos indicus	02	01	01	01	02	01	02	02	02	02	02	02
Bubalus bubalis	--	--	--	--	01	--	--	01	01	01	--	--
Bos/Bubalus	--	--	--	--	02	02	02	01	02	02	02	01
Capra/Ovis	--	--	--	--	01	--	--	--	01	01	01	--
Sus domesticus	--	--	--	--	--	--	--	--	01	--	01	--
Canis lupus	--	--	--	--	--	--	--	--	--	01	--	--
Antilope cervicapra	01	01	02	--	01	01	01	01	01	--	01	01
Gazella bennetti	--	--	--	--	--	--	--	--	01	--	--	--
Boselaphus tragocamelus	--	--	--	--	--	--	--	--	--	01	--	--
Cervus unicolor	--	--	--	--	--	--	--	01	--	--	--	--

Table 4.27. Minimum Number of Individual animals in **Houses of layer 12** at Inamgaon

Species	**74**	**75**	**76**	**77**	**78**	**79**	**80**	**81**	**83**	**84**	**85**
Bos indicus	01	02	--	01	02	03	02	02	03	04	03
Bubalus bubalis	01	--	--	--	--	01	--	--	01	--	01
Bos/Bubalus	01	03	01	--	02	03	01	01	02	--	03
Capra/Ovis	--	-1	--	--	--	--	--	01	--	01	--
Sus domesticus	--	--	--	--	--	--	01	--	--	--	--
Canis familiaris	--	01	--	--	--	01	--	01	--	--	--
Antilope cervicapra	01	01	01	--	--	02	01	01	01	01	01
Gazella bennetti	--	01	--	--	--	--	--	--	--	--	--
Boselaphus tragocamelus	--	01	--	--	--	01	--	--	--	--	--

Table 4.28. Minimum Number of Individual animals in **Houseno. 50** of **layer 13** at Inamgaon

Species	**50**
Bos indicus	01
Bos/Bubalus	01
Capra hircus	02
Capra/Ovis	01
Canis sp.	01
Antilope cervicapra	02
Gazella bennetti	01
Axis axis	01
Cervus unicolor	01

Table 4.29. Minimum Number of Individual animals in **Houses of layer 16** at Inamgaon

Species	**86**	**87**	**88**	**89**	**91**	**93**	**94**	**95**	**96**	**98**	**99**
Bos indicus	01	02	--	--	--	01	--	03	--	--	--
Bubalus bubalis	--	--	--	--	--	--	--	01	--	--	--
Bos/Bubalus	--	02	--	--	--	--	--	03	--	--	--
Sus domesticus	--	01	--	--	--	--	--	--	--	--	--
Canis familiaris	--	01	--	--	--	--	--	--	--	--	--
Canis lupus	--	01	--	--	--	--	--	--	--	--	--
Canis sp.	--	01	--	--	--	--	--	--	--	--	--
Antilope cervicapra	01	02	--	01	--	01	--	--	01	01	--
Tetracerus quadricornis	--	01	--	--	--	--	--	--	--	--	--
Boselaphus tragocamelus	--	01	--	--	--	--	--	01	--	--	--

Table 4.30. Minimum Number of Individual animals in Different layers and houses at Inamgaon

Species	01	H	02	H	03	H	04	01	05	H	06	H	07	H	08	H	09	H	10	H	11	H	12	H	13	H	14	H	15	H	16	H
Bos indicus	11	01	16	09	21	09	12	20	08	--	08	01	08	12	25	20	11	--	04	--	08	--	09	23	07	01	19	--	13	--	04	07
Bubalus bubalis	01	--	03	01	04	--	02	05	01	--	02	--	02	01	04	04	02	--	02	--	03	--	02	04	01	--	02	--	04	--	01	01
Bos/Bubalus	03	--	08	03	08	06	05	08	03	--	05	01	04	02	20	14	03	--	03	--	05	--	03	17	03	01	06	--	03	--	--	05
Capra hircus	02	01	03	--	03	--	01	01	--	--	01	--	01	--	02	--	01	--	01	--	01	--	01	--	--	02	02	--	02	--	--	--
Ovis aries	01	--	02	--	01	--	--	--	01	--	01	--	01	--	01	--	--	--	--	--	02	--	01	--	--	--	--	--	01	--	--	--
Capra/Ovis	04	--	04	02	03	01	03	04	02	--	02	--	01	--	04	04	02	--	02	--	03	--	02	03	--	01	02	--	02	--	--	--
Sus domesticus	--	--	04	--	02	--	02	--	--	--	01	--	--	--	01	02	01	--	--	--	01	--	02	01	01	--	02	--	02	--	--	01
Canis familiaris	01	--	03	01	03	01	02	--	02	--	01	--	02	--	01	--	01	--	01	--	01	--	01	03	01	--	01	--	02	--	--	01
Equus caballus	01	--	01	--	01	01	01	--	--	--	01	--	--	--	01	--	--	--	--	--	--	--	--	--	--	--	--	--	--	--	--	--
Equus asinus	--	--	01	--	--	--	--	--	--	--	--	--	--	--	--	--	--	--	--	--	--	--	--	--	--	--	--	--	--	--	--	--
Elephas maximus	--	--	--	--	--	--	--	--	--	--	--	--	--	--	--	--	--	--	--	--	--	--	--	--	--	--	01	--	--	--	--	--
Canis lupus	01	--	02	--	02	--	--	--	01	--	01	--	--	--	--	01	01	--	--	--	--	--	--	--	--	--	01	--	--	--	01	01
Canis aueus	--	--	01	--	01	--	01	--	--	--	--	--	--	--	01	--	--	--	--	--	--	--	--	--	--	--	--	--	--	--	--	--
Canis sp.	--	--	01	--	01	--	--	--	--	--	01	--	01	--	--	--	01	--	--	--	--	--	--	--	--	01	--	--	--	--	--	01
Felis cahus	02	--	01	--	--	--	01	--	--	--	01	--	01	--	--	--	--	--	--	--	--	--	--	--	--	--	--	--	--	--	--	--
Panthera pardus	--	--	01	--	--	--	--	--	--	--	--	--	--	--	--	--	--	--	--	--	01	--	01	--	--	--	--	--	--	--	--	--
Panthera tigris	--	--	01	--	--	--	--	--	--	--	--	--	--	--	--	--	--	--	--	--	--	--	--	--	--	--	--	--	--	--	--	--
Sus scrofa	02	--	03	--	01	--	--	--	--	--	01	--	--	--	01	--	--	--	--	--	--	--	--	--	01	--	--	--	--	--	--	--
Bos gaurus	--	--	--	--	01	--	01	--	--	--	--	--	--	01	--	--	--	--	--	--	--	--	--	--	--	--	01	--	--	--	--	--
Antilope cervicapra	30	06	79	10	53	12	30	29	07	--	05	01	03	08	05	11	02	--	01	--	04	--	06	10	04	02	04	--	04	--	01	07
Gazella bennetti	01	--	02	--	01	--	01	--	--	--	--	--	--	--	01	01	--	--	--	--	--	--	01	01	--	01	01	--	--	--	--	--
Tetracerus quadricornis	01	--	01	--	02	--	01	--	01	--	01	--	01	--	01	--	--	--	--	--	01	--	01	--	--	--	--	--	03	--	--	01
Boselapus tragocamelus	03	02	06	02	09	03	05	06	02	--	02	--	01	--	03	01	02	--	01	--	01	--	02	02	01	--	03	--	03	--	01	02
Axis axis	02	01	02	01	02	--	01	01	--	--	01	--	--	02	01	--	01	--	--	--	01	--	01	--	--	01	02	--	01	--	01	--
Axis porcinus	01	--	--	--	--	--	--	--	--	--	--	--	--	--	--	--	--	--	--	--	--	--	--	--	--	--	--	--	--	--	--	--
Axis sp	--	--	01	--	--	--	--	--	--	--	--	--	--	--	--	--	--	--	--	--	--	--	--	--	--	--	--	--	--	--	--	--
Cervus unicolor	01	--	01	--	02	--	01	--	--	--	--	--	01	--	--	01	--	--	--	--	--	--	--	--	--	01	01	--	01	--	01	--
Muntiacus muntjak	01	--	--	--	--	--	--	--	--	--	--	--	--	--	01	--	--	--	--	--	--	--	--	--	--	--	--	--	02	--	--	--
Deer sp.	01	--	01	--	02	--	--	--	01	--	--	--	--	--	--	--	--	--	--	--	--	--	--	--	--	--	--	--	--	--	--	--

Table 4.31. The Minimum Number of Individuals in the three cultural phases at Inamgaon

Species	Late Jorwe	%	Early Jorwe	%	Malwa	%
Bos indicus	107	17.83	97	32.44	83	32.30
Bubalus bubalis	17	02.83	20	06.69	15	05.84
Bos/Bubalus	44	07.33	57	19.06	38	14.78
Capra hircus	11	01.83	7	02.34	7	02.72
Ovis aries	5	00.83	5	01.67	2	00.78
Capra/Ovis	23	03.83	18	06.02	10	03.89
Sus domesticus	8	01.33	6	02.01	9	03.50
Canis familiaris	13	02.17	7	02.34	9	03.50
Equus caballus	5	00.83	2	00.67	--	--
Equus asinus	1	00.17	--	--	--	--
Elephas maximus	1	0.17	--	--	1	00.39
Canis lupus	10	01.67	3	01.01	3	01.17
Cani aureus	3	00.51	1	00.33	--	--
Canis sp.	2	00.33	3	01.01	2	00.78
Felis chaus	4	00.67	2	00.67	--	--
Panthera pardus	1	00.17	1	00.33	1	00.39
Panthera tigris	1	00.17	--	--	--	--
Sus srofa	6	01.00	2	00.67	1	00.39
Bos gaurus	2	00.33	1	00.33	1	00.39
Antilope cervicapra	256	42.67	40	13.30	38	14.78
Gazella bennetti	5	00.83	2	00.67	4	01.56
Tetracerus quadricornis	6	01.00	4	01.34	5	01.94
Boselaphus tragocamelus	38	06.33	11	03.68	14	05.45
Axis axis	11	01.83	6	02.01	6	02.33
Axis porcinus	1	00.17	--	--	--	--
Cervus unicolor	5	0083	2	00.67	4	01.56
Muntiacus muntjak	1	00.17	1	00.33	2	00.78
Deer sp.	5	00.83	--	--	--	--
Herpestes edwardsi	4	00.67	1	00.33	--	--
Lepus nigricollis	4	00.67	--	--	1	00.39
Hystrix indica	--	--	--	--	1	0039
Total	600	100%	299	100%	257	100%

Table 4.32. Weight of meat probably consumed in various stratigraphical units at Inamgaon

Species	01	02	03	04	05	06	07	08	09	10	11	12	13	14	15	16
Bos indicus	2400.0	5000.0	6000.0	6400.0	1600.0	1800.0	4000.0	9000.0	2200.0	800.0	1600.0	6400.0	1600.0	3800.0	2600.0	2200.0
Bubalus bubalis	250.0	1000.0	1000.0	1750.0	250.0	500.0	750.0	2000.0	500.0	500.0	750.0	1500.0	250.0	500.0	1000.0	500.0
Bos/Bubalus	675.0	2475.0	3150.0	2925.0	675.0	1350.0	1350.0	7650.0	675.0	675.0	1125.0	4500.0	900.0	1350.0	675.0	1125.0
Capra hircus	37.5	37.5	37.5	25.0	00.0	12.5	12.5	25.0	12.5	12.5	12.5	12.5	25.0	25.0	25.0	00.0
Ovis aries	12.5	25.0	12.5	00.0	12.5	12.5	12.5	12.5	00.0	00.0	25.0	12.5	00.0	00.0	12.5	00.0
Capra/Ovis	50.0	75.0	50.0	87.5	25.0	25.0	12.5	100.0	25.0	25.0	37.5	62.5	12.5	25.0	25.0	00.0
Sus domesticus	00.0	80.0	40.0	40.0	00.0	20.0	00.0	60.0	20.0	00.0	20.0	60.0	20.0	40.0	40.0	20.0
Canis familiaris	7.5	30.0	30.0	15.0	15.0	7.5	15.5	7.5	7.5	7.5	7.5	30.0	7.5	7.5	15.0	7.5
Equus caballus	150.0	150.0	300.0	150.0	00.0	150.0	00.0	150.0	00.0	00.0	00.0	00.0	00.0	00.0	00.0	00.0
Canis lupus	9.0	18.0	18.0	00.0	9.0	9.0	00.0	9.0	9.0	00.0	00.0	00.0	00.0	9.0	00.0	18.0
Canis aureus	00.0	4.0	4.0	4.0	00.0	00.0	00.0	4.0	00.0	00.0	00.0	00.0	00.0	00.0	00.0	00.0
Sus scrofa	200.0	300.0	100.0	00.0	00.0	100.0	00.0	100.0	00.0	00.0	00.0	00.0	100.0	00.0	00.0	00.0
Antilope cervicapra	540.0	1335.0	975.0	885.0	105.0	90.0	165.0	240.0	30.0	15.0	60.0	240.0	90.0	60.0	60.0	120.0
Gazella bennetti	10.0	20.0	10.0	10.0	00.0	00.0	00.0	20.0	00.0	00.0	00.0	20.0	10.0	10.0	00.0	00.0
Tetracerus quadricornis	12.5	12.5	25.0	12.5	12.5	12.5	12.5	12.5	00.0	00.0	12.5	12.5	00.0	00.0	37.5	12.5
Boselaphus tragocamelus	500.0	800.0	1200.0	1100.0	200.0	200.0	100.0	400.0	200.0	100.0	100.0	400.0	100.0	300.0	300.0	300.0
Axis axis	112.5	112.5	75.0	75.0	00.0	37.5	75.0	37.5	37.5	00.0	37.5	37.5	37.5	75.0	37.5	37.5
Cervus unicolor	125.0	125.0	250.0	125.0	00.0	00.0	125.0	125.0	00.0	00.0	00.0	00.0	125.0	125.0	125.0	125.0
Muntiacus muntjak	20.0	00.0	00.0	00.0	00.0	00.0	00.0	20.0	00.0	00.0	00.0	00.0	00.0	00.0	40.0	00.0
Herpestus edwardsi	00.0	00.75	2.25	00.0	00.0	00.0	00.0	00.0	00.0	00.0	00.75	00.0	00.0	00.0	00.0	00.0
Lepus nigricollis	00.0	1.0	2.0	1.0	00.0	00.0	00.0	00.0	00.0	00.0	00.0	1.0	00.0	00.0	00.0	00.00

Table 4.33. Distribution of animals in the Chalcolithic culture of Maharashtra

Species	Nasik	Nevasa	Apegaon	Daimabad	Inamgaon	Kaothe	Walki	Tuljapur garhi	Adam
Domestic Mammals									
Bos indicus	+	+	+	+	+	+	+	+	+
Bubalus bubalis	+	+	+	+	+	+	+	-	+
Bos/Bubalus	-	+	-	-	-	-	+	-	-
Bos sp.	-	+	-	-	-	+	-	-	-
Capra hircus	-	+	+	+	+	-	+	-	+
Ovis aries	+	-	+	+	+	+	+	-	+
Sus domesticus	+	+	-	+	+	+	+	-	+
Sus sp.	-	-	-	-	-	-	-	-	-
Equus caballus	-	+	-	+	+	-	+	-	-
Equus asinus	-	+	-	+	+	-	+	+	-
Equus sp.	-	+	-	-	-	-	-	-	-
Canis familiaris	+?	+	-	+	+	+	+	+	+
Felis catus	-	+	-	+	-	-	+	-	-
Wild Mammals									
Presbytis entellus	-	-	-	-	-	+	-	-	-
Seminopithecus sp.	+	-	-	-	-	-	-	-	-
Bubalus arnee	-	-	-	-	-	-	+	-	+
Bos gaurus	-	-	-	-	+	-	+	-	+
Elephas maximus	-	+	-	+	+	-	-	-	+
Rhinoceros sp.	-	+	-	-	-	-	-	-	-
Boselaphus tragocamelus	+	+	+?	+	+	+	+	-	+
Antilope cervicapra	-	+	+?	+	+	+	+	-	+
Gazella bennetti	-	-	-	-	+	-	+	-	-
Tetracerus quadricornis	+	+	-	+	+	+	+	+	-
Cervus unicolor	+	+	-	+	+	+	+	+	+
Cervus duvauceli	-	-	-	+	-	-	-	-	+
Cervus sp.	-	+	+	-	-	-	-	-	-
Cervus/ Axis	-	+	-	-	-	-	-	-	-
Axis axis	+	-	+	+	+	+	+	+	+
Axis porcinus	-	-	-	-	+	+	+	+	-
Muntiacus muntjak	-	-	-	-	+	+	+	-	-
Sus scrofa	-	-	-	-	+	+	+	+	+
Panthera pardus	-	-	-	-	+	-	+	-	-
Panthera tigris	-	-	-	-	+	-	+	-	-
Felis chaus	-	-	-	-	+	-	+	-	+
Canis lupus	-	-	-	-	+	-	+	-	+
Canis aurus	-	-	-	-	+	-	+	-	+
Vulpes bengalensis	-	-	-	-	-	+	-	-	-
Vulpes sp.	-	+	-	-	-	-	-	-	-
Melursus ursinus	-	-	-	-	-	+	-	-	-
Viverricula indica	-	-	-	-	-	+	-	-	-
Hystrix indica	-	+	-	-	+	+	-	-	-
Herpestes edwardsi	-	-	+	-	+	-	-	+	-
Lepus nigricollis	-	+	-	-	+	+	+	-	-
Rattus rattus	+	+	-	+	+	+	-	+	+
Bandicota indica	-	+	-	-	-	-	-	-	-
Ratufa indica	-	-	-	-	-	-	+	-	-
Birds									
Gallus domesticus	-	+	-	-	+	+	+	-	+
Gallus gallus	-	-	-	-	+	-	-	-	-
Gallus sp.	-	-	-	+	-	-	-	+	-
Anser anser	-	-	-	-	+	+	-	-	-
Pavo cristatus	-	-	-	-	+	+	-	-	-
Ardea sp.	-	-	-	-	-	+	-	-	-
Ardeotis nigriceps	-	-	-	-	-	-	-	+	-
Reptiles									
Trionyx gangeticus	+	+?	+	+	+	+	+	-	-
Batagur baska	-	-	-	-	-	-	+	-	-
Chitra indica	-	+?	-	-	+	-	+	+	-

Kachuga sp.	-	-	-	-	+	-	+	-	-
Lissemys punctata	-	+?	-	-	-	+	-	-	+
Crocodylus palustris	-	-	-	-	+	-	-	-	+
Gavialis gangeticus	-	-	-	-	-	-	+	-	-
Varanus monitor	-	-	-	-	+	-	+	-	+
Snake	-	-	-	-	+	-	-	-	-
Mollusca									
Shells	-	-	-	+	-	-	-	-	-
Lamellidens sp.	+	-	-	-	+	+	+	+	+
Corbicula sp.	-	-	-	-	-	+	+	-	-
Ariophanta sp.	-	-	-	-	-	-	+	-	-
Digoniostoma pulchella	-	-	-	-	+	-	+	-	-
Pila globosa	-	-	-	-	+	-	+	-	-
Melania striatella tuberculata	-	-	-	-	-	+	-	-	-
Sabulina octona	-	-	-	-	-	-	-	-	-
Cryptozona belangeri	-	-	-	-	-	-	-	-	-
Melamia striatella scaba	-	-	-	-	-	+	-	-	-
Glessula sp.	-	-	-	-	-	+	-	-	-
Parreyssia sp.	+	-	-	-	+	+	-	+	-
Viviparus bengalensis	+	-	-	-	+	-	-	+	-
Xesta semirugata	+	-	-	-	-	-	-	+	-
Cyrena impressa	+	-	-	-	-	-	-	+	-
Paphia gallus	-	-	-	-	+	-	-	-	-
Marcia recens	-	-	-	-	+	-	-	-	-
Turbo coronatus	-	-	-	-	+	-	-	-	-
Ornamentaria annulus	+	-	-	-	-	-	-	+	-
Nerita chameleon	+	-	-	-	-	-	-	+	-
Dostia crepidularia	+	-	-	-	-	-	-	+	-
Melanoides pyramis	+	-	-	-	-	-	-	+	-
Indonaia caerulea	+	-	-	-	-	-	-	+	-
Conus sp.	+	-	-	-	-	-	-	+	-
Telescopium telescopium	+	-	-	-	-	-	-	+	-
Cantharus undosus	+	-	-	-	-	-	-	+	-
Gratiadusta pallida	+	-	-	-	-	-	-	+	-
Turbinella pyrum	+	-	-	-	+	+	+	+	-
Cypraea arabica	-	-	-	-	+	+	-	-	-
Meretrix sp.	-	-	-	-	+	-	-	-	-
Murex sp.	+	-	-	-	-	-	-	+	-

+ = Presence, - = Absence, ? = Species doubtful

The evidence of sharing of meat between people of two different houses in the Malwa phase is very significant. The bones of a three years old bull were found on the floors of the houses 83 and 79. Interestingly house number 83 had the right half of the body and house number 79, the left half. These two structures were situated in different locales. The degree of charring and the metrical analyses reveal that, these bones belong to one and the same animal. People from these houses might have butchered this animal at some special occasion. This evidence throws light on the socio-cultural aspects of the people. Cattle bones were also collected from some of the human burials, which may be related to the ceremony of offering to the dead person (Thomas 1988).

Age of cattle at death

The dentition as well as the fused and unfused epiphyses of long bones (Pl. III B) from different layers was studied in detail to find out the age at death of cattle (Table 4.34; Fig. 4.1). In all the three cultural phases at Inamgaon, around 70% cattle were slaughtered between the ages of 1.5-3.5 years. Killing of animals below the age of 1.5 years is represented very less in the Early Jorwe (0.5%) and Late Jorwe phase (0.42%), as compared to the Malwa phase (14.11%) (Table 4.34 & 4.35).Very adult and senile animals were represented in lesser proportion, in the Malwa phase compared to the other two phases. Probably the agricultural activities were of lesser magnitude in the Malwa phase as compared to the other two phases. More representation of adult and senile animals in the Early Jorwe and Late Jorwe phases reveals that a few of these animals might have been also kept for the agricultural operations. The maximum selection of animals between the age group of 1.5- 3.5 years indicates the primary importance of this animal at the site was as meat producer. Thus, stock-raising of cattle was mainly carried out at Inamgaon for regular supply of meat.

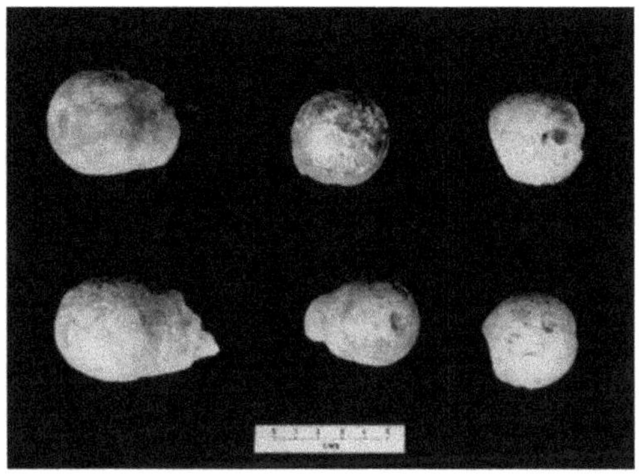

Plate III B. ***Bos indicus*****:** Unfused femur heads

Fig. 4.1. Inamgaon: Age of cattle at death

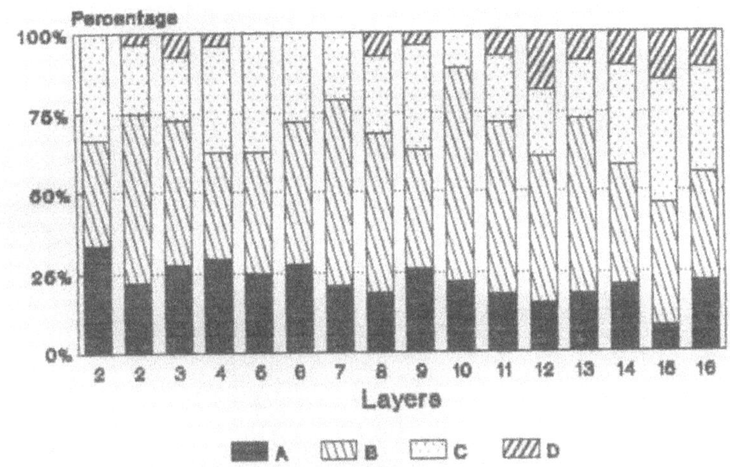

Table 4.34. Age at death of cattle in different layers at Inamgaon.

Layer	A	B	C	D	Total	A%	B%	C%	D%
01	03	03	03	00	09	33.33	33.33	33.33	00.00
02	07	17	07	01	32	21.88	53.12	21.88	03.12
03	11	18	08	03	40	27.50	45.00	20.00	07.50
04	07	08	08	01	24	29.17	33.33	33.33	04.17
05	02	03	03	00	08	25.00	37.50	37.50	00.00
06	05	08	05	00	18	27.78	44.44	27.78	00.00
07	04	11	04	00	19	21.05	57.89	21.05	00.00
08	15	40	20	06	81	18.52	49.38	24.69	07.41
09	07	10	09	01	27	25.93	37.04	33.33	03.70
10	02	06	01	00	09	22.22	66.67	11.11	00.00
11	05	15	06	02	28	17.86	53.57	21.43	07.14
12	05	15	07	06	33	15.15	45.45	21.21	18.18
13	02	06	02	01	11	18.18	54.55	18.18	09.09
14	04	07	06	02	19	21.05	36.84	31.58	10.53
15	01	05	05	02	13	7.15	38.46	38.46	15.38
16	02	03	03	01	09	22.22	33.33	33.33	11.11

A- Age more than 3.5 years
B- Age less than 3.5 years and more than 2.5 years
C- Age less than 2.5 years and more than 1.5 years
D- Age less than 1.5 years

Table 4.35. Age at death of cattle in the three cultural phases at Inamgaon

Cultural phases	A	B	C	D	Total	A%	B%	C%	D%
Malwa phase	14	36	23	12	085	16.47	42.35	27.05	14.11
Early Jorwe phase	38	80	45	09	173	21.96	46.24	26.01	00.05
Late Jorwe phase	30	49	29	05	113	26.54	43.36	25.66	00.42

Bubalus bubalis (buffalo)

Bones of buffalo, though scanty, were recovered from all the 16 layers at Inamgaon (Pl. IV A). Fragments of bones both from appendicular and axial skeletons were identified (Table 4.36). The remains of this animal were also found on some house floors in layers 2, 4, 7, 8, 12 and 16 (Table 4.19-4.31). Cut marks are found on all the long bones, metacarpals, metatarsals, pelvis and scapula. The butchering pattern is the same as it is seen on the cattle bones. Charring marks are also noticed on the bone fragments. The bones of this animal are very sparsely represented in all the Chalcolithic sites of Maharashtra.

Plate IV A. *Bubalus bubalis*: 1. Metacarpal proximal, 2. Metacarpal distal, 3. Metatarsal distal, 4. Second phalanx, 5. First phalanx, 6. Centrotarsal and 7. Astragalus.

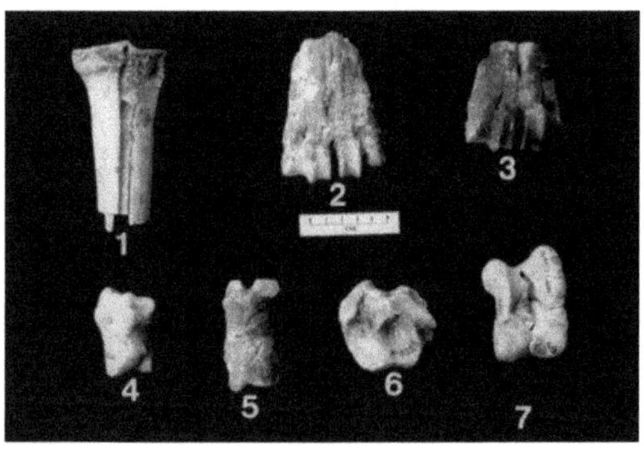

The percentage frequency of bone fragments shows maximum representation in the Early Jorwe phase, i.e. 1.58% (Table 4.17). The number of identified fragments are 129 (0.87%) in the Late Jorwe, 110 (1.58%) in the Early Jorwe and 35 (0.96%) in the Malwa phase. The minimum number of individuals are 17 (2.83%), 20 (6.69%) and 15 (5.84%) in the Late Jorwe, Early Jorwe and the Malwa phases, respectively. In the layer-wise distribution, the maximum representation of bones and individual animals are noticed in layer 8. The amount of meat consumed of this animal at Inamgaon in all the cultural phases shows that buffaloes were the second most important animal in the food economy (Table 4.32).

The quantity of meat consumed of this animal was 3,750 kg, 5,000 kg and 4,250 kg, in the Malwa, Early Jorwe and Late Jorwe phases, respectively. It is interesting to note that, comparatively there is a decline in the representation of bones of buffalo in the Late Jorwe phase.

Morphologically, the bones of cattle and buffaloes are very similar. The fragments of ribs, vertebrae and other very fragmented bones could not be identified to species level. Such fragments were recorded under one group called *Bos/Bubalus*. Since cattle bones are dominating in the overall faunal assemblage a majority of bones from this group may belong to the cattle (Table 4.37). The number of identified fragments of this group accounts for more than 40% in the faunal assemblage at Inamgaon, which is apparently more than that of cattle and buffalo. However, the weight of these fragments is meager, shows the high degree of fragmentation of bones.

Table 4.36. Bone distribution of *Bubalus bubalis* in different layers at Inamgaon

Bones	01	02	03	04	05	06	07	08	09	10	11	12	13	14	15	16
HC/AN	--	--	--	--	--	--	--	--	--	--	--	--	--	--	--	--
CF	--	--	--	--	--	--	--	--	--	--	--	--	--	--	--	--
MN	01	01	--	--	--	--	--	--	--	--	--	01	--	--	--	--
SC	01	02	01	--	--	--	--	--	--	--	--	--	--	01	--	--
HU	--	01	--	--	--	--	--	--	--	--	--	--	--	--	--	--
RA	--	--	01	--	--	--	--	--	--	--	--	--	--	--	--	--
UL	--	--	--	--	--	--	--	--	--	--	--	--	--	--	--	--
MC	01	05	09	--	--	--	01	01	01	--	04	01	--	01	--	01
C/T	--	--	--	--	--	--	--	--	--	--	--	--	--	--	--	--
MP	--	01	--	01	--	--	--	--	--	--	--	--	--	--	--	--
PV	--	--	--	--	--	--	--	--	--	--	--	--	--	--	--	--
FE	--	01	--	--	--	--	--	--	--	--	--	--	--	--	--	--
TI	01	--	--	--	--	--	02	02	--	01	--	--	--	--	--	--
PA	--	01	--	01	--	--	--	--	--	01	--	--	--	--	--	--
MT	--	02	01	--	--	--	--	--	--	--	--	01	--	--	--	--
AS	--	02	02	--	--	--	03	03	02	--	--	02	--	--	03	--
CA	--	--	04	--	01	--	--	--	--	--	--	--	--	01	--	--
CT	--	--	04	04	--	01	--	--	01	--	02	02	01	--	04	--
H1	06	05	10	03	--	01	04	04	01	02	02	--	--	--	--	01
H2	03	10	19	08	01	03	03	03	06	03	05	03	01	04	04	--
H3	02	06	03	01	--	01	--	--	01	--	--	01	--	--	--	--
VB	--	--	--	--	--	--	--	--	--	--	--	01	--	--	--	--
RI	--	--	--	--	--	--	--	--	--	--	--	--	--	--	--	--
UJT	02	--	01	--	--	--	--	--	--	--	--	--	--	--	--	--
LJT	--	--	--	--	--	--	--	--	--	--	--	--	--	--	--	--
OTHERS	--	--	--	--	--	--	--	--	--	--	--	--	--	--	--	--
TOTAL	17	37	55	18	02	06	13	13	12	07	13	12	02	07	11	02

Table 4.37. Bone distribution of *Bos/Bubalus* in different layers at Inamgaon

Bones	**01**	**02**	**03**	**04**	**05**	**06**	**07**	**08**	**09**	**10**	**11**	**12**	**13**	**14**	**15**	**16**
HC/AN	--	02	01	01	01	--	--	02	--	01	02	--	03	01	--	--
CF	46	97	93	59	14	42	20	47	23	13	20	25	05	08	20	02
MN	15	34	22	09	--	06	09	36	08	07	11	19	07	12	08	03
SC	11	18	14	08	02	06	02	20	04	02	03	04	--	--	02	--
HU	10	22	16	17	01	10	05	43	03	03	20	18	02	10	03	03
RA	07	10	03	13	02	09	03	21	10	01	16	14	06	02	03	01
UL	01	08	10	05	04	09	03	10	05	04	05	08	03	11	03	01
MC	03	10	04	09	01	02	01	18	01	--	02	06	02	06	04	02
C/T	05	47	27	40	05	10	05	51	21	04	17	26	15	15	19	02
MP	20	57	40	29	05	19	05	80	28	05	25	44	08	20	16	02
PV	11	21	22	24	03	17	08	61	10	02	07	15	03	17	12	05
FE	13	38	18	13	06	05	15	54	08	05	18	15	09	08	13	01
TI	03	14	11	10	--	06	03	15	02	04	07	07	03	01	03	01
PA	--	--	03	03	--	07	02	09	03	--	04	07	03	05	03	--
MT	06	18	04	09	01	03	03	12	05	06	10	05	03	04	05	04
AS	03	09	04	04	02	02	02	10	04	--	02	10	04	09	08	--
CA	07	17	11	14	02	05	05	26	09	04	09	12	03	11	11	--
CT	--	12	09	07	04	--	01	09	05	03	04	06	--	09	06	01
H1	03	16	06	19	02	01	02	16	06	04	05	10	07	02	07	03
H2	02	09	01	07	01	--	01	02	02	01	05	08	--	01	--	--
H3	02	07	06	09	01	03	--	02	03	02	03	04	07	01	02	01

VB	103	272	233	229	55	107	113	354	144	40	111	191	55	101	80	28
RI	49	155	176	96	25	107	68	134	45	17	78	94	49	53	44	10
UJT	05	24	15	20	02	04	02	43	10	02	17	37	08	05	13	04
LJT	01	05	14	06	02	10	02	30	02	04	16	20	08	04	11	02
OTHERS	15	15	10	12	04	09	05	14	07	04	07	16	05	02	07	--
TOTAL	341	943	765	687	145	394	289	1130	368	138	424	621	230	320	332	78

***Capra/Ovis* (goat/sheep)**

As compared to other Chalcolithic sites of Western Maharashtra goat/sheep bones are meagerly represented at Inamgaon. Altogether, the identified bone fragments of sheep and goat are only 1.41%, in the total faunal assemblage. Skeletal remains are sparsely distributed in all the layers (Pl. IV B). The remains of goat/sheep are absent in layer 16 (Table 4.38-4.40). Bones are also collected from the house floors in the layers 1, 2, 3, 4, 8, 12, and 13 (Table 4.19-4.31).

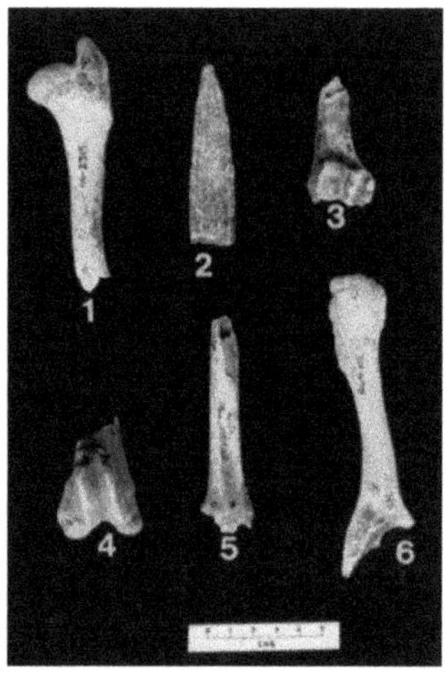

Plate IV B. *Capra/Ovis*: 1. Humerus, 2. Horn core, 3. Humerus distal, 4. Femur distal, 5. Tibia distal and 6. Part of pelvis.

Table 4.38. Bone distribution of *Capra/Ovis* in different layers at Inamgaon

Bones	**01**	**02**	**03**	**04**	**05**	**06**	**07**	**08**	**09**	**10**	**11**	**12**	**13**	**14**	**15**	**16**
HC/AN	--	--	--	--	--	--	--	--	--	--	--	--	--	01	--	--
CF	01	--	--	--	--	--	--	--	--	--	--	--	--	05	--	--
MN	05	05	01	01	01	--	--	-	03	03	02	01	--	--	02	--
SC	--	02	01	--	--	--	--	01	--	01	04	01	01	--	--	--
HU	02	03	02	03	--	--	--	05	01	--	--	02	01	01	--	--
RA	--	01	03	06	--	01	01	02	03	04	03	04	--	02	01	--
UL	--	01	--	01	--	--	--	--	--	--	--	--	--	--	--	--
MC	01	02	01	--	--	--	--	01	01	--	01	02	02	--	01	--
C/T	--	--	--	--	--	--	--	--	--	--	--	--	--	--	--	--
MP	--	01	01	02	03	01	--	--	--	01	--	01	01	--	--	--
PV	01	--	02	--	--	--	--	02	01	08	03	03	01	--	01	--
FE	--	03	--	--	01	01	--	01	01	--	--	03	02	--	--	--
TI	--	--	--	02	--	02	--	01	--	--	01	--	--	01	--	--
PA	--	--	--	--	--	--	--	--	--	--	--	--	--	--	--	--
MT	01	01	01	--	--	--	--	01	--	--	01	--	--	--	01	--
AS	--	--	02	--	--	--	--	--	01	--	--	01	--	--	01	--
CA	--	02	--	--	--	--	--	--	--	02	--	01	01	--	--	--
CT	--	--	--	--	--	--	--	--	--	--	--	--	01	--	--	--
H1	--	--	01	02	01	--	--	--	01	01	03	02	01	01	02	--
H2	--	--	01	--	--	--	---	--	--	---	--	--	01	--	--	--
H3	--	--	--	--	--	--	--	--	--	--	--	01	--	--	--	--
VB	--	02	--	01	01	04	--	03	01	04	01	01	--	--	--	--
RI	--	--	--	--	--	--	--	--	--	02	--	--	--	--	--	--
UJT	01	02	--	03	--	02	--	--	--	--	02	01	01	02	--	--
LJT	--	01	06	--	--	--	--	04	--	03	05	--	02	--	02	--
OTHERS	--	--	--	--	--	--	--	--	--	--	--	--	--	--	--	--
TOTAL	12	26	22	21	07	11	01	21	13	29	26	24	15	13	11	56

The number of identified bones are 154 (1. 02%), 130 (1.87%) and 78 (2.15%) in the Late Jorwe, Early Jorwe and the Malwa phases, respectively (Table 4.1-4.17). The minimum number of individuals represented are 39 (6.49%), 30 (10.03%) and 19 (7.39%) in the Late Jorwe, Early Jorwe and the Malwa phases, respectively. In almost all the Chalcolithic sites the sheep and goat is the second most important animal in the food economy, except at Inamgaon. The probable weight of meat consumed by inhabitants at Inamgaon is 487.50 kg in the Late Jorwe, 375.5kg in Early Jorwe and 237.50 kg in Malwa phase. The maximum amount of meat of this animal has been consumed in the Late Jorwe phase, especially in layer 2 and also in the Early Jorwe phase, in layer 8 (Table 4.32).

A majority of bones of this group belong to the goat (0.34%). The skeletal remains of the sheep are only 0.09% in the collection (Table 4.17).Because of the morphological similarities, some fragments of bones could not be identified to species level which are grouped together as *Capra/Ovis* which accounted for about (0.98 %) in the collection.

Table 4.39. Bone distribution of *Capra hircus* in different layers at Inamgaon

Bones	**01**	**02**	**03**	**04**	**05**	**06**	**07**	**08**	**09**	**10**	**11**	**12**	**13**	**14**	**15**	**16**
HC/AN	--	--	01	--	--	--	--	--	--	--	--	--	--	--	--	--
CF	--	--	01	--	--	--	--	--	--	--	--	--	--	--	--	--
MN	02	03	01	--	--	--	--	02	01	--	01	--	--	--	--	--
SC	--	--	--	--	--	--	01	--	--	--	--	--	--	01	--	--
HU	01	--	03	--	--	--	--	--	--	--	02	--	--	--	--	--
RA	01	--	02	--	--	01	01	02	--	--	--	--	01	01	--	--
UL	--	--	--	--	--	01	--	01	--	--	01	--	--	--	--	--
MC	01	03	02	01	--	--	--	--	--	--	--	--	--	--	--	--
C/T	--	--	--	--	--	--	--	--	--	--	--	--	--	--	--	--
MP	01	01	--	--	--	--	--	--	--	--	--	--	--	--	--	--
PV	--	01	--	--	--	--	--	--	01	--	--	--	--	--	--	--
FE	01	02	--	--	--	--	--	--	--	--	--	--	--	01	--	--
TI	01	02	02	--	--	--	--	01	--	--	--	--	--	--	--	--
PA	--	--	--	--	--	--	--	--	--	--	--	--	--	--	--	--
MT	--	--	--	--	--	--	--	--	--	--	--	--	--	01	--	--
AS	01	--	01	--	--	--	--	01	--	--	--	--	--	--	--	--
CA	--	01	--	--	--	--	--	--	--	--	--	01	--	--	--	--
CT	--	--	01	--	--	--	--	--	--	--	--	--	01	--	--	--
H1	01	06	01	--	--	--	--	--	01	01	--	--	01	--	--	--
H2	--	04	--	01	--	--	--	--	--	--	01	--	--	--	--	--
H3	--	--	--	--	--	--	--	--	--	--	--	--	--	--	--	--
VB	--	--	01	--	--	--	--	--	--	--	--	--	--	--	--	--
RI	--	--	--	--	--	--	--	--	--	--	--	--	--	--	--	--
UJT	--	02	--	--	--	--	--	--	--	--	01	01	--	--	02	--
LJT	--	--	--	--	--	--	--	--	--	--	01	01	--	--	--	--
OTHERS	--	--	--	--	--	--	--	--	--	--	--	--	--	--	--	--
TOTAL	10	25	16	02	00	02	02	07	03	01	07	03	03	04	02	00

Table 4.40. Bone distribution of *Ovis aries* in different layers at Inamgaon

Bones	**01**	**02**	**03**	**04**	**05**	**06**	**07**	**08**	**09**	**10**	**11**	**12**	**13**	**14**	**15**	**16**
HC/AN	--	--	--	--	--	--	--	--	--	--	--	--	--	--	--	--
CF	--	01	--	--	--	--	--	--	--	--	--	--	--	--	--	--
MN	--	02	--	--	--	--	--	--	--	--	--	--	--	--	--	--
SC	--	--	--	--	--	01	--	--	--	--	03	--	--	--	--	--
HU	--	01	01	--	01	--	01	01	--	--	01	--	--	--	--	--
RA	--	01	--	--	--	--	--	--	--	--	--	--	--	--	--	--
UL	--	--	--	--	--	--	--	--	--	--	--	--	--	--	--	--

MC	--	--	--	--	--	--	--	--	--	--	--	01	--	--	--	--
C/T	--	--	--	--	--	--	--	--	--	--	--	--	--	--	--	--
MP	--	--	--	--	--	--	--	--	--	--	--	--	--	--	--	--
PV	--	01	--	--	--	--	--	--	--	--	--	--	--	--	--	--
FE	01	01	01	--	--	--	--	--	--	--	--	--	--	--	--	--
TI	--	--	--	--	--	--	--	--	--	--	--	--	--	--	--	--
PA	--	--	--	--	--	--	--	--	--	--	--	--	--	--	--	--
MT	--	--	--	--	--	--	--	--	--	--	--	--	--	--	--	--
AS	--	--	--	--	--	--	--	--	--	--	--	--	--	--	--	--
CA	--	--	--	--	--	--	--	--	--	--	--	--	--	--	--	--
CT	--	01	--	--	--	--	--	--	--	--	--	--	--	--	--	--
H1	--	--	--	--	--	--	--	--	--	--	--	--	--	--	01	--
H2	--	--	--	--	--	--	--	--	--	--	--	--	--	--	--	--
H3	--	--	--	--	--	--	--	--	--	--	--	--	--	--	--	--
VB	--	01	--	--	--	--	--	--	--	--	--	--	--	--	--	--
RI	--	--	--	--	--	--	--	--	--	--	--	--	--	--	--	--
UJT	--	--	--	--	--	--	--	--	--	--	--	--	--	--	--	--
LJT	--	--	--	--	--	--	--	--	--	--	--	01	--	--	--	--
OTHERS	--	--	--	--	--	--	--	--	--	--	--	--	--	--	--	--
TOTAL	01	09	02	00	01	01	01	01	00	00	04	02	00	00	01	00

Sus domesticus **(domestic pig)**

Domestic pigs are reported from a large number of Chalcolithic sites in western India, except Apegaon (Table 4.33). This animal was also utilized for meat, by early farmers. However, at Inamgaon the representation of bones of pig does not show that it was exploited for food purpose. In the total faunal assemblage from the site only 62 bones of this animal are found. A few skeletal remains are found at Inamgaon in all the layers, except layers 1, 5, 7 and 10 (Table 4.1-4.16), (Pl.V A). The distribution of bone fragments is scanty (Table 4.41). Pig bones are also found in the houses from layers 8 and 12 (Table 4.26 & 4.27). The phase-wise distribution of identifiable bones shows that they are represented 32 (0.21 %) in the Late Jorwe phase, 10 (0.14 %) in the Early Jorwe phase and 20 (0.55 %) in the Malwa phase (Table 4.17). The maximum representation is observed in the Malwa phase with 9 individual animals followed by 8 and 6 in the Late Jorwe and Early Jorwe phase, respectively (Table 4.19-4.31). The possibly consumed meat weight is calculated for different layers and the maximum consumption was noticed in layer 2, i.e. 80 kg (Table 4.32).

Plate V A. ***Sus domesticus*****:** 1. Mandible, 2. Scapula, 3. Ulna, 4. Radius, 5. Femur, 6. Tibia, 7. Astragalus, 8. Second phalanx, 9. Radius and 10. Calcaneum.

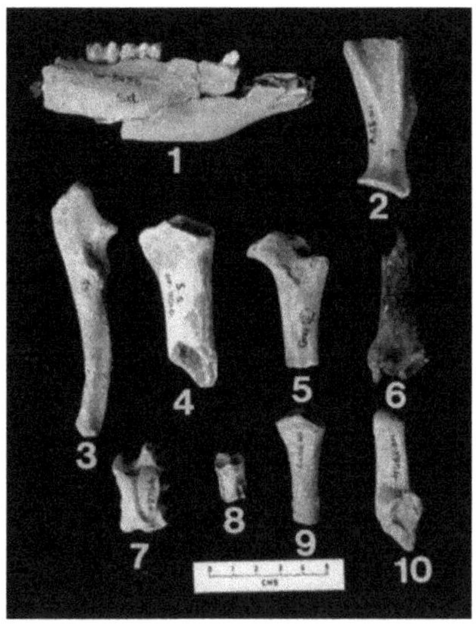

Table 4.41. Bone distribution of *Sus domesticus* in different layers at Inamgaon

Bones	01	02	03	04	05	06	07	08	09	10	11	12	13	14	15	16
CF	--	04	01	--	--	--	--	--	--	--	--	--	--	--	--	--
MN	--	01	--	--	--	--	--	--	01	--	--	01	--	--	--	--
SC	--	01	01	--	--	--	--	--	--	--	--	01	--	01	--	01
HU	--	03	01	--	--	--	--	01	--	--	--	--	--	--	--	--
RA	--	02	01	01	--	--	--	01	--	--	01	--	--	01	--	--
UL	--	01	01	--	--	01	--	02	--	--	--	--	--	01	--	--
MC	--	--	--	--	--	01	--	--	--	--	--	--	--	--	--	--
C/T	--	--	--	--	--	--	--	--	--	--	--	--	--	--	--	--
MP	--	--	--	--	--	--	--	--	--	--	--	--	--	--	--	--
PV	--	--	--	01	--	--	--	--	--	--	--	--	--	--	--	--
FE	--	01	--	--	--	--	--	--	--	--	--	--	--	--	--	--
TI	--	--	01	01	--	--	--	01	--	--	--	--	--	--	--	--
PA	--	--	--	--	--	--	--	--	--	--	--	--	--	--	--	--
MT	--	--	--	--	--	--	--	--	--	--	--	--	--	--	--	--
AS	--	01	--	01	--	--	--	--	--	--	--	01	--	--	01	--
CA	--	01	--	--	--	--	--	--	--	--	--	01	--	--	--	--
CT	--	--	--	--	--	--	--	--	--	--	--	--	--	--	--	--
H1	--	--	--	--	--	--	--	--	--	--	--	--	--	01	--	--
H2	--	01	--	--	--	--	--	--	--	--	01	--	--	01	--	--
H3	--	--	--	--	--	--	--	--	--	--	--	--	--	--	--	--
VB	--	02	01	01	--	--	--	--	--	--	--	02	01	--	01	--
RI	--	--	--	--	--	--	--	--	--	--	--	--	--	--	03	--
UJT	--	02	--	--	--	--	--	--	--	--	--	--	--	--	01	--
LJT	--	--	--	--	--	--	--	--	--	--	--	--	--	--	01	--
OTHERS	--	--	--	--	--	--	--	--	--	--	--	--	--	--	--	--
TOTAL	00	20	07	05	00	02	00	05	01	00	02	06	01	05	07	01

Canis familiaris (dog)

From very early times dog is known as a companion of man and has been reported from a number of archaeological sites (Table 4.33). At Inamgaon the dog bones are found in all the 16 layers. Bones are sparsely distributed in all the cultural phases (Table 4.42). The remains are recovered from the houses in layers 2, 3, 12 and 16 (Table 4.21, 4.22, 4.27 & 4.29).

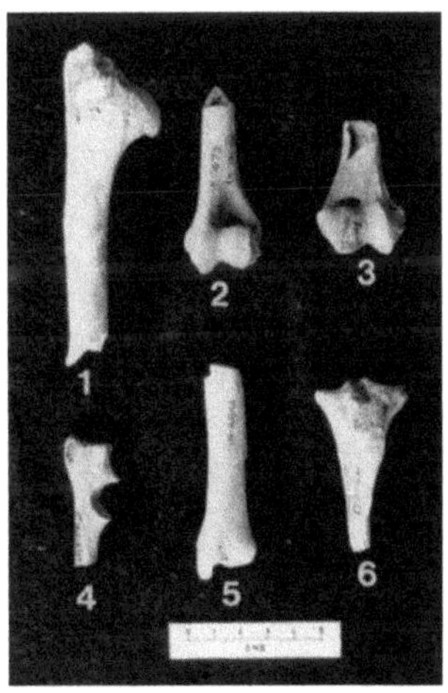

Plate V B. ***Canis familiaris*:** 1. Humerus, 2. &3. Humerus distal, 4. Ulna, 5. Tibia distal and 6. Tibia proximal

Table 4.42. Bone distribution of *Canis familiaris* in different layers at Inamgaon

Bones	01	02	03	04	05	06	07	08	09	10	11	12	13	14	15	16
CF	02	03	--	--	--	01	02	--	01	--	--	--	--	--	--	--
MN	01	03	03	05	--	--	--	01	--	01	--	01	--	--	01	--
SC	02	01	--	--	--	01	--	--	--	--	01	01	--	01	--	--
HU	01	03	05	02	01	--	-	--	--	--	--	--	01	01	02	--
RA	01	--	02	01	--	--	--	--	--	--	--	01	--	01	01	--
UL	--	02	01	01	--	--	--	--	--	--	--	--	--	01	--	--
MC	--	--	--	--	--	--	--	--	--	--	--	--	--	--	--	--
C/T	--	--	--	--	--	--	--	--	--	--	--	--	--	--	--	--
MP	01	01	01	--	01	--	--	03	05	--	--	02	--	01	--	--
PV	01	02	03	02	--	--	--	01	--	--	--	01	--	--	--	01
FE	--	04	01	--	--	--	--	--	--	--	--	--	--	--	01	--
TI	01	07	--	02	01	--	--	--	--	--	--	--	--	--	--	--
FI	--	--	--	--	--	--	--	01	--	--	--	--	--	--	--	--
PA	--	--	--	--	--	--	--	--	--	--	--	--	--	--	--	--
MT	--	--	--	--	01	--	--	--	--	--	--	--	--	--	--	--
AS	--	01	--	--	--	01	--	--	--	--	--	--	--	--	--	--
CA	--	01	01	--	--	01	--	--	--	--	--	--	--	--	--	01
CT	--	--	--	--	01	--	--	--	--	--	--	--	--	--	--	--
H1	--	03	--	--	--	--	01	01	01	--	--	--	--	--	--	--
H2	--	--	--	--	--	--	--	--	--	--	--	--	--	--	--	--
H3	--	--	--	--	--	--	--	--	--	--	--	--	--	--	--	--
VB	--	02	01	--	--	--	--	--	--	--	--	--	--	--	--	--
RI	--	--	--	01	--	--	--	05	--	--	--	--	--	--	--	--
UJT	--	--	--	--	--	--	--	--	--	--	--	--	--	--	--	--
LJT	--	--	01	--	--	--	--	--	--	--	--	--	--	--	--	--
OTHERS	--	--	--	--	--	--	--	--	--	--	--	--	--	--	--	--
TOTAL	10	33	19	14	05	04	03	12	07	01	01	06	01	05	05	02

The bone fragments recovered from the Early Jorwe and the Malwa phases do not show any trace of human activity on them. But in the Late Jorwe phase, cut marks and charring marks are observed on some of the long bones and scapula (Pl. V B). It suggests consumption of dog meat by the inhabitants in this phase. Archaeological evidence is lacking for the consumption of dog meat in India, but an ethnographic parallel is available of a tribe *Katodi*, from Maharashtra who kills the domestic dog for food at the time of scarcity (Thomas 1984a, 1988, 1989). A large number of bone fragments are identified from the Late Jorwe phase. The number of identified fragments are 81 (0.53 %) in the Late Jorwe phase, 28 (0.40 %) in Early Jorwe and 19 (0.52 %) in Malwa phases (Table 4.1-4.16). The numbers of individuals are 12, 6 and 8 in the Late Jorwe phase, Early Jorwe phase and Malwa phase, respectively (Table 4.19-4.31).

***Equus caballus* (horse)**

Horse is introduced towards the end of the Chalcolithic culture at Inamgaon. This animal is also reported from other Chalcolithic sites of this region such as Nevasa, Daimabad and Walki (Table 4.33) (Clason, 1979; Badam 1986; Joglekar 1992; Thomas and Joglekar 1994; Joglekar and Thomas 1995). At Inamgaon the bones are identified from the terminal layers of Early Jorwe phase and maximum are in the Late

Jorwe phase (Table 4.1-4.16). Fragments of dentition like incisor, premolar and molar and the lower limb extremities such as distal end of the tibia, metapodials and the first phalanges were recovered from Inamgaon (Pl. VI A). Also found a centrum portion of the last lumber vertebra. Cut marks are found on the lower limb bones. The Late Jorwe phase is contemporary with the Iron Age culture of this region. From the Megalithic burials of Vidarbha region of Maharashtra, similar kind of occurrence of bones was found (Thomas 1992 a & b). Cut marks and charring on the distal tibia suggests the consumption of meat of this animal probably at some ceremony.

Plate VI A. *Equus caballus*: 1. Lumber vertebra, 2. Metapodial distal, 3. Tibia distal, First phalanx and 5-7. Teeth.

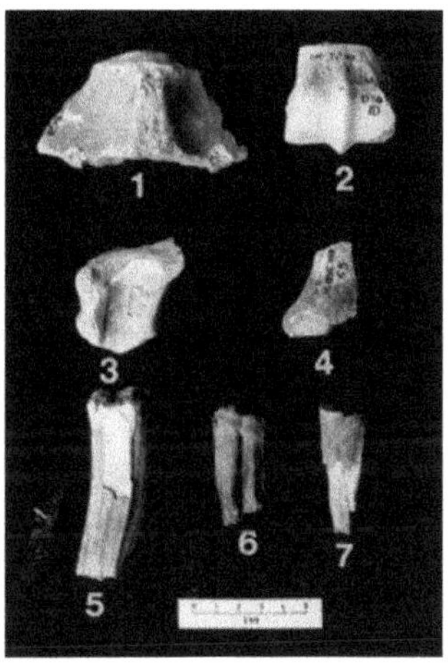

Equus asinus (ass)

Ass is reported from sites such as Nevasa, Daimabad, Inamgaon and Walki (Table 4.33). At Inamgaon this animal is represented by only one bone – a third phalanx. From Walki, only one cervical vertebra is identified (Joglekar 1992; Thomas and Joglekar 1994; Joglekar and Thomas 1995). Ass has also been reported from Daimabad, Nevasa and Tuljapur Garhi (Clason 1979; Badam 1986 and Thomas 1992 a). Since this animal is a beast of burden, it is likely that a small herd of ass might have been associated with this early farming community.

Elephas maximus (elephant)

The presence of this animal is identified from the sites Nevasa, Daimabad and Inamgaon. Ivory objects are found in the Late Jorwe phase and a piece of ivory tusk is collected from the Malwa phase. An Ivory dice is rectangular in shape and is cut on all sides and polished. On the four flat surfaces double circles are incised in numbers ranging from 1 – 4 (Dhavalikar and Ansari 1988).

Canis lupus (wolf)

This animal has not been represented in all the layers at Inamgaon (Table 4.1 – 4.16). A majority of bones of wolf are found in the Late Jorwe phase and almost all the bones of this animal are identified from this phase. In the Early Jorwe phase, this animal is represented by humerus, radius, ulna, and metapodials and in the Malwa phase by one fragment of each of humerus, femur and astragalus and two fragments of tibia. Comparatively the distribution of skeletal remains is very scanty in all the three cultural phases (Table 4.43).

Table 4.43. Bone distribution of *Canis lupus* in different layers at Inamgaon

Bones	01	02	03	04	05	06	07	08	09	10	11	12	13	14	15	16
CF	01	--	--	--	--	--	--	--	--	--	--	--	--	--	--	--
MN	--	01	--	--	--	--	--	--	--	--	--	--	--	--	--	--
SC	--	02	--	--	--	--	--	--	--	--	--	--	--	--	--	--
HU	--	01	01	--	--	--	--	01	--	--	--	--	--	--	--	01
RA	--	01	01	--	--	--	--	--	01	--	--	--	--	--	--	--
UL	--	03	--	--	--	01	--	--	--	--	--	--	--	--	--	--
MC	--	--	--	--	--	--	--	--	--	--	--	--	--	--	--	--
C/T	--	--	--	--	--	--	--	--	--	--	--	--	--	--	--	--
MP	--	--	--	--	01	01	--	--	--	--	--	--	--	--	--	--

PV	--	--	01	--	--	--	--	--	--	--	--	--	--	--	--	--
FE	--	--	01	--	--	--	--	--	--	--	--	--	--	--	--	01
TI	01	--	--	--	--	--	--	--	--	--	--	--	--	02	--	--
FI	--	--	--	--	--	--	--	--	--	--	--	--	--	--	--	--
PA	--	--	--	--	--	--	--	--	--	--	--	--	--	--	--	--
MT	--	--	--	--	--	--	--	--	--	--	--	--	--	--	--	--
AS	--	01	--	--	--	--	--	--	--	--	--	--	--	--	--	01
CA	--	01	02	--	--	--	--	--	--	--	--	--	--	--	--	--
CT	--	--	--	--	--	--	--	--	--	--	--	--	--	--	--	--
H1	--	--	--	--	--	--	--	--	--	--	--	--	--	--	--	--
H2	--	--	--	--	--	--	--	--	--	--	--	--	--	--	--	--
H3	--	--	--	--	--	--	--	--	--	--	--	--	--	--	--	--
VB	01	--	--	--	--	--	--	--	--	--	--	--	--	--	--	--
RI	--	01	--	--	--	--	--	--	--	--	--	--	--	--	--	--
UJT	--	--	--	--	--	--	--	--	--	--	--	--	--	--	--	--
LJT	--	--	--	--	--	--	--	--	--	--	--	--	--	--	--	--
OTHERS	--	--	--	--	--	--	--	--	--	--	--	--	--	--	--	--
TOTAL	03	11	06	00	01	02	00	01	01	00	00	00	00	02	00	03

Altogether 21 bone fragments are found in the Late Jorwe phase, 4 in the Early Jorwe and 5 in the Malwa phase (Table 4.1 – 4.17). Minimum Number of Individuals is 6, 3 and 3, in the Late Jorwe, Early Jorwe and the Malwa phases, respectively (Table 4.19 – 4.31).

Canis aureus (jackal)

Bones of this animal are very sparsely distributed. The skeletal remains are absent in the Malwa phase and are represented by only two bones the scapula and radius, in the Early Jorwe phase. In the Late Jorwe phase seven bones are found such as mandible, ulna, femur one each, and humerus and tibia two each (Table 4.44).The remains of this animal are not identified from any other Chalcolithic sites of Maharashtra, except Walki (Joglekar 1992; Joglekar and Thomas 1995) (Table 4.33).

Table 4.44. Bone distribution of *Canis aureus* in different layers at Inamgaon

Bones	01	02	03	04	05	06	07	08	09	10	11	12	13	14	15	16
CF	--	--	--	--	--	--	--	--	--	--	--	--	--	--	--	--
MN	--	01	--	--	--	--	--	--	--	--	--	--	--	--	--	--
SC	--	--	--	--	--	--	--	01	--	--	--	--	--	--	--	--
HU	--	02	--	--	--	--	--	--	--	--	--	--	--	--	--	--
RA	--	--	--	--	--	--	--	01	--	--	--	--	--	--	--	--
UL	--	--	01	--	--	--	--	--	--	--	--	--	--	--	--	--
MC	--	--	--	--	--	--	--	--	--	--	--	--	--	--	--	--
C/T	--	--	--	--	--	--	--	--	--	--	--	--	--	--	--	--
MP	--	--	--	--	--	--	--	--	--	--	--	--	--	--	--	--
PV	--	--	--	--	--	--	--	--	--	--	--	--	--	--	--	--
FE	--	--	--	01	--	--	--	--	--	--	--	--	--	--	--	--
TI	--	01	01	--	--	--	--	--	--	--	--	--	--	-	--	--
FI	--	--	--	--	--	--	--	--	--	--	--	--	--	--	--	--
PA	--	--	--	--	--	--	--	--	--	--	--	--	--	--	--	--
MT	--	--	--	--	--	--	--	--	--	--	--	--	--	--	--	--
AS	--	--	--	--	--	--	--	--	--	--	--	--	--	--	--	-
CA	--	--	--	--	--	--	--	--	--	--	--	--	--	--	--	--
CT	--	--	--	--	--	--	--	--	--	--	--	--	--	--	--	--
H1	--	--	--	--	--	--	--	--	--	--	--	--	--	--	--	--
H2	--	--	--	--	--	--	--	--	--	--	--	--	--	--	--	--
H3	--	--	--	--	--	--	--	--	--	--	--	--	--	--	--	--
VB	--	--	--	--	--	--	--	--	--	--	--	--	--	--	--	--

RI	--	--	--	--	--	--	--	--	--	--	--	--	--	--	--	--
UJT	--	--	--	--	--	--	--	--	--	--	--	--	--	--	--	--
LJT	--	--	--	--	--	--	--	--	--	--	--	--	--	--	--	--
OTHERS	--	--	--	--	--	--	--	--	--	--	--	--	--	--	--	--
TOTAL	00	04	02	01	00	00	00	02	00	00	00	00	00	00	00	00

***Panthera pardus* (leopard)**

Leopard bones were identified from layers 2, 11 and 12. Only a few bone fragments were found and they are - three complete first phalanges, one proximal part of femur, three fragments of metapodials, etc. (Pl. VI B). Epiphyses of all the bones were fused. There is no trace of charring and cut marks to suggest human activity on these bone fragments. Therefore, it is doubtful that these animals may have contributed to the subsistence economy at Inamgaon. The only other site where the skeletal remains of this animal were reported was at Walki (Joglekar 1992; Joglekar and Thomas 1995) very near to Inamgaon, in the Pune district of Maharashtra.

Plate VI B. *Panthera pardus*: 1. Metapodia, 2&3. Scapula, 4-6. First phalanges.

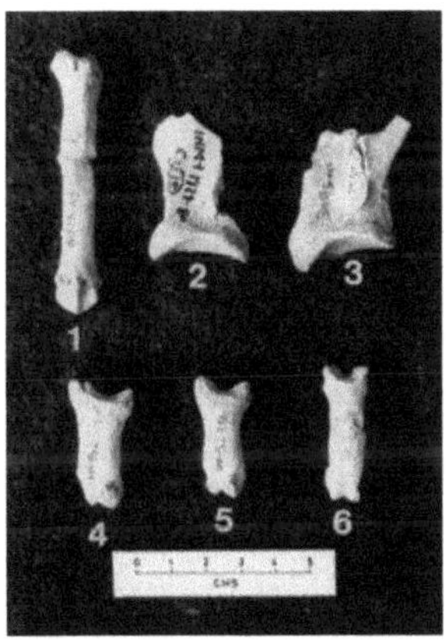

***Panthera tigris* (tiger)**

Only one fragment of the distal part of a radius was identified from layer 2. The distal end is unfused and the lack of any human or animal activity on the bone suggests that this animal was not in the diet of the people. The skeletal remains of this animal are also reported from Walki (Joglekar 1992 and Joglekar and Thomas 1995).

***Felis chaus* (jungle cat)**

This animal is identified from the layer 1, 2, 3, 4, 6 and 7. Fragments of scapula, humerus, radius, pelvis, femur, tibia, calcaneum and metapodia are found with cut marks and charring (Pl. VII A). A majority of the remains are recovered from the Late Jorwe phase, i.e. 10, (0.06 %) (Table 4.17). This animal has also been reported from Kaothe (Thomas and Joglekar 1990), Walki (Joglekar 1992; Joglekar and Thomas 1995) and Adam (Thomas and Joglekar in press) (Table 4.33).

Plate VII A. *Felis chaus*: 1&2. Pelvis, 3. Tibia and 4. Femur proximal.

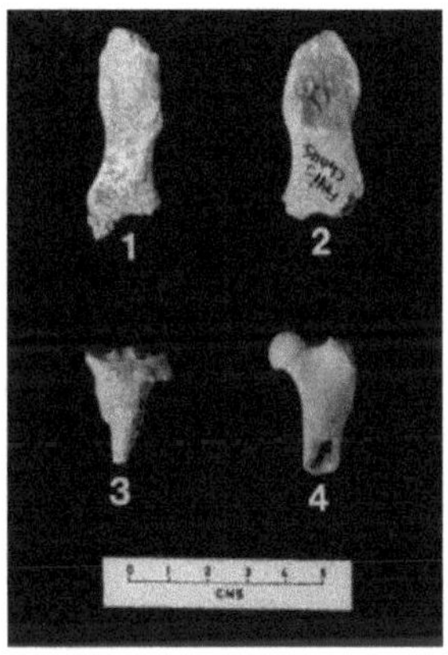

***Bos gaurus* (gaur)**

Only two first phalanges, one third phalanx (Pl. VII B) and a fragment of calcaneum are identified at Inamgaon (Table 4.1- 4.16). It is interesting to note that these bones constitute the non-meat bearing parts of the skeleton of an animal and are found well preserved. Probably these bones have reached the site by scavenging activities of human or non- human agencies. Gaur has also been reported from the Chalcolithic sites such as Walki (Joglekar 1992 and Joglekar and Thomas 1995) in western Maharashtra and Adam (Thomas and Joglekar in press) in eastern Maharashtra (Table 4.33).

Plate VII B. *Bos gaurus*: 1. Third phalanx, 2&3. First phalanges.

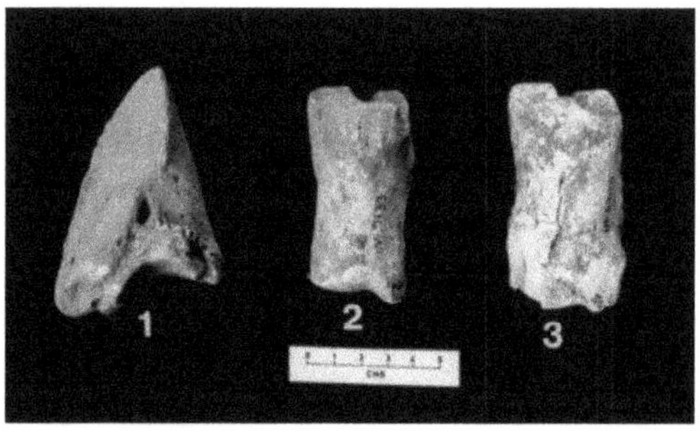

Sus scrofa (wild pig)

Wild pig is identified from Kaothe, in the Dhule district (Thomas and Joglekar 1990) and Walki, in the Pune district (Joglekar 1992 and Joglekar and Thomas 1995) (Table 4.33). At Inamgaon remains of this animal are represented sparsely in all the cultural phases. From layer 13 of the Malwa phase, only one femur is identified. In the Early Jorwe phase, bones are found in layers 6 and 8. However, maximum representation is noticed in the Late Jorwe phase. The identified fragments include cranial, mandible, humerus, tibia, metacarpal, calcaneum and some vertebral fragments (Table 4.45). Altogether 16 bone fragments are recovered from the Late Jorwe phase, 7 from the Early Jorwe phase and one from the Malwa phase (Table 4.1-4.16). The bones of wild pig account for only 0.09 % in the entire faunal assemblage (Table 4.17). The Minimum Number of Individuals is 6, 2 and 1, in the Late Jorwe, Early Jorwe and Malwa phases, respectively (Table 4.19-4.31). Approximately 600 kg of meat of this animal has been consumed in the Late Jorwe phase (Table 4.34).

Table 4.45. Bone distribution of *Sus scrofa* in different layers at Inamgaon

Bones	01	02	03	04	05	06	07	08	09	10	11	12	13	14	15	16
CF	01	--	--	--	--	--	--	--	--	--	--	--	--	--	--	--
MN	--	03	--	--	--	--	--	--	--	--	--	--	--	--	--	--
SC	--	01	--	--	--	--	--	--	--	--	--	--	--	--	--	--
HU	01	--	01	--	--	01	--	--	--	--	--	--	--	--	--	--
RA	--	--	--	--	--	--	--	01	--	--	--	--	--	--	--	--
UL	02	02	--	--	--	--	--	--	--	--	--	--	--	--	--	--
MC	--	--	--	--	--	--	--	01	--	--	--	--	--	--	--	--
C/T	--	--	--	--	--	--	--	--	--	--	--	--	--	--	--	--
MP	--	--	--	--	--	--	--	--	--	--	--	--	--	--	--	--
PV	--	--	01	--	--	--	--	--	--	--	--	--	--	--	--	--
FE	--	--	--	--	--	--	--	--	--	--	--	--	01	--	--	--
TI	--	--	--	--	--	--	--	01	--	--	--	--	--	-	--	--
PA	--	--	--	--	--	--	--	--	--	--	--	--	--	--	--	--
MT	--	--	--	--	--	--	--	--	--	--	--	--	--	--	--	--
AS	--	--	--	--	--	--	--	--	--	--	--	--	--	--	--	-
CA	--	--	01	--	--	--	--	--	--	--	--	--	--	--	--	--
CT	--	--	--	--	--	--	--	--	--	--	--	--	--	--	--	--
H1	--	--	--	--	--	--	--	--	--	--	--	--	--	--	--	--
H2	--	--	--	--	--	--	--	--	--	--	--	--	--	--	--	--
H3	--	--	--	--	--	--	--	--	--	--	--	--	--	--	--	--
VB	--	02	01	--	--	--	--	02	--	--	--	--	--	--	--	--
RI	--	--	--	--	--	--	--	--	--	--	--	--	--	--	--	--
UJT	--	--	--	--	--	01	--	--	--	--	--	--	--	--	--	--
LJT	--	--	--	--	--	--	--	--	--	--	--	--	--	--	--	--
OTHERS	--	--	--	--	--	--	--	--	--	--	--	--	--	--	--	--
TOTAL	04	08	04	00	00	02	00	05	00	00	00	00	01	00	00	00

Gazella bennetti (chinkara)

This animal is presently distributed in the plains and low hills of northern, western and central India, extending through the open lands of Deccan to a little south of the river Krishna (Prater 1965).

Chinkara is identified from all the three cultural phases. However, they are absent in many of the stratigraphical units (Table 4.1-4.16). Bones are also collected from the structures in the layers 8, 12 and 13 (Table 4.26-4.28). Compared to the Malwa and the Early Jorwe phases, the skeletal remains are found more in the Late Jorwe phase. Fragments of horn core, scapula, humerus, radius, metacarpal, metatarsal, pelvis, femur, tibia, astragalus, calcaneum, first, second and third phalanges and a few fragments of vertebrae are identified in the assemblage (Table 4.46). Altogether 29 bones are recovered in the Late Jorwe phase, 5 in the Early Jorwe phase and 4 in the Malwa phase (Table 4.17). The Minimum Number of Individuals is 5, 2 and 4 in the Late Jorwe, the Early Jorwe and in the Malwa phases, respectively. The probable meat consumption from this animal is 50 kg, 20 kg and 40 kg in the Late Jorwe, Early Jorwe and the Malwa phases, respectively (Table 4.32).

Table 4.46. Bone distribution of *Gazella bennetti* in different layers at Inamgaon

Bones	**01**	**02**	**03**	**04**	**05**	**06**	**07**	**08**	**09**	**10**	**11**	**12**	**13**	**14**	**15**	**16**
HC/AN	01	--	01	--	--	--	--	--	--	--	--	--	--	--	--	--
CF	--	--	--	--	--	--	--	--	--	--	--	--	--	--	--	--
MN	--	--	--	--	--	--	--	--	--	--	--	--	--	--	--	--
SC	--	02	--	--	--	--	--	--	--	--	--	--	--	--	--	--
HU	--	01	--	--	--	--	--	01	--	--	--	--	--	--	--	--
RA	--	01	--	--	--	--	--	01	--	--	--	--	--	--	--	--
UL	--	--	--	--	--	--	--	--	--	--	--	--	--	--	--	--
MC	01	--	--	--	--	--	--	--	--	--	--	--	--	--	--	--
C/T	--	--	--	--	--	--	--	--	--	--	--	--	--	--	--	--
MP	--	--	01	--	--	--	--	--	--	--	--	--	01	01	--	--
PV	--	01	--	--	--	--	--	--	--	--	--	--	--	--	--	--
FE	01	--	--	--	--	--	--	--	--	--	--	--	--	--	--	--
TI	01	--	01	--	--	--	--	02	--	--	--	--	--	-	--	--
PA	--	--	--	--	--	--	--	--	--	--	--	--	--	--	--	--
MT	--	01	--	--	--	--	--	--	--	--	--	--	--	--	--	--
AS	--	--	01	--	--	--	--	--	--	--	--	01	--	--	--	-
CA	--	--	--	--	--	--	--	01	--	--	--	--	--	--	--	--
CT	--	--	--	--	--	--	--	--	--	--	--	--	--	--	--	--
H1	02	--	01	02	--	--	--	--	--	--	--	--	01	--	--	--
H2	02	01	01	01	--	--	--	--	--	--	--	--	--	--	--	--
H3	--	01	--	--	--	--	--	--	--	--	--	--	--	--	--	--
VB	02	--	--	--	--	--	--	--	--	--	--	--	--	--	--	--
RI	--	--	--	--	--	--	--	--	--	--	--	--	--	--	--	--
UJT	--	--	--	--	--	--	--	--	--	--	--	--	--	--	--	--
LJT	--	02	--	--	--	--	--	--	--	--	--	--	--	--	--	--
OTHERS	--	--	--	--	--	--	--	--	--	--	--	--	--	--	--	--
TOTAL	10	10	06	03	00	00	00	05	00	00	00	01	02	01	00	00

This animal also identified at Walki (Table 4.33) in the Pune district (Joglekar 1992). However, they are not presently distributed in the Pune district but are found in the adjacent districts such as Solapur, Ahmadnagar, Dhule, Nasik, Bid, Osmanabad, Nanded, Yavatmal, Wardha, Nagpur, Bhandara and Chandrapur of Maharashtra (Fig.4.2).

Fig. 4.2. Map showing past and present distribution of Gazella bennetti in Maharashtra.

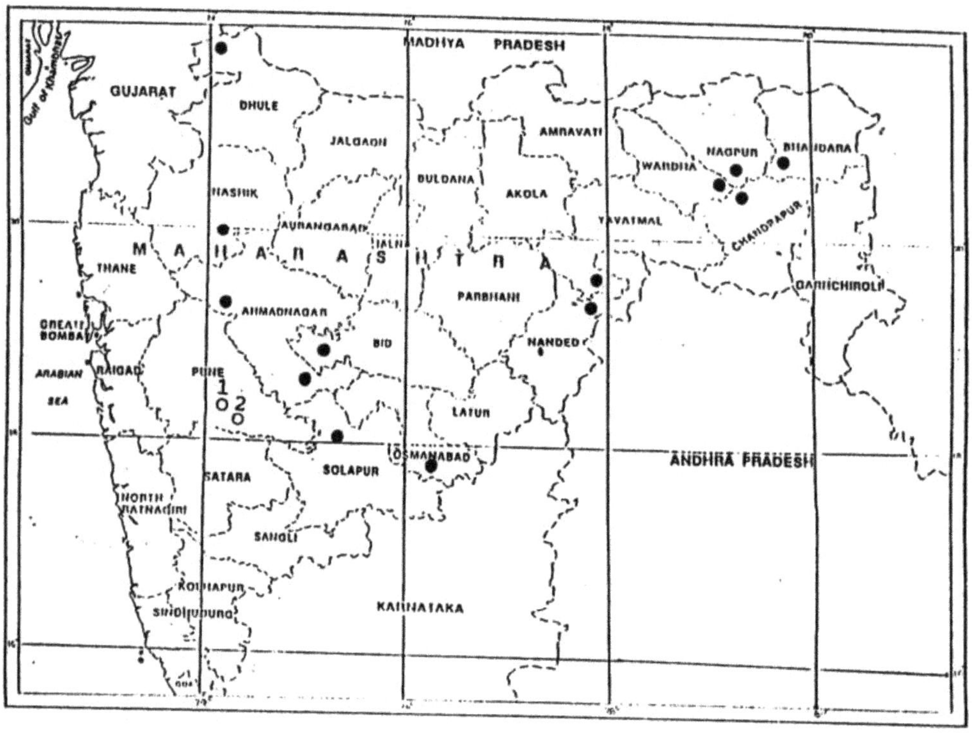

○ Past distribution ● Present distribution

1. Walki, 2. Inamgaon.

Tetracerus quadricornis **(four-horned antelope)**

Four-horned antelope shelters in tall grass and open jungle and is distributed in Peninsular India. This animal is not found in too densely forested area. Four-horned antelope is characterized by two pairs of horns and the presence of a pair of well developed glands between the false hooves of the hind legs in both males and females (Prater 1965).

The bones of this animal are identified (Pl. VIII A) from all the layers, except layers 9, 10, 13 and 14. The distribution of bones is very scanty. Only 11 bone fragments are identified from Malwa, 7 from the Early Jorwe and 30 from the Late Jorwe phase. Fragments of horn core, cranium, scapula, humerus, radius, metacarpal, pelvis, femur, tibia, metatarsal, astragalus, calcaneum, first and second phalanges and isolated upper and lower jaw teeth are found in the collection (Table 4.47).

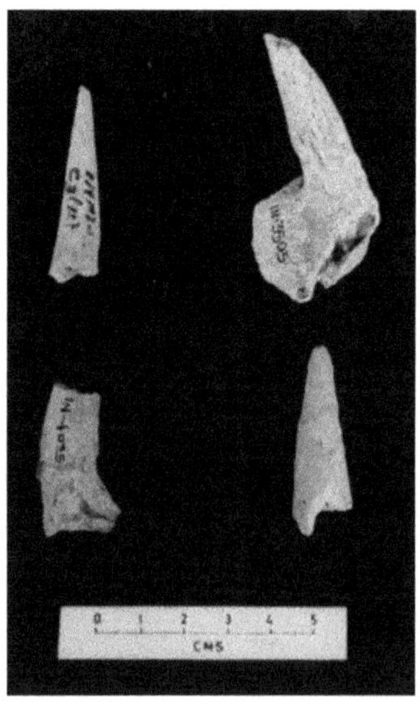

Plate VIII A. ***Tetracerus quadricornis*:** Horn cores.

Table 4.47. Bone distribution of *Tetracerus quadricornis* in different layers at Inamgaon

Bones	**01**	**02**	**03**	**04**	**05**	**06**	**07**	**08**	**09**	**10**	**11**	**12**	**13**	**14**	**15**	**16**
HC/AN	--	02	04	--	--	--	--	--	--	--	01	--	--	--	--	--
CF	01	01	--	--	--	--	--	--	--	--	--	--	--	--	--	--
MN	--	--	--	--	--	--	--	--	--	--	--	--	--	--	--	--
SC	--	01	--	01	--	--	--	--	--	--	--	--	--	--	--	01
HU	--	02	01	--	--	01	--	--	--	--	--	--	--	--	--	--
RA	--	--	01	--	--	--	--	--	--	--	--	--	--	--	01	--
UL	--	--	--	--	--	--	--	--	--	--	--	--	--	--	--	--
MC	--	--	02	--	--	--	--	--	--	--	--	--	--	--	01	--
C/T	--	--	--	--	--	--	--	--	--	--	--	--	--	--	--	--
MP	--	--	--	--	--	--	--	--	--	--	--	--	--	--	--	--
PV	--	--	01	--	--	--	--	--	--	--	--	--	--	--	01	--
FE	01	--	--	--	--	--	--	01	--	--	--	--	--	--	--	--
TI	--	02	--	--	01	--	--	01	--	--	--	01	--	-	--	--
PA	--	--	--	--	--	--	--	--	--	--	--	--	--	--	--	--
MT	--	02	01	--	--	--	--	--	--	--	--	02	--	--	--	--
AS	--	--	02	01	--	--	--	--	--	--	01	--	--	--	02	-
CA	--	--	01	--	--	--	--	--	--	--	--	--	--	--	--	--
CT	--	--	--	--	--	--	--	--	--	--	--	--	--	--	--	--
H1	--	01	--	--	--	--	--	--	--	--	--	--	--	--	--	--
H2	--	--	01	--	--	--	--	--	--	--	--	--	--	--	--	--
H3	--	--	--	--	--	--	--	--	--	--	--	--	--	--	--	--
VB	--	--	--	--	--	--	--	--	--	--	--	--	--	--	--	--
RI	--	--	--	--	--	--	--	--	--	--	--	--	--	--	--	--
UJT	--	--	--	--	--	01	--	--	--	--	--	--	--	--	02	--
LJT	--	--	--	--	--	--	01	--	--	--	--	--	--	--	--	--
OTHERS	--	--	--	--	--	--	--	--	--	--	--	--	--	--	--	--
TOTAL	02	11	14	02	01	02	01	02	00	00	02	03	00	00	07	01

From Layer 16, only one bone fragment is recovered from house number 87. The number of individuals are 6 (1.00 %), 4 (1.34 %) and 5 (1.94 %) in the Late Jorwe, Early Jorwe and Malwa phases, respectively (Table 4.19-4.31). The maximum consumption of meat is noticed in layer 15, (37.50 kg) (Table 4.32). A gradual decline in the representation of this animal is found towards the end of the cultural phase.

Four-horned antelopes are also reported from Nasik (George 1955), Nevasa (Clason 1979) and Daimabad (Badam 1986) in the Ahmadnagar district, Kaothe in the Dhule district (Thomas and Joglekar 1990), Walki in the Pune district (Joglekar 1992) and Tuljapur Garhi in the Amravati district (Thomas 1992)(Table 4.31). In Pune, the present day distribution is reported towards the north of this district and is absent in Ahmadnagar district. The distribution in Dhule and Amravati districts is found near the location of the Chalcolithic sites. At present in Maharashtra, this species is reported from the districts of Dhule, Nasik, Thane, Pune, Raigadh, Ratnagiri, Satara, Sangli, Kolhapur, Sindhudurga, Amravati, Bhandara, Chandrapur and Gadchiroli (Fig. 4.3).

Fig. 4.3. Map Showing Past and Present Distribution of *Tetracerus quadricornis* in Maharashtra

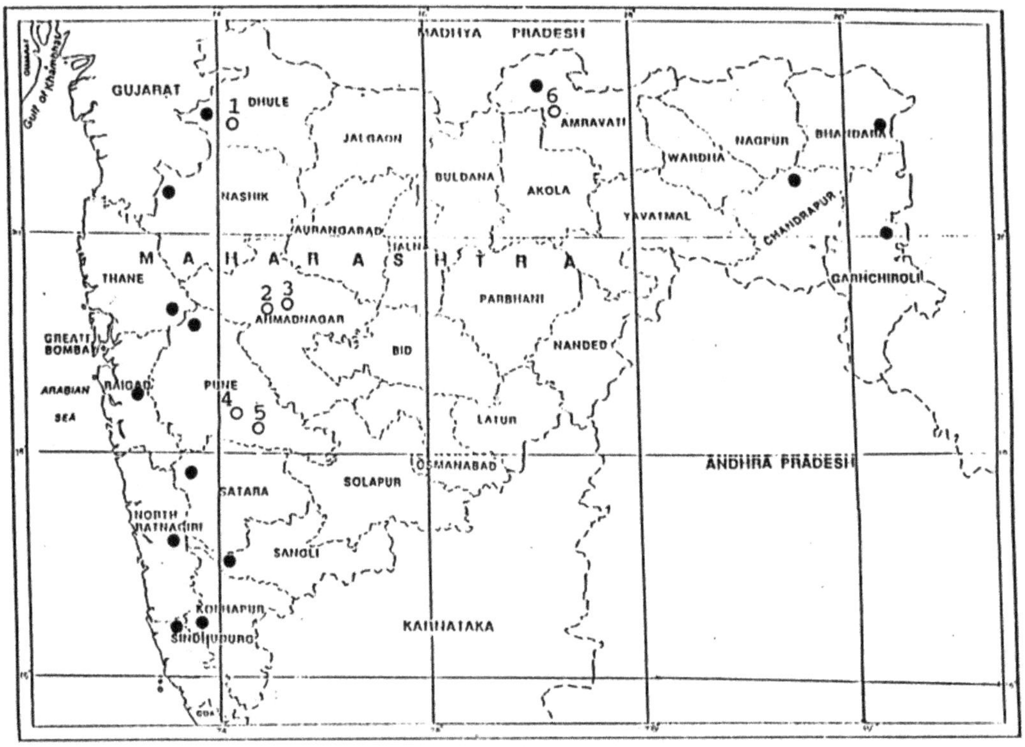

○ Past distribution   ● Present distribution

1. Kaothe, 2. Daimabad, 3. Nevasa, 4. Walki, 5. Inamgaon and 6. Tuljapur Garhi

Antilope cervicapra (blackbuck)

Blackbuck is usually found in the plains and avoids forest and hilly tracts. They feed on grass and various cereal crops (Prater 1965). This is the most predominant animal in the Late Jorwe phase and the second most important animal in the Early Jorwe and Malwa phases. Both appendicular and axial skeletal parts are identified from all the 16 layers (Table 4.48). A majority of bone fragments constitute phalanges (Pl. VIII B) and vertebrae. From the Late Jorwe phase 6331 (41.64 %), the Early Jorwe 307 (4.39 %) and the Malwa phase 230 (6.34 %) bone fragments are identified (Table 4.1-4.17). The total bone fragments of blackbucks are recovered at Inamgaon account for 26.60 % in the faunal assemblage. In layer 1, about 52 % bone fragments belong to this animal (Table 4.1-4.17). Bones are also identified from the house floors. The Minimum Number of Individuals is worked out for different layers and houses. The number of individuals is in the Late Jorwe phase 257 (42.67 %), Early Jorwe phase 40 (13.38 %) and in the Malwa phase 38 (14.78 %) (Table 4.19 - 4.31). The maximum amount of meat of this animal is consumed in layer 2 i.e. 1335 kg. The amount of meat of this animal consumed by the inhabitants varies in three phases – in Late Jorwe phase 3840 kg, Early Jorwe phase 600 kg and 1170 kg in the Malwa phase. Numerous cut marks and charring marks are observed on the skeletal remains. Some skulls are found with cut marks on the occipital region suggesting the removal and consumption of brain of this animal. Some of these skulls are also found with charring (Pl IX A). Cut marks are also found on the mandible, near the angle of ascending ramus (Pl. IX B), on the long bones near the proximal and distal ends (Pl. X A), scapula near the glenoid cavity, pelvis near the acetabulum, ilium and ischium. The metacarpals and the metatarsals also had cut marks near the proximal and distal parts (Pl. X B). Bones of this animal were also used for making tools. Blackbuck was predominant in the Late Jorwe phase and was the second important animals in the Early Jorwe and the Malwa phases.

Plate VIII B. ***Antilope cervicapra*:** 1st row- First phalanges, 2nd row- Second phalanges and 3rd row- Third phalanges

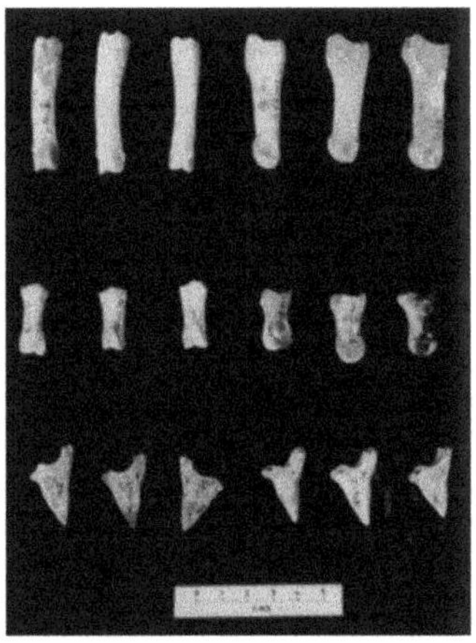

Plate IX A. ***Antilope cervicapra*:** Skull fragments with cut marks and charring marks.

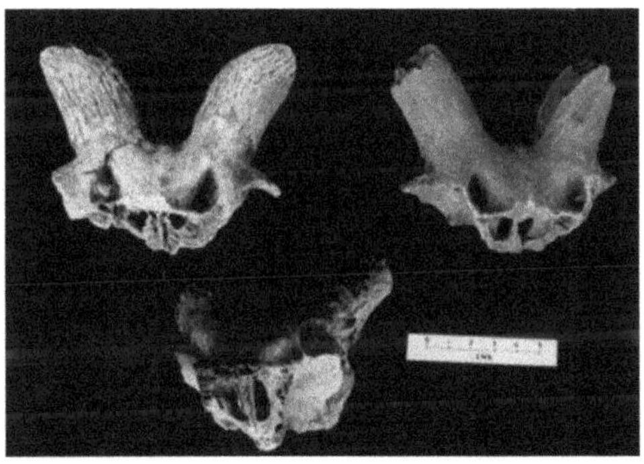

Plate IX B. ***Antilope cervicapra*:** Mandibles

Plate X A. ***Antilope cervicapra*:** Proximal epiphyses of 1. Humerus, 2. Radius, 3. Femur, 4. Tibia and Distal epiphyses of 5. Humerus, 6. Radius, 7. Femur, 8. Tibia.

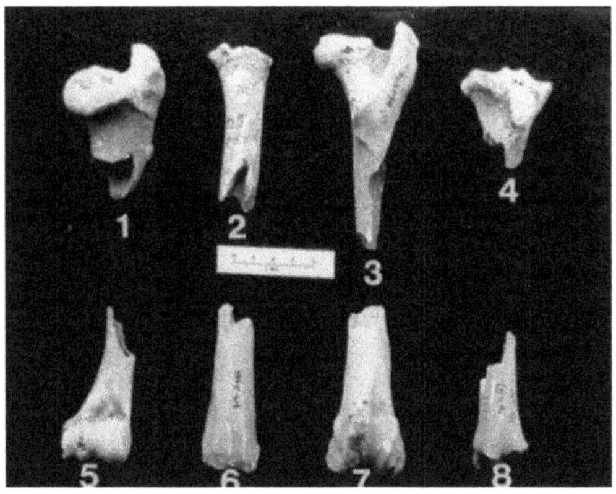

Plate X B. ***Antilope cervicapra*:** 1&2. Proximal and distal ends of Metatarsal, 3&4. Proximal and distal ends of Metacarpal, 5&6. Astragalus, 7&8. Calcaneum, 9&10. Centrotarsal.

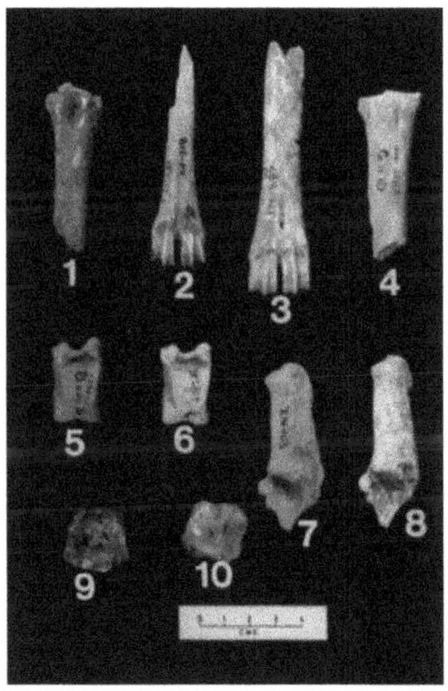

Blackbucks are identified from a large number of Chalcolithic sites of Maharashtra except Nasik (George 1955), Tuljapur Garhi (Thomas 1992a) and Apegaon (Table 4.33). At Apegaon, the presence of this species is doubtful (Badam 1979). The present distribution of this animal shows locales almost closer to the archaeological sites except at Dhule and Nagpur districts (Fig.4.4).

Fig. 4.4. Map Showing Past and Present Distribution of *Antilope cervicapra* in Maharashtra.

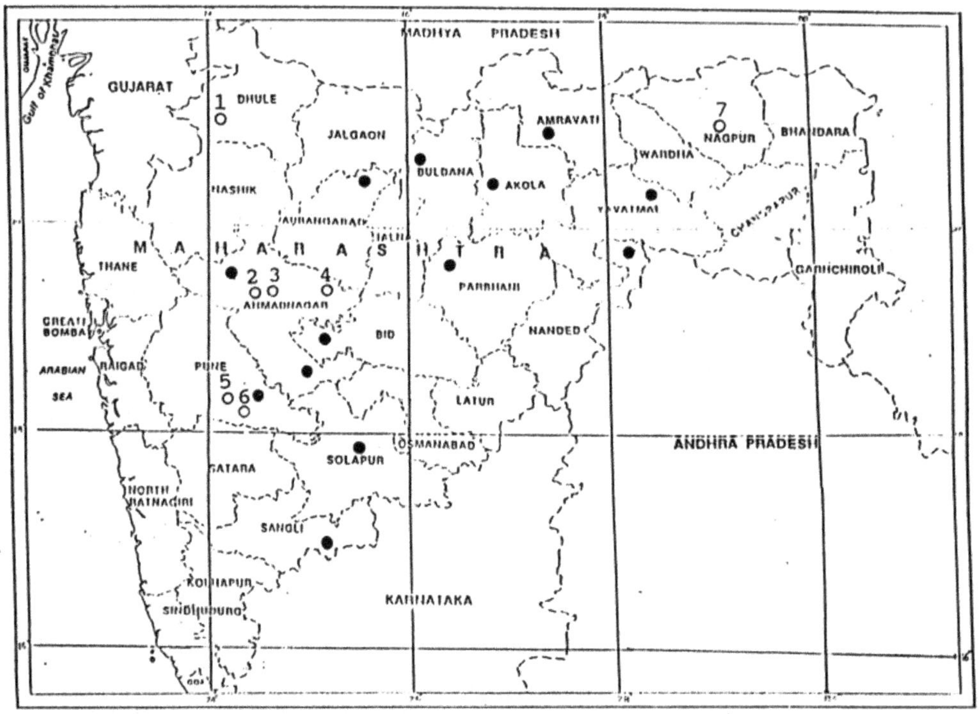

○ Past distribution   • Present distribution

1. Kaothe, 2. Daimabad, 3. Nevasa, 4.Apegaon, 5. Walki, 6. Inamgaon and 7. Adam.

Table 4.48. Bone distribution of *Antilope cervicapra* in different layers at Inamgaon

Bones	01	02	03	04	05	06	07	08	09	10	11	12	13	14	15	16
HC/AN	33	65	48	24	01	01	--	01	--	--	07	02	--	01	02	--
CF	45	53	54	17	--	--	01	01	--	--	--	01	--	01	01	--
MN	77	138	83	37	03	04	03	03	--	--	02	03	02	01	03	01
SC	25	61	45	22	02	03	02	02	01	--	04	03	--	02	02	01
HU	56	102	59	36	03	02	03	04	--	--	01	03	02	02	01	--
RA	40	75	54	37	01	01	--	02	02	--	01	02	01	--	04	01
UL	13	18	15	06	01	04	01	03	01	--	01	03	--	--	01	--
MC	94	225	126	52	10	11	01	03	01	--	03	04	--	02	--	03
C/T	04	14	19	08	--	01	--	02	--	--	--	--	--	01	--	--
MP	37	95	126	85	06	03	03	02	--	--	02	02	02	--	01	--
PV	37	117	100	51	02	03	01	01	03	--	02	03	02	01	02	01
FE	27	66	36	17	03	02	03	05	--	01	--	01	--	04	02	01
TI	30	113	63	40	04	07	02	03	--	--	01	02	02	02	03	01
PA	--	--	05	05	--	01	01	01	--	--	--	01	--	--	01	--
MT	81	193	158	105	08	03	05	07	01	01	04	04	05	01	03	--
AS	40	95	79	60	04	08	02	08	01	--	05	08	02	03	04	01
CA	61	170	129	76	10	11	05	09	01	--	03	11	10	02	09	02
CT	11	21	37	26	01	03	02	02	--	--	--	04	02	--	--	--
H1	113	316	235	129	08	07	03	03	03	--	08	09	06	02	03	01
H2	25	93	93	64	05	02	04	--	--	01	01	05	02	02	01	--

H3	11	36	33	31	--	04	01	--	--	--	03	02	02	--	--	--
VB	116	221	152	78	02	03	07	14	01	01	01	02	02	--	--	01
RI	--	03	05	03	--	--	01	--	--	--	--	--	--	--	03	--
UJT	28	44	19	06	--	--	--	02	03	--	02	04	04	01	04	03
LJT	21	43	46	19	03	06	04	04	--	--	05	08	--	01	03	--
OTHERS	--	--	01	--	--	--	01	--	--	--	--	--	--	--	01	--
TOTAL	1425	2377	1820	1034	75	91	56	82	18	04	56	87	43	29	54	17

Some very fragmentary bones of small sized ruminants which could not identified to species are included in one group called small ruminants. A majority of these fragments may belong to blackbucks. The Number of Identified Specimens are 917 (6.03 %), 455 (6.51 %) and 322 (8.87 %) in the Late Jorwe, Early Jorwe and Malwa phases, respectively (Table 4.1- 4.17). The distribution of bone fragments in different layers is given in Table 4.49. Among these fragments of vertebrae constitute the maximum.

Table 4.49. Bone distribution of Small Ruminants in different layers at Inamgaon

Bones	**01**	**02**	**03**	**04**	**05**	**06**	**07**	**08**	**09**	**10**	**11**	**12**	**13**	**14**	**15**	**16**
HC/AN	--	--	--	--	--	--	--	--	02	--	--	--	--	--	--	--
CF	06	12	08	02	08	04	04	02	09	01	--	06	02	03	09	--
MN	05	13	10	06	02	02	01	07	03	04	03	09	05	01	02	01
SC	04	11	04	06	--	01	02	05	03	03	--	04	01	--	--	01
HU	13	18	08	06	01	01	--	04	05	02	04	06	05	02	03	01
RA	05	11	06	02	01	--	--	01	06	02	03	04	01	--	01	--
UL	01	01	--	02	--	01	01	01	--	01	--	04	--	--	02	--
MC	--	04	01	01	--	--	--	--	01	--	01	02	--	--	02	--
C/T	--	--	--	03	--	--	--	--	--	--	02	01	--	--	--	--
MP	08	12	04	01	--	--	01	01	02	01	01	06	02	02	01	--
PV	06	18	12	09	05	06	01	04	05	02	03	05	03	01	--	02
FE	07	28	10	06	03	04	01	05	04	01	05	03	05	--	02	--
TI	05	06	07	04	--	--	--	05	05	02	02	07	02	--	02	01
PA	--	--	--	01	--	--	--	--	--	--	--	--	01	--	--	--
MT	02	02	01	--	--	--	--	--	--	--	--	02	01	--	--	--
AS	--	--	--	01	--	--	--	--	--	--	01	--	01	02	--	--
CA	--	01	04	03	--	--	--	--	--	01	--	02	05	--	--	--
CT	--	--	--	--	01	--	--	--	--	--	--	--	--	--	--	--
H1	02	06	--	--	--	--	--	--	--	--	--	--	01	--	--	--
H2	--	01	--	--	--	--	--	--	--	--	--	--	--	--	--	--
H3	--	--	--	--	--	--	--	--	--	--	--	--	--	--	--	--
VB	34	72	71	37	19	17	10	11	29	07	21	33	11	05	09	--
RI	02	11	19	14	04	09	04	01	03	02	18	05	24	--	--	01
UJT	--	02	--	03	--	--	--	--	--	--	03	05	01	02	--	--
LJT	--	01	01	01	--	--	04	02	02	--	01	03	03	--	--	--
OTHERS	--	03	04	01	01	--	01	02	04	01	01	01	01	--	02	--
TOTAL	100	234	164	109	45	44	30	51	83	30	69	108	75	18	35	07

***Boselaphus tragocamelus* (nilgai)**

The present day distribution shows the occurrence of nilgai in the Indian Peninsula. Generally nilgai is found on the hills with sparse vegetation or undulating plains covered with grass and patches of scrub (Prater 1965).

The remains of this animal are represented in all the layers (Pl. XI A & B). However, maximum bone fragments are found in the Late Jorwe phase. Compared to the Late Jorwe (303, 1.99 %) and the Malwa (79, 2.18 %) phases, the number of bone fragments is very less in the Early Jorwe phase (29, 0.42 %) (Table 4.1-4.17). Except horn core, skull, scapula, patella and rib fragments, all the other skeletal remains are identified.

However, the non-meat-bearing parts were much more than the meat-bearing (Table 4.50). The Minimum Number of Individuals are 38 (6.33 %) in the Late Jorwe phase, 11 (3.68 %) in the Early Jorwe phase and 14 (5.45 %) in the Malwa phase (Tables 4.19-4.31).

Plate XI A. ***Boselaphus tragocamelus*:** 1st row- Metacarpals, 2nd row- Metatarsals.

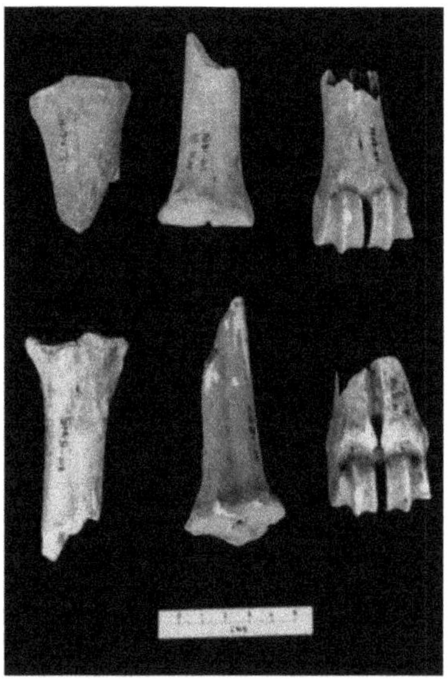

Plate XI B. ***Boselaphus tragocamelus*:** 1. Calcaneum, 2. Astragalus, 3. Centrotarsal, 4. First phalanx, 5. Second phalanx and 6. Third phalanx.

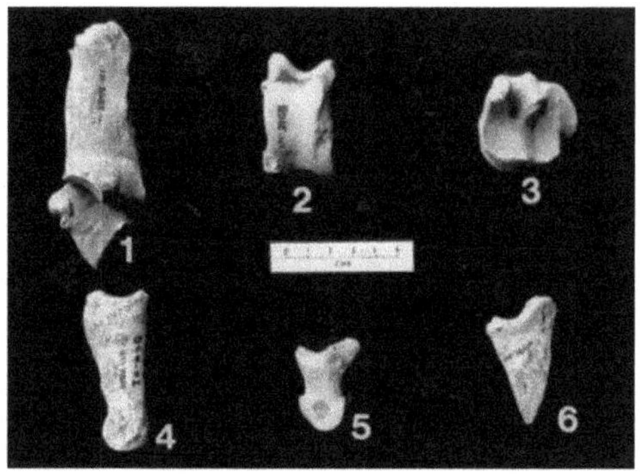

The skeletal remains are also recovered from the floors of structures. The maximum amount of meat consumed of this animal in the Late Jorwe, Early Jorwe and the Malwa phases are 3800 kg, 1100 kg and 1400 kg, respectively (Table 4.32).

Table 4.50. Bone distribution of *Boselaphus tragocamelus* in different layers at Inamgaon

Bones	**01**	**02**	**03**	**04**	**05**	**06**	**07**	**08**	**09**	**10**	**11**	**12**	**13**	**14**	**15**	**16**
HC/AN	--	--	--	--	--	--	--	--	--	--	--	--	--	--	--	--
CF	--	--	--	--	--	--	--	--	--	--	--	--	--	--	--	--
MN	--	--	--	--	--	--	--	--	--	--	--	--	--	01	--	--
SC	--	--	--	--	--	--	--	--	--	--	--	--	--	--	--	--
HU	--	01	01	--	--	--	--	--	--	--	--	--	--	--	--	--
RA	--	--	01	01	01	--	--	--	--	--	--	01	--	--	01	--
UL	--	03	--	01	--	--	--	--	02	--	--	01	01	--	02	--
MC	01	02	02	01	--	--	--	--	--	--	--	--	--	--	--	--
C/T	--	--	01	07	--	--	--	--	--	--	--	--	--	--	01	--
MP	--	--	02	01	02	--	--	02	--	--	--	--	01	01	01	--
PV	01	--	03	--	--	--	--	02	--	--	--	--	--	--	--	--
FE	01	01	--	--	--	--	--	--	--	--	--	--	--	--	--	--
TI	02	02	01	--	--	--	--	--	--	--	--	--	--	--	01	--
PA	--	--	--	--	--	--	--	--	--	--	--	--	--	--	--	--
MT	03	06	03	01	--	--	--	01	--	--	--	--	01	03	03	--
AS	01	07	12	08	--	--	--	01	02	--	01	--	01	03	02	02
CA	--	03	04	05	--	--	--	--	--	--	--	--	01	01	03	--
CT	--	04	02	01	--	--	--	--	--	01	01	01	--	--	01	--
H1	10	25	21	27	04	--	02	06	01	--	01	01	0	05	10	01
H2	08	13	31	12	02	--	--	01	--	--	01	03	02	09	06	01
H3	04	12	12	14	--	02	--	--	--	--	--	--	--	--	05	01
VB	02	01	--	--	--	--	--	01	--	--	--	--	--	--	--	--
RI	--	--	--	--	--	--	--	--	--	--	--	--	--	--	--	--

UJT	--	01	01	--	--	01	--	--	--	--	--	--	--	--	--	--
LJT	01	01	01	--	--	--	--	--	--	--	--	--	--	--	--	--
OTHERS	--	--	--	01	--	--	--	--	--	--	--	--	--	--	--	--
TOTAL	34	82	98	80	09	03	02	14	05	01	04	07	08	23	36	05

Nilgai is reported from the Chalcolithic sites in the districts of western Maharashtra like Pune, Ahmadnagar, Nasik, Dhule and also Nagpur of eastern Maharashtra. The present distribution shows the absence of nilgai in all these districts except Dhule and Nagpur. The occurrence in Dhule district is reported from extreme north of this region. Presently this species is distributed in districts Gadchiroli, Bhandara, Nagpur, Wardha, Yavatmal, Jalgaon and Dhule in Maharashtra (Fig. 4.5).

Fig. 4.5. Map Showing Past and Present Distribution of *Boselaphus tragocamelus* in Maharashtra.

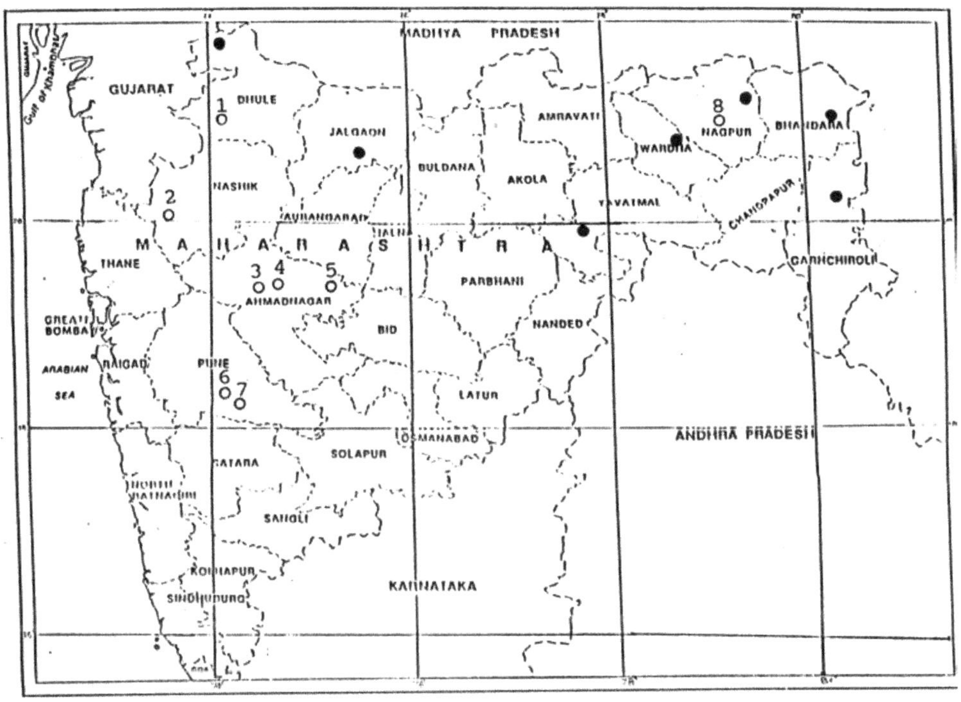

○ Past distribution ● Present distribution

1. Kaothe, 2. Nasik, 3. Daimabad, 4. Nevasa, 5. Apegaon, 6. Walki, 7. Inamgaon and 8. Adam.

Axis axis (chital)

The habitat of this animal is in the forest with good grazing and a plentiful supply of water. Chital are distributed through out the peninsular India (Prater 1965). At Inamgaon this animal is identified sparsely from all the layers (Table 4.51). The remains of both the appendicular and the axial skeletons were recovered (Pl. XII A & B). Some of the bones were also used for making tools, especially the antlers for picks, points, chisels and punches. The maximum bone representation is found in the Late Jorwe phase. The number of identified fragments in the Late Jorwe, Early Jorwe and the Malwa phases are 38 (0.26 %), 20 (0.26 %) and 11 (0.30), respectively (Table 4.1-4.17). Bone fragments are also recovered from the houses. The minimum number of individuals are 11 (1.83 %), 6 (2.01 %) and 6 (2.33 %) from the Late Jorwe, Early Jorwe and Malwa phases, respectively (Table 4.19-4.31). The meat weight shows the maximum amount of consumption in the Late Jorwe phase i.e. 375 kg and in the earlier two phases it is 225 kg each (Table 4.32). However, a gradual decline in the representation is noticed towards the upper layers.

Plate XII A. ***Axis axis*:** 1&2. First phalanges, 3&4. Second phalanges, 5. Centrotarsal, 6. Astragalus and 7. Tibia distal.

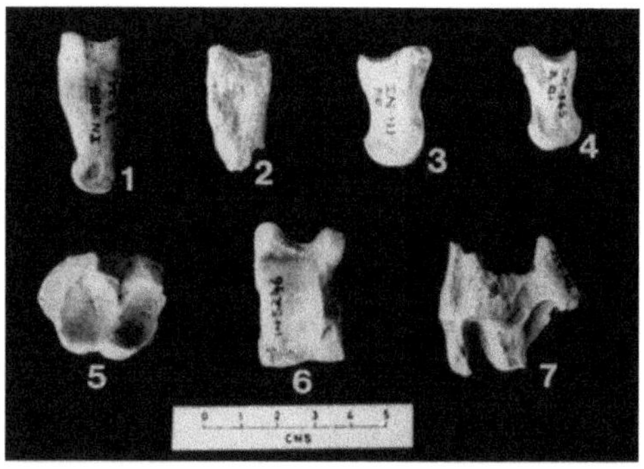

Plate XII B. ***Axis axis*:** Antlers

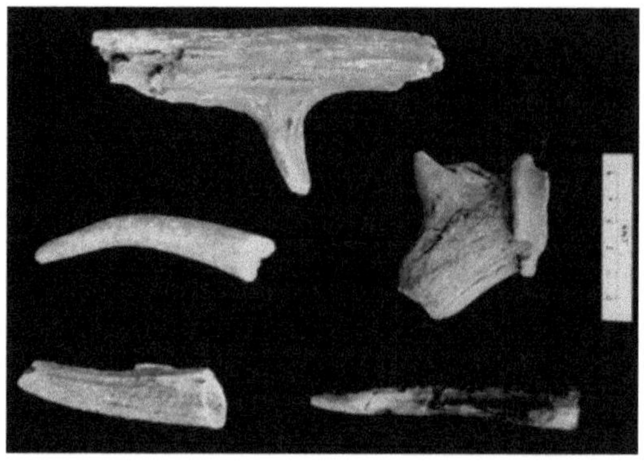

Table 4.51. Bone distribution of *Axis axis* in different layers at Inamgaon

Bones	01	02	03	04	05	06	07	08	09	10	11	12	13	14	15	16
HC/AN	--	01	02	05	--	--	07	01	--	--	02	--	--	--	01	--
CF	--	--	--	--	--	--	--	01	--	--	--	--	--	--	--	--
MN	--	--	--	01	--	--	--	--	--	--	--	--	--	--	--	--
SC	--	--	--	--	--	--	--	--	--	--	--	--	--	--	--	--
HU	01	--	01	--	--	--	--	--	--	--	--	--	--	01	--	--
RA	--	--	01	--	--	--	--	--	01	--	01	01	--	--	--	--
UL	--	--	--	--	--	--	--	--	--	--	--	--	--	--	--	--
MC	--	02	--	--	--	--	--	--	--	--	--	--	--	--	--	--
C/T	--	--	--	--	--	--	--	--	--	--	--	--	--	--	--	--
MP	--	--	--	--	--	--	--	--	--	--	--	--	--	--	--	--
PV	--	--	--	--	--	--	--	--	--	--	--	--	--	--	--	--
FE	--	--	01	--	--	--	--	--	01	--	01	--	--	--	--	--
TI	01	--	--	--	--	01	--	--	--	--	--	--	--	--	--	--
PA	--	--	--	--	--	--	--	--	--	--	--	--	--	--	--	--
MT	--	--	--	--	--	01	--	--	--	--	--	01	--	--	--	--
AS	--	--	--	01	--	--	--	--	--	--	01	--	--	02	--	01
CA	--	01	--	02	--	--	--	--	--	--	--	--	01	--	01	--
CT	--	--	02	--	--	--	--	--	--	--	--	--	--	--	--	--
H1	--	01	03	01	--	--	--	--	--	--	--	--	--	--	--	--
H2	--	02	--	02	--	--	--	--	--	--	--	--	--	--	--	--
H3	--	--	--	--	--	--	--	--	--	--	--	--	--	--	--	--
VB	01	01	--	01	--	--	--	--	01	--	--	--	--	--	--	--
RI	--	--	--	--	--	--	--	--	--	--	--	--	--	--	--	--
UJT	04	--	--	--	--	--	--	01	--	--	--	01	--	--	--	--
LJT	--	--	--	--	--	--	--	--	--	--	--	--	--	01	--	--
OTHERS	--	--	--	--	--	--	--	--	--	--	--	--	--	--	--	--
TOTAL	07	08	10	13	00	02	07	03	03	00	05	03	01	04	02	01

This animal is distributed from all the Chalcolithic sites of western Maharashtra (Table 4.33). At present chital is distributed in the north-eastern districts of the state such as Gadchiroli, Chandrapur, Bhandara, Nagpur, Akola and Amravati (Fig. 4.6).

Fig. 4.6. Map Showing Past and Present Distribution of *Axis axis* in Maharashtra.

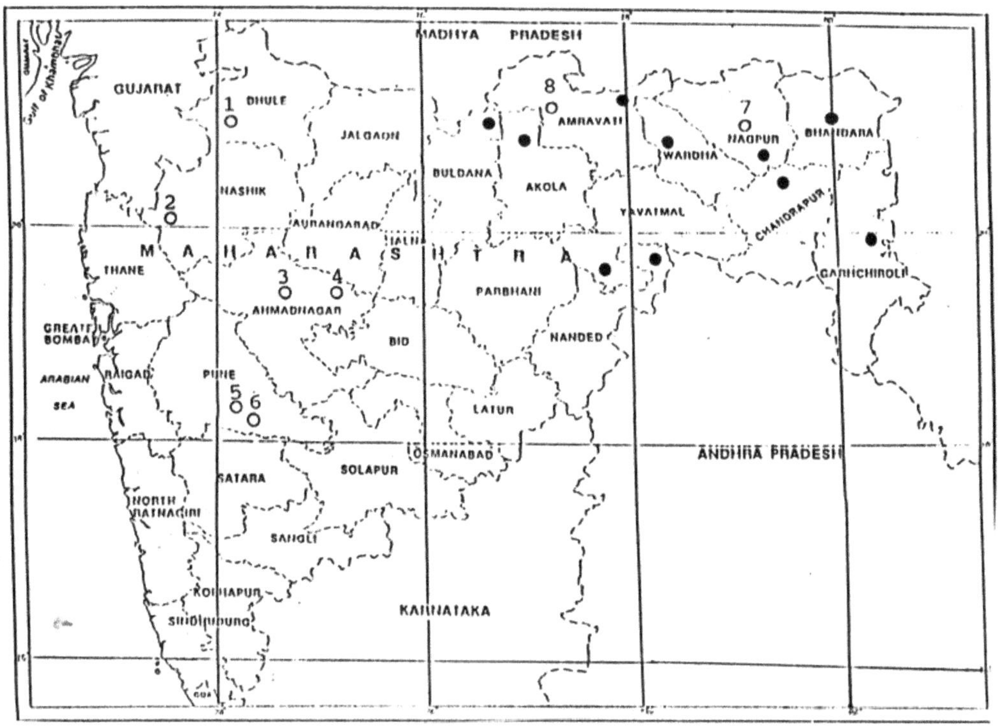

○ Past distribution ● Present distribution

1. Kaothe, 2. Nasik, 3. Nevasa, 4. Apegaon, 5. Walki, 6. Inamgaon, 7. Adam and 8. Tuljapur Garhi.

Axis porcinus (hog deer)

Hog deer generally live in grass jungles by the river banks. This animal occurs in north India, from Sind Punjab to Assam. It is absent in the Indian Peninsula (Prater 1965).

The archaeological evidence of this animal is very poor. This animal was identified from Kaothe (Thomas and Joglekar 1990) and Walki (Joglekar 1992; Thomas and Joglekar 1995) (Table 4.33). The number of bone fragments in the assemblage is very less. From Kaothe only a fragment of radius, From Walki three bone fragments and from Inamgaon only one second phalanx is identified. These are the only evidence for the representation of this animal in the Chalcolithic culture of Maharashtra.

Cervus unicolor (sambar)

Sambar is the biggest in Indian deer. The height of this animal at shoulder is approximately over 1.5 m. This animal is distributed in the wooded districts of India. Their food mainly consists of grass, leaves and various kinds of fruits. Sambar sheds its antler between the end of March and the mid April (Prater 1965).

At Inamgaon this animal is not identified from all the layers. The skeletal remains of this animal are absent in layers 5, 6, 9, 10, 11 and 12. Compared to the Late Jorwe phase, bones are distributed very sporadically in the Early Jorwe and the Malwa phases (Table 4.52) (Pl. XIII A). The Number of Identified Specimens are 30 (0.20 %),

3 (0.04 %) and 4 (0.11 %) in the Late Jorwe, Early Jorwe and the Malwa phases, respectively (Table 4.1-4.17). The Minimum Number of Individuals are calculated for different layers and phases which are as follows: in the Late Jorwe phase 5 (0.83 %), Early Jorwe phase 2 (0.67 %) and in the Malwa phase 4 (1.56 %) (Table 4.19-4.31). The inhabitants at Inamgaon have consumed the meat of this animal which is 625 kg in the Late Jorwe, 250 kg in Early Jorwe and 500 kg in the Malwa phases (Table 4.32).

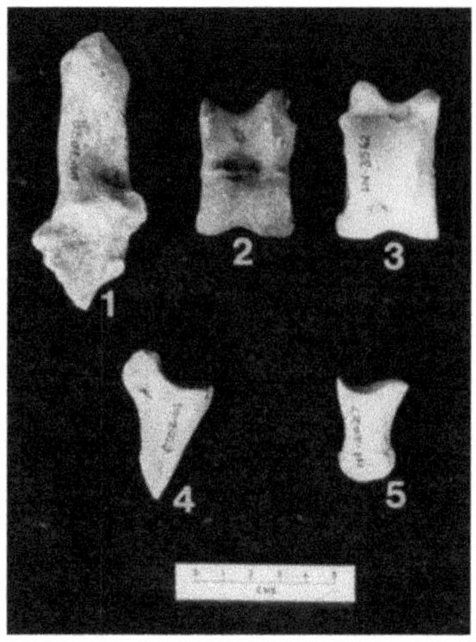

Plate XIII A. Cervus unicolor: 1. Calcaneum, 2&3. Astragalus, 4. Third phalanx and 5. Second phalanx.

Table 4.52. Bone distribution of *Cervus unicolor* in different layers at Inamgaon

Bones	**01**	**02**	**03**	**04**	**05**	**06**	**07**	**08**	**09**	**10**	**11**	**12**	**13**	**14**	**15**	**16**
HC/AN	01	--	03	08	--	--	--	--	--	--	--	--	--	--	--	--
CF	--	--	--	--	--	--	--	--	--	--	--	--	--	--	--	--
MN	01	--	--	--	--	--	--	--	--	--	--	--	--	--	--	--
SC	--	--	--	--	--	--	--	--	--	--	--	--	--	--	--	--
HU	--	--	01	--	--	--	--	01	--	--	--	--	--	--	--	--
RA	01	--	--	--	--	--	--	--	--	--	--	--	--	--	01	--
UL	01	--	01	--	--	--	--	--	--	--	--	--	--	--	--	--
MC	--	--	--	--	--	--	--	--	--	--	--	--	--	--	--	--
C/T	--	--	--	--	--	--	--	--	--	--	--	--	--	--	--	--
MP	--	02	--	--	--	--	--	--	--	--	--	--	--	--	--	--
PV	--	--	--	--	--	--	--	--	--	--	--	--	--	--	--	--
FE	--	--	01	--	--	--	--	--	--	--	--	--	--	--	--	--
TI	--	--	01	--	--	--	--	--	--	--	--	--	--	--	--	--
PA	01	--	--	--	--	--	--	--	--	--	--	--	--	01	--	--
MT	--	01	--	--	--	--	01	--	--	--	--	--	--	--	--	--
AS	--	01	--	--	--	--	--	--	--	--	--	--	--	--	--	--
CA	--	--	--	--	--	--	--	--	--	--	--	--	--	--	--	--
CT	--	--	--	--	--	--	--	--	--	--	--	--	01	--	--	--
H1	--	--	--	--	--	--	01	--	--	--	--	--	--	--	--	--
H2	01	--	--	--	--	--	--	--	--	--	--	--	--	--	--	--
H3	01	01	01	01	--	--	--	--	--	--	--	--	--	--	--	01
VB	--	--	--	--	--	--	--	--	--	--	--	--	--	--	--	--
RI	--	--	--	--	--	--	--	--	--	--	--	--	--	--	--	--
UJT	--	--	--	--	--	--	--	--	--	--	--	--	--	--	--	--
LJT	--	--	01	--	--	--	--	--	--	--	--	--	--	--	--	--
OTHERS	--	--	--	--	--	--	--	--	--	--	--	--	--	--	--	--
TOTAL	07	05	09	09	00	00	02	01	00	00	00	00	01	01	01	01

This animal is reported from almost all the Chalcolithic sites of western Maharashtra, but at present sambar is totally absent in this region. However, they are found in some parts of the eastern, northern and southern Maharashtra (Fig. 4.7).

Fig. 4.7. Map Showing Past and Present Distribution of *Cervus unicolor* in Maharashtra

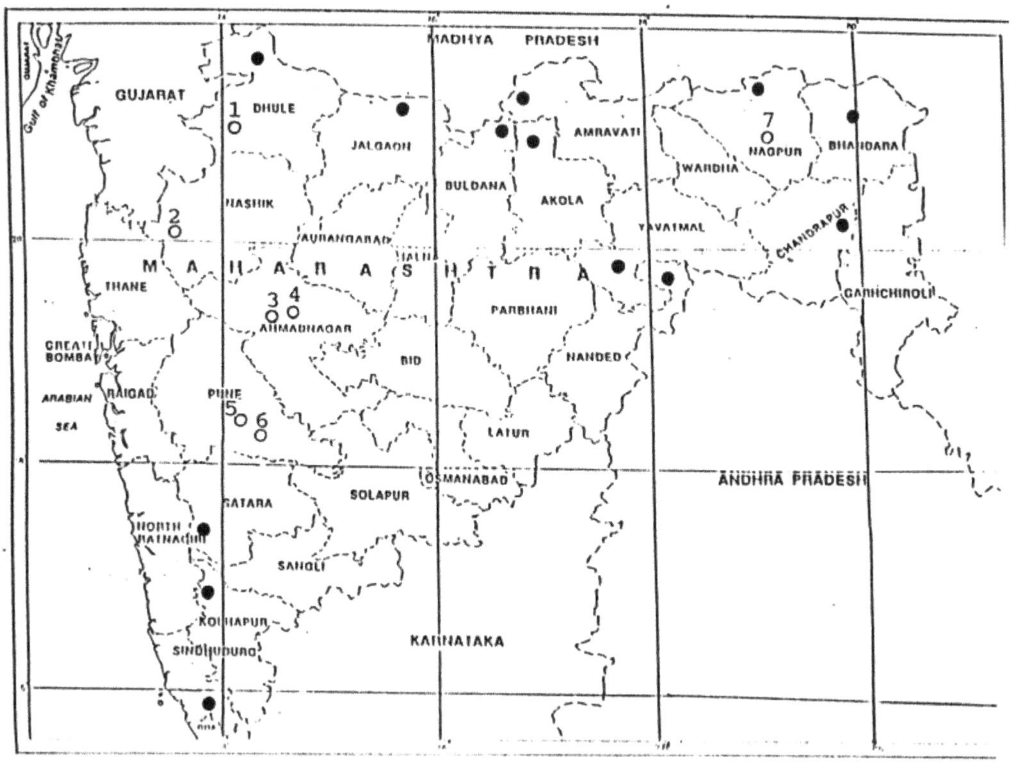

○ Past distribution ● Present distribution

1. Kaothe, 2. Nasik, 3. Daimabad, 4. Nevasa, 5. Walki, 6. Inamgaon and 7. Adam.

Muntiacus muntjak (barking deer)

This animal is usually found in the thickly wooded hills and come out to graze in the outskirts of forests. They are diurnal in habit (Prater 1965).

Table 4.53. Bone distribution of *Muntiacus muntjak* in different layers at Inamgaon

Bones	01	02	03	04	05	06	07	08	09	10	11	12	13	14	15	16
HC/AN	--	--	--	--	--	--	--	--	--	--	--	--	--	--	--	--
CF	--	--	--	--	--	--	--	--	--	--	--	--	--	--	--	--
MN	--	--	--	--	--	--	--	--	--	--	--	--	--	--	01	--
SC	--	--	--	--	--	--	--	--	--	--	--	--	--	--	--	--
HU	--	--	--	--	--	--	--	--	--	--	--	--	--	--	--	--
RA	--	--	--	--	--	--	--	--	--	--	--	--	--	--	--	--
UL	--	--	--	--	--	--	--	--	--	--	--	--	--	--	01	--
MC	--	--	--	--	--	--	--	01	--	--	--	--	--	--	--	--
C/T	--	--	--	--	--	--	--	--	--	--	--	--	--	--	--	--
MP	--	--	--	--	--	--	--	--	--	--	--	--	--	--	--	--
PV	--	--	--	--	--	--	--	--	--	--	--	--	--	--	01	--
FE	--	--	--	--	--	--	--	--	--	--	--	--	--	--	--	--
TI	--	--	--	--	--	--	--	--	--	--	--	--	--	--	01	--

PA	--	--	--	--	--	--	--	--	--	--	--	--	--	--	--	--
MT	01	--	--	--	--	--	--	--	--	--	--	--	--	--	--	--
AS	--	--	--	--	--	--	--	--	--	--	--	--	--	--	--	--
CA	--	--	--	--	--	--	--	--	--	--	--	--	--	--	--	--
CT	--	--	--	--	--	--	--	--	--	--	--	--	--	--	--	--
H1	--	--	--	--	--	--	--	--	--	--	--	--	--	--	--	--
H2	--	--	--	--	--	--	--	--	--	--	--	--	--	--	--	--
H3	--	--	--	--	--	--	--	--	--	--	--	--	--	--	--	--
VB	--	--	--	--	--	--	--	--	--	--	--	--	--	--	--	--
RI	--	--	--	--	--	--	--	--	--	--	--	--	--	--	--	--
UJT	--	--	--	--	--	--	--	--	--	--	--	--	--	--	01	--
LJT	--	--	--	--	--	--	--	--	--	--	--	--	--	--	--	--
OTHERS	--	--	--	--	--	--	--	--	--	--	--	--	--	--	--	--
TOTAL	01	00	00	00	00	00	00	01	00	00	00	00	00	00	05	00

At Inamgaon, very few bones of this animal are found (Table 4.1-4.17). Only seven bones are recovered at Inamgaon from layers 1, 8 and 15 Table 4.53). The fragments of mandible, ulna, metacarpal, pelvis, tibia, metatarsal and upper jaw teeth are identified. The Minimum Number of Individuals are 1, 1 and 2 in the Late Jorwe, Early Jorwe and Malwa phases, respectively (Table 4.19-4.31, 4.33). The remains of this animal are also identified from Walki in the Pune district (Joglekar 1992; Thomas and Joglekar 1995). The present distribution of this animal in Maharashtra has shown in the map (Fig.4.8).

Fig. 4.8. Map Showing Past and Present Distribution of *Muntiacus muntjak* in Maharashtra.

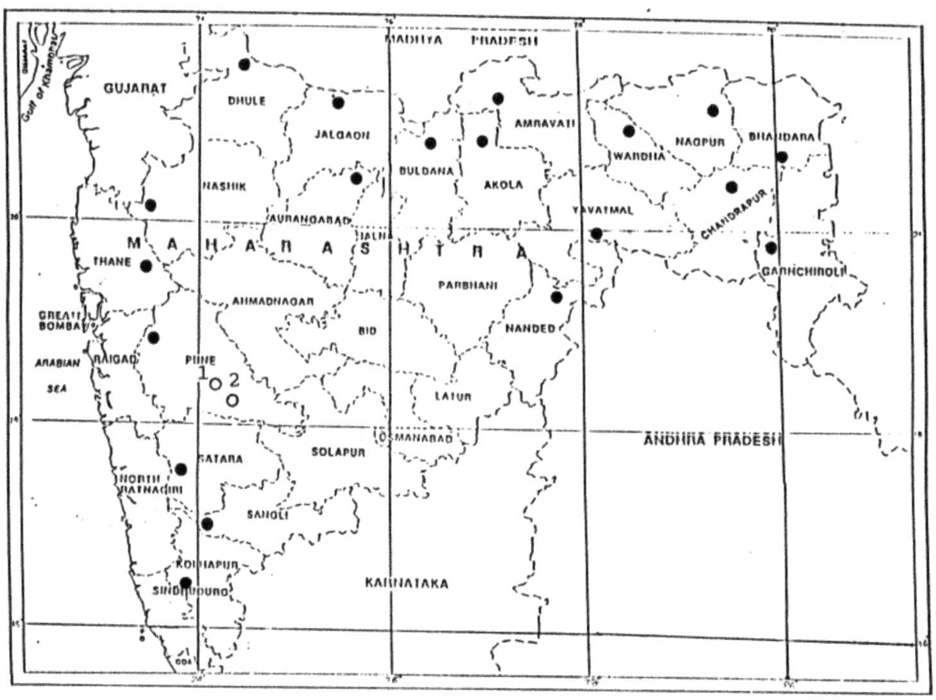

○ Past distribution ● Present distribution

1. Walki and 2. Inamgaon.

Hystrix indica (porcupine)

Only two fragments of femur, the proximal and the distal ends (Pl. XIII B) are found in layer 15 (Table 4.15). This animal is reported from Nevasa (Clason 1979) and from Kaothe by one molar tooth (Thomas and Joglekar 1990). Porcupine is generally found in rocky and hill sides and inhabits in both open as well as forest (Prater 1965).

Plate XIII B. ***Hystrix indica*****:** Proximal and Distal ends of Femur

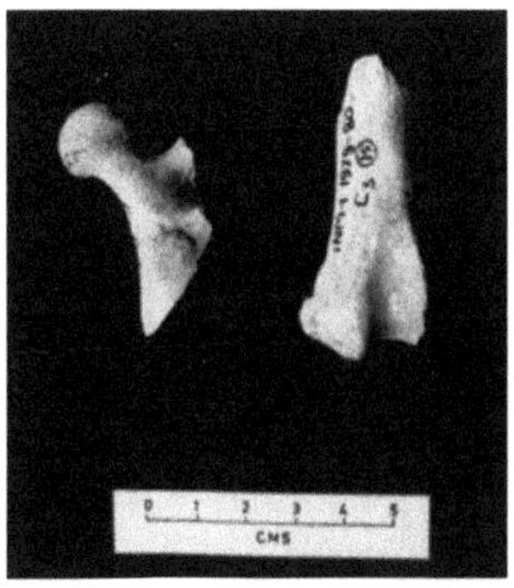

Lepus nigricollis (hare)

In India this animal is widely distributed. The concentration of the skeletal remains was less. Bone fragments of this animal are identified from the Late Jorwe and the Malwa phases (Pl. XIV A). The number of identified fragments are 6 (0.04 %) and 1 (0.03 %), in the Late Jorwe and Malwa phases, respectively. Hare was also reported from Nevasa (Clason 1079), Kaothe (Thomas and Joglekar 1990) and Walki (Joglekar 1992; Thomas and Joglekar 1995).

Plate XIV A. ***Lepus nigricollis*****:** 1. Pelvis, 2. Humerus distal, 3. Femur proximal, 4&5. Humerus distal and 6. Femur head.

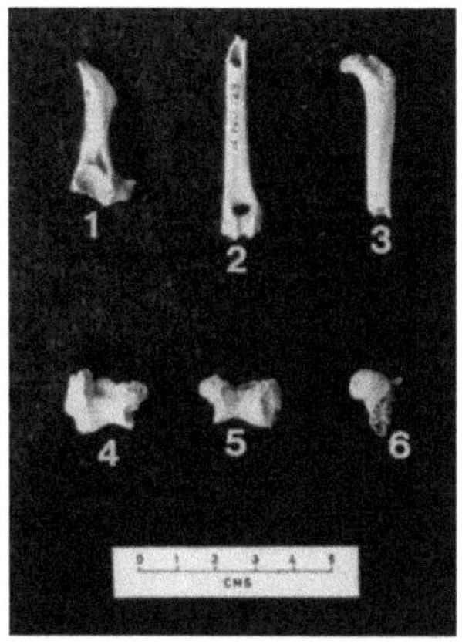

Herpestes edwardsi (mongoose)

Mongoose is also widely distributed in India. This has been very sparsely identified from all the cultural phases. A few charred scapulae and mandibles are found (Pl. XIV B). The number of identified specimens are 6 (0.04 %), 1 (0.01 %) and 1 (0.03 %), in the Late Jorwe, Early Jorwe and the Malwa phases, respectively (Table 4.1-4.17, 4.33). The skeletal remains of this animal are also reported from Apegaon (Badam 1979) and Tuljapur garhi (Thomas 1992).

Plate XIV B. ***Herpestes edwardsi*****:** 1-4. Mandibles, 5. Femur and 6. Scapula.

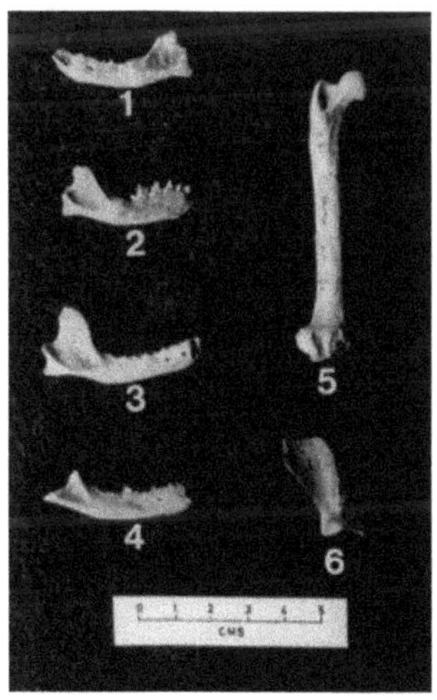

Rattus rattus **(rat)**

Rat bones are found in all the cultural phases. Bones are generally found complete. There is no trace of human activity on these bones. In the cultural deposit at Inamgaon, many rat holes were found while the site was being excavated. It shows that they were definitely later intrusion in the cultural deposit.

Birds

At Inamgaon four species of birds are identified. *They are Gallus gallus, Gallus domesticus, Anser indicus* and *Pavo cristatus* (Pl. XV A). Bird bones are mainly recovered from the Late Jorwe phase and only one fragment from the Early Jorwe phase. A few charred bone fragments are also found. A majority of the collection comprises the bones of *Gallus* species. Fragments of long bones, metapodials and phalanges are identified. The bone fragments which could not be identified specifically were put together in a general group called Bird sp. The Number of identified Specimens are 23 (0.15 %) and 1 (0.01 %) in Late Jorwe and Early Jorwe phases. Bird bones are absent in the Malwa phase (Table 4.1-4.17).

Plate XV A. ***Pavo cristatus*****:** Long bones.

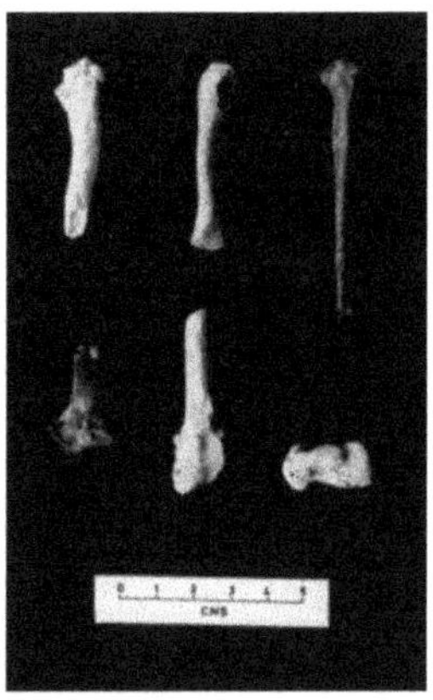

Reptiles

Six species of reptiles are identified which include turtle, crocodile, varanus and snake. Snake bones could not be identified to the species level. Two lower jaw fragments of snake, where one is charred (Pl. XV B) are identified in the Late Jorwe phase. The turtle bone and carapace fragments (Pl. XVI A & B) are collected from all the layers except in layers 6, 9, 11 and 13. The three species of turtle identified are *Trionyx gangeticus, Chitra indica* and *Kachuga* sp. Fragments of carapace, humerus, scapula, pelvis and a few vertebrae are found in the collection. Charring marks are found on some of these bones. A majority of the bones are found in the Late Jorwe phase compared to the other two cultural phases – the Early Jorwe and Malwa. The bone fragments of crocodile, snake and varanus are found only in the Late Jorwe phase. The lower jaw and vertebral fragments are the only evidence of crocodile at the site. A few vertebrae of varanus are also identified.

Plate XV B. Reptiles: 1st row- mandibles of snake, 2nd row- mandible of crocodile and 3rd row- vertebrae of crocodile

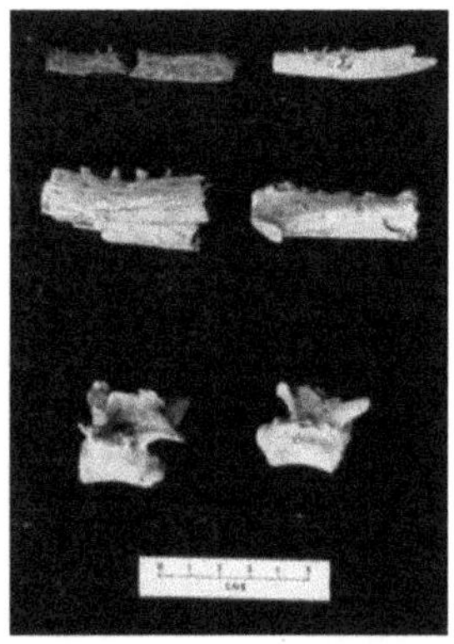

. **Plate XVI A. Turtle/Tortoise:** 1. Pelvis, 2. Femur, 3. Humerus, 4&5 Vertebrae, 6. Femur and 7. Phalanx.

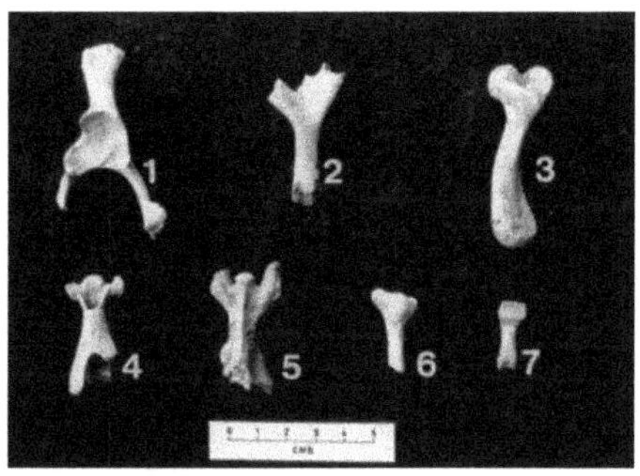

Plate XVI B. Carapace: 1, 2 &5. *Trionyx gangeticus*, 3&4. *Kachuga* sp. and 6. *Chitra indica*

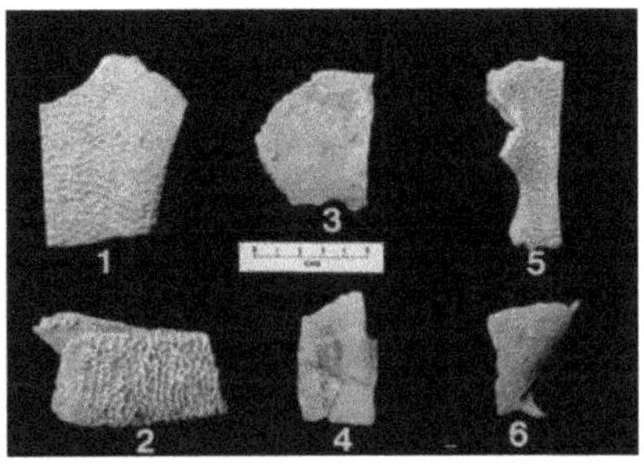

Fish

Fish bones are recovered from the Late and Early Jorwe phases (Pl. XVII A). The species identified from the collection include *Labeo rohita, Wallago wallago, Ritha ritha* and cat fish. Only jaws and vertebrae are found in the collection. Some vertebrae have perforation in the centre of the disc. These perforated vertebral discs might have been used as pendants. One of them may belong to the Shark family. The availability of marine fish throws light on the trade contacts of the Chalcolithic population at Inamgaon. The number of identified fragments are 37 (0.24%) and 3 (0.04%) in the Late Jorwe and Early Jorwe phases, respectively.

Plate XVII A. Fish bones.

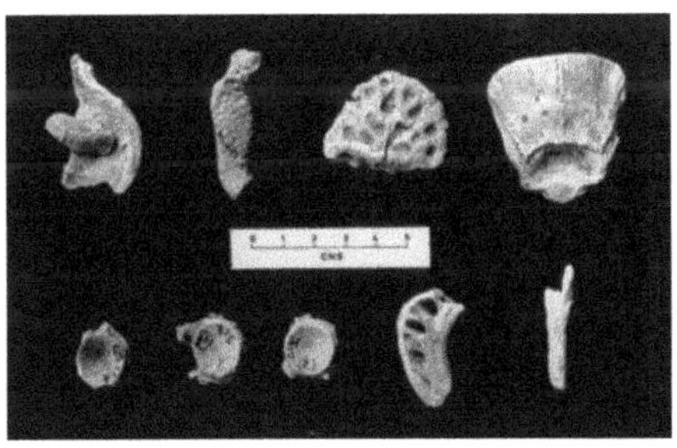

Molluscs

Lamellidens marginalis, Parreyssia corrugate, Turbo, coronatus, Digoniostoma pulchella, Viviparus bengalensis, Turbinella pyrum, Paphia gallus, Meretrix sp., *Marcia, recene, Pila globosa, Turritella* sp., *Melania striatella tuberculata*, etc. are the molluscan species identified in different layers at Inamgaon. This constitutes both marine (Pl. XVII B & XVIII A) and fresh water (Pl. XVIII B & XIX) shells. Charred shells are also noticed in the collection. Some shells with thick rolled edges indicate that people might have used them for scraping. Shell bangles and shell beads are collected. Shells are also found with perforations which might have been used as pendants. The number of identified shells are 520 (3.42%) and 43(1.18%) in the Late Jorwe, Early Jorwe and Malwa phases, respectively.

Plate XVII B. Marine Shells: 1, 4-6, *Meretrix* sp., 2. *Paphia* sp., 3. *Marcia recens.*

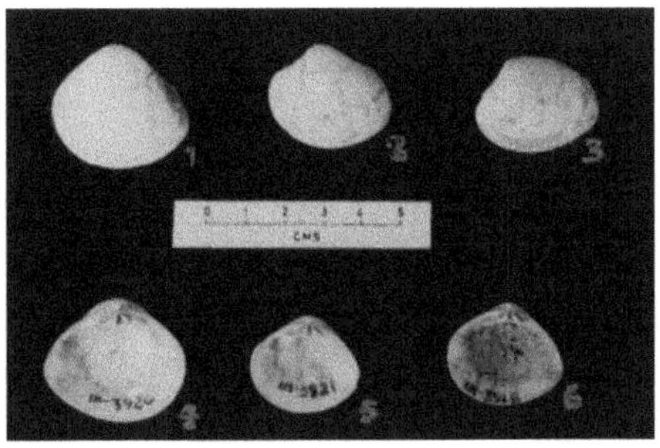

Plate XVIII A. Marine Shells: 1&2. *Turbo coronatus*, 3&5. *Turbinella pyrum*, 4. *Turritella* sp.

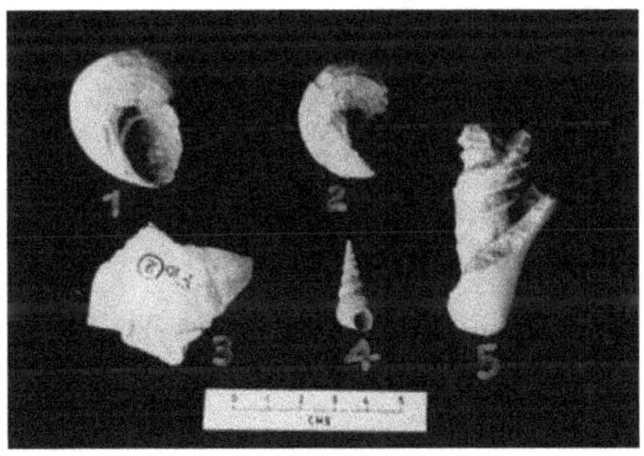

Plate XVIII B. Fresh water bivalves: 1st and 2nd row- *Parreyssia corrugate*, 3rd row- *Lamellidens marginalis.*

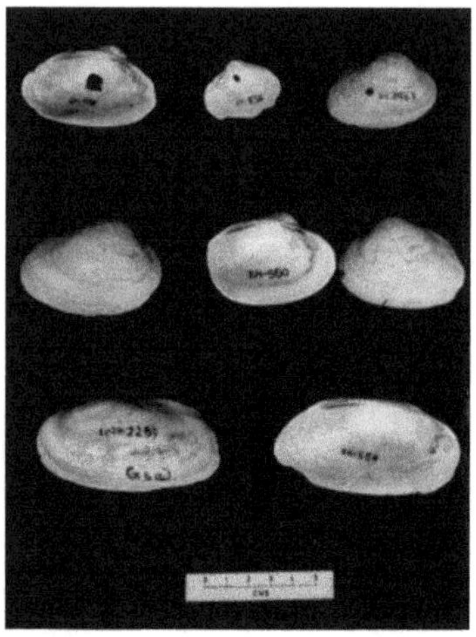

Plate XIX. Fresh water gastropods: 1-3 & 5. *Viviparus bengalensis*, 6. *Melania striatella* tuberculata, 4. Fresh water bivalve *Parreyssia corrugata*

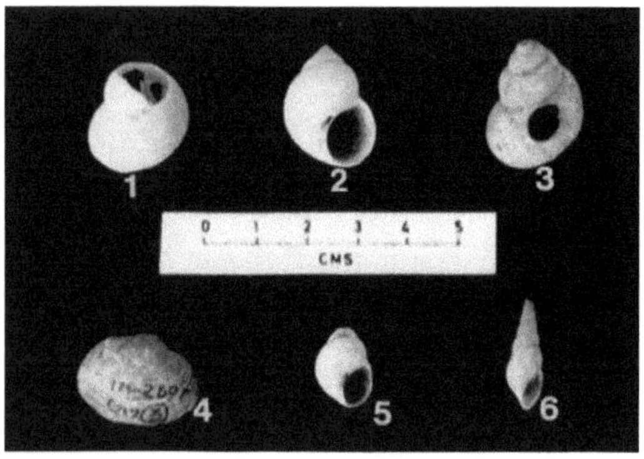

The presence of marine shell and fish at Inamgaon throws light on the trading contacts of the Chalcolithic people to outsiders. Especially the Chank shell was used for making bangles, pendants, etc. This species of marine shell is available only on the coast of Saurashtra, Tamil Nadu and Bay of Bengal. The nearest sea coast to Inamgaon is Konkan coast which is at least 200 km. to the west. The Chank shell is not available on this coast. Considering the distance to Tamil Nadu coast and the Bay of Bengal, it is possible that the Saurashtra coast could have been the source for the procurement of Chank shell. Shell industry is reported from a majority of the Harappan sites of Saurashtra. The cultural affinity between the late Harappan of Gujarat and the Chalcolithic cultures of Deccan is evident from the late Harappan phase at Daimabad in the Tapi valley (Sali 1986; Dhavalikar 1988 b; Thomas 1988).

Crustaceans

Only four mobile and immobile finger of crab are found in the Late Jorwe and the Malwa phases. In the Early Jorwe phase the remains are absent.

Further Investigation into Human Behaviour

I. Degree of fragmentation

a. Level of identification

The faunal assemblage from Inamgaon have been analysed using some improved methods in Archaeozoology. Of the 66, 847 bone fragments 25,928 were identified, whereas 40,919 fragments could not be identified. Thus, only 38.79% bone fragments were identifiable when the bones were counted but the weight of which accounted for 69.12%. Each and every bone could not be identified to the species level as some of them were very fragmentary, without any qualitative markers. Thus, in this dissertation the bones were classified into three levels of identification. NISP Level I, includes the bone fragments which could be identified to the species and genus; NISP Level II includes the bones which could not identified even to the generic levels but can be broadly classified to large bovid, small ruminants, etc. and Level III comprises all the unidentified bones without diagnostic characters. This classification has been done to determine the degree of fragmentation of bones in different stratigraphical units. The bone fragments from all the 16 layers are classified accordingly in table 5:1 and also their weight is given in table 5.2. The percentage of unidentified fragments is more in the layers of Early Jorwe phase, i.e. in layers 6-11.

Table 5.1. Level of Identification: Fragment count

Layers	NISP I	%	NISP II	%	UF	%	Total
01	1600	54.13	374	12.65	982	33.22	2956
02	4118	45.40	1316	14.51	3637	40.09	071
03	3471	46.89	942	17.73	2989	40.38	7402
04	2136	33.86	791	12.54	3381	53.60	6308
05	402	21.02	81	04.23	1430	74.75	1913
06	655	19.78	183	05.52	2474	74.75	3312
07	728	19.28	150	03.97	2899	76.75	3777
08	2173	34.94	493	07.92	3554	57.14	6220
09	672	10.46	426	06.63	5325	82.91	6423
10	213	04.39	379	07.82	4257	87.79	4849
11	734	19.64	246	06.58	2758	73.78	3738
12	1153	17.47	302	04.58	5144	77.95	6599
13	368	38.61	148	15.53	437	45.86	953
14	649	41.36	67	04.27	853	54.37	1569
15	561	37.15	183	12.12	766	50.73	1510
16	170	74.89	24	10.57	33	14.54	227
Total	19803	100	6125	100	40919	100	66847

Table 5.2. Level of Identification: Weight of fragments in grams

Layers	NISP I Wt. (gms)	%	NISP II Wt. (gms)	%	UF Wt. (gms)	%	Total Wt. (gms)
01	20265	68.36	3833	12.93	5545	18.71	29643
02	39624	59.67	9829	14.80	16949	25.53	66402
03	42689	68.15	7129	11.38	12820	20.46	62638
04	22667	60.54	4866	12.99	9910	26.47	37443
05	5312	51.67	649	06.31	4320	42.02	10281
06	8414	50.27	1186	07.09	7137	42.64	10281
07	9497	49.82	1010	05.30	8555	44.88	19062
08	33850	70.89	2868	06.01	11030	23.10	47748
09	10066	40.94	1820	07.40	12700	51.66	24586
10	2201	17.70	1215	09.77	9020	72.53	12436
11	9801	49.72	1511	07.67	8399	42.61	19711
12	15568	51.95	1379	04.60	13020	43.44	29967
13	4865	63.54	1042	13.61	1750	22.85	7657
14	10207	71.94	591	14.17	3390	23.89	14188
15	8533	59.26	2126	14.76	3740	25.98	14399
16	3010	83.36	271	07.50	330	09.14	3611
Total	246569	100	41325	100	128615	100	416509

The Level of Identification and the weight of bone fragments reveal high degree of bone fragmentation in the Early Jorwe phase (Fig. 5.1 & 5.2). The number of unidentified fragments is very high in layers 4-12. However, the peaks are noticed in layers 9 and 12, but the weight of these unidentified fragments is very less. The fragmentation of bones depends on many factors like taphonomy, the pattern of butchery, carcass utilization, cooking, the secondary uses of bones, excavations, transport, etc.

Fig. 5.1. Level of Identification of bones at Inamgaon: Fragment count

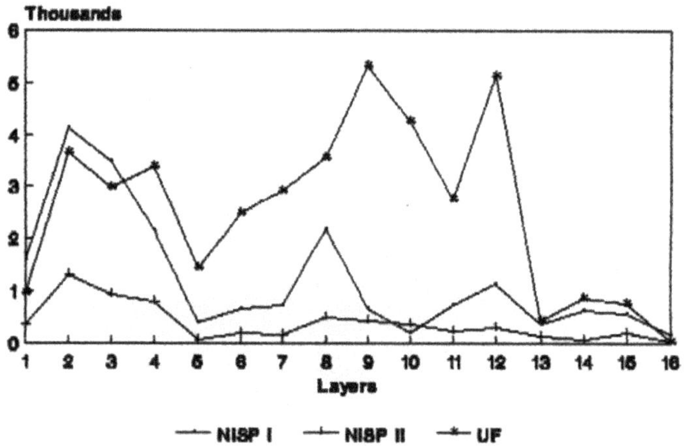

Fig. 5.2. Level of Identification of bones at Inamgaon: Fragment Weight

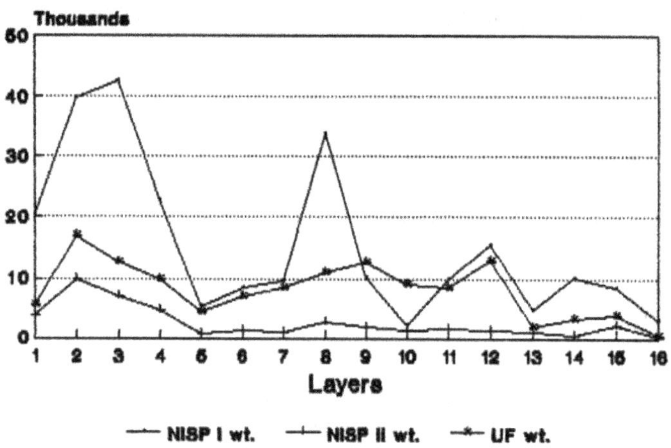

b. Identified versus unidentified bone fragments

The ratio between the number of identified specimens and the unidentified fragments, in different layers also reveals high degree of fragmentation of bones in the Early Jorwe phase (Fig.5.3). The maximum number of unidentified fragments is noticed in layer 9. A similar pattern has also been seen in layer 12 of the Malwa phase. The peaks of the number of identified fragments are seen in layer 2 of the Late Jorwe, layer 8 of Early Jorwe and in layer 15 of the Malwa phases.

Fig. 5.3. Identified (NISP) versus Unidentified (UF) bone fragments

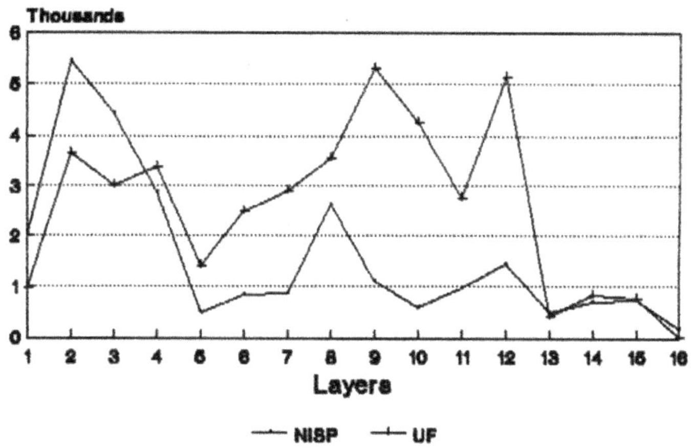

c. Measurability index (MSI)

Measurability index is the percentage of the total measurable bone fragments in NISP level I, which includes the complete and partially complete bones. This index has been calculated for different layers (Table 5.3). The result of the Measurability Index also supports the point of high degree of fragmentation of bones in the Early Jorwe phase. The highest percentage of measurable bones is found in layer 16, i.e. 74.18 %. In the layers of Late Jorwe phase (1-5), the index is higher than the earlier two phases, the Early Jorwe and the Malwa (Fig. 5.4).

Though, the Measurability Index is high in the Late Jorwe phase it does not mean that there was less human activity on the bones. A large number of bone fragments with cut marks, charring and butchering marks are found in the Late Jorwe phase (Table 5.4).

Fig. 5.4. Measurability Index of bones at Inamgaon.

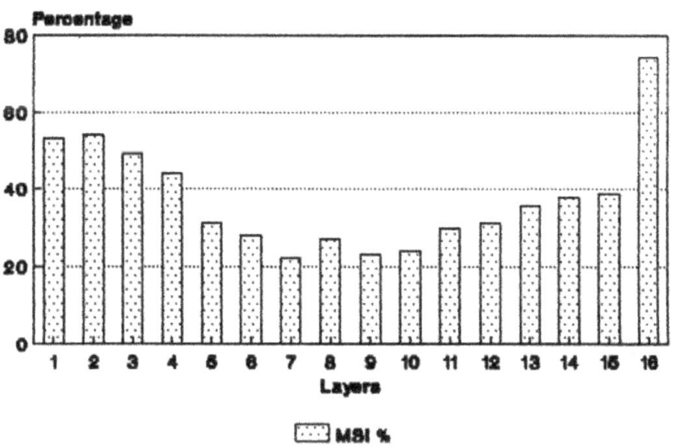

Table 5.3. Measurability Index of Bones at Inamgaon

Layers	Total Measurable Bone	NISP I	MSI %
01	815	1600	53.19
02	2218	4118	53.86
03	1714	3471	49.38
04	948	2136	44.38
05	125	402	31.09
06	183	655	27.94

07	162	728	22.35
08	592	2173	27.24
09	156	672	23.21
10	51	213	23.94
11	221	734	30.10
12	360	1153	31.22
13	131	368	35.59
14	246	649	37.90
15	217	561	38.68
16	126	170	74.18

Table 5.4. Layer-wise distribution of bones with human activity at Inamgaon

Layers	A	B	C	D	E
01	30	165	02	08	07
02	210	370	01	40	170
03	235	189	03	46	200
04	90	222	00	34	193
05	25	09	00	03	184
06	68	22	00	09	361
07	93	50	00	19	589
08	123	202	00	27	208
09	38	64	00	05	352
10	36	03	00	05	352
11	82	26	01	18	426
12	110	61	00	32	709
13	64	21	01	14	46
14	47	62	00	14	155
15	41	39	00	12	58
16	14	05	00	04	02

A : Charred bone fragments
B : Bone fragments with cut marks
C : Butchering marks
D : Charred bone fragments with cut marks
E : Charred unidentified fragments

II. Exploitation of domestic and wild animals

The taxonomic abundance of domestic and wild animals calculated from both the NISP and MNI reveals almost the same results. Thus, it can be assumed that there is some relationship between these two quantitative methods (Fig.5.5 & 5.6). Both the graphs show three peaks for the domestic and wild animals in layers 12, 8 and 4. It is interesting to note that in the food economy the over dependence on animals is evidenced after an interval of every three layers.

In the Late Jorwe phase, there is a remarkable change in the composition of animals which is unique. The exploitation of wild animals increased in this phase. In the Malwa and the Early Jorwe phases the proportion of domestic and wild animals exploited is 80.11%:19.89% and 87.28%:12.72%, respectively. Whereas, in the Late Jorwe phase; the percentage of the wild animals is 54.12 % (Table 5.5). Hunting of wild animals is seen from the beginning of the settlement. However, the people of the Malwa and Early Jorwe phases gave more emphasis on the exploitation of domestic animals. In the Late Jorwe phase the hunting of wild animals increased gradually till layer 2. Hunting has played a vital role in the subsistence system of the Late Jorwe people. The other associated archaeological evidence of the phase shows high mortality rates of the populations, change in the settlement structures from rectangular to circular, decline in farming activity and plant remains and other cultural materials, etc. This overall change in many cultural aspects probably can be attributed to the local environmental degradation. This region falls in the semi-arid zone where the average rainfall is 400-1000 mm. The vegetation of this area is of dry deciduous type. After every 2-3 years the region experiences drought and famine. Risk and uncertainty in the subsistence system play an important role in making decisions by human about new subsistence strategies. To cope up with the situation, the Late Jorwe people might have hunted wild animals to substantiate their food economy.

Fig. 5.5. NISP of Domestic and Wild Animals at Inamgaon.

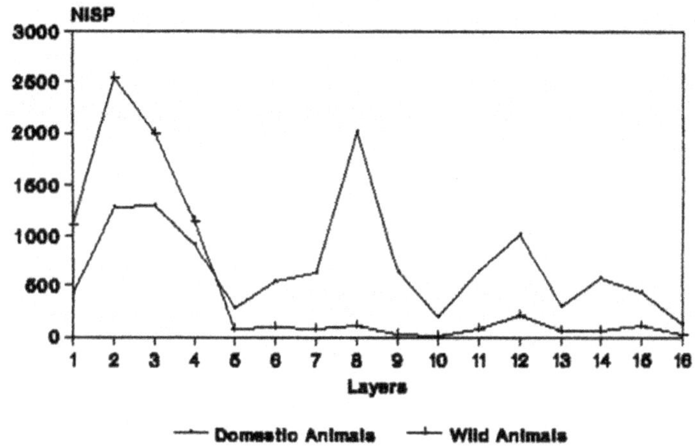

Fig. 5.6. MNI of Domestic and Wild Animals at Inamgaon.

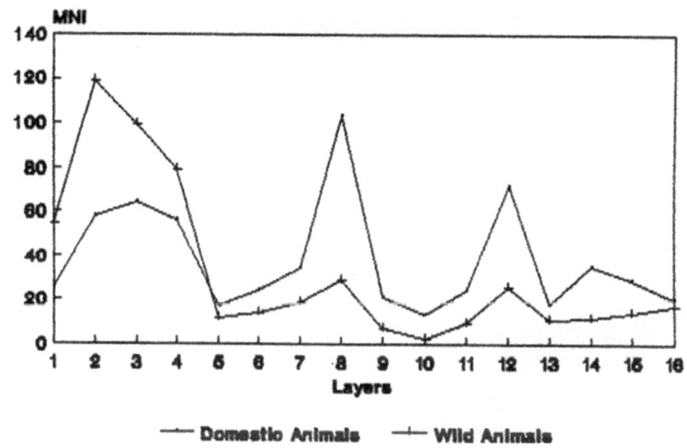

Table 5.5 Proportion of domestic and wild animals in the three cultural phases at Inamgaon

Species	Late Jorwe Phase	%	Early Jorwe Phase	%	Malwa Phase	%	Total	%
Domestic	6737	45.88	6015	87.28	2873	80.11	15625	62.10
Wild	7946	54.12	877	12.72	713	19.89	9536	38.90
Total	14623	100	6892	100	3586	100	25161	100

II. Standardization of faunal assemblage

At Inamgaon, the excavators concentrated more on the Late Jorwe phase than the Early Jorwe and the Malwa phases and excavated more area of this cultural phase. The volume of the excavated soil from the Late Jorwe phase is 28195 m^3, the Early Jorwe 1611 m^3 and the Malwa 977.625 m^3.Thus naturally the faunal remains were more from the Late Jorwe phase. An attempt has been made to standardize the results of faunal studies at one level with an assumption that the distribution of faunal remains in the site for the excavated and the unexcavated area is uniform. The results of the Number of Identified Specimens, Minimum Number of Individuals and the Meat Weight for mammalian species in the faunal assemblage are standardized (Pawankar and Thomas 1997) (Table 5.6-5.8). The graphs of the projected NISP and the MNI for the domestic and the wild animals show four peaks in layers 14, 12, 8 and 4 (Fig. 5.7 &5.8). The two peaks in the Malwa phase in layers 14 and 12 reveal maximum exploitation of domestic animals. The layer 12 peak point is the highest amongst the four peaks of the three cultural phases. The increase and decrease in the exploitation of domestic and wild animals in different layers is similar both in the actual Number of Identified Specimens and the Minimum Number of Individuals. The results of the actual and the projected Number of Identified Specimens and the Minimum Number of Individuals have revealed a definite increase in the exploitation of wild animals in the Late Jorwe phase. To illustrate the difference between the actual and the projected Number of Identified Specimens a few species of animals such as cattle, sheep/goat, pig, dog, antelope, deer, wolf and jackal are selected (Fig. 5.9-5.15).

Fig. 5.7. Projected NISP of Domestic and Wild Animals at Inamgaon.

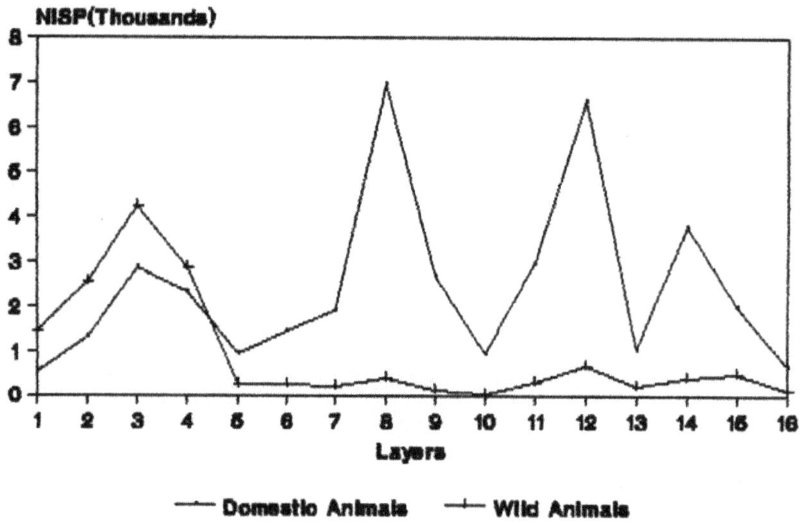

Fig. 5.8. Projected MNI of Domestic and Wild Animals at Inamgaon.

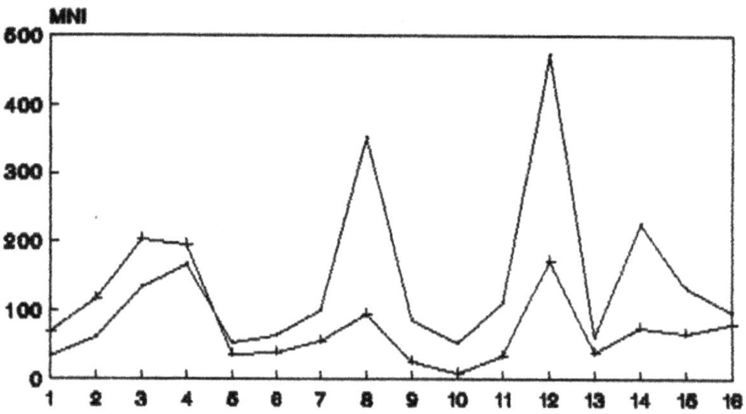

Fig. 5.9. Actual and Projected NISP of Cattle at Inamgaon.

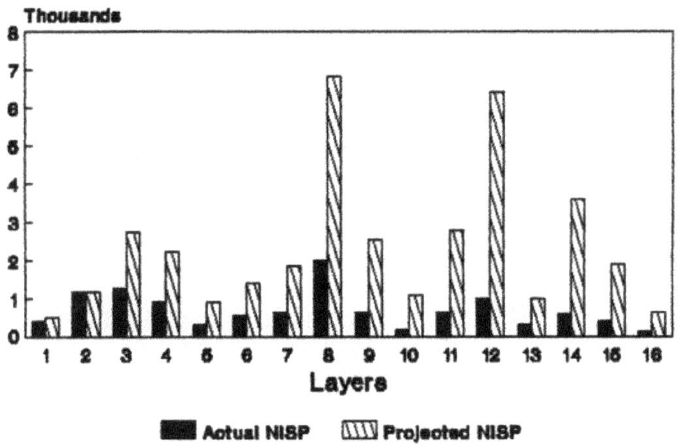

Fig. 5.10. Actual and Projected NISP of Sheep and Goat at Inamgaon.

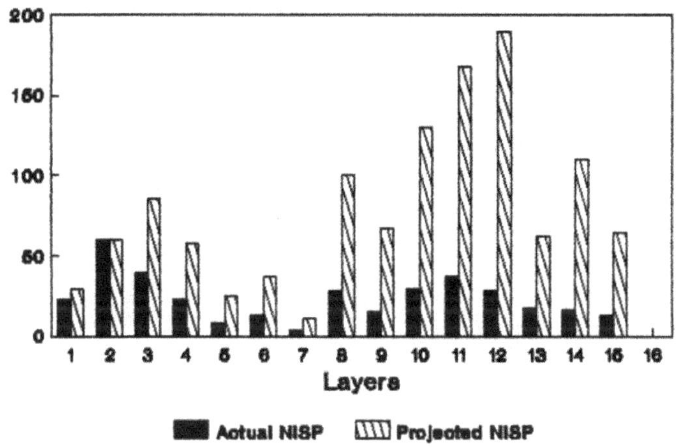

Fig. 5.11. Actual and Projected NISP of Pig at Inamgaon.

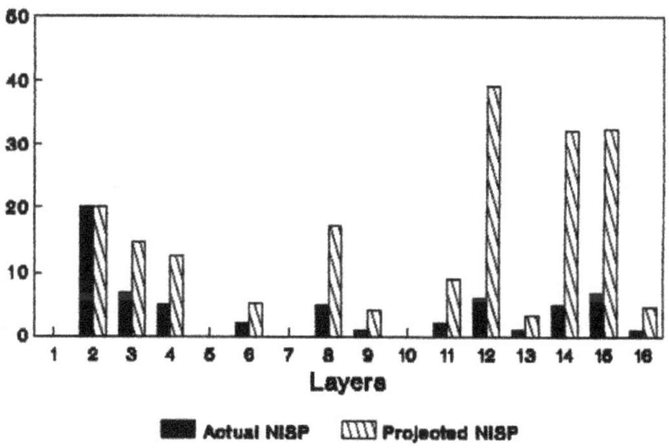

Fig. 5.12. Actual and Projected NISP of Dog at Inamgaon.

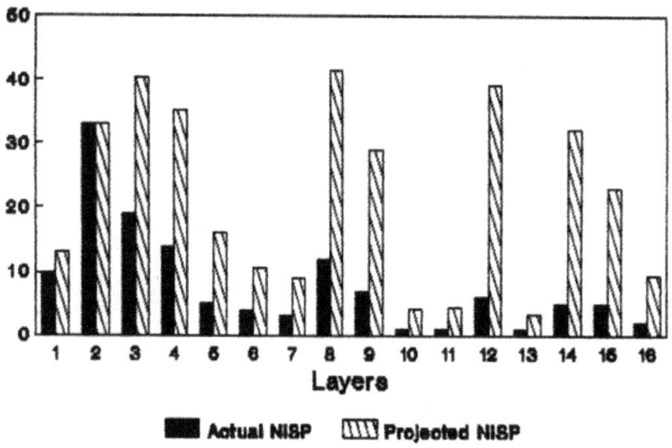

Fig. 5.13. Actual and Projected NISP of Antelope at Inamgaon.

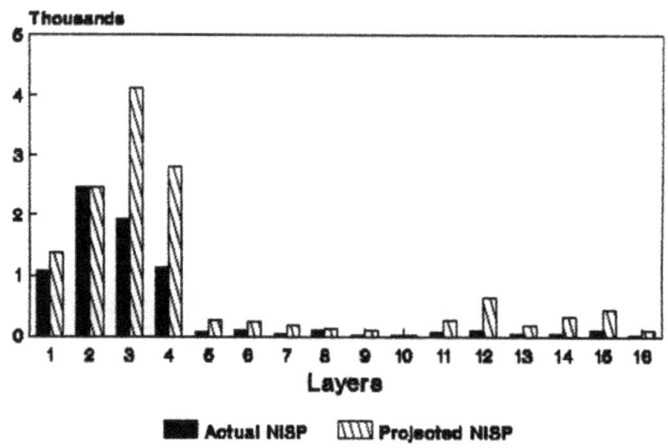

Fig. 5.14. Actual and Projected NISP of Deer at Inamgaon.

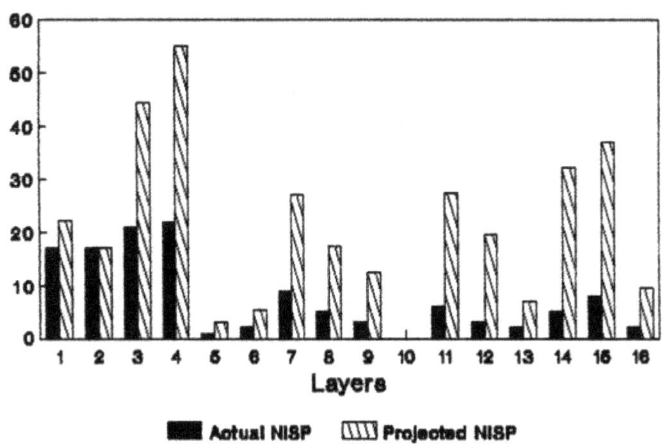

Fig. 5.15. Actual and Projected NISP of Wolf and Jackal at Inamgaon.

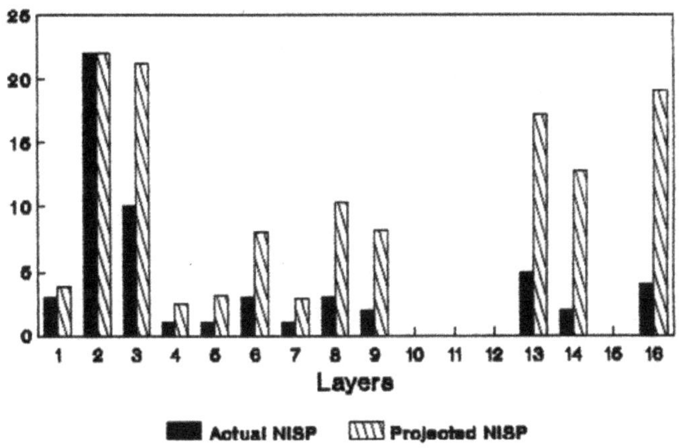

The graph of meat weight of the domestic and wild animals reveals that the maximum amount of meat of domestic animals was consumed in all the three cultural phases (Fig. 5.16). Layer 8 of the Early Jorwe phase shows maximum consumption of meat of domestic animals. In Late Jorwe phase the percentage of meat of domestic animal is decreased compared to other two phases, but it is not less than the meat weight of wild animals since the hunted animal was smaller in size. It shows that the people of Late Jorwe phase hunted more wild animals to substantiate the food economy. The result of meat weight of the domestic and the wild animals is also standardized (Fig. 5.17; Table 5.8).

Fig. 5.16. Meat Weight of Domestic and Wild Animals at Inamgaon.

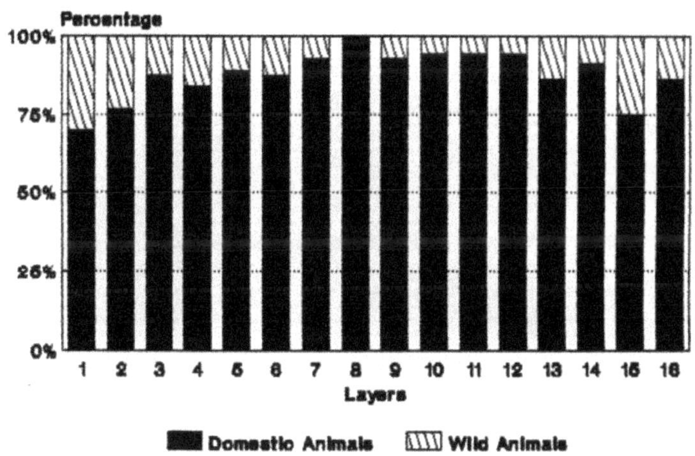

Fig. 5.17. Projected Meat Weight of Domestic and Wild Animals at Inamgaon.

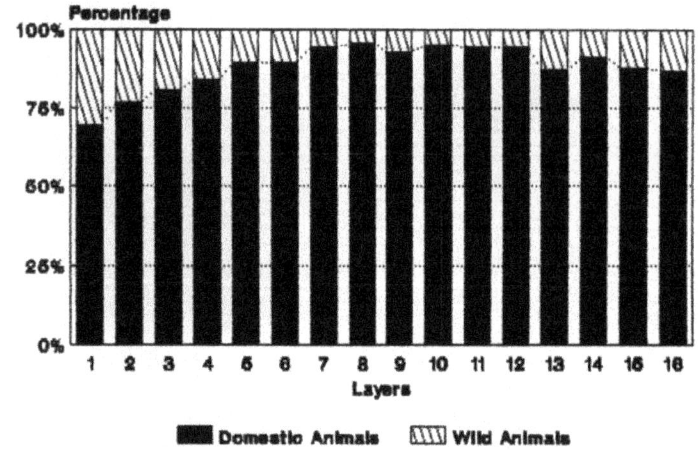

Table 5.6. Projected Number of Identified Specimens in all the layers at Inamgaon

Species	01	02	03	04	05	06	07	08	09	10	11	12	13	14	15	16
Bos indicus	334	777	2007	1308	489	549	1108	3295	1309	387	1033	2724	396	1779	815	350
Bubalus bubalis	22	37	117	45	06	16	39	204	50	30	59	18	07	52	51	10
Bos/Bubalus	138	351	573	862	396	811	708	3289	1192	373	1688	3600	568	1747	1010	283
Capra hircus	13	25	34	05	--	05	06	24	12	04	31	20	10	26	09	--
Ovis aries	01	09	04	--	02	03	03	03	--	--	18	13	--	--	04	05
Capra/Ovis	16	26	47	53	22	29	03	72	54	126	118	157	52	34	51	--
Sus domesticus	--	20	15	13	--	05	--	17	04	--	09	39	03	32	25	05
Canis familiaris	13	33	40	35	16	11	09	41	29	04	05	33	03	32	23	10
Equus caballus	04	04	09	02	--	03	--	03	--	--	--	--	--	--	--	--
Equus asinus	--	01	--	--	--	--	--	--	--	--	--	--	--	--	--	--
Elephas maximus	--	--	--	--	--	--	--	--	--	--	--	--	--	06	--	--
Canis lupus	04	11	13	--	03	05	--	03	04	--	--	--	--	13	--	14
Canis aureus	--	04	04	--	--	--	--	07	--	--	--	--	--	--	--	--
Canis sp.	--	04	04	--	--	03	03	--	04	--	--	--	03	--	--	05
Felis chaus	03	05	04	--	--	03	03	--	--	--	--	--	--	--	--	--
Panthera pardus	--	02	--	--	--	--	--	--	--	--	09	20	--	--	--	--
Panthera tigris	--	01	--	--	--	--	--	--	--	--	--	--	--	--	--	--
Sus scrofa	05	08	08	--	--	05	--	17	--	--	--	--	03	--	--	--
Bos gaurus	--	--	02	01	--	--	03	--	--	--	--	--	--	06	--	--
Antilope cervicapra	1332	2377	3864	2600	240	244	169	283	75	17	255	568	148	187	250	82
Gazella bennetti	13	10	13	07	--	--	--	17	--	--	--	07	07	06	--	--
Tetracerus quadricornis	03	11	30	05	03	05	03	07	--	--	09	20	--	--	32	05
Boselaphus tragocamelus	44	82	208	201	29	08	06	48	21	04	18	46	28	148	167	24
Axis axis	09	08	21	33	--	--	21	10	12	--	23	20	03	26	09	05
Axis porcinus	01	--	--	--	--	--	--	--	--	--	--	--	--	--	--	--
Axis sp.	--	01	--	--	--	--	--	--	--	--	--	--	--	--	--	--
Cervus unicolor	09	05	19	23	--	--	06	03	--	--	--	--	03	06	05	05
Muntiacus muntjak	01	--	--	--	--	--	--	03	--	--	--	--	--	--	23	--
Deer sp.	01	03	04	--	03	--	--	--	--	--	--	--	--	--	--	--
Herpestus edwardsi	--	01	11	--	--	--	--	--	--	--	05	--	--	--	--	--
Lepus nigricollis	--	01	08	02	--	--	--	--	--	--	--	07	--	--	--	--

Table 5.7. Projected Minimum Number of Individuals in all the layers at Inamgaon

Species	01	02	03	04	05	06	07	08	09	10	11	12	13	14	15	16
Bos indicus	16	25	64	80	26	24	60	155	46	17	36	209	28	122	60	53
Bubalus bubalis	01	04	08	18	03	05	09	28	08	09	14	39	03	13	18	10
Bos/Bubalus	04	11	30	33	10	16	18	117	12	13	23	131	14	39	14	24
Capra hircus	04	03	06	05	--	03	03	05	04	04	05	07	07	13	09	--
Ovis aries	01	02	02	--	03	03	02	03	--	--	09	07	--	--	05	--
Capra/Ovis	05	06	08	18	06	05	03	28	08	07	14	33	03	13	09	--
Sus domesticus	--	04	04	05	--	03	--	10	04	--	05	20	03	13	09	05
Canis familiaris	01	04	08	05	06	03	06	03	04	04	05	26	03	06	09	05
Equus caballus	01	01	04	03	--	03	--	03	--	--	--	--	--	--	--	--
Equus asinus	--	01	--	--	--	--	--	--	--	--	--	--	--	--	--	--
Elephas maximus	--	--	--	--	--	--	--	--	--	--	--	--	--	06	--	--
Canis lupus	01	02	04	--	03	03	--	03	04	--	--	--	--	06	--	10
Canis aureus	--	01	02	02	--	--	--	03	--	--	--	--	--	--	--	--
Canis sp.	--	01	02	--	--	03	03	--	04	--	--	--	03	--	--	05
Felis chaus	03	01	--	02	--	03	03	--	--	--	--	--	--	--	--	--
Panthera pardus	--	01	--	--	--	--	--	--	--	--	01	07	--	--	--	--
Panthera tigris	--	01	--	--	--	--	--	--	--	--	--	--	--	--	--	--
Sus scrofa	03	03	02	--	--	03	--	03	--	--	--	--	03	--	--	--
Bos gaurus	--	--	02	02	--	--	03	--	--	--	--	--	--	--	--	--
Antilope cervicapra	47	89	138	148	22	16	33	55	08	04	18	105	21	26	19	38
Gazella bennetti	01	02	02	02	--	--	--	05	--	--	--	13	03	06	--	--
Tetracerus quadricornis	01	01	04	02	03	03	03	03	--	--	01	07	--	--	14	05
Boselaphus tragocamelus	07	08	25	28	06	05	03	14	08	04	05	26	03	19	14	14
Axis axis	04	03	04	05	--	03	06	03	04	--	04	07	03	13	05	05
Axis porcinus	01	--	--	--	--	--	--	--	--	--	--	--	--	--	--	--
Axis sp.	--	01	--	--	--	--	--	--	--	--	--	--	--	--	--	--
Cervus unicolor	01	01	04	02	--	--	03	03	--	--	--	--	03	06	05	05
Muntiacus muntjak	01	--	--	--	--	--	--	03	--	--	--	--	--	--	09	--
Deer sp.	01	01	04	--	03	--	--	--	--	--	--	--	--	--	--	--
Herpestus edwardsi	--	01	06	--	--	--	--	--	--	--	05	--	--	--	--	--
Lepus nigricollis	--	01	04	02	--	--	--	--	--	--	--	07	--	--	--	--

Table 5.8. Projected Meat Weight in all the layers at Inamgaon

Species	01	02	03	04	05	06	07	08	09	10	11	12	13	14	15	16
Bos indicus	3200.0	5000.0	12800.0	16000.0	5200.0	4800.0	12000.0	31000.0	9200.0	3400.0	7200.0	41800.0	5600.0	24400.0	12000.0	10600.0
Bubalus bubalis	250.0	1000.0	2000.0	4500.0	750.0	1250.0	2250.0	7000.0	2000.0	2250.0	3500.0	9750.0	750.0	3250.0	4500.0	2500.0
Bos/Bubalus	900.0	2475.0	6750.0	7425.0	2250.0	3600.0	4050.0	26325.0	2700.0	2925.0	5175.0	29475.0	3150.0	8775.0	3150.0	5400.0
Capra hircus	50.0	37.5	75.0	62.5	00.0	37.5	37.5	62.5	50.0	50.0	62.5	87.5	87.5	162.5	112.5	00.0
Ovis aries	12.5	25.0	25.0	00.0	37.5	37.5	25.0	37.5	00.0	00.0	112.5	87.5	00.0	00.0	62.5	00.0
Capra/Ovis	62.5	75.0	100.0	225.0	75.0	62.5	37.5	350.0	100.0	87.5	175.0	412.5	37.5	162.5	112.5	00.0
Sus domesticus	00.0	80.0	80.0	100.0	00.0	60.0	00.0	200.0	80.0	00.0	100.0	400.0	60.0	260.0	180.0	100.0
Canis familiaris	7.5	30.0	60.0	37.5	45.0	22.5	45.0	22.5	30.0	30.0	37.5	195.0	22.5	45.0	67.5	37.5
Equus caballus	150.0	150.0	600.0	450.0	00.0	450.0	00.0	450.0	00.0	00.0	00.0	00.0	00.0	00.0	00.0	00.0
Canis lupus	9.0	18.0	36.0	00.0	27.0	27.0	00.0	27.0	36.0	00.0	00.0	00.0	00.0	54.0	00.0	90.0
Canis aureus	00.0	4.0	8.0	8.0	00.0	00.0	00.0	12.0	00.0	00.0	00.0	00.0	00.0	00.0	00.0	00.0
Sus scrofa	300.0	300.0	200.0	00.0	00.0	300.0	00.0	300.0	00.0	00.0	00.0	00.0	300.0	00.0	00.0	00.0
Antilope cervicapra	705.0	1335.0	2070.0	2220.0	330.0	240.0	495.0	825.0	120.0	60.0	270.0	1575.0	315.0	390.0	285.0	570.0
Gazella bennetti	10.0	20.0	20.0	20.0	00.0	00.0	00.0	50.0	00.0	00.0	00.0	130.0	30.0	60.0	00.0	00.0
Tetracerus quadricornis	12.5	12.5	50.0	25.0	37.5	37.5	37.5	37.5	00.0	00.0	12.5	87.5	00.0	00.0	175.0	62.5
Boselaphus tragocamelus	700.0	800.0	2500.0	2800.0	600.0	500.0	300.0	1400.0	800.0	400.0	500.0	2600.0	300.0	1900.0	1400.0	1400.0
Axis axis	150.0	112.5	150.0	187.5	00.0	112.5	225.0	112.5	150.0	00.0	150.0	262.5	112.5	487.5	187.5	187.5
Cervus unicolor	125.0	125.0	500.0	250.0	00.0	00.0	375.0	375.0	00.0	00.0	00.0	00.0	375.0	750.0	625.0	625.0
Muntiacus muntjak	20.0	00.0	00.0	00.0	00.0	00.0	00.0	60.0	00.0	00.0	00.0	00.0	00.0	00.0	180.0	00.0
Herpestus edwardsi	00.0	00.75	4.5	00.0	00.0	00.0	00.0	00.0	00.0	00.0	3.75	00.0	00.0	00.0	00.0	00.0
Lepus nigricollis	00.0	1.0	4.0	2.0	00.0	00.0	00.0	00.0	00.0	00.0	00.0	7.0	00.0	00.0	00.0	00.0

IV. Determination of Species, Sex and Stature based on Osteometric analysis

a. Species separation

Some of the species which are morphologically very similar can be separated out with the help of certain measurements of bones. Morphologically the bones of blackbuck, sheep and goat are very similar. To differentiate these species the bones are studied with the help of the reference skeletons. These species could easily be separated out with the help of measurements of scapula, humerus, radius and femur (Fig. 5.18-5.21). Cattle, buffalo and nilgai could be separated out with the help of measurements such as proximal breadth and the thickness of the first phalanx (Fig. 5.22). Similarly the measurements of the distal breadth and the thickness of humerus have been used to differentiate between dog, wolf and jackal (Fig. 5.23). Wild and domestic pig can also separate out by taking the measurements of the proximal breadth and thickness of the radius (Fig. 5.24).

Fig. 5.18. Species separation based on Scapula measurements (GLP vs. BG)

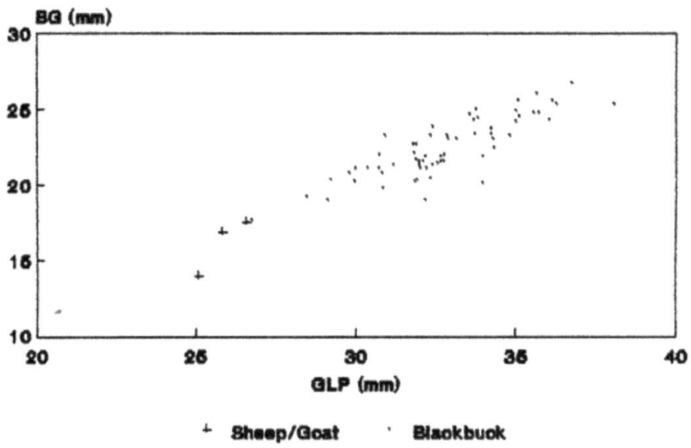

Fig. 5.19. Species separation based on Humerus measurements (BD vs. TD)

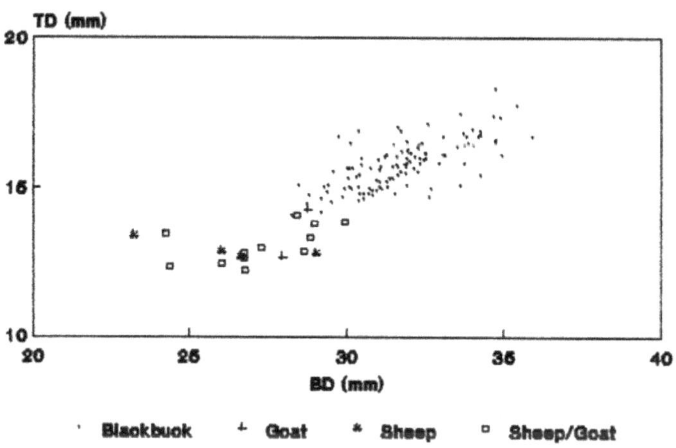

Fig. 5.20. Species separation based on Radius measurements (BP vs. TP).

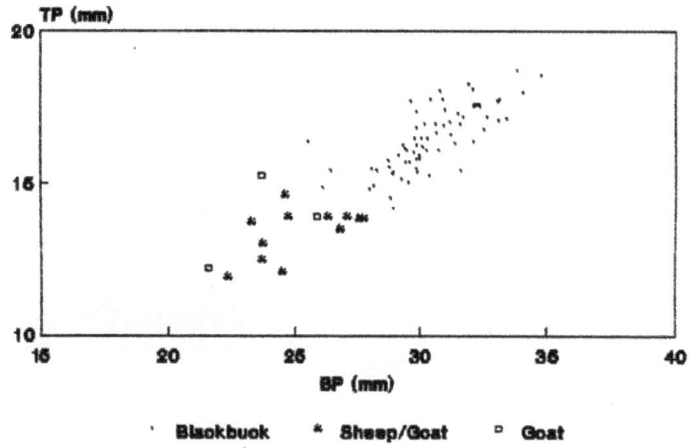

Fig. 5.21. Species separation based on Femur measurements (BP vs. TP).

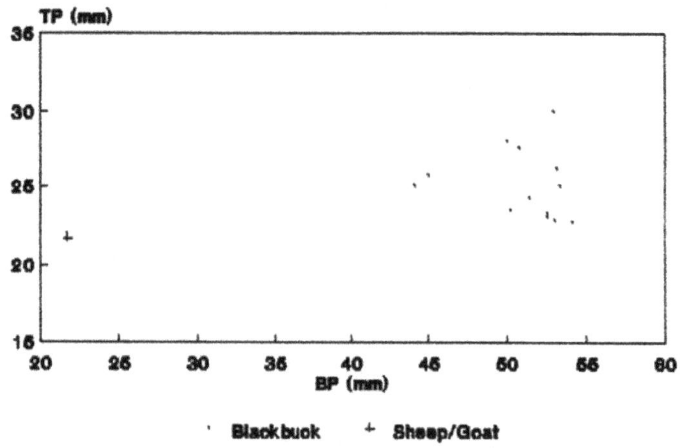

Fig. 5.22. Species separation based on First Phalanx measurements (BP vs. TP).

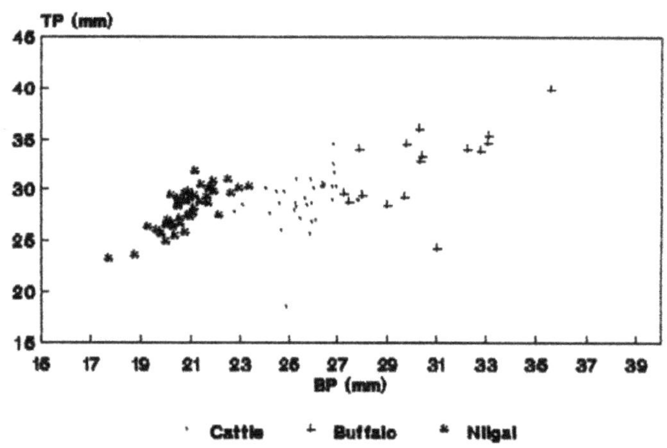

Fig. 5.23. Species separation based on Humerus measurements (BD vs. TD).

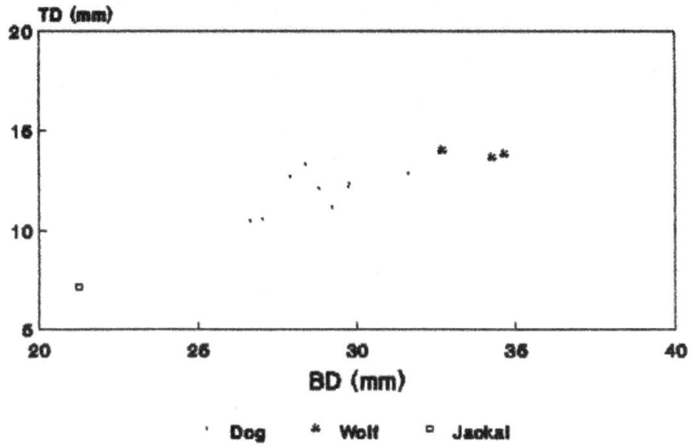

Fig. 5.24. Species separation based on Radius measurements (BP vs. TP).

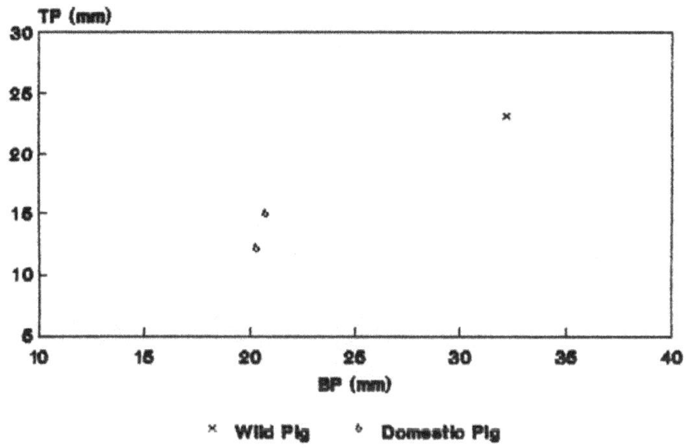

b. Sex determination

Qualitatively it was difficult to distinguish the bones of male and female cattle at Inamgaon, as the bones were very fragmentary. Hence, metrical analyses have been carried out on well preserved bones like astragalus and metapodials (Fig. 5.25 & 5.26). This has given some very good evidence for male to female ratio, which was 2:1. In stock breeding, more number of males is not required. As a result of this, there is a tendency to kill a large number of males when they are young. Such selection of animals shows the complete knowledge of the early farming population about the stock-raising aspects.

Fig. 5.25. Sex Determination of Cattle at Inamgaon: Astragalus (GL vs.BD).

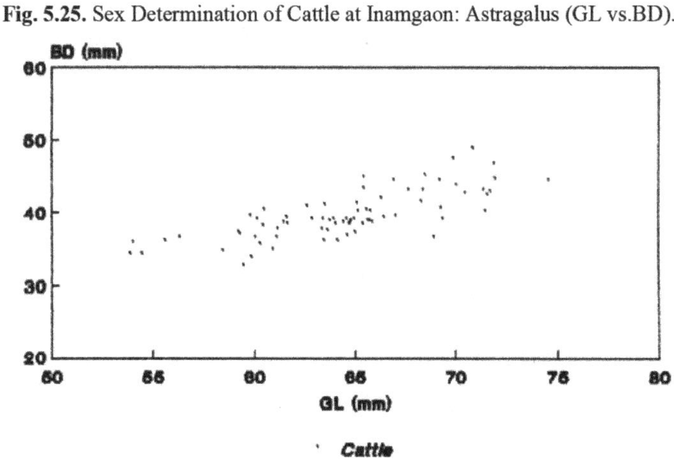

Fig. 5.26. Sex Determination of Cattle at Inamgaon: Metacarpal (BD vs.WT).

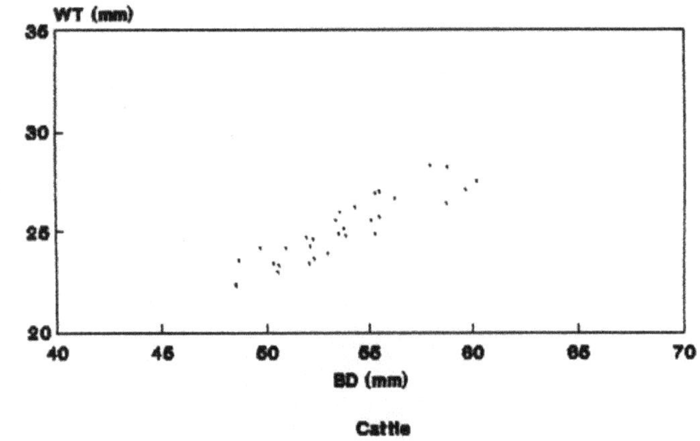

Metrical analyses of bones of blackbuck such as astragalus, first and the second phalanx reveal that about 70-75% of the blackbuck hunted were males (Fig. 5.27-5.30). This is significant, because while hunting animals, this kind of a selection is generally not possible. Thus probably these animals were killed during their mating season. Blackbucks breed at all seasons, but the main rut is noticed between February and the March. This time the bucks fight for the possession of the does. The unsuccessful males are driven out from the territory and probably such animals were killed at Inamgaon.

Fig. 5.27. Sex Determination of Blackbuck at Inamgaon: Astragalus (GLI vs.BD).

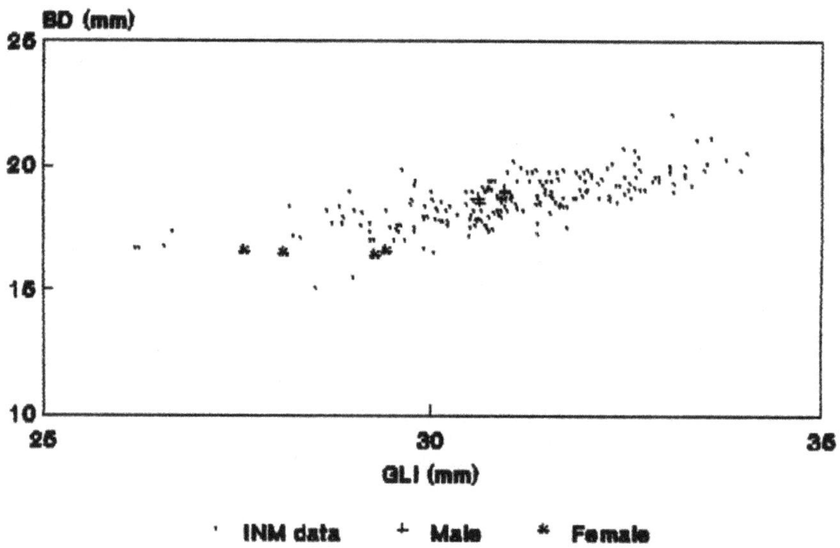

Fig. 5.28. Sex Determination of Blackbuck at Inamgaon: First Phalanx (GL vs. BD).

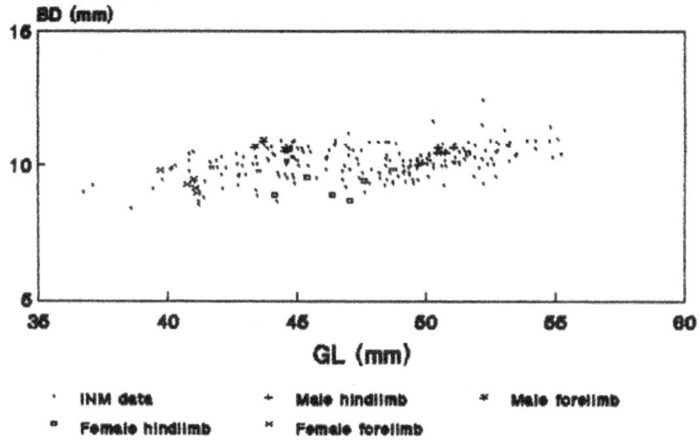

Fig. 5.29. Sex Determination of Blackbuck at Inamgaon: Second Phalanx (GL vs.BD).

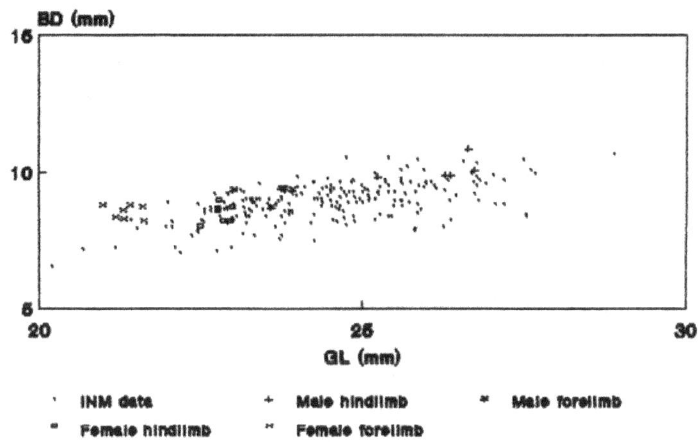

Fig. 5.30. Male to Female Ratio of Blackbuck in Various Phases at Inamgaon.

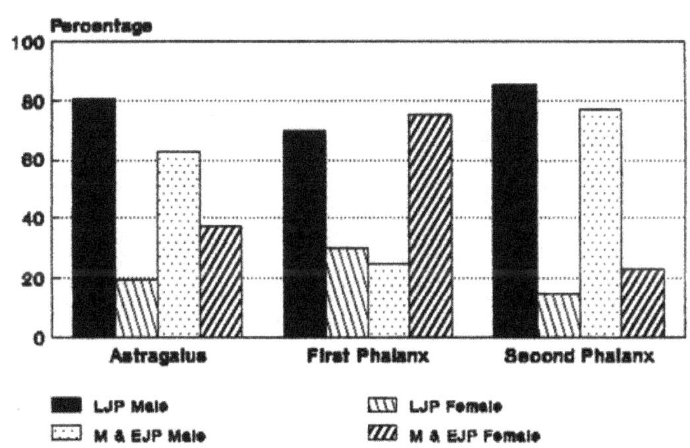

c. Stature of cattle at Inamgaon

The withers height of cattle remained almost static between 106 to113 cm in all the phases. However, a close scrutiny shows a gradual decrease in the size of the animal in the transitional layers between the Malwa, Early Jorwe and Late Jorwe phases and there after a continuous increase in size. When Malwa people from Central India migrated towards south to settle in the Deccan, they might have brought their live- stock with them to start their life in this region. As this was an agro-pastoral community the use of cattle was manifold. Inbreeding of animals within the same stock for a longer period of time, results in the size reduction. But in the Malwa phase size has increased, which shows either the selection within the stock or probably in the beginning itself these people had some contact and exchange of cattle with other local cultural complexes of the area, which resulted in an increase in the size of cattle during this phase. The Malwa phase at Inamgaon continued for approximately 200 years. No further contact and the continuous inbreeding for such a longer period of time may have resulted in the size reduction of animals towards the end of this phase. This is followed by the Early Jorwe phase at the site and contact with the Jorwe occupation continued uninterrupted for about 4 centuries at the site. Hence, like in the Malwa phase, the process occurred till about the beginning of the Late Jorwe phase. The Late Jorwe phase at Inamgaon was more or less contemporary with the Iron Age cultures of Maharashtra. In the Late Jorwe phase probably because of the association with Iron Age people the size of cattle again increased (Pawankar and Thomas 1997). A burial site of the Iron Age period has been reported at Pimpalsuti, hardly 6 km to the west of Inamgaon (Ansari and Dhavalikar 1975). One of the burials had yielded a typical spouted vessel of painted Jorwe type, which throws light on the cultural contact with Jorwe people (Fig. 5.31).

Fig. 5.31. Stature Analysis of Cattle based on Astragalus at Inamgaon.

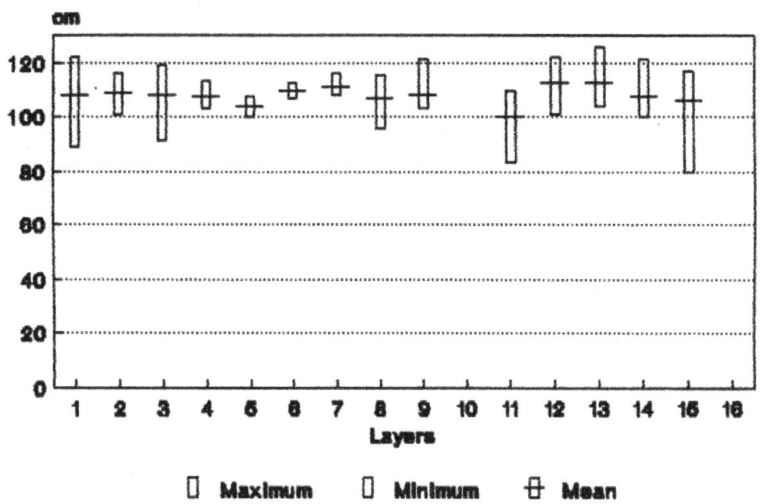

V. Horizontal distribution of animal remains

Layer 16

A majority of the settlement structures in layer 16 are oriented in northeast-southwest direction. Animal remains are found only in houses 84, 87 and 95. Of this, houses 87 and 95 are littered with non-meat-bearing parts whereas; house number 84 has normal distribution of both meat-bearing and non-meat-bearing parts. Non-meat-bearing parts are also found adjacent to the structure 87 in trench B5 (Fig. 5.32).

Fig. 5.32. Horizontal Distribution of Bones in Layer 16 at Inamgaon.

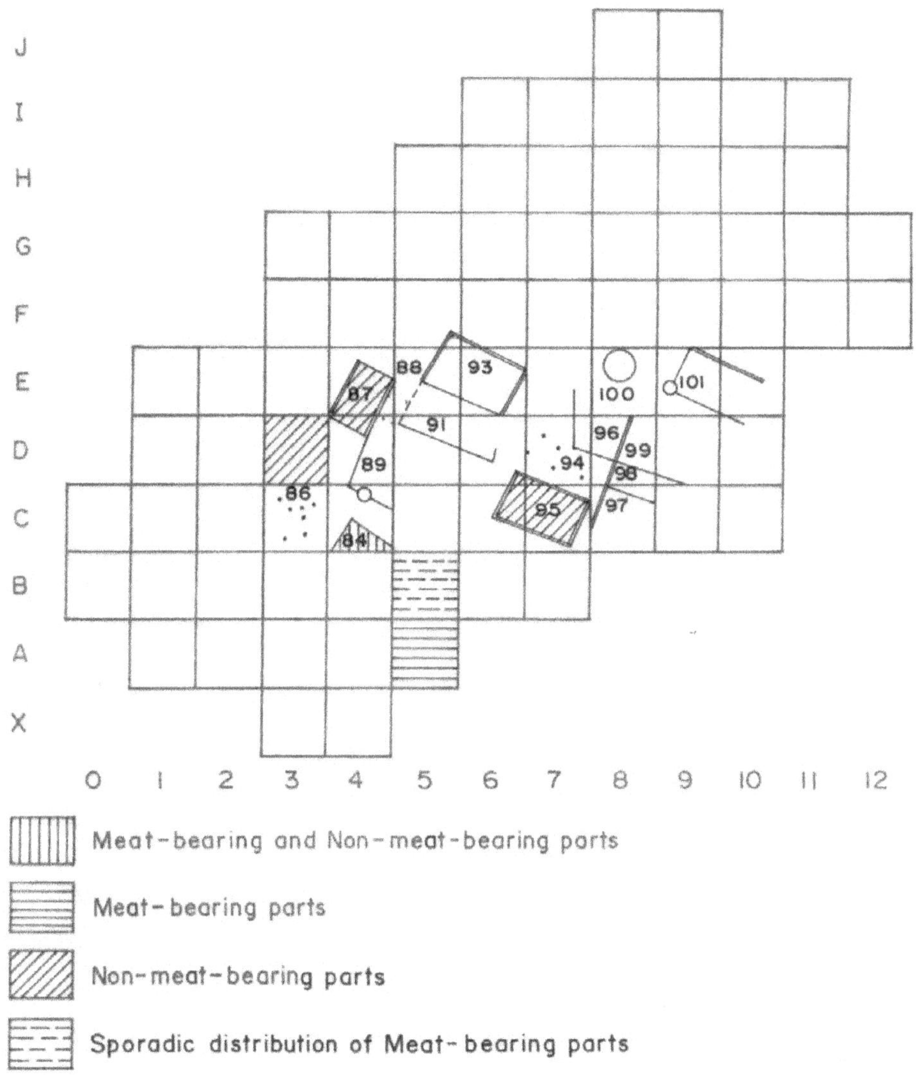

Layer 15

Only one structure is identified in layer 15, i.e. house number 90 which is devoid of non-meat bearing parts is noticed in the four corners of structure 90. However, meat-bearing and non-meat-bearing parts are found along the periphery towards the northwest and southwest of structure 90, in trenches E3, E4, OB1, B3, A5 and B5 (Fig.5.33).

Fig. 5.33. Horizontal Distribution of Bones in Layer 15 at Inamgaon.

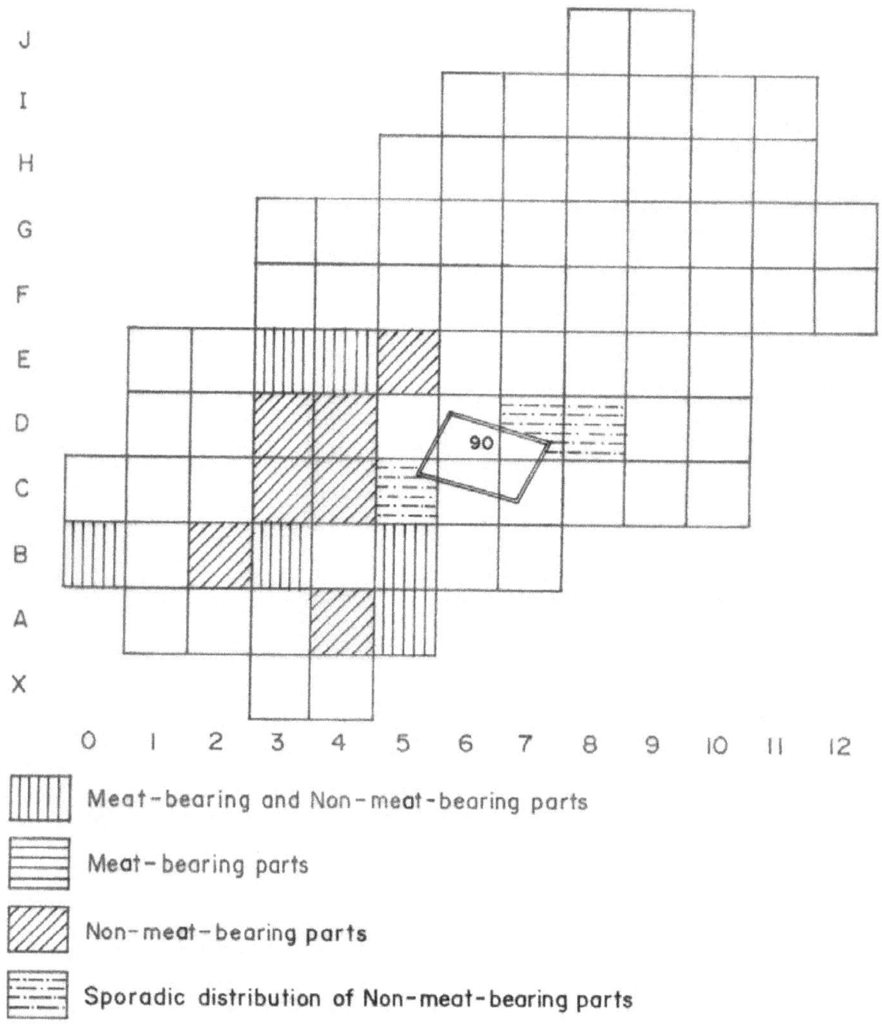

Layer 14

No structure has been identified in layer 14. The concentration of non-meat-bearing parts is found towards the eastern part and non-meat-bearing parts towards the western part of the excavated area. This indicates that the butchering of animals may have been in the eastern side and what one gets in the living area. In the previous layer 15, the trenches C3, C4, C5, D3, D4 and D5 which were concentrated with non-meat-bearing parts are devoid of any anima remains in Layer 14. Probably some structures may have been associated with these places in layer 14 (Fig. 5.34).

Fig. 5.34. Horizontal Distribution of Bones in Layer 14 at Inamgaon.

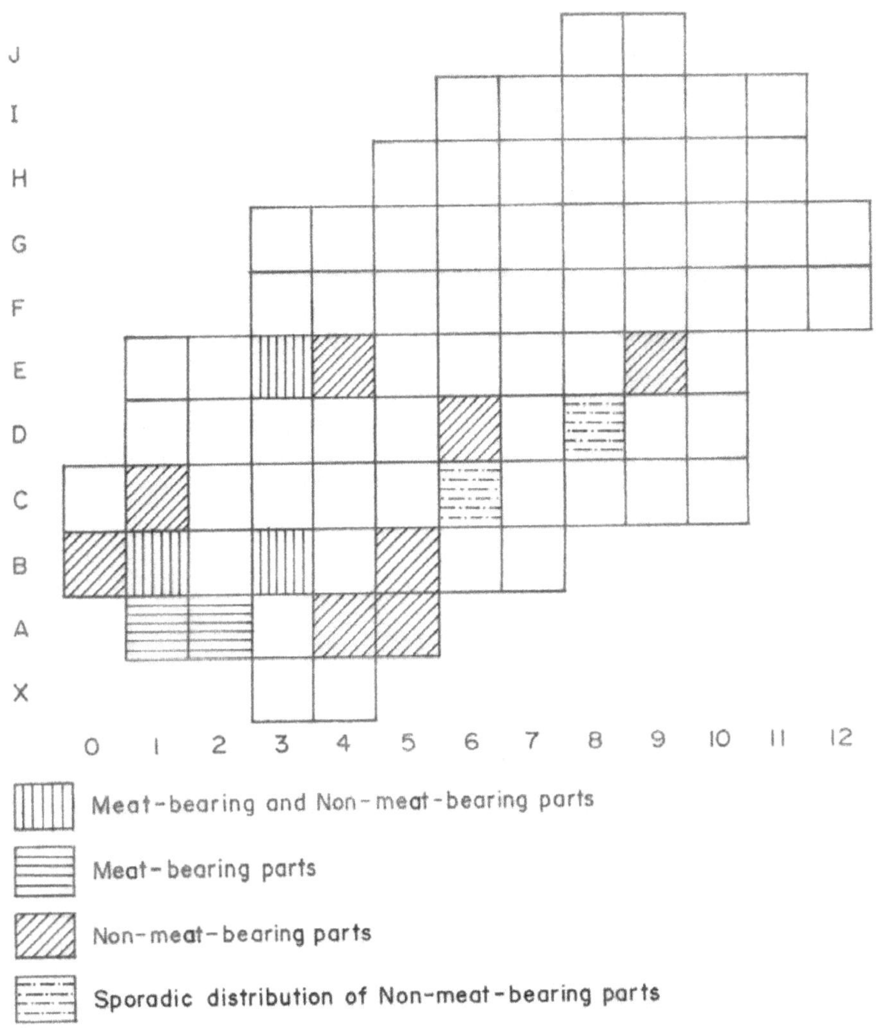

Layer 13

Again layer 13 is devoid of any structure. Much of the non-meat-bearing parts are found towards the eastern side whereas, the concentration of normal distribution, as well as the meat-bearing parts are noticed in the southern part of the excavated area. Probably those were the food refuse thrown from the living area. Non-meat-bearing parts are found in trenches B0, D3, E4, E5 and C4 and forms a regular pattern in layer 13, as it was in layer 14 (Fig. 5.35).

Fig. 5.35. Horizontal Distribution of Bones in Layer 13 at Inamgaon.

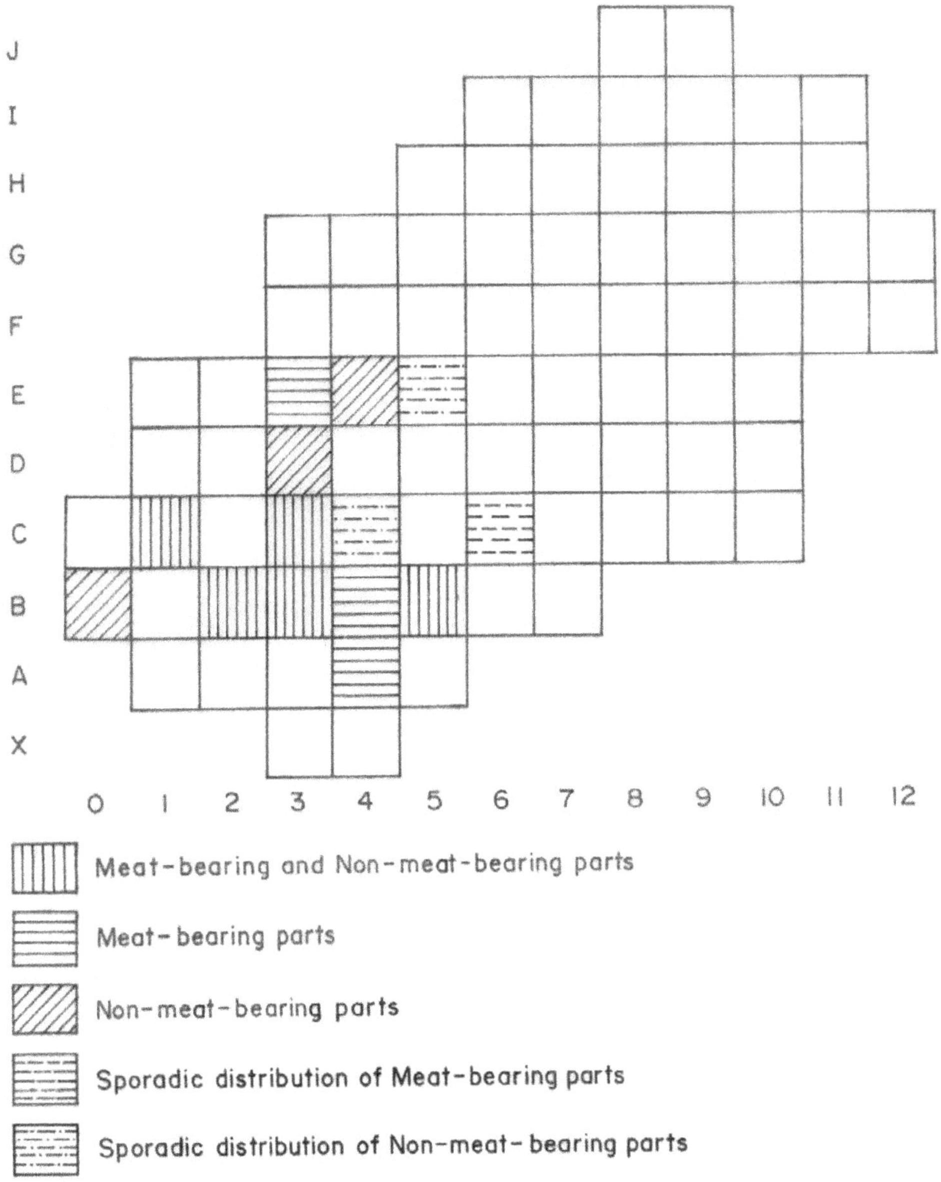

Layer 12

A large number of rectangular structures are found in southwest-northeast direction as has been in layer 16. The non-meat-bearing parts are found in the centre of the excavated area and also in a few structures like house numbers 74, 75, 76, 77 and 85. The two largest structures (75 and 85) have yielded a number of non-meat-bearing parts. It appears that there is something unusual about these two houses (Dhavalikar 1988). In house number 85, a burial (No. 115) and potters kiln were found. Thus, this house does not fit into the conventional term of living place. House number 75 also gives a similar impression with Chalcedony blades, roundels, grass patches, etc. (Dhavalikar 1988). The grass patch in house number 75 probably was not the remains of a fallen roof but intentionally kept vegetable matter on the floor to protect the meat from mixing it with soil. Thus, these structures can be identified as the activity areas of the settlement. Also the concentration of non-meat-bearing parts in these structures, suggests butchering of the animal within these structures. However, some of the structures (78, 79, 80, 81, 83 and 84) in the southeast area are with meat-bearing parts. The concentration of non-meat-bearing parts in the central part of the excavated area suggests that much of the activities were within the area of the settlement as has been noticed in the earlier layers. In this layer, the refuse dump areas are excavated in the southwestern region in trench B0, C0, B1, C1, B2 and C2. Structures 25, 82 and 82a are devoid of any faunal remains (Fig. 5.36).

Fig. 5.36. Horizontal Distribution of Bones in Layer 12 at Inamgaon.

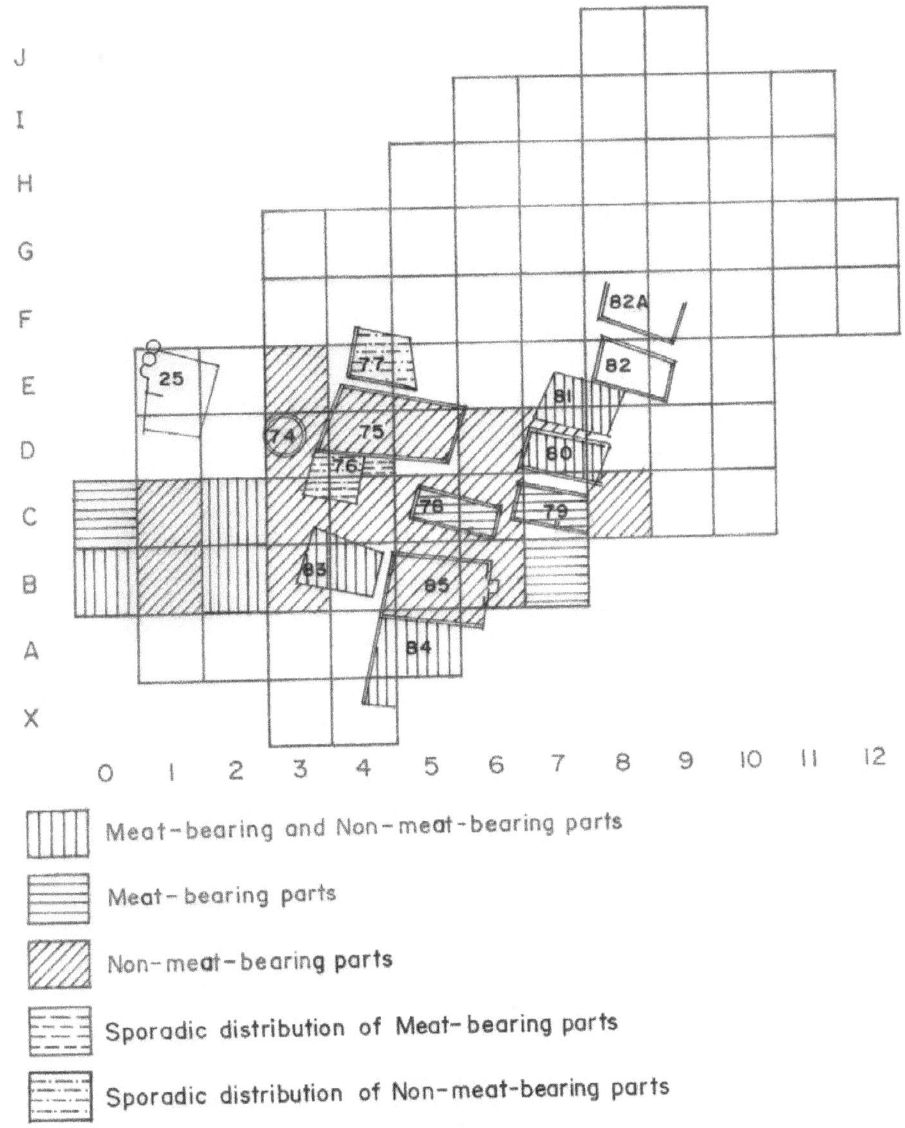

Layer 11

The concentration of non-meat-bearing parts in layer 11 appears to be slightly shifted to the eastern side in trenches B5-B8 and C8. Much of the western part of the excavated area is littered with bones of meat-bearing parts. The blank spaces in between the trenches with bone distribution probably indicate the structural remains of the previous layers (Fig. 5.37).

Fig. 5.37. Horizontal Distribution of Bones in Layer 11 at Inamgaon.

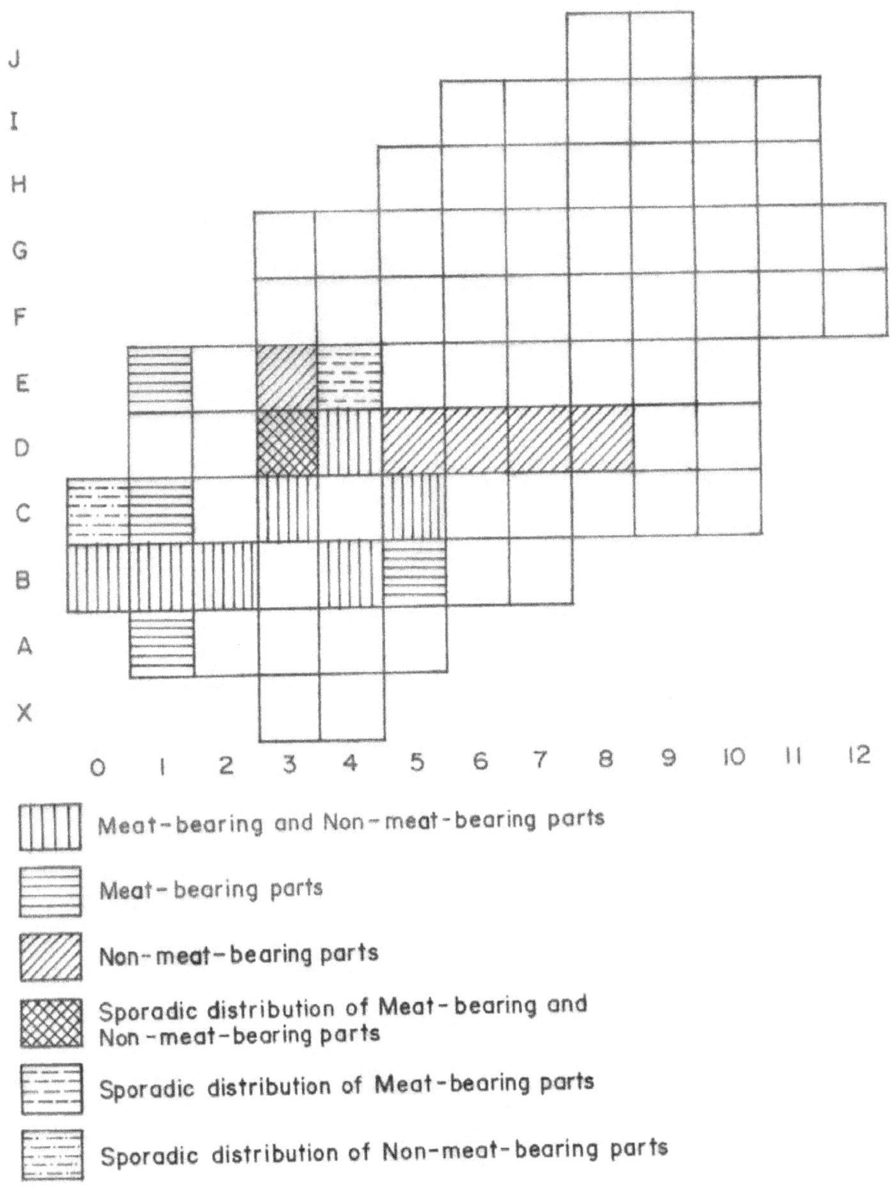

Layer 10

This layer is devoid of any structure. Bones are only found in the southwestern part of excavated area. The non-meat-bearing parts are noticed only in trench C1. Around this trench the meat-bearing parts and the normal distribution are also found (Fig. 5.38).

Fig. 5.38. Horizontal Distribution of Bones in Layer 10 at Inamgaon.

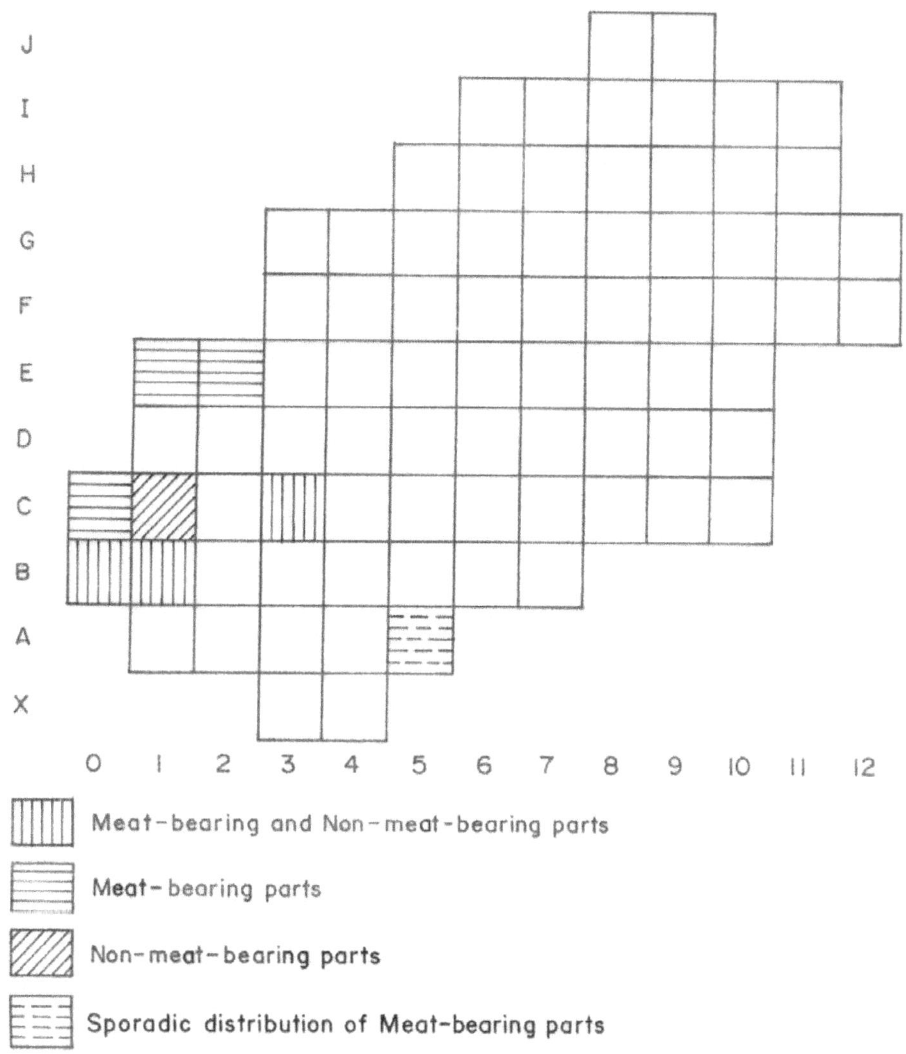

Layer 9

Structures 40 and 41 are identified in layer 9, which are free from any faunal material. These structures are located in the northeastern part of the excavated area whereas; the concentration of bones is on the southwestern side. Non-meat-bearing parts are found only in trenches C4 and C5 and could be identified as animal butchering area. This is surrounded by the normal distribution of bones and high meat-bearing parts (Fig. 5.39).

Fig. 5.39. Horizontal Distribution of Bones in Layer 9 at Inamgaon.

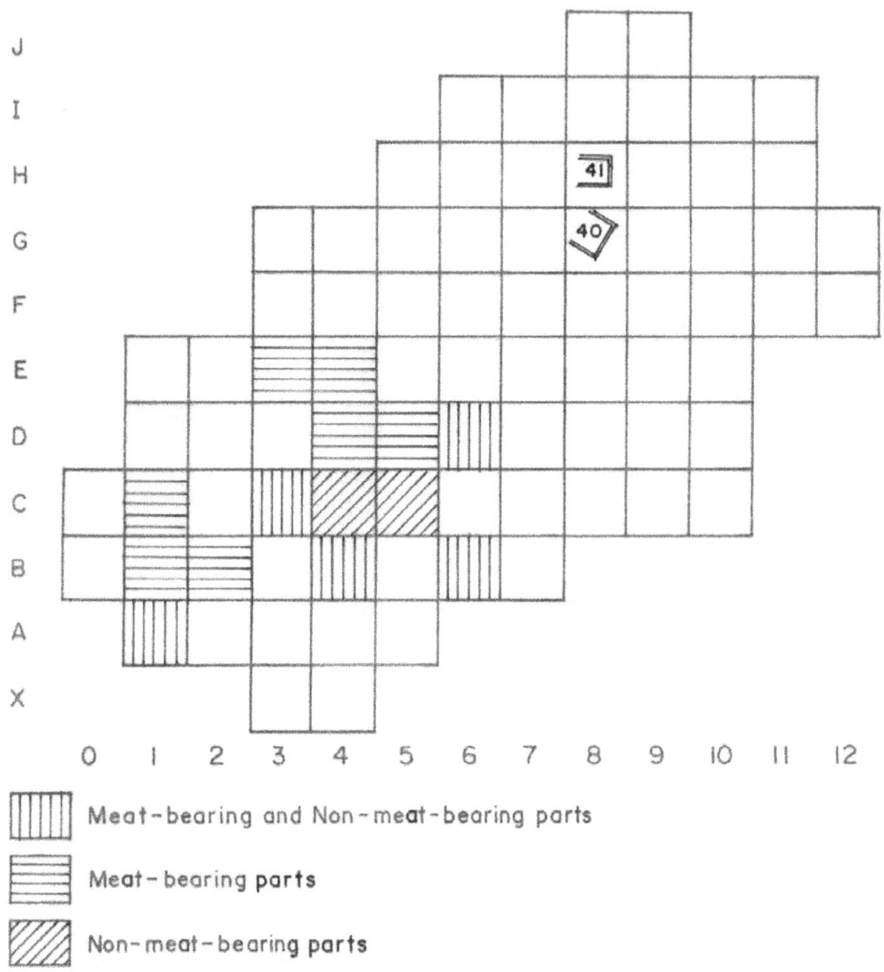

Layer 8

The structures in layer 8 are more or less identical with those in layer 12. It appears that only some modifications were made to the earlier structures. Once again the bulk of non-meat-bearing parts are found in the central part of excavated area, as it has been in layer12. It is also interesting to note that structures 61, 64, 69, 71, 71a and 72 have yielded non-meat-bearing parts. In this structural complex structure 69 is one of the largest. The meat-bearing parts are found in the southwestern region and also found in structures 60, 62, 63, 65, 67, 68, 70 and 73.

The high meat bearing parts are found on the southwestern side of the excavated area outside the structures; probably indicate that it was formed as a sort of refuse clearance from structures as well as outside the structures. Probably the structures with non-meat-bearing parts were associated with animal butchering activity (Fig. 5.40).

Fig. 5.40. Horizontal Distribution of Bones in Layer 8 at Inamgaon.

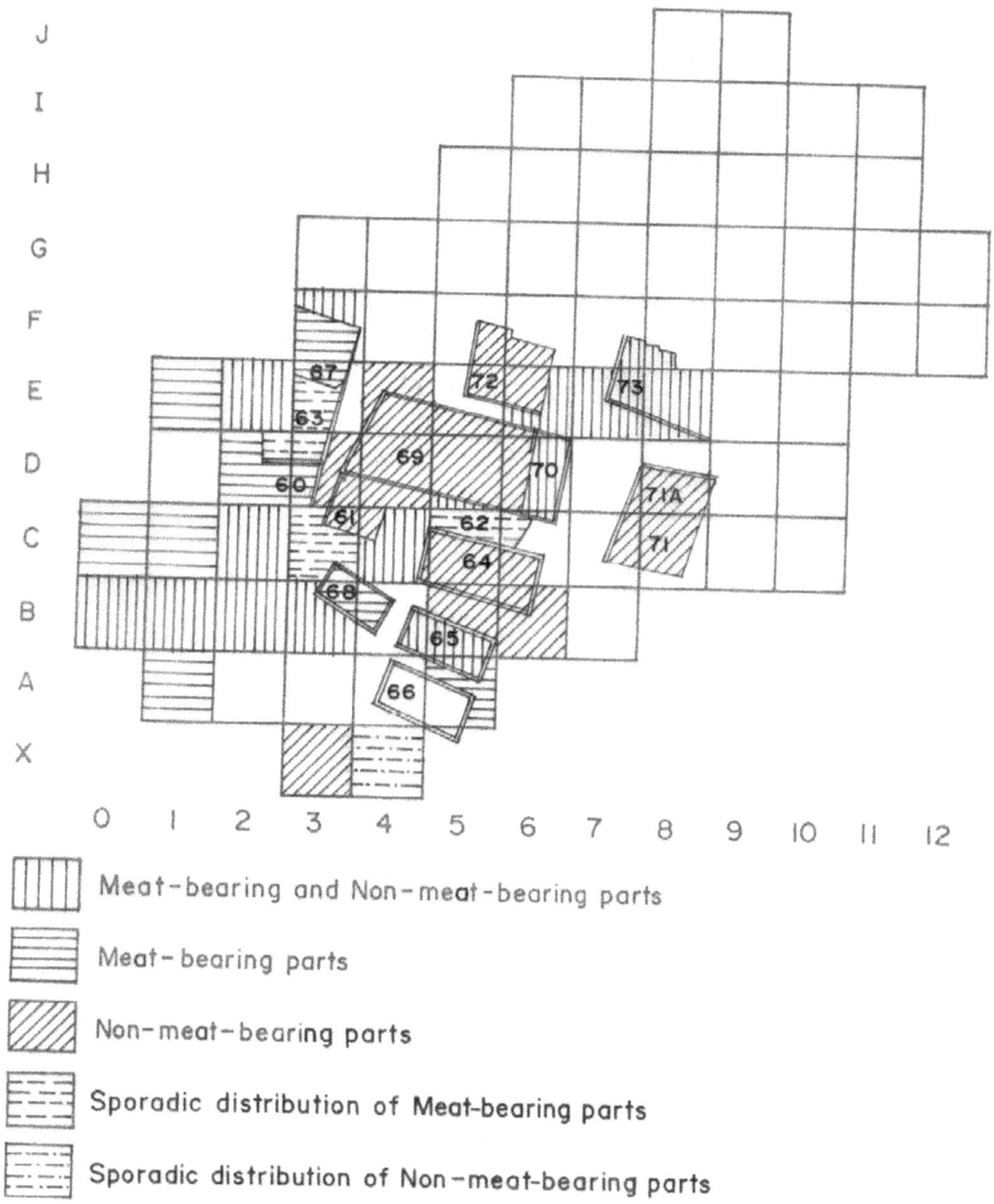

Layer 7

In layer 7, an appreciable number of meat bearing parts are observed in the southern part of the excavated area. The two big structures 53 and 54 are located almost at the centre of the trenches excavated. The circular structure 57 in trench E7 and on the northern side trench G4, structure 119 and 42 has yielded non-meat-bearing parts. Probably all these areas were associated with animal butchering activities. One interesting thing noticed is that the concentration of non-meat-bearing parts was found in two different localities suggest that involvement of more than one family in the butchering activity. Some of the structures such as 55, 56, 58, 59, 38 and 39 are free from animal remains. Structures 51, 51a and 52 have yielded meat-bearing parts. In general the distribution of meat-bearing parts is found in the southern and northern part of the structures, probably indicates refuse dumps (Fig. 5.41).

Fig. 5.41. Horizontal Distribution of Bones in Layer 7 at Inamgaon.

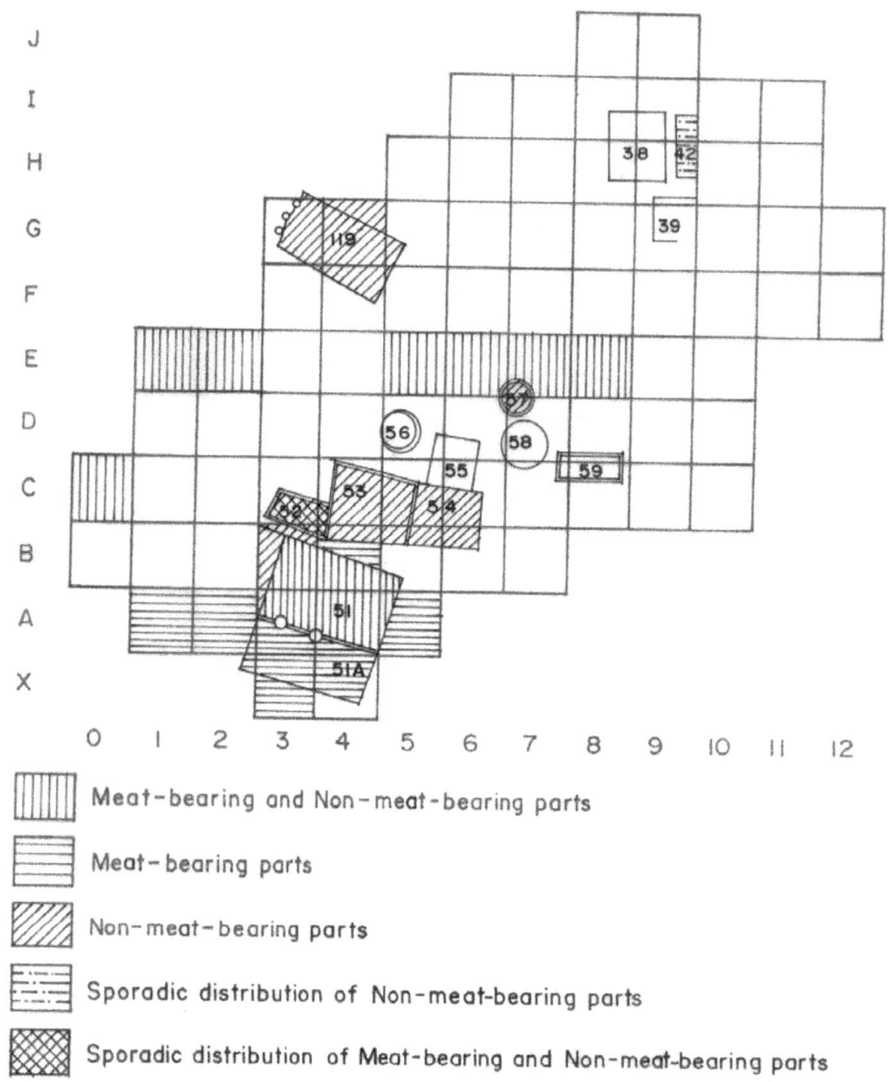

Layer 6

Structure number 18 is identified in layer 6, which is devoid of any animal remains. The concentration of meat-bearing parts is found in the southern, western and eastern part of the excavated area. The areas where the structures were identified in the previous layer 7 remain blank as far as the bones were concerned. The non-meat-bearing parts are noticed in the northwestern part of the excavated area, where in layer 7, structure 119 was located. This area can be identified as a butchering place in layer 6 (Fig.5.42).

Fig. 5.42. Horizontal Distribution of Bones in Layer 6 at Inamgaon.

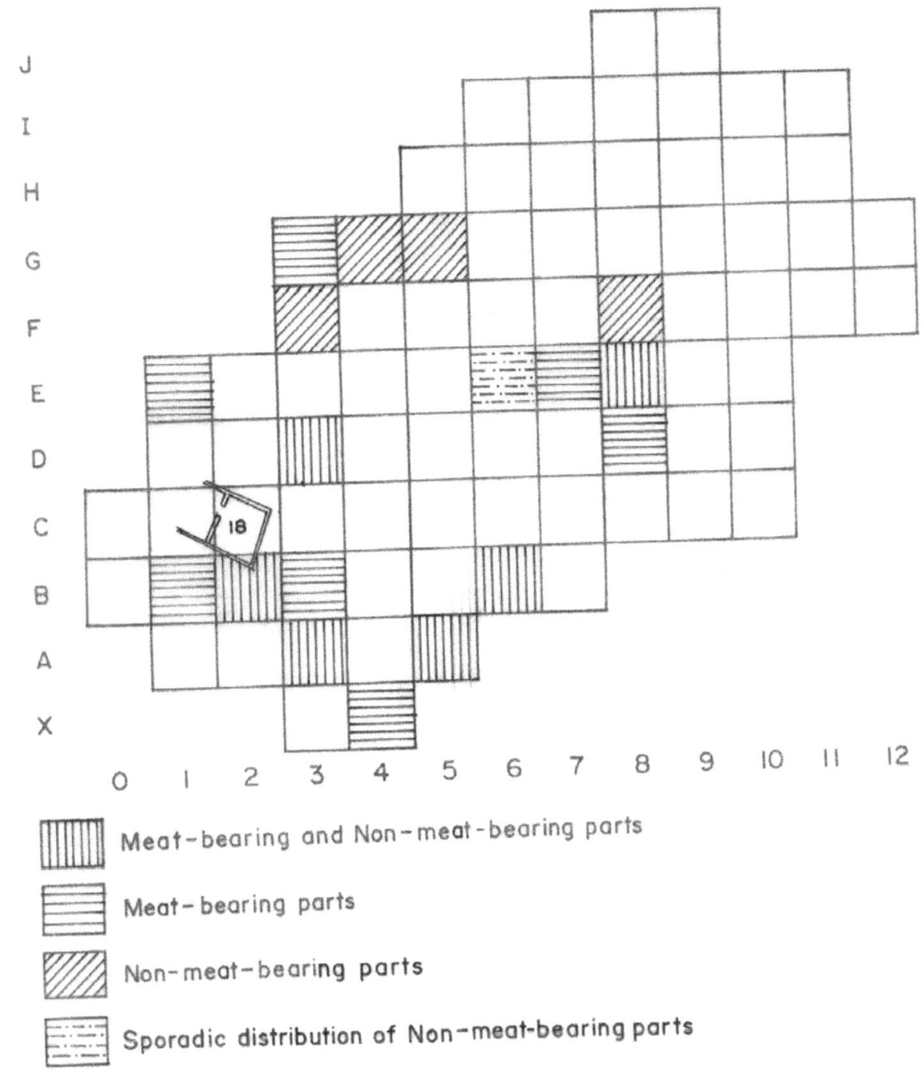

Layer 5

This layer is characterized by structure 15, 16 and 17 and located in the southwestern side of the excavated area. The non-meat-bearing parts are noticed in excavated area. The non-meat-bearing parts are noticed in trenches B0, E5 and F8 and in structures 15 and 16. Structure 17 is without any animal remains. The meat-bearing parts are noticed in trenches B and C, in the southern side of excavated area and this can be identified as a refuse dump (Fig. 5.43).

Fig. 5.43. Horizontal Distribution of Bones in Layer 5 at Inamgaon.

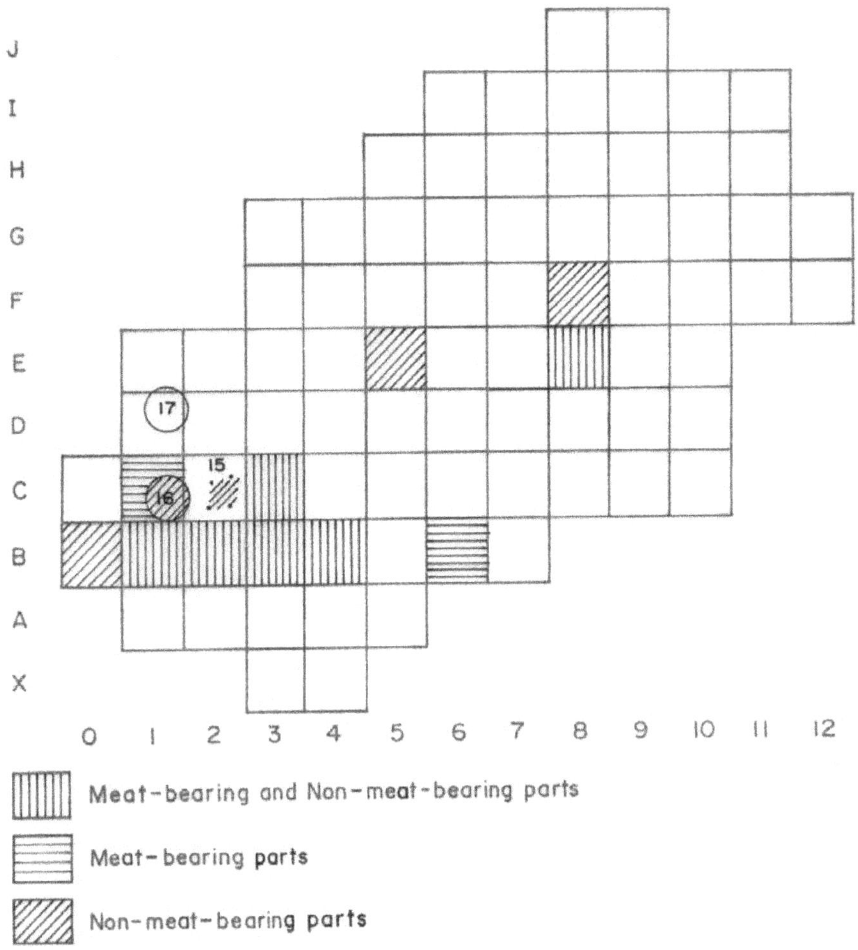

Layer 4

All the rectangular structures of the Malwa and Early Jorwe phases were replaced with smaller circular structures in the Late Jorwe phase. However, the orientation of the structures and locales are similar to the early Jorwe and Malwa phases. A majority of these circular structures are devoid of faunal remains. The trenches around the circular structures are also not yielding any animal remains. In addition to the previous excavated area, trenches C, D, E, F and G (23-28) have been extended eastwards. There is very scanty representation of bones in layer 4, except in the newly excavated area (23-28 trenches. A few bones of normal distribution are found in the southwestern region. Another cluster is noticed in trenches D10 and E10. In the northern periphery, in trenches H5 and H7 more number of non-meat-bearing parts are found. A concentration of non-meat-bearing parts is found associated with structures 115, 116, 117 and 120. A few of the circular structures have also yielded faunal material. Extension trenches on the eastern side C, D, E, F and G (23-28) have yielded a vast number of animal remains. Interestingly, a couple of rectangular structures (105 and 110) were identified from this area. However, the type of bones identified from the floor of these structures appears to be different. House number 110 is concentrated with non-meat-bearing parts whereas 105, with high meat-bearing parts. High concentration of non-meat-bearing parts is found in the northwest (E24, F24-25 and G24-26) corner of the extended trenches whereas meat-bearing parts are found in a few places in the southeast.

From the description given above, it is clear that the butchering activities may have been in at least two places in Trench H5 and H7 in one area and trenches E, F and G (24-26) in another area. It is important to note that the rectangular structures and the concentration of bones are found in the eastern side. As already stated a majority of the circular structures and the areas around them are devoid of animal remains, probably indicate a change in the functional aspect of these structures. Much of the activities concerning food habits may have been in the eastern side of the mound. This shift of the activity area probably starts in layer 7. In the northeastern part of the excavated area, a few rectangular structures are identified. These structures may have continued in layers 6 and 5, thereby animal remains are few in this area whereas, from layer 4 upwards concentration of faunal remains is seen in the eastern side (Fig. 5.44).

Fig. 5.44. Horizontal Distribution of Bones in Layer 4 at Inamgaon.

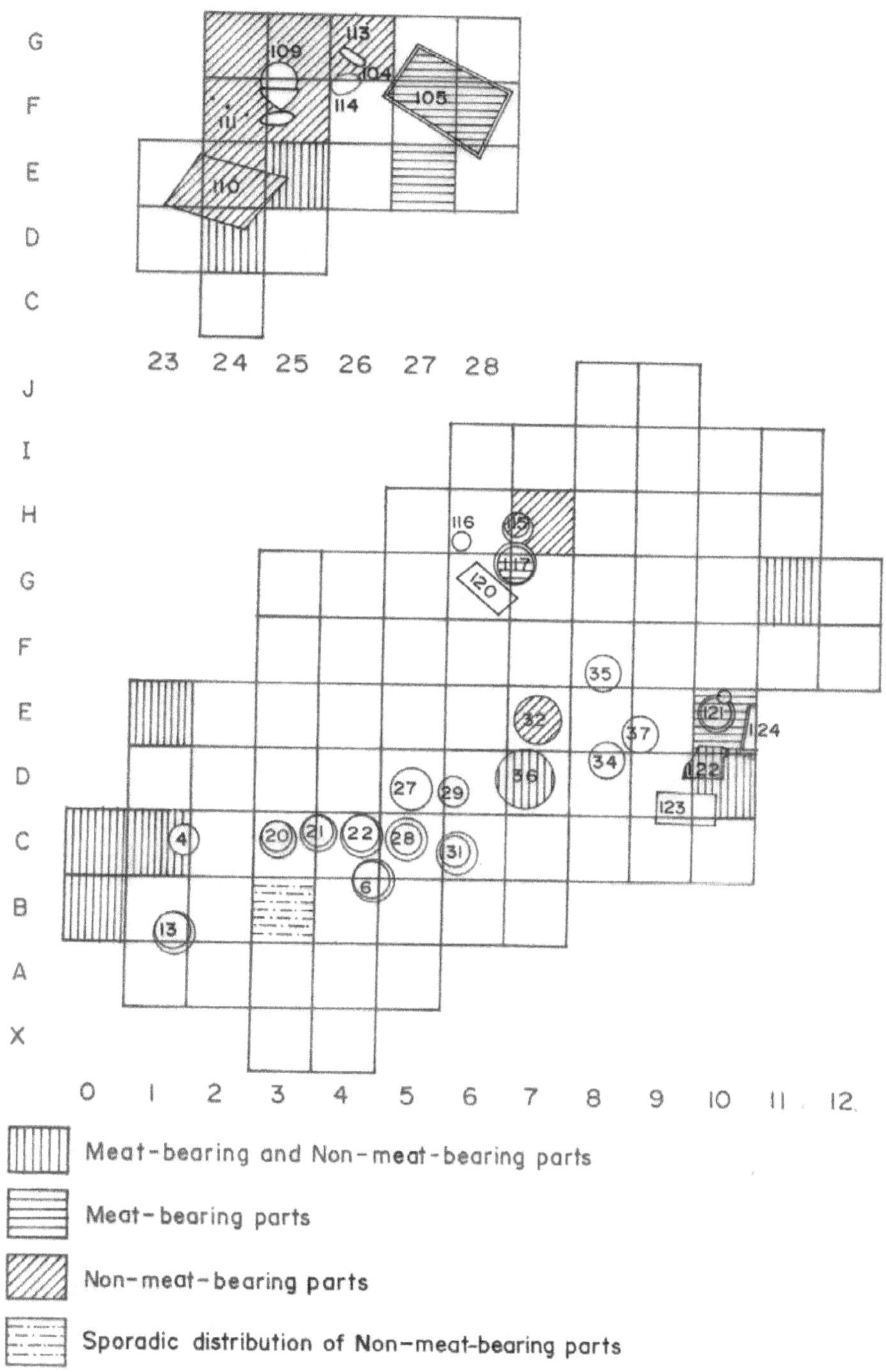

Layer 3

Compared to layer 4, only very few structures are identified which are mostly concentrated in the southwest region of the excavated area. A few structures have also been found in the northeastern region. Maximum amount of meat-bearing parts are mainly found in two areas i.e. G10, H10 and E, F, G (25027). However, sporadic distribution of non-meat-bearing parts is noticed in the G4, H7 and D8. These two areas are identified as butchering areas. Unlike in layer 4, much of the meat-bearing parts were found in the eastern side of the excavated area. This area can be identified as a refuse dumping area in layer 3. So far in all the other layers from 16-4, this area remains vacant or devoid of bones. In all the earlier layers the concentration of refuse dump was in the southern as well as western part of the excavated region. This shift of activity area probably has to do with an expansion of the settlement (Fig. 5.45).

Fig. 5.45. Horizontal Distribution of Bones in Layer 3 at Inamgaon.

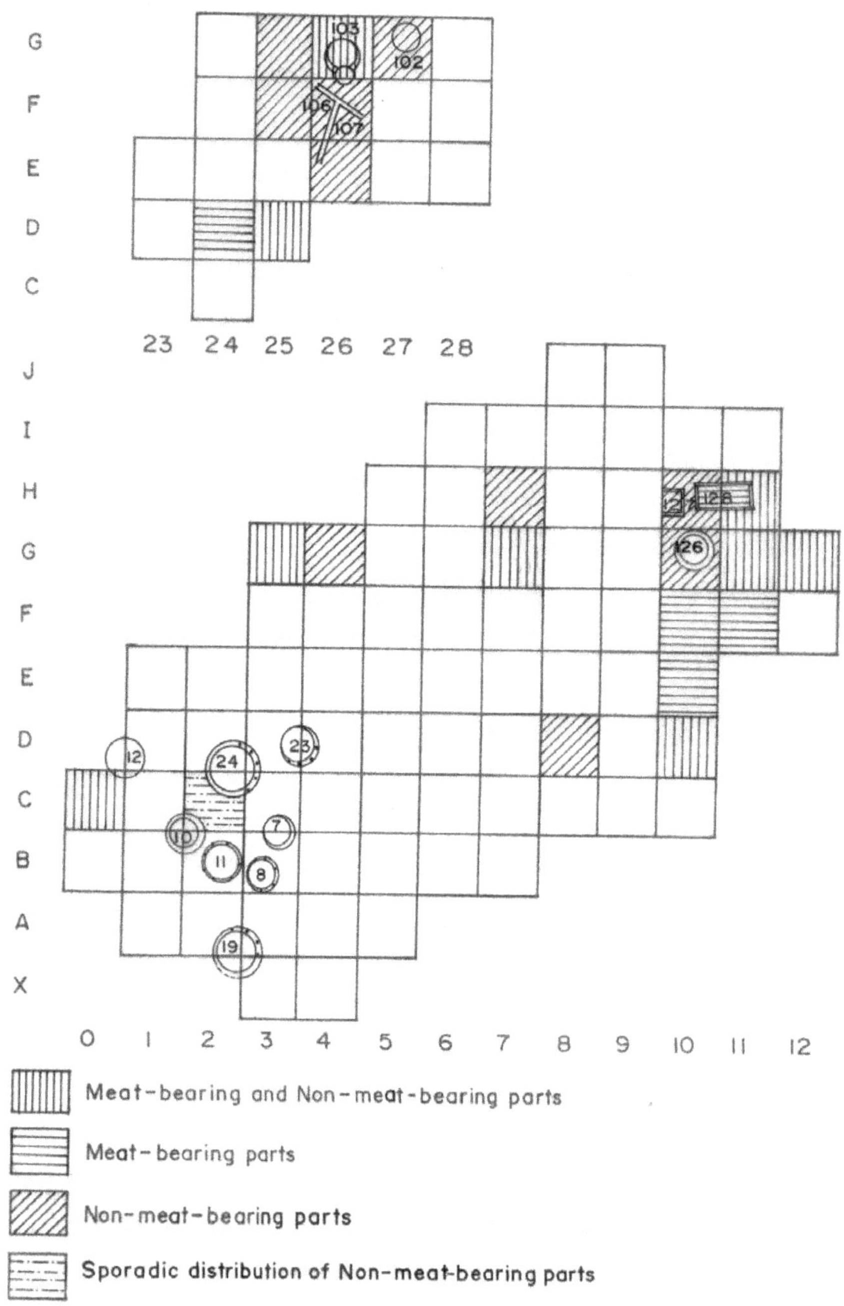

Layer 2

Structures 108, 125, 129 and 130 are identified in layer 2. The concentration of meat-bearing parts is found in all the corners of the excavated area, however, more in the southern side. It is significant that the non-meat-bearing parts are mainly concentrated in the northern part of the excavated area. Distinctly four clusters of non-meat-bearing are found (Clusterl – H5, H6, G6, G7 and I6; Cluster 2- F11, G11, H11 and I12; Cluster 3 – E24 and D25 and Cluster 4- G25 and F26). Structures 125, 129 and 130 are with meat-bearing parts whereas; structure 108 is with non-meat-bearing parts. Like in the earlier layers 3 and 4, the central part of the excavated area is devoid of animal remains indicating probably the existence of structures of the previous layer (Fig. 5.46).

Fig. 5.46. Horizontal Distribution of Bones in Layer 2 at Inamgaon.

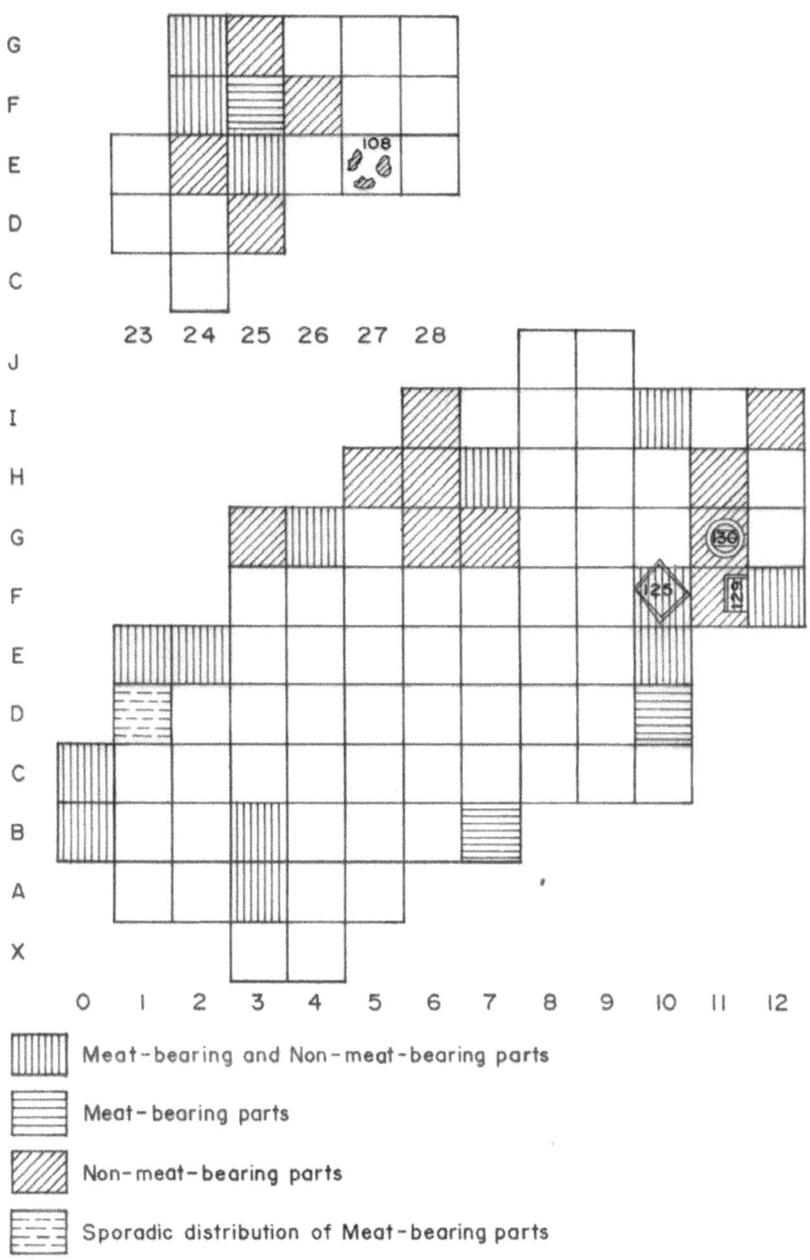

Layer 1

No structure is identified in this layer. However, the central blank spaces without bones continued to occur even in this layer suggesting the continuation of at least some of the earlier structural complexes. The meat-bearing as well as non-meat-bearing parts form distinct clusters in the eastern and northwestern part of the excavated area. Thus, they can be identified as refuse-dumping and butchering areas. The refuse dump clusters are Cluster 1- F11, G11, H11 and I11; Cluster 2- G4 and H5 and Cluster 3 is F26-27. The butchering areas are located in trenches E10, F10, G6, G7 and D, E, F, G (23-25) (Fig.5.47).

Fig. 5.47. Horizontal Distribution of Bones in Layer 1 at Inamgaon.

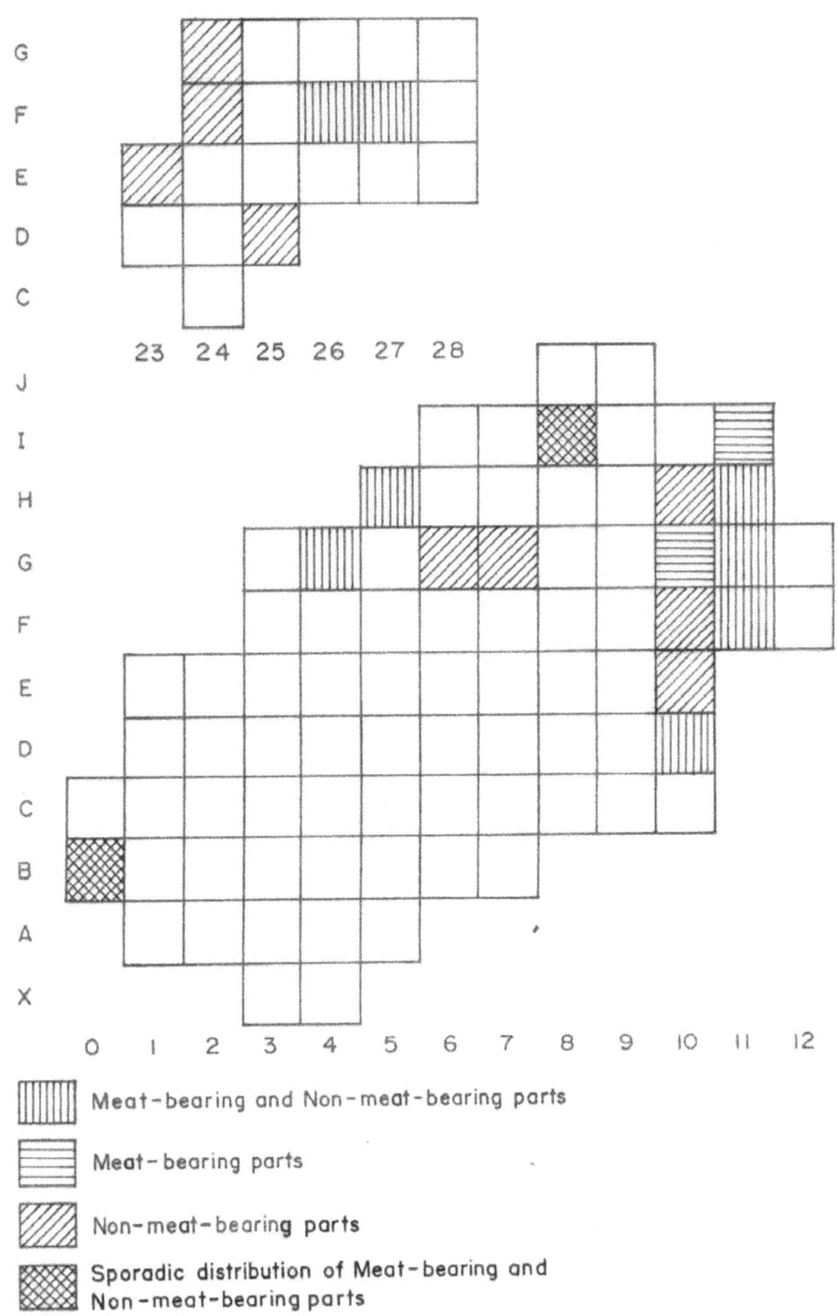

Summary

A majority of the structures are oriented in the north-east, south-west direction in different layers. The horizontal distribution of bones in different layers shows that much of the butchering activity was taking place in the central parts of the structures in the Malwa and Early Jorwe phases. The structures in layer 8 of the Early Jorwe and 12 of the Malwa phase are more or less identical, only some minor modifications and expansion could be seen in the Early Jorwe phase. The high concentration of meat-bearing parts was noticed in the south –west part of the excavated area and outside the structures. In the Late Jorwe phase all the rectangular structures of the Early Jorwe and Malwa phases were replaced with circular structures. However, the orientation of the structures was similar. A majority of these circular structures were devoid of faunal remains, probably indicating a change in the functional aspect of these huts. Interestingly, a couple of rectangular structures were also identified in the extended area towards the eastern side. The concentration of bones was more in and around these structures. Much of the activities concerning food may have been in the eastern side of the mound in the Late Jorwe phase. In the earlier two phases the concentration of refuse dump was in the southern as well as in the western sides.

VI. Carcass utilization

a. Relative representation of skeletal parts

1. *Bos indicus*

The axial skeleton of Bos indicus was well represented in the Late Jorwe phase compared to the earlier Early Jorwe and Malwa phases. The representation of proximal and distal ends of humerus and tibia is more or less identical in all the three cultural phases. However, compared to the proximal, the distal ends of humerus are more that would indicate the differential preservation of these parts. The distal ends of humerus are hard and sturdy while the proximal ends are more vulnerable to human as well as natural destruction. A similar sort of portrayal has been noticed for the proximal ends of femur and radius. The pattern is more or less consistent in all the three cultural phases. Compared to scapula the pelvic bones are numerous. This could be attributed to the structural and size differences of these bones. The carpals and tarsals are always found in abundance and they are better preserved since they are sturdy, small in size and have no meat value. Because of the non-meat value there is less human activity on the bones. To an extent this may be true for the phalanges also. Compared to the first and the second phalanges scanty occurrence of third phalanges is noticed. Probably this is an indication for the use of hide at the site. When the hide is removed from the carcass of an animal, generally the third phalanx and occasionally the second phalanx are attached to the hide. The representation of metapodials shows that comparatively proximal ends are more than the distal. One of the reasons could be that while butchering the animals proximal ends of metapodials were also cut along with other tarsals and carpals and get preserved along with other non-meat-bearing bones. It is also noticed that the shaft of the distal metapodials are used as tools. This portion of the metapodia is preferred because of the distal trochleas which could be used as a firm handles (Fig. 5.48-5.50).

Fig. 5.48. Relative Representation of Skeletal Parts of *Bos indicus* in the Malwa Phase at Inamgaon.

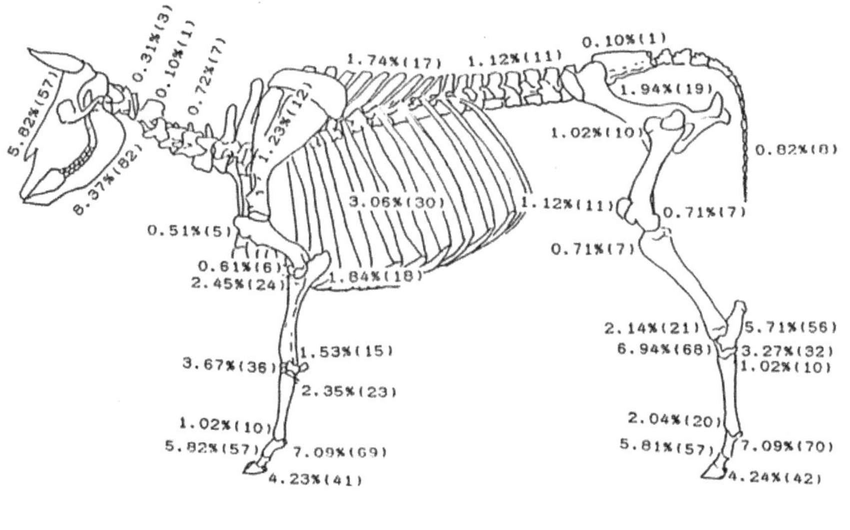

Fig. 5.49. Relative Representation of Skeletal Parts of *Bos indicus* in the Early Jorwe Phase at Inamgaon.

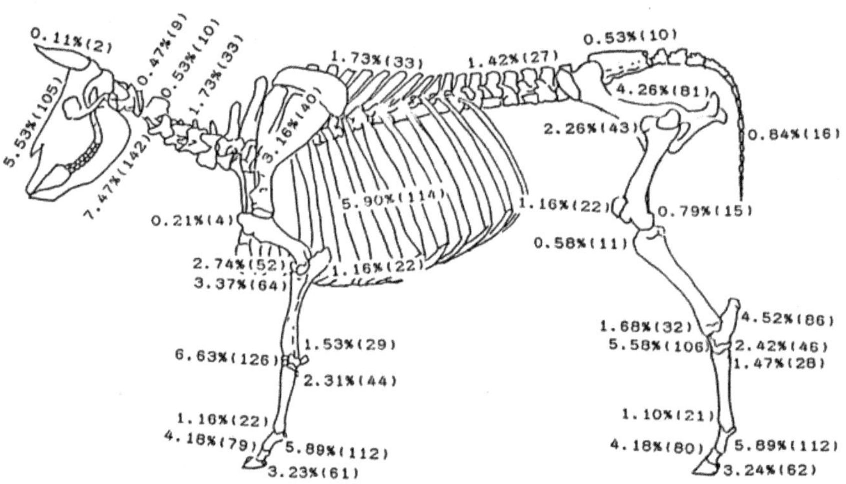

Fig. 5.50. Relative Representation of Skeletal Parts of *Bos indicus* in the Late Jorwe Phase at Inamgaon.

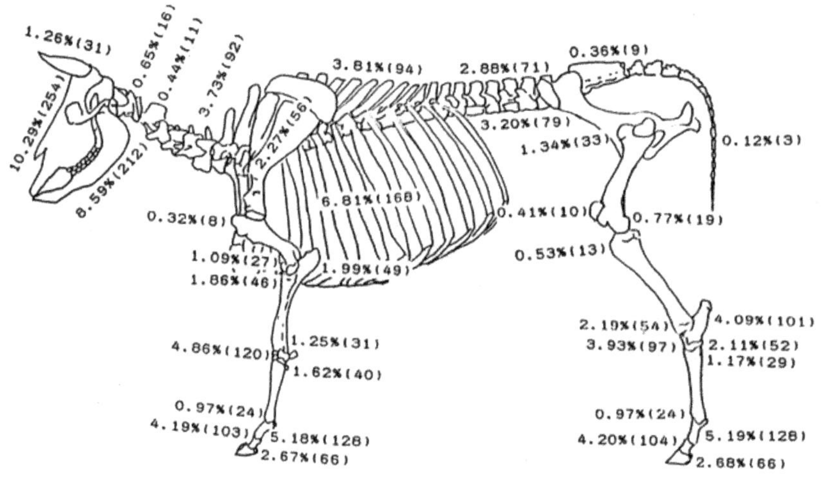

2. *Bubalus bubalis*

Buffalo bones are more in the Late Jorwe phase compared to the Malwa and Early Jorwe phases. The axial skeletal parts are scantly represented in all the three cultural phases. To distinguish between the bones of *Bos indicus* and *Bubalus bubalis* is difficult from this part of the skeleton except for few bones from cranium, mandible, atlas and axis vertebrae. Thus, a large number of fragments which could not identify to the species level were classified together in another group called *Bos/Bubalus*. Compared to *Bos indicus* the representation of bones is very scanty. The preservation of carpal and tarsal bones is more or less identical with those of *Bos indicus*, and is found in sizable numbers. However, the portrayal of phalanges of *Bubalus bubalis* compared to *Bos indicus* is different. Generally the first phalanges are more in *Bubalus bubalis*. It is intriguing that the second phalanges are more than the first phalanges. From the faunal assemblage no secondary use of first phalanges could be identified except for a couple of hunting whistles (Fig. 5.51-5.53).

Fig. 5.51. Relative Representation of Skeletal Parts of *Bubalus bubalis* in the Malwa Phase at Inamgaon.

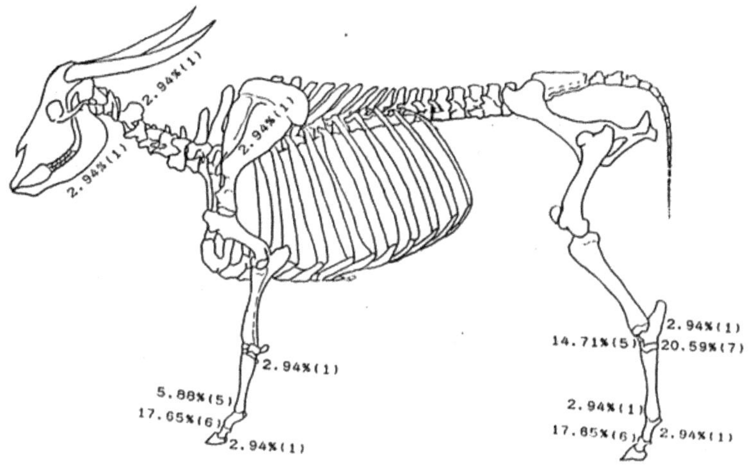

Fig. 5.52. Relative Representation of Skeletal Parts of *Bubalus bubalis* in the Early Jorwe Phase at Inamgaon.

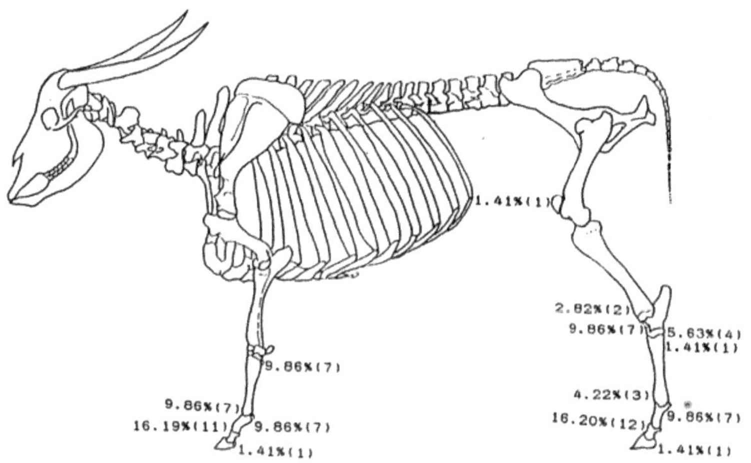

Fig. 5.53. Relative Representation of Skeletal Parts of *Bubalus bubalis* in the Late Jorwe Phase at Inamgaon.

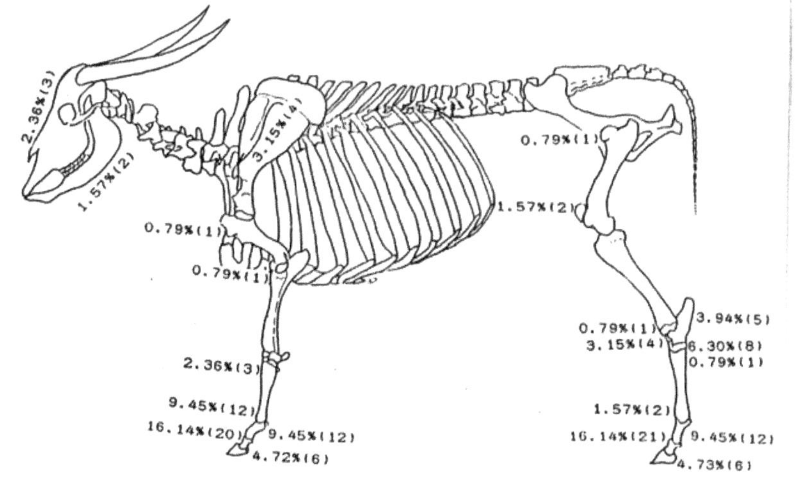

3. *Capra/Ovis*

Representation of bones of goat/sheep is found in the Early Jorwe and the Late Jorwe phases. In the Malwa phase a few bone fragments belonging to the cranial and mandibular regions were found. Fairly a good number of the meat-bearing parts are found in the Early and Late Jorwe phases. However, the percentage frequency of the carpal and tarsal bones is comparatively less in the Early Jorwe phase. The proximal parts of the humerus are not represented in the Early Jorwe and Malwa phases. Absence of proximal part can be attributed to the differential preservation of this part. Distal ends of humerus are equally distributed in all the three phases. Likewise, the proximal ends of radius and distal ends of tibia are found more compared to their distal and proximal counterparts. Proximal and distal ends of femur are more or less equally represented in the Malwa phase whereas; it varies in other two phases. The proximal and the distal ends of metapodials also show erratic representation. Among the phalanges, first phalanges are more compared to the second and third. The scarcity of the third phalanges is noticed for this species also, probably again an indication of the hide being used for some purpose (Fig. 5.54-5.56).

Fig. 5.54. Relative Representation of Skeletal Parts of *Capra/Ovis* in the Malwa Phase at Inamgaon.

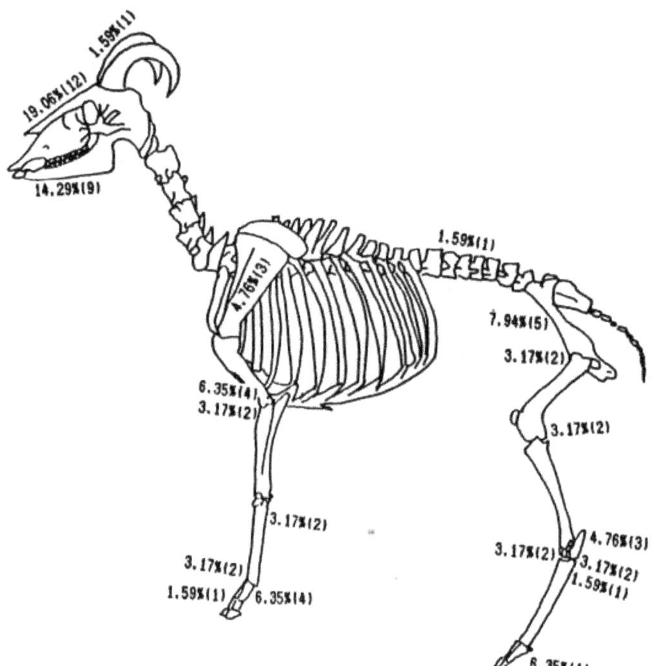

Fig. 5.55. Relative Representation of Skeletal Parts of *Capra/Ovis* in the Early Jorwe Phase at Inamgaon.

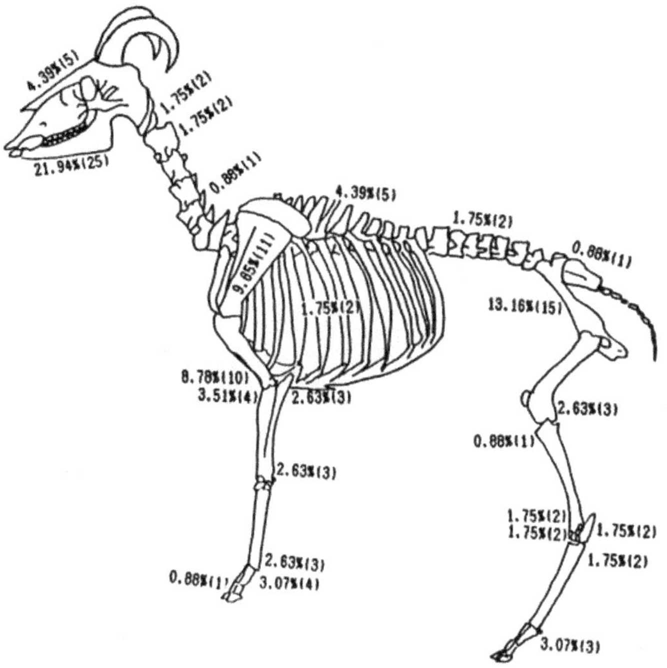

Fig. 5.56. Relative Representation of Skeletal Parts of *Capra/Ovis* in the Late Jorwe Phase at Inamgaon.

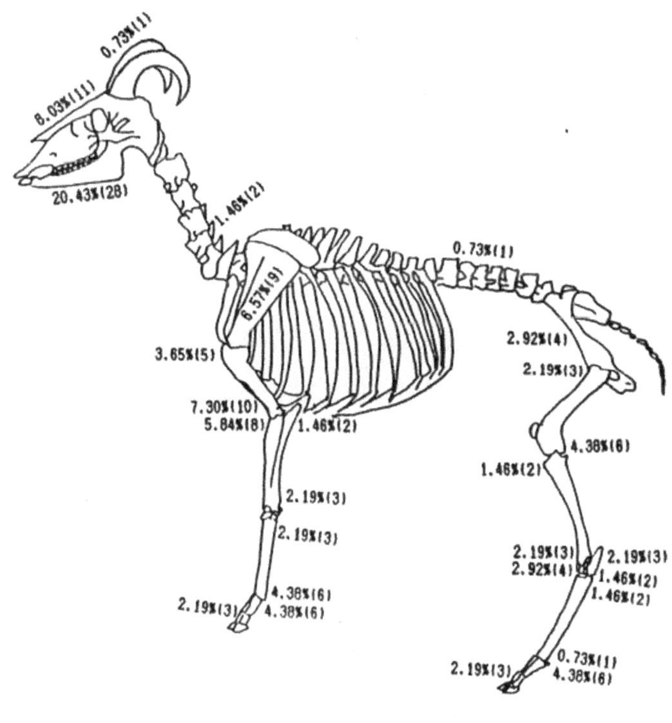

4. *Canis familiaris*

Evidence suggests the consumption of dog meat in Late Jorwe phase as a number of bones are found with cut marks and charring. A few bones found in the Early Jorwe and the Malwa phases, probably indicate predator or scavenging activities at the site. However, the representation of skeletal parts of dog in the Late Jorwe phase is more or less identical with other species of animals which were regularly butchered for food purposes. From the axial skeleton a large number of cranial and mandibular fragments have been found. The number of distal ends of humerus and tibia are comparatively higher than the proximal ends; probably again suggests the differential preservation of these parts. Similarly, the proximal ends of femur, radius and ulna are found more in number than the distal ends. This also can be attributed to differential preservation. The carpal bones are completely absent while, a good collection of tarsal bones is seen in the Late Jorwe phase (Fig. 5.57-5.59).

Fig. 5.57. Relative Representation of Skeletal Parts of *Canis familiaris* in the Malwa Phase at Inamgaon.

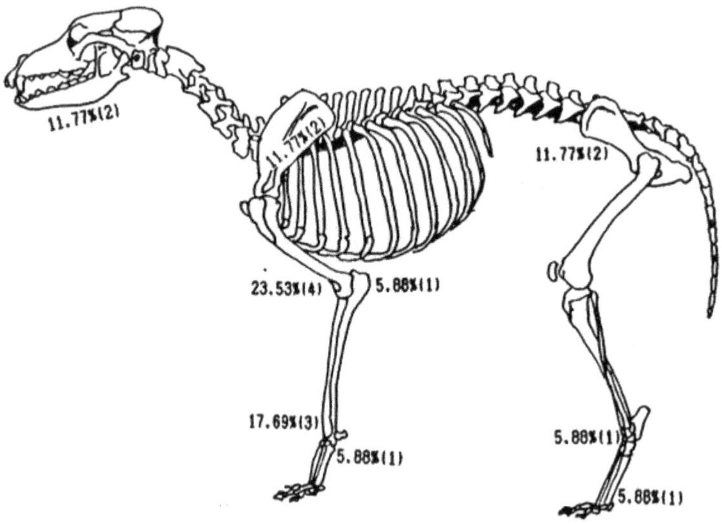

Fig. 5.58. Relative Representation of Skeletal Parts of *Canis familiaris* in the Early Jorwe Phase at Inamgaon.

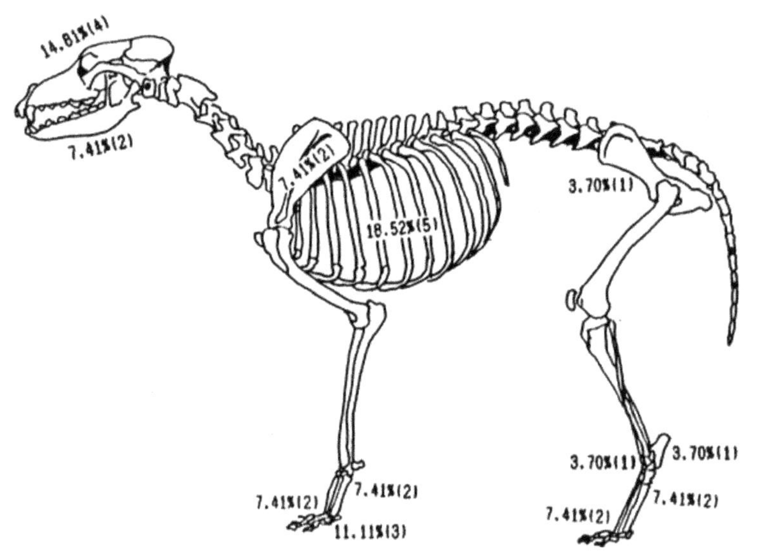

Fig. 5.59. Relative Representation of Skeletal Parts of *Canis familiaris* in the Late Jorwe Phase at Inamgaon.

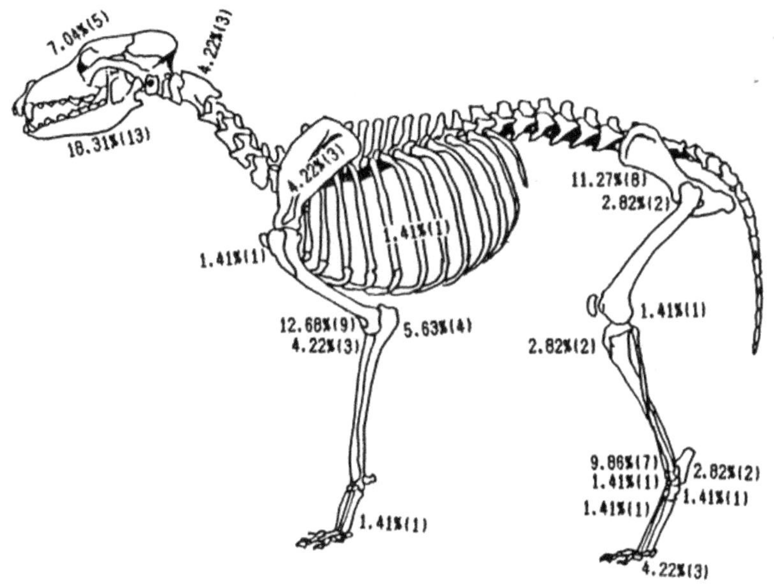

5. *Sus domesticus*

Bones of domestic pigs are scantly represented in all the cultural phases. However, they are slightly more in the Late Jorwe. Axial skeleton is completely absent in the Early Jorwe phase whereas, a few bones of this part are represented in the Malwa as well as in the Late Jorwe phases. The hind limb portions with meat-bearing parts (pelvis, femur and tibia) are absent in the Malwa and the Early Jorwe phases but a couple of bones from this part of the body are found in the Late Jorwe phase. From the distribution of bones it appears that there was no regular butchering of this animal for food purpose. However, a few bones represented may have reached the site probably from outside as a sort of delicacy. This animal is also found in the wild state in abundance in this region even today. The non-exploitation of this species shows probably a taboo on pork (Fig. 5.60-5.62).

Fig. 5.60. Relative Representation of Skeletal Parts of *Sus domesticus* in the Malwa Phase at Inamgaon.

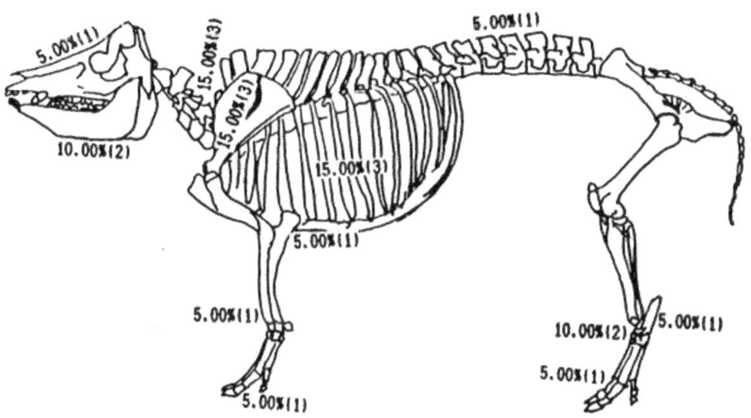

Fig. 5.61. Relative Representation of Skeletal Parts of *Sus domesticus* in the Early Jorwe Phase at Inamgaon.

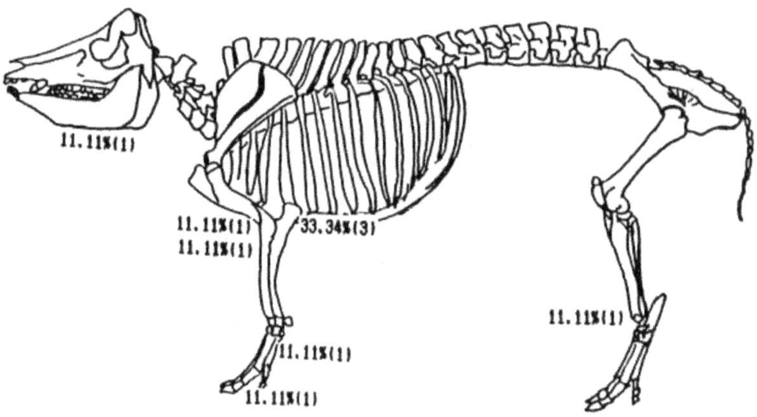

Fig. 5.62. Relative Representation of Skeletal Parts of *Sus domesticus* in the Late Jorwe Phase at Inamgaon.

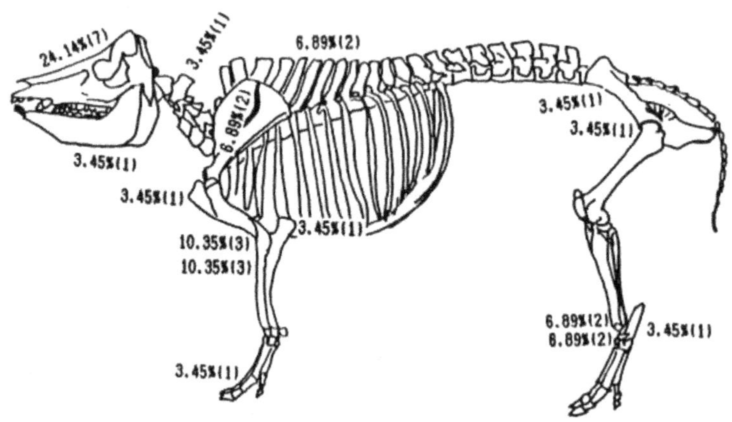

6. *Antilope cervicapra*

An appreciable number of bones of Blackbucks have been found in the Late Jorwe phase compared to the other two phases. The portrayal of axial part of the skeleton is more or less identical in all the three cultural phases at Inamgaon. However, individual bones such as cranial fragments, rib fragments, etc. may differ slightly in different phases. Scapula, pelvis and humerus are some what equally represented in all the three phases, though a slight increase in percentage is noticed in the Late Jorwe phase. In the stratigraphical units, the distal ends of humerus are found at least eight times more than the proximal ends, probably indicates the hardness of the distal ends. Femur fragments are erratically found in all the three phases. Distal ends of radius are meagerly represented suggests the differential preservation of both proximal and distal ends. The proximal and distal ends of tibia are also behaving the same way where the distal ends are always found more in number. The frequency of ulna fragments is more in the Early Jorwe and the Malwa phases whereas, in the Late Jorwe phase they are less represented. Probably the proximal ends of ulna may have been used as bone tools in the Late Jorwe phase. Compared to tarsal bones, carpals are found less in number. The smaller size of the carpals may have the reason for the meager quantity at the site. Both proximal and distal ends of metapodials are almost equally represented at Inamgaon. Some individual variations are noticed in the Malwa as well as in Early Jorwe phases. Ratio of phalanges first, second and third are more or less similar in all the cultural phases. The first phalanges are always found in abundance followed by the second and then third. The scanty representation of second and third phalanges compared to the first may have been due to the use of hide of this animal (Fig. 5.63-5.65).

Fig. 5.63. Relative Representation of Skeletal Parts of *Antilope cervicapra* in the Malwa Phase at Inamgaon.

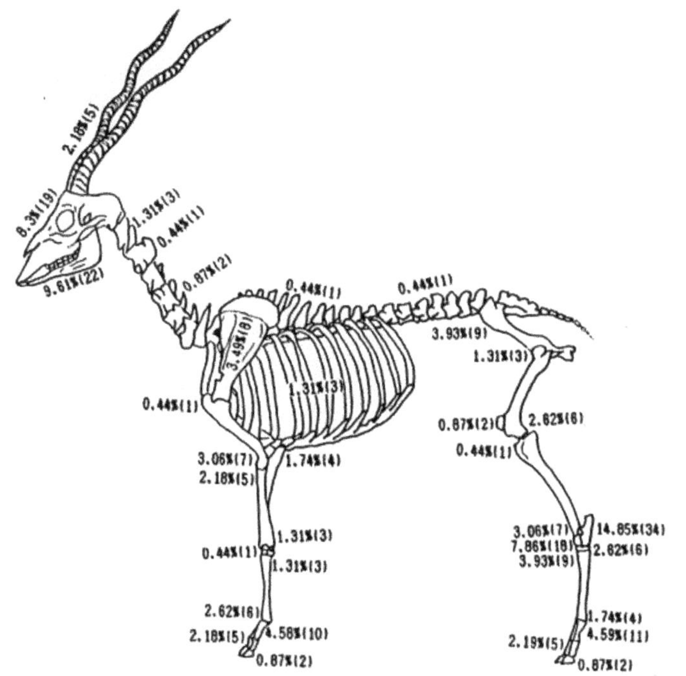

Fig. 5.64. Relative Representation of Skeletal Parts of *Antilope cervicapra* in the Early Jorwe Phase at Inamgaon.

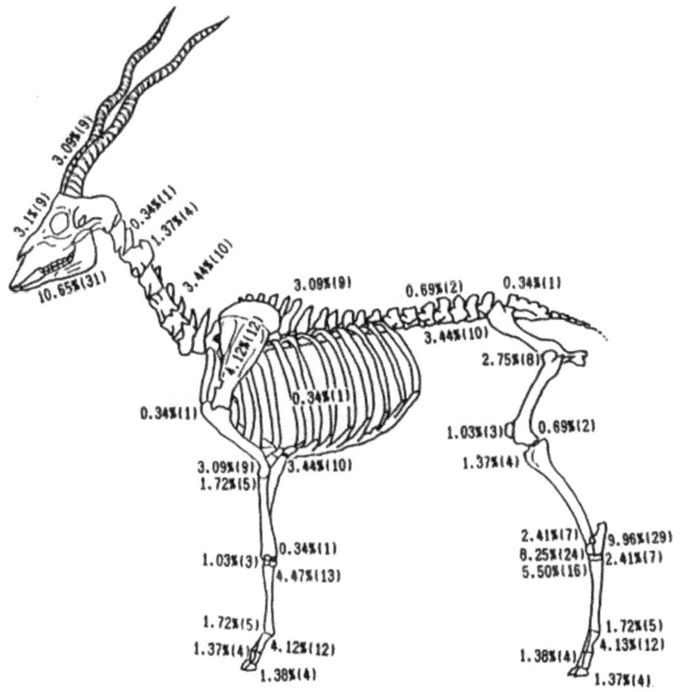

Fig. 5.65. Relative Representation of Skeletal Parts of *Antilope cervicapra* in the Late Jorwe Phase at Inamgaon.

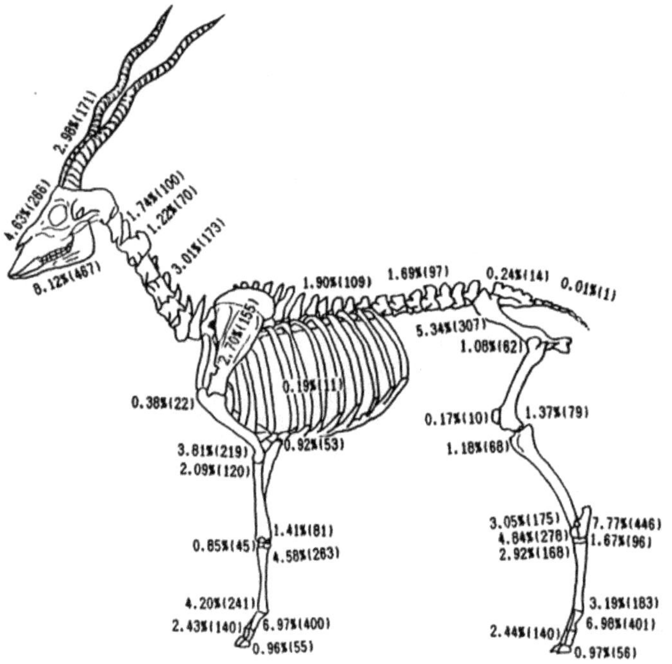

7. *Boselaphus tragocamelus*

Maximum number of bones of this animal is found in the Malwa and the Late Jorwe phases. Bones are mostly represented by lower extremities of hind and fore limbs. The axial skeleton is very poorly represented. Among the bones identified tarsals and phalanges are found more in number. Surprisingly, the meat-bearing parts of this animal are scanty in all the three cultural phases. Considering the amount of lower extremity bones represented in the collection, a few more bones of meat-bearing parts are expected. The absence of meat-bearing parts to this extend suggest probably the transport of meat of this animal to some other place (Fig. 5.66-5.68).

Fig. 5.66. Relative Representation of Skeletal Parts of *Boselaphus tragocamelus* in the Malwa Phase at Inamgaon.

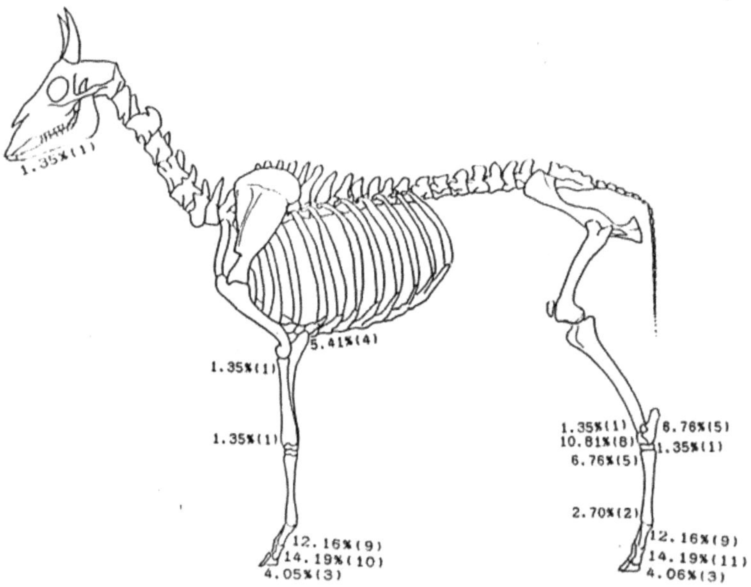

Fig. 5.67. Relative Representation of Skeletal Parts of *Boselaphus tragocamelus* in the Early Jorwe Phase at Inamgaon.

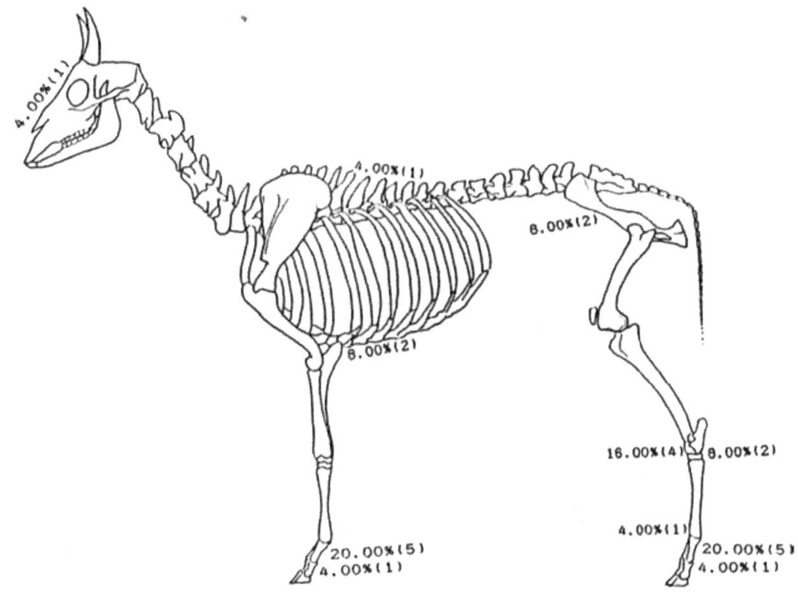

Fig. 5.68. Relative Representation of Skeletal Parts of *Boselaphus tragocamelus* in the Late Jorwe Phase at Inamgaon.

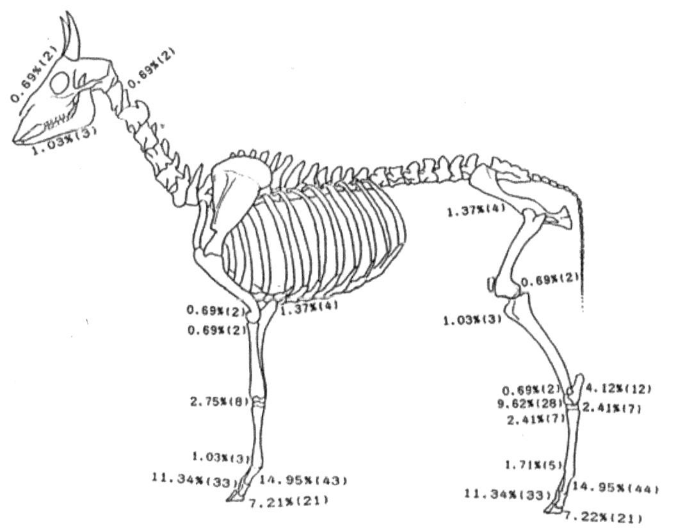

b. Phalangial ratio

The ratio of first, second and third phalanges of animals can throw light on the utilization of hide at a site. While removing the hide of an animal, generally the second and the third phalanx go away with the hide. Hide is preserved in archaeological context only in very favourable climatic conditions. Because of its perishable nature, it is not found in any of the Indian archaeological sites. In the normal distribution of bones the phalangial ratio is expected to be one.

Table 5.9. Ratio of Phalanges of *Bos indicus* at Inamgaon

Layers	H1	H2	H3	H2/H1	H3/H2
01	26	14	10	0.53	0.71
02	75	56	36	0.74	0.64
03	70	77	44	1.1	0.57
04	73	41	35	0.56	0.85
05	12	19	07	1.58	0.36
06	26	18	17	0.69	0.94
07	35	22	21	0.62	0.95
08	101	73	45	0.72	0.60
09	17	19	20	1.11	1.05
10	14	09	04	0.64	0.44
11	31	18	16	0.58	0.88
12	60	42	30	0.7	0.71
13	19	17	15	0.89	0.88
14	34	34	21	1.00	0.61
15	19	16	13	0.84	0.81
16	07	05	04	0.71	0.80

Table 5.10. Ratio of Phalanges of *Bubalus bubalis* at Inamgaon

Layers	H1	H2	H3	H2/H1	H3/H2
01	06	03	02	0.5	0.66
02	05	10	06	2.00	0.60
03	10	19	03	1.9	0.15
04	03	08	01	2.66	0.12
05	00	01	00	0.00	0.00
06	01	03	01	3.00	0.33
07	04	03	00	0.75	0.00
08	13	22	02	1.69	0.09
09	01	06	01	6.00	0.16
10	02	03	00	1.50	0.00
11	02	05	00	2.50	0.00
12	00	03	01	0.00	0.33
13	00	01	00	0.00	0.00
14	00	04	00	0.00	0.00
15	00	04	00	0.00	0.00
16	01	00	00	0.00	0.00

Table 5.11. Ratio of Phalanges of *Boselaphus tragocamelus* at Inamgaon

Layers	H1	H2	H3	H2/H1	H3/H2
01	10	08	04	0.8	0.5
02	25	13	12	0.52	0.92
03	21	31	13	1.47	0.41
04	27	12	14	0.44	1.16
05	04	02	00	0.5	0.00
06	00	00	02	0.00	0.00
07	02	00	00	0.00	0.00
08	06	01	00	0.16	0.00
09	01	00	00	0.00	0.00
10	01	00	00	0.00	0.00
11	01	01	01	1.00	1.00
12	01	01	03	1.00	3.00
13	01	02	00	2.00	0.00
14	05	09	00	1.8	0.00
15	10	06	05	0.65	0.83
16	01	01	01	1.00	1.00

Table 5.12. Ratio of Phalanges of *Antilope cervicapra* at Inamgaon

Layers	H1	H2	H3	H2/H1	H3/H2
01	113	25	11	0.22	0.44
02	316	93	36	0.29	0.38
03	235	93	33	0.39	0.35
04	129	64	31	0.49	0.48
05	08	05	00	0.62	0.00
06	07	02	04	0.28	2.00
07	03	04	01	0.75	0.25
08	03	00	00	0.72	0.00
09	03	00	00	0.00	0.00
10	00	01	00	0.00	0.00
11	08	01	03	0.12	3.00
12	09	05	02	0.55	0.40
13	06	02	00	0.33	0.00
14	02	02	00	1.00	0.00
15	03	01	00	0.33	0.00
16	01	00	00	0.00	0.00

The ratios were calculated for species like cattle (Table 5.9), buffalo (Table 5.10), nilgai (Table 5.11) and blackbuck (Table 5.12) where enough data was available for comparison. The ratio calculated for these different animals shows considerable loss of third and second phalanges compared to the first one. Therefore, it can be assumed that the hide of a majority of animals were utilized for different purposes like clothing, shelter, containers, etc.

c. Bone modification

1. Human activity

Faunal remains recovered from most of the archaeological sites represent the garbage of ancient people's meal. Some of these bones are also found in modified form by human as well as non-human agencies. During the course of excavation at Inamgaon, a number of bone tools were found. Most of them were made on long bones like humerus, radius, ulna, femur, tibia, metapodials and the flat bones like scapula and pelvis; and also ribs, horn core and antlers of animals like cattle, buffalo, antelopes and deer. An exhaustive account of these tools is dealt with in the excavation report (Dhavalikar and Ansari 1988 c; Thomas 1988).

Bone objects recovered include points which are found in large number in the Late Jorwe phase, are usually prepared from all types of bones and antlers. Scrapers are generally made on scapula, mandible, rib and long bones of cattle and blackbucks. However, punches are made on sturdy antlers and thick cattle bones. A majority of chisels are found in the Late Jorwe phase and are prepared on the thin bone splinters, and thick bone or antlers which were suitable for preparing hides. Tow picks, one each from Early and the Late Jorwe phase are unique and made on bifurcated antler where the short shaft of antler (brow tine) is grouped into a sharp point and the longer one serves the purpose of handle (Dhavalikar and Ansari 1988). The methodology used in making bone tools includes cutting, chipping, grinding and fire hardening. Probably a majority of these tools were used for skinning the animal and in the preparation of hides (Thomas 1988).

In addition to the already reported bone tools and objects, a few more have been encountered while analyzing the faunal assemblage of Inamgaon. A punch made of antler is ground and polished on all sides and at the tip (Fig.5.69 a). This is also charred, probably for hardening. Rib fragments with serrated edge generally with a couple of points are found in the collection (Fig. 5.70). Probably they were used for pottery designs. The distance between the points of the serrated edge and the width of the lines on some of the pottery pieces matches well. Rows of incised vertical strokes have been found on the pottery, e.g. No. D137 and T132. In another tool the serrated points appear to be blunt. Rib bone has also been used for making chisels.

Fig. 5.69. Bone Tools.

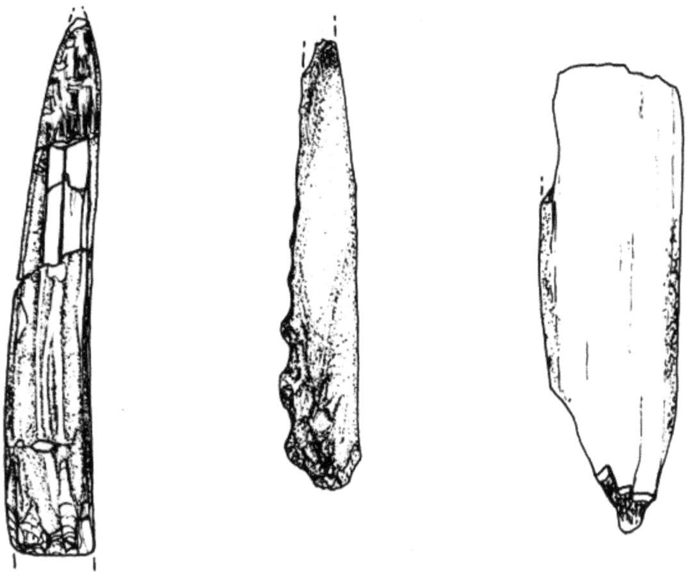

Fig. 5.70. Bone tools used for pottery designing at Inamgaon.

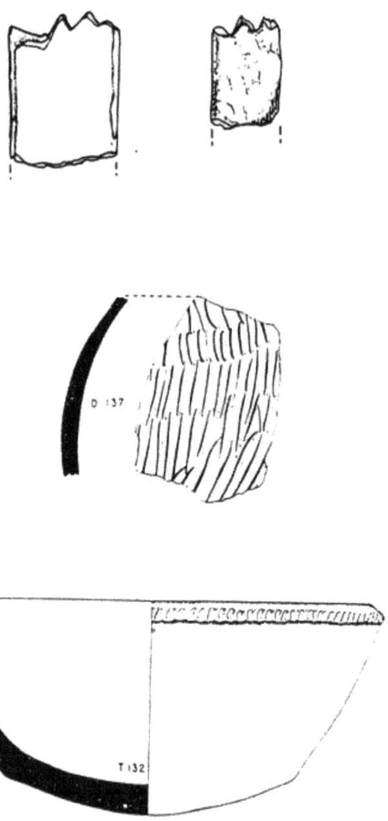

A few first phalanges of cattle were recovered with square shaped hole on the dorsal side, near the distal end (Pl. XX A). These holes appear to be intentionally and regularly made. This part of the skeleton does not have much marrow to assume that they were made for the extraction of marrow. The hole is only on one side of the bone, so they could not be used as pendants or beads. Such phalanges are reported to be hunting whistles. Similar evidence has also been found at the Chalcolithic sites of Prabhas Patan (Gujarat), Ramapurm (Andhra Pradesh), Neolithic/Megalithic site of Veerapuram (Andhra Pradesh) and the Megalithic site of Naikund (Maharashtra) (Thomas 1979, 1981, 1984 a and 1984 B, 1988).

Plate XX A. Hunting Whistles made on First Phalanges of *Bos indicus*.

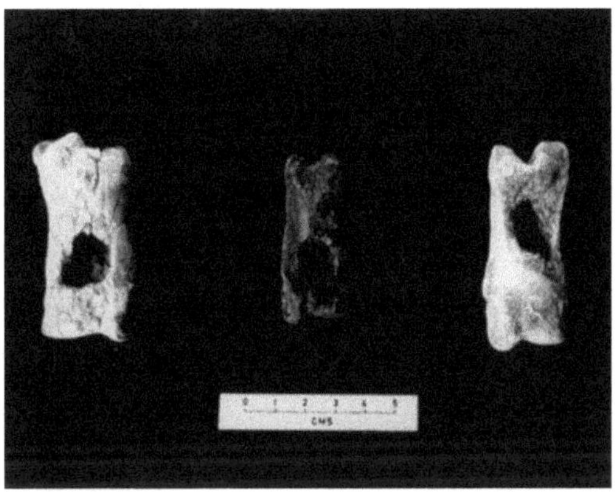

However, a few first phalanges and tarsal bones of smaller ruminants (blackbuck) have been found with perforations (Pl. XX Band XXI A). These perforations are comparatively smaller than those of cattle phalanges; rather they appear to be more like small drilling marks. Generally the holes are found on both side of the phalanges but occasionally only on one side. These phalanges and calcaneum were probably used as pendants or beads.

Plate XX B. Drilling marks on calcaneum, tibia and metapodia of *Antilope cervicapra*.

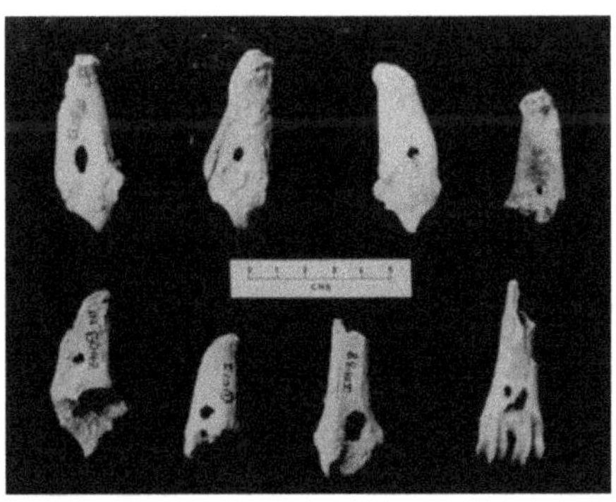

Plate XXI A. Driling marks on phalanges of *Antilope cervicapra*.

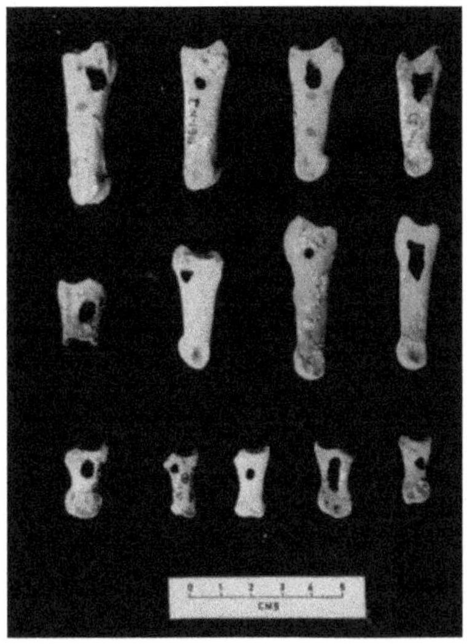

2.Carnivore and rodent activity

A few fragments of astragali, phalanges, centrotarsals and the distal ends of metapodial are found deformed in the collection (Pl. XXI B). The external cover or the cortex of the bones are comparatively damaged with etched irregular surface. The bones have become thinner and shapeless. In the butchering activity these non-meat-bearing parts are cut and thrown out. Such bones are meal to the scavengers like dog and they swallow these bones. The changes noticed on the bones may have been due to the reaction of gastric juice of carnivores. Payne and Munson (1986) have observed the reaction of stomach juices of carnivores like do on different bones. Dogs are able to swallow the bone fragments smaller than 2.5 cm in diameter. The swallowed bone fragments are heavily attacked by the stomach juices. The smaller fragments result in their complete disappearance. The Surface of the digested bones is etched, often shiny, wafer-thin and sharp. In the archaeological context such evidence is reported of phalanges and tarsals of gazelle from the Natufian culture period (10,300-8,500 B. C.)(Davis 1985 and 1987).

Plate XXI B. Effect of gastric juice of carnivores on bones.

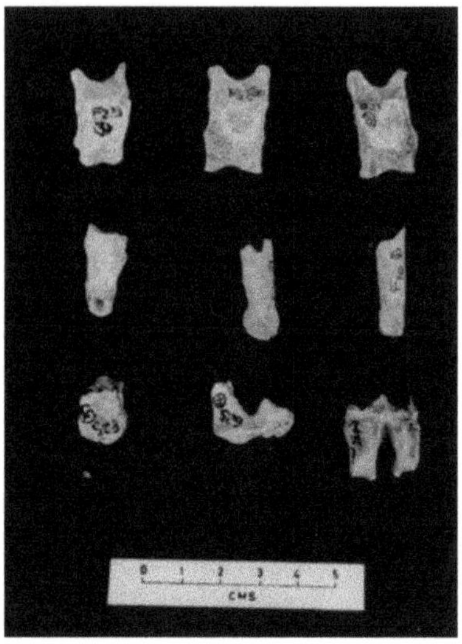

The other carnivore activity such as puncture marks is also found on bones like radius, ulna and pelvis of blackbucks. The observed tooth marks are apparently the result of the canine and the incisor teeth of animals like dog. However, at the site only very few bones are found with such modifications (Fig. 5.71).

Fig.5.71. Activities of carnivores on bones.

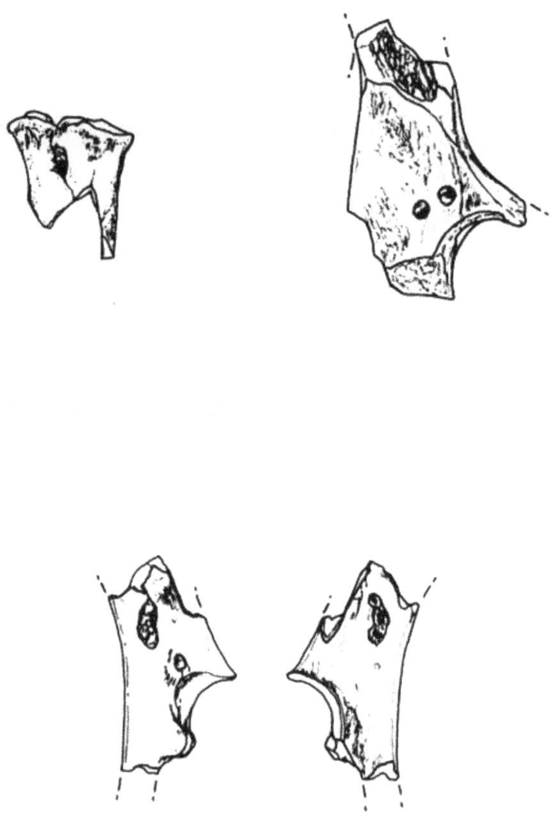

Some bones are also found with gnawing marks. The observations on bones with gnawing marks suggest involvement of more than one animal species for these incisions. The smaller incisions could be attributed to animals like rodents while the large ones to the carnivores (Fig. 5.72 a). Mainly the gnawing marks are observed on the proximal and distal epiphyses of bones.

3. Insect and Bacterial activity

The mandibular condyle of the blackbuck has anterior-posterior groove formation on the articular surface (Fig. 5.72b). Being the condyle of a young animal this is very spongy and the formation appears to be post-depositional. Probably bacterial or insect activity may have been the reason for these grooves (Baker and Brothwell 1980). Similarly one of the first phalanges of cattle has very irregularly made small holes on both epiphyses. These appear to be more like insect holes (Fig. 5.72 c).

Fig. 5.72. Activities of a. Rodent, b.Bacteria and c. Insect on bones at Inamgaon.

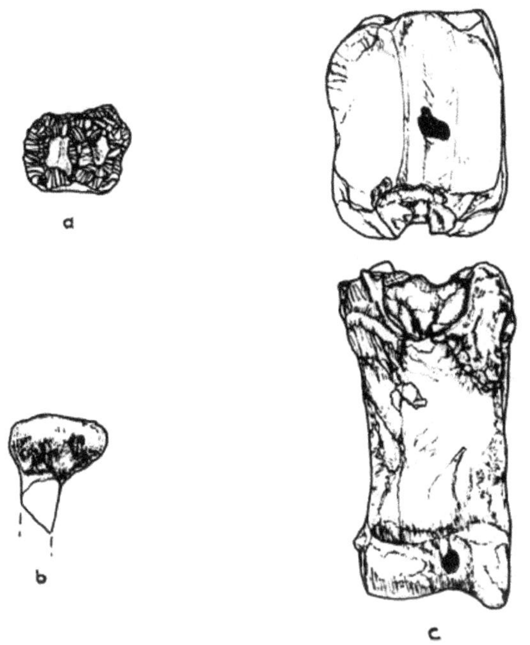

4. Pathological and non-pathological lesions

The spine of the thoracic vertebra has two natural holes very near to the posterior zygapophyses (Fig. 5.73 a) and seems to be only a deformity which may not be pathological in character. This has also gnawing marks on the posterior zygapophysis by rodents.

An example for trauma has also been found on the mid-shaft of the metacarpal (IN 4517) of blackbuck (Fig. 5.73 b). The fracture of the bone have got healed up with extra growth of bone around the injure area (Baker and Brothwell 1980).

Some of the first and second phalanges of cattle show pathological lesions. In one of the first phalanges (IN 2177) around the proximal articular surface expansion of bones is seen on both lateral and medial sides (Fig. 5.72 c). Similarly another first phalanx (IN 4851) has a depression and extra growth of bone on the medial side. The fossa may have been formed due to the extra growth of bone on the other phalanx in the pair (Fig. 5.73 c). In yet another first phalanx the shape and size are completely altered and extra growth of bone is seen. These extra growths of bones or osteoperiotitis may have been due to pathological reasons (Fig.5.73 d). In one of the second phalanges of cattle, the bone has become very porous and extra growth of bone is noticed on the lateral side of the proximal epiphysis. This could be due to osteoporosis, another pathological condition generally seen in domestic animals (Fig. 5.73 e) (Pawankar 2001).

Fig. 5.73. Pathological and non-pathological lesions.

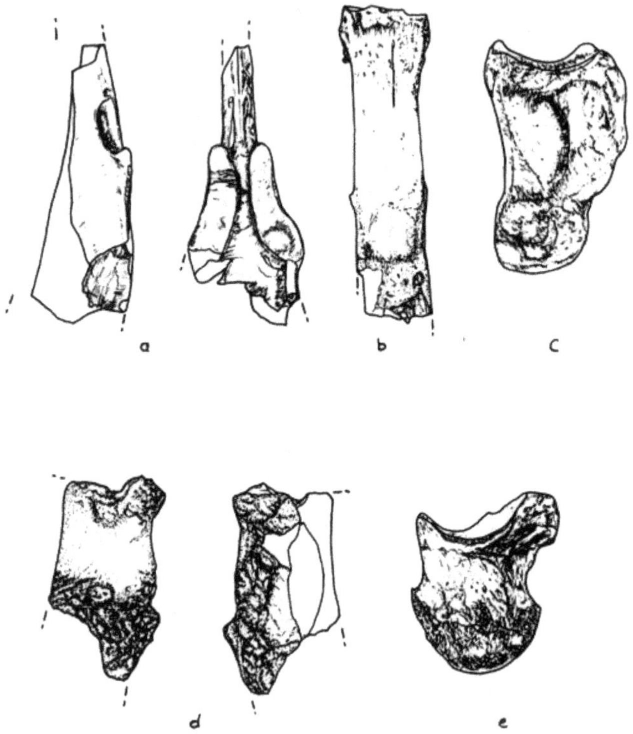

Conclusion

The objective of this study was to find out the man-animal relationship in the ancient past especially among the early farmers of western India. The main focus of this work was the re-analysis of the faunal remains from the Chalcolithic site, Inamgaon (district Pune, Maharashtra), housed in Archaeozoology laboratory of Deccan College, Pune. This was necessary since after the earlier work done many methodological changes have taken place in the field of Archaeozoology.

The faunal assemblage at Inamgaon consisted of over 66,847 bone fragments. Only about 38.79 % of the total collection could be identified, weight wise which accounted for 69.12 %. This shows a high degree of fragmentation of bones at Inamgaon. The results of measurability index and the level of identification of bones suggest more fragmentation of bones in the Early Jorwe phase than in the Late Jorwe and Malwa phases. The fragmentation of bones depends on factors such as taphonomy, butchering techniques, carcass utilization, cooking, secondary uses of bones, excavation, transport, etc. One or many of these factors may have been responsible for the high degree of fragmentation.

The archaeofauna at Inamgaon comprises more than 50 species of animals consisting of mammals, birds, reptiles, fish, molluscs and crustaceans. Some of the fish and birds could not be identified to the species. The diversity and the wide spectrum of the animal species represented at the site suggest the significance of these animals in the promotion of the cultural activities at Inamgaon. In the faunal remains, cattle bones predominate in the Early Jorwe and Malwa phases. Almost everywhere in the protohistoric cultures of India, cattle bones are abundant which indicate the importance of this animal both in the food economy as well as in agricultural operations. It was observed that about 70 % cattle were killed between the ages of 1.5 to 3.5 years. However, in the Malwa phase, a few very young animals below the age of 1.5 years were found butchered as against old animals in the Early and Late Jorwe phases. This probably reveals the lesser magnitude of agricultural activities in the Malwa phase. Large number of senile and adult animals in the Early Jorwe and Late Jorwe phases suggests that some of these animals may have been kept also for agricultural operations. Likewise, more number of males was killed; the male and female ratio of cattle butchered at Inamgaon was 2:1. Such selection shows the stock-raising skill of the early farmers at Inamgaon.

The importance of this animal can also be deduced from the evidence of sharing meat of this animal possibly at some ceremony between the occupants of two houses in the Malwa phase. These two houses 79 and 83 were situated in different locales. The degree of charring and the metrical analyses reveal that, these bones belong to same animal. People from these two houses might have butchered this animal on some special occasion. This evidence throws light on the socio-cultural aspects of the people. Cattle bones have also been found from the human burials as remains of offering to dead persons.

Terracotta figurines of humped cattle and terracotta wheeled toy carts are reported from a vast majority of protohistoric sites of India and these animals are also depicted on pottery and rock shelters. Inamgaon has also yielded an engraving of a cart drawn by bullocks on a storage jar. This representation suggests the use of cattle as draught animals in the Chalcolithic period.

Cattle by-products also must have been of significance to the Chalcolithic population. For a vast majority of items like milk, hide and cow dung as fuel; direct evidence is lacking for the utilization of these resources at the site. However, based on certain methods used in this research programme it may be possible to postulate the use of some of these items. The sex ratio and the age group studies have shown that female cattle were killed when they were old. This suggests the possible use of milk at the site. The phalangial ratio between the first, second and third phalanges shows the meager representation of second and third phalanges compared to the first. Probably this is an indication of the use of hide, because in the preparation of hide the third and the second phalanges of an animal are generally lost with the hide. The cow dung was extensively used for plastering the floors and walls of the settlement at Inamgaon. However, its use as a fuel is not certain. Cattle dung is the main source of domestic cooking fuel even today in Indian villages. Burnt cattle dung is reported from a number of protohistoric sites. Cattle dung could also be used as manure in agriculture.

A large number of bone tools made of cattle bones which are comparatively hard and sturdy are found among the collection of bones at Inamgaon. A few first phalanges have also been recovered with square shaped hole on the dorsal side of the phalanx, near the distal end which may have been used as hunting whistles. Parallel evidences are found from the Chalcolithic sites of Prabhas Patan in Gujarat, Ramapuram in Andhra Pradesh and Iron Age sites such as Veerapuram in Andhra Pradesh and Naikund in Maharashtra.

The withers height of cattle remained almost static in between 106-113 cm in all the phases. However, a close scrutiny shows a gradual decrease in the size of animals in the transitional layers, between cultural phases and there after a continuous increase in size. Inbreeding of animals within the same stock for a longer period of time; results in the reduction of size. When Malwa people from Central India migrated to settle in the Deccan, they might have brought their live-stock with them. As this was an agro-pastoral community the use of cattle was manifold. Probably in the beginning itself the Malwa people had some contact and exchange of cattle with other local cultural complexes of the area which resulted in an increase in the size of cattle during this phase, and the same may have happened in the Jorwe phase again. Increase in the stature of cattle also occurred in the Late Jorwe phase, probably because of the cultural contacts with contemporary Iron Age people.

Scanty representation of buffalo bones was found in the faunal assemblage which may account for about 2-3 % in the total collection. Because of close morphological similarity with cattle, some of the bones could not be identified precisely. The skeletal representation of bones of buffalo was more or less identical with those of cattle. One notices almost the same degree of bone representation at other Chalcolithic sites like Nevasa, Nasik, Jorwe, Daimabad, Kaothe and Walki. Unlike cattle, buffalo is not adapted to very harsh climatic conditions hence, the meager amount of buffalo bones at these sites.

In all the Chalcolithic sites, the second most important animals in the food economy were the sheep and the goat. However, at Inamgaon, while considering the number of animals killed for food purpose, it was blackbuck. Thus, it can be presumed that the sheep and goat herd structure at Inamgaon was not very significant as it was elsewhere.

The skeletal representation of pigs at Inamgaon shows that these animals were not butchered at the site. A few body parts retrieved from excavation may have been brought from outside as delicacy. Pigs were also available in the wild state in and around the site of Inamgaon. The non-exploitation of both the wild and domestic pigs at Inamgaon probably indicates a taboo on this animal.

Horse was introduced towards the end of the Jorwe culture. A majority of bones were retrieved from layers 1-6, which is mainly a part of the Late Jorwe phase. The occurrence of bones was identical with those found the megalithic burials of Vidarbha. Mainly the lower extremities of limb bones and cranial fragments including teeth were found in the collection. A distal end of tibia was found with cut marks and charring, suggesting possible consumption of horse meat at some ceremony, as it was practiced in the Iron Age culture of Vidarbha, Chronologically also, the Late Jorwe phase matches with the Iron Age culture of this region.

Another beast of burden represented in the faunal spectrum of Inamgaon was the ass, which was identified by a solitary bone in layer 2. Since this animal was not certainly in the diet of the people, one does not expect a large number of bones of ass among the food refuse.

Dog is known as a companion of man right from very early times. Dog bones were identified from all the phases at Inamgaon. However, they were scanty in Malwa and Early Jorwe phases. The skeletal representation also shows scanty and erratic portrayal of the dog bones which had no human activity on them. However, the dog bones from the Late Jorwe depict human activity such as charring and cut marks. The skeletal representation also shows a more or less uniform pattern as seen on the bones of other animals utilized for food purpose. The evidence certainly suggests the consumption of dog meat in the Late Jorwe phase. This is unparalleled in any other archaeological site in India. Here, it is interesting to note that a tribe, *Katodi* in Maharashtra eats dog meat at the time of scarcity and famine.

The subsistence economy at Inamgaon was not only based on agriculture and animal husbandry. Wild animals consisting of terrestrial, avian and aquatic fauna were exploited in large number throughout the cultural history at Inamgaon, though the maximum exploitation was in the Late Jorwe phase. The bones of wild animals accounted for more than 54 % in the Late Jorwe phase, whereas in the Malwa and Early Jorwe phases they were about 20 % and 13 %, respectively. Among the bones of wild animals, the blackbuck bones alone constituted 41 %, in the Late Jorwe phase. At Inamgaon, blackbucks were the second most important animal in the food economy of the Malwa and the Jorwe people and numerically the bone fragments recovered were more than the cattle bones in the Late Jorwe phase. Almost all types of bones of blackbuck were found at the site. This suggests that these animals were killed within the vicinity and the entire body was brought into the settlement and butchered at the site. Evidence is also available for the consumption of blackbucks' brain as a number of skulls were found with cut marks on the occipital region of the cranium. Metrical analyses revealed that about 70-75 % of the blackbucks killed were males. This is significant, because while hunting this sort of selection is generally not possible. Thus, probably these animals were killed mainly in the mating season. Blackbucks breed at all seasons, but the main rut is noticed between February and March. This time the bucks fight for the possession of does. The unsuccessful males are driven away from the territory and probably such animals were killed at Inamgaon.

A number of phalanges and calcaneum were found with perforations on the lateral side, near the proximal epiphyses. Probably these were beads or pendants in making. In a few bones like astragali, phalanges, centrotarsals and the distal ends of metapodials of blackbucks, the cortex of the bones have been found removed and became shapeless. Probably this is due to the scavenging activity of animals like dog. The bones which are swallowed undergo a lot of changes in the gastric juice of carnivores, resulting in the modification. Rodent activity like gnawing has also been noticed on a few bones.

The skeletal representation of bones of Nilgai shows the presence of a large number of non-meat-bearing parts like the lower extremities of limb bones. The absence of meat-bearing parts is intriguing since no taphonomical factor could be attributed to this animal was butchered at the site. Only the possible explanation could be that the meat of this animal was priced and given in exchange to some outside community.

Other antelopes like chinkara, four-horned antelope and deer species such a sambar, chital, barking deer, hog deer, etc. were found at Inamgaon. Almost the same species of animals were identified at many of the Chalcolithic sites of western Maharashtra from districts Pune, Ahmadnagar, Nasik and Dhule. If one looks at the present day distribution of animals like sambar, chital, barking deer, nilgai and chinkara, they are not found in these districts, except nilgai which is found in the extreme north of Dhule districts whereas, the past distribution was in the south west part of the district. Presently, the blackbucks are absent in Dhule district, but are present in Pune and Ahmadnagar districts, very near to the Chalcolithic sites. Past and present distribution of these animals shows the environmental change from the Chalcolithic period to the present time.

A large number of bone tools and objects made of antlers, bones, shells, ivory, etc. are reported from Inamgaon. Points, punches, chisel and scarpers are the common types of bone tools. These tools were probably used in animal butchering, wood working, pottery designing, etc. Chipping, grinding, polishing and fire hardening are some of the techniques used in the manufacture of these tools.

The presence of fish and the shells of marine origin at Inamgaon suggest trading activities with outsiders. Mainly bangles, beads, pendants, etc. were made of Chank shell. The waste products, finished and unfinished shell objects at the site indicate that these ornaments were manufactured at the site. Perforated fish vertebrae also might have been used as pendants. To the west of the site, the nearest sea coast is the Konkan coast which is at least 200 km away. The Chank shell is not available on this sea coast. This species of shell is found on the Saurashtra coast, Tamil Nadu coast and the Bay of Bengal. Considering the long distance to Tamil Nadu coast and the Bay of Bengal, probably the Saurashtra coast could be the source for the procurement of Chank shell. In Saurashtra, shell industry is reported from a majority of the Harappan sites. The Late Harappans of Gujarat had trading contacts with the Chalcolithic cultures of the Deccan which is evident from the Late Harappan phase at Daimabad in the Tapi valley.

A majority of the structures in different layers are oriented in the north-east, south-west direction. Spatial analysis of bones from different layers shows that much of the butchering activity was taking place in the central part of the settlement in the Malwa and Early Jorwe phases, as the concentration of non-meat-bearing parts was found in this area. A very high concentration of meat-bearing parts was noticed in the south-western region of the excavated area, outside the structures. This could be identified as a food refuse dumping area. The extent of non-meat-bearing parts in the central part of the settlement and the concentration of the same in two different localities in layer 7, probably suggest the involvement of more families in the butchering activities in the Malwa and Early Jorwe phases have also given evidences for butchering activity.

In the Late Jorwe phase, all the rectangular structures of the earlier two phases were replaced with circular structures. However, the orientation of the structures was similar to the earlier phases. A majority of these circular structures were devoid of faunal remains, probably indicating a change in the functional aspect of these huts. In this phase the extension trenches on the eastern side C, D, E, F, G (23-28) have yielded a vast number of animal remains. Interestingly, a couple of rectangular structures were also identified in this area. Much of the activities concerning food habits may have been in the eastern side of the mound. In this phase the refuse dumping area was identified in the eastern side of the excavated area. In the earlier phases the concentration of refuse dump was in the southern as well as in the western part. This shift of activity areas probably was related to the expansion of the settlement.

The most outstanding and unique find from Inamgaon was the change in the consumption of animals in the Late Jorwe phase. In the first two phases, cattle were predominant over other animals in the food economy. However, in the Late Jorwe phase greater emphasis was given to hunting wild animals, especially the blackbucks. Culturally this shift in the subsistence economy from the domestic stock to the wild animals is very significant. If one analyses the cultural processes of the Late Jorwe phase, it is quite different from the earlier two phases. The rectangular structures of the settlement were replaced increasingly by circular structures. The botanical evidence indicates greater decline in the agricultural activities. Also the trace element studies on the human skeletal remains from this phase reveal the concentration of copper, zinc, strontium and manganese, suggesting greater dependence on animal food. With the available evidence it is reasonable to think that there was considerable environmental degradation in the Late Jorwe phase and a general decline in the economy which resulted in the hunting of more animals in this phase. Mixed economy of early farmers of Maharashtra was based on subsistence agriculture, animal husbandry, hunting and fishing.

Inamgaon is an important site for the Late Jorwe cultural phase. In all the other Chalcolithic sites of Central India, Western India and the Deccan, the activities of the early farming communities came to an end around 1000 B.C. and the settlements remained unoccupied until the beginning of the Early Historic period. But in the Bhima valley, Inamgaon continued till the end of 700 B.C.

References

Agrawal, D. P. and Sheela Kusumgar 1968. Tata Institute Radiocarbon Date List V, *Radiocarbon* 10 (1): 131-143.

Ansari, Z. D. and M. K. Dhavalikar 1975. Megalithic burials at Pimpalsuti, *Bulletin of the Deccan College Postgraduate and Research Institute* 36 (1-4): 84-88.

Badam G. L. 1977. Faunal Remains from Chalcolithic Inamgaon, *Man and Environment* 1: 58-60.

Badam, G. L. 1979. Bone remains, *Apegaon Excavations 1976* (S. B. Deo, M. K. Dhavalikar and Z. D. Ansari Eds.), pp.40-49. Pune: Deccan College Postgraduate and research Institute.

Badam, G. L. 1984. Holocene faunal material from India with special reference to domesticated animals, *Animals and Archaeology* (J. Clutton-Brock and C. Grigson Eds.), 3:339-353. England: British Archaeological Report 202.

Badam, G. L. 1986. Preliminary Report on the Faunal Remains from Chalcolithic Daimabad, Maharashtra, *Archaeological studies*, pp.393-100. Varanasi: Bharat Kala Bhavan.

Baker, J. and D. Brothwell 1980. *Animal Diseases in Archaeology*. London: Academic Press.

Bergstrom, P. L. and L.H. von Wijngaarden-Bakker 1983. The metapodia as the predictors of size and weight in cattle, *Report No. B 206*, Instituut voor Veeteelkudfig Onderzoek Schoonord, Zeist.

Binford, L. R. 1968. Post-pleistocene adaptations in *New Perspectives in Archaeology* (S. R. Binford and L. R. Binford Eds.), pp. 313-341. Chicago: Aldine.

Boessneck, J. 1956. Ein Beitrag zur Errechnung der Widerristhohe nach metapodienmassen bel Rindern, *Zeitschrift fur Tierzuchtg und Zuchtgsbiol* 68: 75-90.

Boessneck, J. 1963. Osteological differences between sheep (*Ovis aries* Linne.) and goat (*Capra hircus* Linne.), *Science in Archaeology* (D. R. Brothwell and E. Higgs Eds.), pp. 331-358. New York: Basic Books.

Bokonyi, S. 1970. A new method for the determination of the number of individuals in animal bone material, *American Journal of Archaeology* 74: 291-292.

Bokonyi, S. 1989. Definitions of animal domestication, *The Walking Larder: Patterns of Domestication, Pastoralism and Predation* (J. Clutton Brock Ed.), pp.22-27. London: UNWIN HYMAN.

Bose, A. B. 1975. Pastoral Nomadism in India: nature, problems and prospects, *Pastoralists and Nomads in South Asia* (L. S. Leshnik and G. D. Sontheimer Eds.), pp. 1-15. Wiebaden.

Casteel, R. W. 1979. Faunal assemblages and the 'wiegermethode' or weight method, *Journal of Field Archaeology* 5: 71-77.

Chaplin, R. E. 1971. *The Study of Animal Bones from Archaeological Sites*. London and New York: Seminar Press.

Child V. G. 1928. *The most ancient East: the Oriental Prelude to European Prehistory*. London: Kegan Paul, Trench, Trubner.

Clason, A. T. 1979. *Wild and Domestic Animals in Prehistoric and Early Historic India*. Lucknow: Ethnographic and Folk Culture Society.

Clutton-Brock, J. 1989. Introduction to domestication, *The Walking Larder: Patterns of Domestication, Pastoralism and Predation* (J. Clutton-Brock Ed.), pp.7-9. London: UNWIN HYMAN.

Cohen, M. N. 1977. *The Food Crisis in Prehistory*. New Haven: Yale University Press.

Cook, C. S. and A. E. Treganza 1950. The quantitative Investigation of Indian Mounds with special reference to the relation of physical components to the probable material culture, *UC Publication of American Archaeology and Ethnology*, 40: 245.

Cornwall, I. W. 1956. *Bones for the Archaeologist*. London: Phoenix House.

Davis, S. J. M. 1985. A preliminary Report of the Fauna from Hatoula: a Natufian-Khiamian (PPNA) site near Latroun, Israel, *Le site Natoufien-Khiamen de Hatoula, pres de Latroun, Israel* (M. Lechevallier and A. Ronen Eds.). Centre de Recherche Francais de Jerusalem. Service de documentation I.

Davis, S. J. M. 1987. *The Archaeology of Animals*. London: B. T. Batsford Ltd.

Deniz, E. and S. Payne 1982. Eruption and wear in the mandibular dentition as a guide to ageing Turkish Angora goats, *Ageing and Sexing animal bones from archaeological sites* (B. Wilson, C. Grigson and S. Payne Eds.), pp. 155-205. Oxford: BAR British Series 109.

Deo, S. B., M. K. Dhavalikar and Z. D. Ansari 1979. *Apegaon Excavations 1976*. Pune: Deccan College Postgraduate and Research Institute.

Deshpande, C. D. 1948. *Western India: A Regional Geography*. Dharwar: Students' Own Book Depot.

Deshpande, C. D. 1971. *Geography of Maharashtra*. New Delhi: National Book Trust of India.

Dhavalikar, M. K., H. D. Sankalia and Z. D. Ansari 1988 a. *Excavations at Inamgaon*. Pune: Deccan College Postgraduate and Research Institute.

Dhavalikar, M. K. 1988 b. Inamgaon- A Chalcolithic Chiefdom, *Excavations at Inamgaon* (M. K. Dhavalikar, H. D. Sankalia and Z. D. Ansari Eds.), 1 (2): 1001-1008. Pune: Deccan College Postgraduate and Research Institute.

Dhavalikar, M. K. and Z. D. Ansari 1988 c. Other Artifacts, *Excavations at Inamgaon* (M. K. Dhavalikar, H. D. Sankalia and Z. D. Ansari Eds.), 1 (2): 553-724. Pune: Deccan College Postgraduate and Research Institute.

Dhavalikar, M. K., V. S. Shinde and S. Atre1990 a. *Excavations at Kaothe.* Pune: Deccan College Postgraduate and Research Institute.

Dhavalikar, M. K., V. S. Shinde and S. Atre 1990 b. Small site archaeology: Excavations at Walki, *Bulletin of Deccan College Postgraduate and Research Institute* 50:197-228.

Driesch von den 1976. *A Guide to the Measurements of Animal Bones from Archaeological sites*, Peabody Museum Bulletin 1. Cambridge, Massachusetts: Harvard University Press.

Ducos, P. (translated by Marie Matthews) 1989. Defining Domestication: a clarification, *The Walking Larder: Patterns of Domestication, Pastoralism and Predation* (J. Clutton-Brock Ed.), pp. 28-30. London: UNWIN HYMAN.

Durest, J. U. 1926. Verleichende Untersuchungs methoden am skelett bei saugern, *Handbuch der Bilogischen Arbeitsmethoden,* 2:125-530. Berlin and Wien.

During, E. 1985. On relative quantification of faunal remains from archaeological sites, *I.S.K.O S. Helsinki.* 5: 227-281.

Eapen, J. 1960. A report on the animal remains of Nevasa excavations, *From History to Prehistory at Nevasa* (H. D. Sankalia, S. B. Deo, Z. D. Ansari and S. Ehahardt Eds.). pp. 531-536. Pune Deccan College Postgraduate Research Institute.

Edwards, J. K., R. L. Marchinton and G. F. Smith 1982. Pelvic girdle criteria for sex determination of white-tailed deer, *Journal of Wildlife Management* 46: 544-547.

Fieller, N. R. J. and A Turner 1982. Estimation of Number in Vertebrate Samples, *Journal of Archaeological Science* 9(1): 49-62.

Flannery K. V. 1969. Origin and Early Domestication in Iran and the Near East, *The domestication and exploitation of plants and animals* (P. J. Ucko and G. W. Dimbleby Eds.). Pp.73-100. Gerald, Duckworth, London.

Fock, J. 1966. *Metrische Untersuchungen an metapodien einiger Europaischer Rinderrassen,* Dissertation, Munich: University of Munich.

Galton, F. 1883. *Inquiry into human faculty and its development*, London: Macmillion.

Gauthier, A. 1984. How do I count you? Let me count the ways, *BAR International Series 277.* pp.237-251.

Gazetteer of India 1965.

Gejvall, N. G. 1969. *Lerna a pre-classical site in the Argoilid,* vol. 1 The fauna, Princeton: American school of classical studies at Athens.

George, J. C. 1955. Identification of bones, *Report on the excavations at Nasik and Jorwe 1950-51* (H. D. Sankalia and S. B. Deo Eds), pp. 142-146. Pune: Deccan College Postgraduate and Research Institute.

Grant, A. 1975. The use of tooth wear as a guide to the age of domestic animals: a brief explanation, *The excavations at Portchester Castle 1* (B. Cunliffe ed.), pp.437-450. London: Society of Antiquaries.

Grant, A. 1978. Variation in dental attrition in mammals and its relevance to age estimation, *Research Problems in Zooarchaeology* (D. R. Brothwell, K. D. Thomas and J. Clutton-Brock Eds.), pp. 103-106. London: Institute of Archaeology.

Grant, A. 1982. The use of tooth wear as a guide to the age of domestic ungulates, *Aging and sexing animal bones from archaeological sites* (B. Wilson, C. Grigson and S. Payne Eds.), pp. 155-205. Oxford: BAR British Series 109.

Grayson, D.K. 1973. On the methodology of Faunal analysis, *American Antiquity* 38: 432-439.

Grayson D. K. 1979. On the quantification of vertebrate archaeofauna, *Advances in Archaeological Methods and Theory* (M. B. Schiffer Ed.), 2: 199-237. New York: Academic Press.

Grayson, D. K. 1984. *Quantitative Zooarchaeology.* Orlando, U.S.A.: Academic Press.

Grigson, C. 1982. Sex and age determination of some bones and teeth of domestic cattle: a review of the literature, *Aging and sexing animal bones from archaeological sites* (B. Wilson, C. Grigson and S. Payne Eds.), pp. 7-24. Oxford: BAR British Series 109.

Grigson, C. 1982. Sexing Neolithic domestic cattle skulls and horn cores, *Aging and sexing animal bones from archaeological sites* (B. Wilson, C. Grigson and S. Payne Eds.), pp. 25-36. Oxford: BAR British Series 109.

Haak, D. 1965. *Metrische Untersuchungen an Rohernknochen Bei Deutchen Merinolandschafen und Heidschnucken* Dissetation. Germany: University of Munich.

Harcourt, R. A. 1974. The dog in prehistoric and early historic Britain, *Journal of Archaeological Science* 1: 151-175.

Herre, W. 1963. The Science and History of Domestic Animals, *Science in Archaeology* (D. R. Brothwell and E. Higgs Eds.), pp. 257-271. New York: Basic Books.

Issac, E. 1962. On the domestication of cattle, *Science* 137: 195-204.

Joglekar, P. P. 1992. *A biometric approach to the faunal remains of western India with special reference to Kaothe and Walki*. Ph. D. Dissertation. Pune: University of Poona.

Joglekar, P. P. 1993-94. Height at Withers: An Osteometric Approach, *Bharati* XX, 31-88.

Joglekar, P. P. and P. K. Thomas 1995. Faunal Diversity at Walki –A Small Chalcolithic Settlement in Western Maharashtra, *Bulletin of the Deccan College Postgraduate and Research Institute* 53 (1993): 75-94.

Kajale, M. D. 1988 a. Plant Economy, *Excavations at Inamgaon* (M. K. Dhavalikar, H. D. Sankalia and Z. D. Ansari Eds.) 1 (2): 727-821. Pune: Deccan College Postgraduate and Research Institute.

Kajale, M. D. 1988 b. Ancient Plant Economy at Tuljapur Garhi, District Amravati, Maharashtra, *Current Science* 57 (7): 377-379.

Khan, E. 1975. The role of man and environment in the development of domestic animal, *Archaeozoological Studies* (A. T. Clason Ed.), pp. 358-361. North Holland, Amsterdam.

Kieswalter, L. 1888. *Skelettmessungen an pferden als Beitrag zur theoritischen Grundlge der Beurteilungslehre der pferdes,* Dissertation. Germany: University of Leipzig.

Klein, R. G. and K. Cruz-Uribe 1984. *The Analysis of Animal Bones from Archaeological site*, Chicago and London: University of Chicago Press.

Krantz, G. S. 1968. A new method of counting mammal bones, *American Journal of Archaeology* 72: 286-288.

Kubasiewicz, M. 1956. O metodyce badan wykipalisko wych szczatkow kostnych sqierzecych, *Materaly Zachodnio-Pomorskie* 2: 235-244.

Lauwerier, R. C. G.M. 1988. *Animals in Roman times in the Dutch Eastern River Area.*Amsterdam: R. O. B.

Matolcsi, J. 1970. Historische Erforschung der Korpergrosse der Rinder auf Grund von Ungarischen Knochen Material, *Zeitschrift fur Tierzuchtung und Zuchtungs biologie* 87: 89-128.

McMeekan, C. P. 1940.Growth and the Development in the Pig, with Special Reference to Carcass Quality Characters, Part I, *Journal of Agricultural Science* 30: 276-343.

McMeekan, C. P. 1940.Growth and the Development in the Pig, with Special Reference to Carcass Quality Characters, Part II. The influence of the Plane Nutrition on Growth and Development, *Journal of Agricultural Science* 30: 387-436.

McMeekan, C. P. 1940.Growth and the Development in the Pig, with Special Reference to Carcass Quality Characters, Part III. Effect of the Plane of Nutrition on the form and composition of the Bacon pig, Journal *of Agricultural Science* 30: 511-569.

Moris, P. 1972. A review of mammalian age determination methods, *Mammal Review* 2:69-104.

Moris, P. 1978. The use of teeth for estimating the age of wild mammals, *Development, function and evolution of teeth* (P. M. Butler and K. A. Joysey Eds.), pp.483-494. New York: Academic Press.

Nath, A. 1988-89. Excavation Adam, District Nagpur *Indian Archaeology 1988-89 –A Review*, pp. 50-62.

Pappu, R. S. 1988. Site Catchment Analysis, Excavations *at Inamgaon* (M. K. Dhavalikar, H. D. Sankalia and Z. D. Ansari Eds.) 1 (1): 107-120. Pune: Deccan College Postgraduate and Research Institute.

Pawankar, S.J. and P. K. Thomas 1997. Fauna and Subsistence Pattern in the Chalcolithic Culture of Western India, With Special Reference to Inamgaon, *Anthropozoologica* Proceedings of the 7th International Conference for Archaeozoology, ICAZ, Constance, Germany, September 1994 (M. Kokabi and J. Wahl Eds.) 25-26, pp. 737-746.

Pawankar, S. J. and P. K. Thomas 2001. Osteological Differences between Blackbuck (*Antilope cervicapra*), Goat (*Capra hircus*) and Sheep (*Ovis aries*), *Man and Environment* XXVI (1) pp.109-126. Pune: Indian Society for Prehistoric and Quaternary Studies

Pawankar, S. J. 2001. Animal Bones Modifications: A case study at Inamgaon, a Chalcolithic Site in India, *Pratnatattva* 5: 37-40. Dhaka: Department of Archaeology, Jahangirnagr University.

Payne, S. and P. J. Munson 1986. *Rubyand how many squirrels ? the destruction of bones by dogs.* Oxford, BAR.

Perkins, D. (Jr.) 1973. A critique on the methods of quantifying faunal remains from archaeological sites, *Domestikationsforschung und Geschichte der Haustiere* (J. Matolcsi Ed.), pp. 367-369. Budapest: Akademiai Kiado.

Pitts, G. C. and T. R. Bullard 1968. Some Inter specific Aspects of Body Composition in Mammals, *Body Composition in Animals and Man,* pp. 49-69. National Research Council, Washington D. C.: National Academy of Science.

Prater, S. H. 1965. *The Book of Indian Animals*, Bombay: Bombay Natural History Society.

Prummel, W. 1983. *Excavations at Dorestad 2, Early Medieval Dorestad an Archaeozoological study*. Amersfoot: R. O. B.

Prummel, W. and Hans-Jorg Frisch 1986. A Guide for the distinction of Species, Sex and Body side in Bones of Sheep and Goat, *Journal of Archaeological Science* 13:567-577.

Rajaguru, S.N. 1973. Late Pleistocene Climatic Change in Western India, *Radiocarbon and Indian Archaeology* (D. P. Agrawal and A. Ghosh Eds.), 87-88. Bombay: Tata Institute of Fundamental Research.

Rajaguru, S. N. 1988. Environment, *Excavations at Inamgaon* (M. K. Dhavalikar, H. D. Sankalia and Z. D. Ansari Eds.) 1 (1): 9-14. Pune: Deccan College Postgraduate and Research Institute.

Reed, C. A. 1963. Osteoarchaeology, *Science in Archaeology* (D. R. Brothwell and E. Higgs Eds.), PP. 215. New York: Basic Books.

Reed, C. A. 1984. The beginning of animal domestication, *Evolution of domesticated animals* (I. L. Mason Ed.), pp. 1-6. England: Longman Group Limited.

Reynolds, W. W. and W. J. Karlotki 1977. The allometric relationship of skeleton weight to body weight in Teleost Fishes: A preliminary comparison with birds and mammals, *Copeia*, pp. 160-163.

Sali, S. A. 1986. *Daimabad 1976-79*. New Delhi: Archaeological Survey of India, Government of India.

Sankalia, H. D. and S. B. Deo 1955. *Report on the Excavations at Nasik and Jorwe 1950-51*. Pune: Deccan College Postgraduate and Research Institute.

Sankalia, H. D., S.B. Deo, Z. D. Ansari and S. Ehahardt 1960. *From History to Prehistory at Nevasa*, Pune: Deccan College Postgraduate and Research Institute.

Schramm, Z. 1967. Long bones and height at withers in goat, *Rocziki Wyzazcj Szkoly Roiniczey W Pozaniu* 36: 89-105.

Silver, I.A. 1963. The Aging of Domestic Animals, *Science in Archaeology* (D. R. Brothwell and E. Higgs Eds.), pp. 283-302. New York: Basic Books.

Singh, R. L. 1971. *India – A Regional Geography*, Varanasi: N. G. S. I.

Taber, R. D. 1956. Characteristics of the pelvic girdle in relation to sex in black-tailed and white-tailed deer, *California Department of Fish and Game* 42:15-21.

Teichert, M. 1990. Withers Height Calculation: Remarks and Experience, paper presented at VI conference of ICAZ, Washington D.C.

Thomas, P. K. 1975. Role of Animals in the Food Economy of the Mesolithic Culture of Western and Central India, *Archaeozoological studies* (A. T. Clason Ed.), pp.322-328. Amsterdam: North Holland.

Thomas, P. K. 1977. *Archaeozoological Aspects of the Prehistoric Culture of Western India,* Ph. D. Dissertation. Pune: University of Poona.

Thomas P. K. 1979. Archaeozoological Aspects of Prehistoric Culture of Prabhas Patan, *Archaeozoology* (M. Kubasiewicz Ed.), 1: 176-186. Szczecin, Poland: Agricultural Academy.

Thomas, P. K. 1980-81. A preliminary report of the animal remains from Ramapuram, Dist. Kurnool, A.P., *Indian Archaeology – A Review 1980-81,* p.7.

Thomas, P. K. 1984 a. Faunal background of the Chalcolithic culture of Western India, *Animals and Archaeology* (J. Clutton-Brock and C. Grigson Eds.),3: 355-361. England: British Archaeological Report 202.

Thomas, P. K. 1984 b. Fauna assemblage of Veerapuram, *Veerapuram a type site for cultural study in the Krishna valley* (T. V. G. Sastri, M. Kasturi Bai, J. Vera Prasad Rao Eds.), pp. i-xi, Hyderabad: Bhavani Printers.

Thomas, P. K. 1988. Faunal assemblage, *Excavations at Inamgaon* (M. K. Dhavalikar, H. D. Sankalia and Z. D. Ansari Eds.) 1 (2): 822-961. Pune: Deccan College Postgraduate and Research Institute.

Thomas, P. K. 1989. Utilization of domestic animals in pre and protohistoric India, *The Walking Larder: Patterns of Domestication, Pastoralism and Predation* (J. Clutton-Brock ed.), pp.108-112. London: UNWIN HYMAN.

Thomas, P. K. and P. P. Joglekar 1990. Faunal remains, *Excavations at Kaothe* (M. K. Dhavalikar, V. Shinde and S. Atre Eds.), pp. 233-264. Pune: Deccan College Postgraduate and Research Institute.

Thomas, P. K. 1992 a. Faunal Assemblage and Subsistence Strategies at Tuljapur Garhi, *Man and Environment* 17 (2):71-74.

Thomas, P. K. 1992 b. Faunal Assemblage of the Iron Age Culture of Maharashtra, *Man and Environment* 17 (2): 75-79.

Thomas, P. K. and P. P. Joglekar 1994. Holocene Faunal Studies in India, *Man and Environment* 18 (1-2):179-203.

Thomas, P. K. (in press) Faunal remains from the Megalithic habitation at Naikund.

Thomas, P. K. and P. P. Joglekar. (in press) Faunal remains from Adam.

Tonapi, G. T. 1980. *Fresh water animals of India.* New Delhi: Oxford – IBH.

Uerpmann, H. P. 1973. Animal bone finds and economic archaeology: a critical study of osteo-archaeological methods, *World Archaeology* 4: 307-322.

White, T. E. 1953. A method of calculating the dietary percentage of various food animals utilized by aboriginal people, *American Antiquity* 18: 396-398.

Wijngaarden-Bakker, L. H. and P.L. Bergstrom 1988. Estimation of the shoulder height of cattle, *Archeozoologia* 2 (1.2): 67-82. The Acts of the fifth International Conference of Archaeozoology Bordeaux- August 1986.

Zalkin, V. I. 1960. Metapodial variation and its significance for the study of ancient cattle, *Bull. Moskova obshch Ispit. Pri. Biol .n.. s.* 65 (1): 109-126.

Zalkin, V. I. 1970. Drevniejsije domasnije zivotnije vostochoj Europy, Moskova.

Zeuner, F. E. 1963. *A history of domesticated animals*. London: Hutchinson.

Appendix A

A.1. Abbreviations used in the tables of measurements of bone

Scapula

- Greatest length of the Processus articularis (glenoid cavity) GLP
- Length of the glenoid cavity LG
- Breadth of the glenoid cavity BG
- Smallest length of the neck of the scapula SLC

Humerus

- Greatest length GL
- Breadth of the diaphysis DD
- Breadth of the proximal epiphysis BP
- Thickness of the proximal epiphysis TP
- Breadth of the distal epiphysis BD
- Thickness of the distal epiphysis TD

Radius

- Greatest length GL
- Breadth of the diaphysis DD
- Breadth of the proximal epiphysis BP
- Thickness of the proximal epiphysis TP
- Breadth of the distal epiphysis BD
- Thickness of the distal epiphysis TD

Ulna

- Greatest length GL
- Breadth of the proximal epiphysis BP
- Smallest depth of the olecranon SDD

Metacarpal and Metatarsal

- Greatest length GL
- Breadth of the diaphysis DD
- Breadth of the proximal epiphysis BP
- Thickness of the proximal epiphysis TP
- Breadth of the distal epiphysis BD
- Thickness of the distal epiphysis TD
- Width of the single distal trochlea WT
- Thickness of the sagital ridge TT
- Thickness of the lateral trochlea TLT

Pelvis

- Length of acetabulum LA

Femur

- Greatest length GL
- Breadth of the diaphysis DD
- Breadth of the proximal epiphysis BP
- Thickness of the proximal epiphysis TP
- Breadth of the distal epiphysis BD
- Thickness of the distal epiphysis TD
- Length of the femur head LC
- Thickness of the femur head TC

Tibia

• Greatest length	GL
• Breadth of the diaphysis	DD
• Breadth of the proximal epiphysis	BP
• Thickness of the proximal epiphysis	TP
• Breadth of the distal epiphysis	BD
• Thickness of the distal epiphysis	TD

Astragalus

• Greatest length	GL
• Lateral length	GLL
• Medial length	GLM
• Lateral thickness	TL
• Medial thickness	TM
• Width of the distal trochlea	BD

Calcaneum

• Greatest length	GL
• Greatest breadth	GB

Centrotarsal

• Greatest Breadth	GB

First and Second Phalanges

• Greatest length	GL
• Breadth of the proximal epiphysis	BP
• Thickness of the proximal epiphysis	TP
• Breadth of the distal epiphysis	BD
• Thickness of the distal epiphysis	TD

Third Phalanx

• Greatest length	GL
• Dorsal length	DL
• Breadth of the proximal epiphysis	BP
• Thickness of the proximal epiphysis	TP

A.2. Measurements of Bones of *Bos indicus* in mm.

Table A.2.1. Measurements of Scapula of *Bos indicus*

Site	Reg. No.	Trench	Layer	GLP	LG	BG
INM-1	1721	G11	1	00.00	50.00	36.96
INM-1	3992	G4	2	00.00	53.00	47.30
INM-1	2826	D10	2	00.00	49.72	40.88
INM-1	2827	D10	2	79.37	61.28	50.23
INM-1	2281	E10	2	58.59	50.39	38.63
INM-1	6663	F24	2	00.00	45.30	36.10
INM-1	6664	F24	2	00.00	48.40	43.65
INM-1	7912	Hno.108	2	67.25	55.50	44.60
INM-1	1895	E10	3	00.00	49.63	40.30
INM-1	1896	E10	3	00.00	62.56	49.72
INM-1	3701	A3	6	68.30	58.15	45.85
INM-1	5213	B3	6	00.00	58.90	48.00
INM-1	3776	F8	6	69.00	57.25	44.80
INM-1	4666	A5	8	00.00	56.80	50.20
INM-1	4997	C2	8	00.00	52.60	4.50
INM-1	4705	F3	8	00.00	49.45	45.10
INM-1	4892	Hno.71	8	00.00	52.00	43.75
INM-1	4893	Hno.71	8	65.05	56.88	44.70
INM-1	2883	D5	9	70.03	56.80	51.25
INM-1	6161	Hno.75	12	00.00	59.60	46.80
INM-1	6288	B3	14	68.00	56.40	43.85

Table A.2.2. Measurements of Humerus of *Bos indicus*

Site	Reg.No.	Trench	Layer	BD	TD
INM-1	1092	H7	2	00.00	37.27
INM-1	1499	G4	6	55.54	43.70
INM-1	3699	A3	6	79.35	36.50
INM-1	3879	B6	6	68.30	30.15
INM-1	3681	C6	8	62.56	29.93
INM-1	4397	B6	8	68.10	32.55
INM-1	4740	C1	8	75.68	00.00
INM-1	680	A1	11	77.31	31.32
INM-1	5336	D5	11	72.40	31.25
INM-1	5337	D5	11	68.55	31.85
INM-1	5434	E13	15	84.15	35.00

Table A.2.3. Measurements of Radius of *Bos indicus*

Site	Reg.No.	Trench	Layer	BP	TP	BD	TD
INM-1	36	LJP	0	78.85	38.11	00.00	00.00
INM-1	60	LJP	0	00.00	00.00	67.60	43.52
INM-1	315	LJP	0	60.95	31.61	00.00	00.00
INM-1	332	LJP	0	75.38	37.59	00.00	00.00
INM-1	386	LJP	0	00.00	00.00	62.64	31.59
INM-1	1018	E10	1	00.00	00.00	55.82	42.84
INM-1	1019	E10	1	00.00	00.00	55.88	42.96
INM-1	8077	F24	1	00.00	00.00	61.20	35.55
INM-1	848	F12	2	63.62	31.81	00.00	00.00
INM-1	2101	G6	2	00.00	00.00	70.06	46.46
INM-1	3027	F10	2	66.67	33.32	00.00	00.00
INM-1	3993	G4	2	72.35	37.25	00.00	00.00
INM-1	3994	G4	2	00.00	00.00	61.70	34.95
INM-1	2256	F11	3	00.00	35.05	00.00	00.00
INM-1	3256	D10	3	00.00	00.00	76.05	50.60
INM-1	6578	F25	3	78.60	36.63	00.00	00.00
INM-1	2967	D10	4	00.00	00.00	47.05	34.70
INM-1	4660	Hno111	4	00.00	00.00	69.05	49.80

INM-1	4121	C3	5	00.00	00.00	62.35	41.35
INM-1	3707	A3	6	00.00	35.25	00.00	00.00
INM-1	584	E6	7	72.95	34.14	00.00	00.00
INM-1	3661	DB1	7	64.84	34.40	00.00	00.00
INM-1	4860	C0	7	83.35	44.95	00.00	00.00
INM-1	4141	XA3B	8	00.00	00.00	53.47	40.95
INM-1	4265	B2	8	73.00	36.30	00.00	00.00
INM-1	34	C1	9	00.00	00.00	65.66	39.78
INM-1	4686	B6	9	79.95	40.95	00.00	00.00
INM-1	5196	C3	9	00.00	00.00	51.95	29.10
INM-1	5223	E1	11	74.00	37.60	00.00	00.00
INM-1	5554	E9	11	00.00	00.00	59.60	40.75
INM-1	5625	D7	12	73.85	36.40	00.00	00.00
INM-1	5511	Hno85	12	00.00	00.00	44.66	29.30
INM-1	5304	E3	14	65.68	32.65	00.00	00.00
INM-1	5461	B1	14	68.18	33.80	00.00	00.00
INM-1	5462	B1	14	72.75	36.08	00.00	00.00
INM-1	5670	A4	14	00.00	00.00	53.25	35.60
INM-1	6079	B5	14	67.35	34.96	00.00	00.00
INM-1	5477	D3	15	00.00	00.00	66.50	48.25
INM-1	5663	A5	15	00.00	00.00	60.30	35.90
INM-1	5702	E9	15	00.00	00.00	50.60	31.10

Table A.2.4. Measurements of Ulna of *Bos indicus*

Site	Reg. No.	Trench	Layer	BP	SDD
INM-1	2109	G6	2	34.19	00.00
INM-1	2560	F11	2	31.36	00.00
INM-1	6113	Hno105	4	38.50	00.00
INM-1	667	Hno110	4	31.33	00.00
INM-1	5892	E25	4	44.05	00.00
INM-1	7685	G25	4	00.00	44.77
INM-1	5912	B3	14	41.42	46.20
INM-1	5478	D3	15	41.20	00.00
INM-1	2508	D3	16	40.61	00.00

Table A.2.5. Measurements of Radius and Ulna of *Bos indicus*

Site	Reg.No.	Trench	Layer	BP	TP	BD	TD
INM-1	2247	F11	3	00.00	00.00	68.74	43.95
INM-1	7140	G24	4	00.00	00.00	80.75	50.65
INM-1	2977	B3	5	69.85	35.00	00.00	00.00
INM-1	3898	B3C3	8	00.00	35.97	00.00	00.00

Table A.2.6. Measurements of Metacarpal of *Bos indicus*

Site	Reg.No.	Trench	Layer	DD	BP	TP	BD	TD	WT	TT	TLT
INM-1	195	LJP	0	00.00	00.00	00.00	58.67	00.00	26.41	32.52	19.84
INM-1	316	LJP	0	00.00	57.65	00.00	00.00	00.00	00.00	00.00	00.00
INM-1	317	LJP	0	00.00	59.53	35.79	00.00	00.00	00.00	00.00	00.00
INM-1	390	LJP	0	31.87	00.00	00.00	55.05	00.00	25.52	31.45	22.58
INM-1	6352	A3B3	0	00.00	00.00	00.00	50.30	31.85	23.40	30.65	23.35
INM-1	1096	G4	1	00.00	00.00	00.00	50.89	27.43	24.18	27.04	22.28
INM-1	2717	G6	1	00.00	52.13	30.67	00.00	00.00	00.00	00.00	00.00
INM-1	2038	G7	1	00.00	52.00	29.70	00.00	00.00	00.00	00.00	00.00
INM-1	3867	D25	1	00.00	00.00	00.00	50.50	27.95	22.95	27.90	21.55
INM-1	1165	G10	1	00.00	00.00	00.00	54.24	30.83	30.32	25.00	21.41
INM-1	3971	G4	2	00.00	00.00	00.00	55.23	35.24	26.86	32.95	24.75
INM-1	2084	G6	2	00.00	44.47	28.34	00.00	00.00	00.00	00.00	00.00
INM-1	1586	H5	2	00.00	51.18	32.16	00.00	00.00	00.00	00.00	00.00
INM-1	1617	H5	2	00.00	00.00	00.00	00.00	00.00	23.62	30.31	21.52
INM-1	2825	D10	2	00.00	52.73	30.11	00.00	00.00	00.00	00.00	00.00
INM-1	3008	F10	2	00.00	39.69	23.61	00.00	00.00	00.00	00.00	00.00

INM-1	3572	I10	2	00.00	00.00	00.00	55.27	30.25	24.85	30.00	20.36
INM-1	4988	C2	3	00.00	40.50	25.95	00.00	00.00	00.00	00.00	00.00
INM-1	2851	G3	3	00.00	41.91	28.31	00.00	00.00	00.00	00.00	00.00
INM-1	3254	D10	3	00.00	00.00	00.00	53.74	28.90	25.10	25.10	20.90
INM-1	4088	D10	3	00.00	50.90	30.85	00.00	00.00	00.00	00.00	00.00
INM-1	4089	D10	3	28.10	00.00	00.00	55.45	31.15	25.65	30.30	22.10
INM-1	1256	D25	3	00.00	00.00	00.00	00.00	00.00	21.17	30.71	23.97
INM-1	1893	E10	3	00.00	62.65	00.00	00.00	00.00	00.00	00.00	00.00
INM-1	2625	H10	3	00.00	00.00	00.00	53.83	29.76	24.78	29.97	22.48
INM-1	4354	C1	4	00.00	38.70	24.70	00.00	00.00	00.00	00.00	00.00
INM-1	1525	E10	4	20.48	00.00	00.00	00.00	00.00	00.00	00.00	00.00
INM-1	5893	E25	4	00.00	47.10	27.20	00.00	00.00	00.00	00.00	00.00
INM-1	5894	E25	4	00.00	54.20	32.45	00.00	00.00	00.00	00.00	00.00
INM-1	5969	F25	4	00.00	00.00	00.00	53.35	27.95	25.55	28.45	18.80
INM-1	7252	Hno104	4	00.00	00.00	00.00	52.25	28.70	24.60	28.42	20.05
INM-1	7139	G24	4	00.00	00.00	00.00	52.10	28.20	24.20	28.10	19.50
INM-1	3811	E8	5	00.00	00.00	00.00	54.28	30.63	26.17	29.94	20.60
INM-1	4337	F8	5	00.00	00.00	00.00	56.15	30.28	26.59	29.90	23.25
INM-1	4419		6	00.00	53.81	34.30	00.00	00.00	00.00	00.00	00.00
INM-1	4420		6	00.00	51.35	30.35	00.00	00.00	00.00	00.00	00.00
INM-1	5048	B1	7	00.00	49.95	28.55	00.00	00.00	00.00	00.00	00.00
INM-1	4426	E7	7	00.00	00.00	00.00	52.95	26.70	23.89	26.60	19.45
INM-1	1838	G4	7	00.00	00.00	00.00	53.52	29.28	24.84	28.98	21.82
INM-1	1844	G4	7	00.00	53.63	31.41	00.00	00.00	00.00	00.00	00.00
INM-1	7930	Hno51	7	27.30	52.80	31.20	00.00	00.00	00.00	00.00	00.00
INM-1	627	B5	8	00.00	00.00	00.00	53.55	28.65	25.96	28.65	22.17
INM-1	628	B5	8	00.00	00.00	00.00	48.50	28.09	22.30	27.27	18.95
INM-1	629	B5	8	00.00	00.00	00.00	59.62	32.21	27.07	31.35	23.45
INM-1	4395	B6	8	00.00	00.00	00.00	58.73	31.55	28.20	31.60	24.05
INM-1	861	D5	8	00.00	00.00	00.00	48.69	26.34	23.52	25.50	20.63
INM-1	5129	D7	8	00.00	00.00	00.00	52.30	30.30	23.65	30.10	23.05
INM-1	4246	E1	8	00.00	00.00	00.00	00.00	00.00	23.52	29.26	20.87
INM-1	4937	E6	8	00.00	00.00	00.00	57.90	33.45	28.25	32.05	25.50
INM-1	4953	E6	8	00.00	52.36	34.98	00.00	00.00	00.00	00.00	00.00
INM-1	3666	E8	8	00.00	44.70	28.93	00.00	00.00	00.00	00.00	00.00
INM-1	3667	E8	8	00.00	00.00	00.00	48.54	26.46	22.32	26.05	21.35
INM-1	3781	X4	8	00.00	00.00	00.00	50.55	19.05	23.30	28.77	20.30
INM-1	3893	B3C3	8	22.05	45.40	26.35	00.00	00.00	00.00	00.00	00.00
INM-1	3894	B3C3	8	00.00	49.00	29.55	00.00	00.00	00.00	00.00	00.00
INM-1	3895	B3C3	8	00.00	00.00	00.00	00.00	00.00	27.00	32.50	23.70
INM-1	3737	Hno.69	8	00.00	51.90	31.63	00.00	00.00	00.00	00.00	00.00
INM-1	2885	D5	9	00.00	00.00	00.00	00.00	00.00	28.91	31.50	23.25
INM-1	4239	E4	9	00.00	00.00	00.00	49.66	29.37	24.15	29.20	22.35
INM-1	4212	B4D13	9	00.00	57.70	33.80	00.00	00.00	00.00	00.00	00.00
INM-1	*4811	B0	10	00.00	20.55	15.35	00.00	00.00	00.00	00.00	00.00
INM-1	5715	B6	11	00.00	53.90	34.65	00.00	00.00	00.00	00.00	00.00
INM-1	5266	C0	11	00.00	58.38	35.05	00.00	00.00	00.00	00.00	00.00
INM-1	3	B2	12	17.88	00.00	00.00	00.00	00.00	00.00	00.00	00.00
INM-1	6211	B6	12	00.00	54.75	32.70	00.00	00.00	00.00	00.00	00.00
INM-1	5788	Hno74	12	00.00	61.30	37.00	00.00	00.00	00.00	00.00	00.00
INM-1	5321	Hno79	12	00.00	54.30	32.26	00.00	00.00	00.00	00.00	00.00
INM-1	5322	Hno79	12	00.00	60.40	37.75	00.00	00.00	00.00	00.00	00.00
INM-1	5323	Hno79	12	00.00	00.00	00.00	51.90	29.67	24.65	29.32	21.75
INM-1	5548		13	00.00	00.00	00.00	55.44	28.90	27.00	28.37	20.40
INM-1	5089	B2	13	00.00	50.75	31.95	00.00	00.00	00.00	00.00	00.00
INM-1	5442	B3	13	00.00	00.00	00.00	00.00	00.00	30.05	33.62	23.30
INM-1	6313	A5	14	00.00	51.70	30.40	00.00	00.00	00.00	00.00	00.00
INM-1	4797	B0	14	00.00	50.15	26.35	00.00	00.00	00.00	00.00	00.00
INM-1	6069	B5	15	00.00	00.00	00.00	52.06	29.75	23.40	29.35	20.40
INM-1	4818	B0	15	00.00	64.55	38.75	00.00	00.00	00.00	00.00	00.00
INM-1	6030	Hno87	15	00.00	57.05	34.05	00.00	00.00	00.00	00.00	00.00
INM-1	6002	A5	16	00.00	58.85	31.55	00.00	00.00	00.00	00.00	00.00
INM-1	2500	D3	16	00.00	64.20	38.81	00.00	00.00	00.00	00.00	00.00

Table A.2.7. Measurements of Metapodium of *Bos indicus*

Site	Reg.No.	Trench	Layer	BD	TD	WT	TT	TLT
INM-1	5281	LJP	0	00.00	00.00	18.50	30.25	23.64
INM-1	5282	LJP	0	47.90	29.45	22.50	29.35	21.50
INM-1	8078	F24	1	00.00	00.00	24.22	28.65	21.35
INM-1	1698	G11	1	00.00	00.00	26.28	30.74	23.34
INM-1	2347		2	00.00	00.00	29.25	34.94	26.94
INM-1	2348		2	00.00	00.00	23.62	29.52	21.16
INM-1	1993	D10	2	00.00	00.00	27.50	33.19	22.40
INM-1	1444	D25	2	00.00	00.00	25.20	28.44	21.57
INM-1	6655	F24	2	00.00	00.00	24.25	29.46	21.20
INM-1	7506	F24	2	00.00	00.00	30.75	32.80	22.50
INM-1	6653	F24	2	00.00	00.00	31.90	33.16	24.55
INM-1	6654	F24	2	00.00	00.00	24.40	29.20	23.15
INM-1	6651	F24	2	50.60	28.00	23.44	28.05	21.70
INM-1	7505	F24	2	49.65	30.65	23.30	30.00	23.35
INM-1	3218	G11	2	00.00	00.00	20.75	26.80	17.92
INM-1	3217	G11	2	00.00	00.00	26.50	33.55	25.86
INM-1	3478	H7	3	55.38	31.00	25.70	30.47	22.00
INM-1	4092	D10	3	00.00	00.00	23.05	29.45	20.65
INM-1	4091	D10	3	00.00	00.00	22.80	30.25	21.75
INM-1	1238	D25	3	00.00	00.00	26.48	31.62	23.61
INM-1	1239	D25	3	00.00	00.00	23.68	30.71	21.04
INM-1	1890	E10	3	54.85	29.15	26.35	27.74	21.82
INM-1	6575	F25	3	00.00	00.00	24.43	29.77	21.90
INM-1	6576	F25	3	00.00	00.00	21.55	27.10	18.90
INM-1	6572	F25	3	00.00	00.00	25.00	29.15	21.00
INM-1	6573	F25	3	00.00	00.00	28.15	31.90	23.93
INM-1	6574	F25	3	00.00	00.00	26.60	30.15	23.05
INM-1	2228	G10	3	00.00	00.00	28.26	33.68	25.53
INM-1	2211	G10	3	00.00	00.00	23.71	22.00	00.00
INM-1	4355	C1	4	00.00	00.00	27.70	29.88	21.53
INM-1	1378	E1	4	53.50	30.94	25.95	29.80	22.05
INM-1	3059	H7	4	00.00	00.00	25.00	28.50	22.36
INM-1	8029	E24	4	00.00	00.00	25.55	31.25	21.86
INM-1	5885	E25	4	00.00	00.00	28.90	30.50	21.80
INM-1	7680	G25	4	51.20	30.20	23.85	29.60	20.65
INM-1	4781	B0	4	00.00	00.00	22.15	28.31	20.55
INM-1	4564	E5&B0	5	00.00	00.00	22.65	27.90	21.20
INM-1	5210	B3	6	46.30	28.05	21.10	27.50	19.05
INM-1	515	F3	6	00.00	00.00	25.05	29.22	22.80
INM-1	1509	G4	6	00.00	00.00	27.45	33.67	25.25
INM-1	1510	G4	6	00.00	00.00	27.10	30.49	24.05
INM-1	1511	G4	6	00.00	00.00	25.44	00.00	21.33
INM-1	1507	G4	6	00.00	00.00	23.82	26.73	19.89
INM-1	4196	E8	7	00.00	00.00	21.40	27.65	21.05
INM-1	4267	B2	8	00.00	00.00	27.10	33.48	25.08
INM-1	4269	B2	8	00.00	00.00	23.66	29.30	21.30
INM-1	4270	B2	8	00.00	00.00	25.13	32.20	22.78
INM-1	4268	B2	8	00.00	00.00	25.55	31.60	24.40
INM-1	4848	B5	8	00.00	00.00	22.80	25.07	19.25
INM-1	4991	C2	8	00.00	00.00	25.35	29.90	22.10
INM-1	4989	C2	8	00.00	00.00	22.35	27.75	19.25
INM-1	4990	C2	8	00.00	00.00	28.95	31.32	25.15
INM-1	4550	C4	8	00.00	00.00	23.30	29.14	20.80
INM-1	4385	C5	8	00.00	00.00	21.80	28.90	19.75
INM-1	4386	C5	8	00.00	00.00	21.78	28.24	20.20
INM-1	3680	C6	8	48.95	27.12	20.44	27.05	21.00
INM-1	4408	C7	8	00.00	00.00	27.80	33.50	23.39
INM-1	4825	D3	8	00.00	00.00	20.05	24.75	19.70
INM-1	860	D5	8	00.00	00.00	24.21	28.33	18.94

INM-1	5130	D7	8	00.00	00.00	23.55	29.30	20.20
INM-1	5074	E5	8	00.00	00.00	23.70	29.40	21.40
INM-1	5064	E5	8	00.00	00.00	22.55	28.55	21.23
INM-1	4944	E6	8	00.00	00.00	24.50	2815	18.75
INM-1	4943	E6	8	00.00	00.00	00.00	31.55	22.55
INM-1	4942	E6	8	00.00	00.00	27.31	30.30	24.90
INM-1	4940	E6	8	00.00	00.00	23.60	27.68	20.57
INM-1	4939	E6	8	00.00	00.00	24.10	31.91	24.55
INM-1	4941	E6	8	00.00	00.00	2.85	31.95	21.85
INM-1	4938	E6	8	47.30	00.00	20.55	27.30	21.10
INM-1	4615	E7	8	00.00	00.00	22.70	28.85	19.95
INM-1	4617	E7	8	00.00	00.00	24.20	30.74	20.65
INM-1	3896	B3&C3	8	00.00	00.00	27.50	31.25	25.05
INM-1	4138	X3&B2	8	00.00	00.00	22.25	28.29	28.80
INM-1	4709	Hno61	8	00.00	00.00	25.60	30.85	21.55
INM-1	4895	Hno71	8	00.00	00.00	27.00	30.62	23.95
INM-1	3068	C4	9	00.00	00.00	23.96	29.53	22.58
INM-1	4501	D2	9	00.00	00.00	23.70	28.25	23.35
INM-1	4213	B4&D13	9	00.00	00.00	26.04	29.60	21.15
INM-1	605	C2	10	00.00	00.00	21.30	26.31	18.83
INM-1	4839	E1	10	00.00	00.00	24.05	30.50	21.70
INM-1	5234	C3	11	00.00	00.00	24.54	31.15	21.45
INM-1	5225	C3	11	00.00	00.00	21.50	26.35	18.21
INM-1	5698	C5	11	00.00	00.00	25.90	31.65	21.65
INM-1	6130	D4	11	00.00	00.00	23.75	30.70	23.20
INM-1	3697	B1	12	52.32	29.60	22.90	29.35	20.33
INM-1	4181	B1	12	00.00	00.00	22.45	26.60	18.85
INM-1	4284	C2	12	00.00	00.00	29.85	00.00	26.40
INM-1	5755	C4	12	51.55	28.70	25.55	28.00	21.90
INM-1	5846	C6	12	00.00	00.00	30.45	34.00	26.35
INM-1	6273	C8	12	00.00	00.00	23.35	31.20	21.55
INM-1	5326	Hno79	12	00.00	00.00	24.00	28.05	19.56
INM-1	5325	Hno79	12	00.00	00.00	25.10	28.60	22.05
INM-1	6142	Hno83	12	00.00	00.00	23.70	30.20	22.60
INM-1	6143	Hno83	12	00.00	00.00	24.35	30.30	21.28
INM-1	5091	B2	13	00.00	00.00	24.07	28.48	20.25
INM-1	4681	B0	13	00.00	00.00	22.30	28.60	20.48
INM-1	5672	A4	14	00.00	00.00	26.36	30.55	22.25
INM-1	5673	A4	14	00.00	00.00	22.50	27.85	20.60
INM-1	5915	B3	14	00.00	00.00	23.35	27.97	20.95
INM-1	6078	B5	14	46.00	26.80	21.80	27.15	19.80
INM-1	6077	B5	14	00.00	00.00	29.65	33.26	25.57
INM-1	6263	E9	14	00.00	00.00	26.85	31.05	20.60
INM-1	4800	B0	14	51.85	31.30	24.56	30.95	23.88
INM-1	4799	B0	14	00.00	00.00	29.85	30.68	24.66
INM-1	4798	B0	14	00.00	00.00	29.53	30.62	23.54
INM-1	5662	A5	15	00.00	00.00	26.10	30.00	22.57
INM-1	5586	E3	15	00.00	00.00	28.80	32.40	24.05
INM-1	5585	E3	15	00.00	00.00	27.45	31.70	20.83
INM-1	5587	E3	15	00.00	00.00	26.00	30.55	23.85
INM-1	6033	Hno87	15	00.00	00.00	26.75	33.50	25.70
INM-1	6031	Hno87	15	00.00	00.00	23.16	29.40	21.80
INM-1	6032	Hno87	15	00.00	00.00	26.05	33.40	23.35

Table A.2.8. Measurements of Pelvis of *Bos indicus*

Site	Reg. No.	Trench	Layer	LA
INM-1	1826	F10	3	65.54
INM-1	4427	E7	7	70.95

Table A.2.9. Measurements of Femur of *Bos indicus*

Site	Reg.No.	Trench	Layer	BP	LC	TC
INM-1	2616	G6	1	00.00	63.50	42.90
INM-1	3337	F11	1	00.00	00.00	44.00
INM-1	3763	I11	1	00.00	55.68	42.85
INM-1	293	F26	2	00.00	53.30	41.29
INM-1	3995	G4	2	94.93	56.75	41.55
INM-1	2453	F25-26	3	00.00	47.40	39.02
INM-1	4218	A3	6	00.00	49.79	40.80
INM-1	4197	E8	7	00.00	00.00	42.15
INM-1	4560		7	00.00	57.88	44.88
INM-1	6120	Hno51	7	00.00	56.65	38.25
INM-1	4398	B6	8	00.00	63.88	46.40
INM-1	4743	C1	8	00.00	65.77	46.55
INM-1	4595	E3	9	00.00	00.00	41.30
INM-1	6216	B6	12	00.00	58.60	41.30
INM-1	5493	C1	14	00.00	64.25	47.86
INM-1	6265	E9	14	00.00	61.45	48.05

Table A.2.10. Measurements of Tibia of *Bos indicus*

Site	Reg.No.	Trench	Layer	DD	BP	TP	BD	TD
INM-1	33	LJP	0	00.00	00.00	00.00	55.72	48.80
INM-1	44	LJP	0	00.00	00.00	00.00	53.99	43.58
INM-1	148	LJP	0	00.00	00.00	00.00	49.82	28.89
INM-1	149	LJP	0	00.00	00.00	00.00	58.21	45.67
INM-1	150	LJP	0	00.00	00.00	00.00	60.73	44.62
INM-1	353	LJP	0	00.00	00.00	00.00	59.84	47.66
INM-1	2163	Hno132	1	00.00	00.00	00.00	52.01	39.78
INM-1	32	F27	2	35.55	00.00	00.00	53.61	42.26
INM-1	2110	G6	2	00.00	00.00	00.00	60.10	00.00
INM-1	3591	I6	2	00.00	58.20	42.65	00.00	00.00
INM-1	6656	F24	2	00.00	00.00	00.00	56.10	42.90
INM-1	2318	E10	3	00.00	00.00	70.19	00.00	00.00
INM-1	3464	H7	3	00.00	00.00	00.00	59.60	45.15
INM-1	4087	D10	3	00.00	00.00	00.00	53.45	40.85
INM-1	2452	F25-26	3	00.00	00.00	00.00	49.57	37.22
INM-1	6114	Hno105	4	00.00	00.00	00.00	62.45	48.77
INM-1	353		5	00.00	00.00	00.00	60.62	46.58
INM-1	3702	A3	6	00.00	00.00	00.00	55.56	43.88
INM-1	4209	D8&E7	6	00.00	00.00	00.00	60.05	44.35
INM-1	4240	E4	6	00.00	00.00	00.00	53.65	42.27
INM-1	4200	E8	7	00.00	00.00	00.00	47.85	38.45
INM-1	4299	E4	7	00.00	00.00	60.85	00.00	00.00
INM-1	4372	E5	7	00.00	00.00	00.00	51.39	38.50
INM-1	381		8	00.00	00.00	00.00	62.68	48.48
INM-1	966	E2	8	00.00	53.37	39.94	00.00	00.00
INM-1	4142	X3B2	8	00.00	65.37	34.45	00.00	00.00
INM-1	4399	B6	8	00.00	00.00	00.00	55.35	40.55
INM-1	4554	C4	8	00.00	00.00	00.00	58.98	44.68
INM-1	4945	E6	8	00.00	00.00	00.00	52.75	00.00
INM-1	5127	D7	8	00.00	00.00	00.00	55.25	45.00
INM-1	5186	D6	8	00.00	00.00	00.00	53.15	43.35
INM-1	700	B1	9	00.00	00.00	00.00	55.90	43.66
INM-1	5197	C3	9	00.00	00.00	00.00	51.65	40.50
INM-1	4288	C2	12	00.00	00.00	00.00	00.00	51.93
INM-1	6299	Hno81	12	00.00	00.00	00.00	63.10	50.15
INM-1	5841	B3	13	00.00	00.00	00.00	61.35	45.70
INM-1	5495	C1	14	00.00	00.00	00.00	42.30	31.80
INM-1	5913	B3	14	00.00	00.00	00.00	57.75	00.00
INM-1	6289	B3	14	00.00	00.00	74.70	00.00	00.00
INM-1	5390	C3	15	00.00	00.00	00.00	53.63	41.75

INM-1	5637	C5	15	00.00	00.00	00.00	49.15	36.75
INM-1	5703	E9	15	16.70	00.00	00.00	36.05	25.65
INM-1	2509	D3	16	00.00	61.41	49.73	00.00	00.00

Table A.2.11. Measurements of Metatarsal of *Bos indicus*

Site	Reg.No.	Trench	Layer	DD	BP	TP	BD	TD	WT	TT	TLT
INM-1	194	LJP	0	28.28	00.00	00.00	53.17	00.00	24.28	30.77	22.84
INM-1	389	LJP	0	27.92	00.00	00.00	00.00	24.80	00.00	00.00	00.00
INM-1	2039	G7	1	00.00	22.49	24.89	00.00	00.00	00.00	00.00	00.00
INM-1	2040	G7	1	00.00	21.80	23.93	00.00	00.00	00.00	00.00	00.00
INM-1	2041	G7	1	00.00	22.98	24.45	00.00	00.00	00.00	00.00	00.00
INM-1	2042	G7	1	00.00	21.75	23.92	00.00	00.00	00.00	00.00	00.00
INM-1	1164	G10	1	00.00	00.00	00.00	51.84	33.17	31.70	23.79	22.12
INM-1	3970	G4	2	00.00	00.00	00.00	49.30	29.20	24.00	28.36	21.76
INM-1	2824	D10	2	00.00	00.00	00.00	52.89	30.15	23.87	30.00	22.55
INM-1	3016	F10	2	00.00	00.00	00.00	53.77	32.31	25.29	32.19	23.17
INM-1	7582	F26	2	00.00	00.00	00.00	00.00	00.00	29.45	34.25	24.55
INM-1	1349	G11	2	00.00	00.00	00.00	49.54	00.00	22.60	28.13	20.64
INM-1	3195	G11	2	00.00	40.08	38.37	00.00	00.00	00.00	00.00	00.00
INM-1	3476	H7	3	00.00	45.00	43.15	00.00	00.00	00.00	00.00	00.00
INM-1	3477	H7	3	00.00	00.00	00.00	55.75	30.55	25.10	30.20	21.19
INM-1	3255	D10	3	00.00	00.00	00.00	00.00	00.00	20.70	29.05	19.85
INM-1	4090	D10	3	00.00	00.00	00.00	51.25	30.32	23.70	30.00	23.05
INM-1	2207	G10	3	00.00	43.48	39.84	00.00	00.00	00.00	00.00	00.00
INM-1	2208	G10	3	00.00	00.00	00.00	50.02	30.63	24.44	29.68	23.11
INM-1	1279	G11	3	00.00	00.00	00.00	00.00	00.00	23.40	31.76	00.00
INM-1	4876	Hno107	3	00.00	40.60	37.90	00.00	00.00	00.00	00.00	00.00
INM-1	4878	Hno107	3	00.00	00.00	00.00	00.00	00.00	23.72	30.60	23.05
INM-1	2961	D10	4	00.00	00.00	00.00	50.49	29.83	23.32	29.77	22.24
INM-1	5968	F25	4	00.00	00.00	00.00	00.00	00.00	22.10	27.50	20.60
INM-1	7138	G24	4	00.00	43.25	43.40	00.00	00.00	00.00	00.00	00.00
INM-1	6842	G26	4	00.00	00.00	00.00	47.15	26.60	21.85	26.55	21.10
INM-1	6843	G26	4	00.00	00.00	00.00	00.00	00.00	23.70	29.50	20.50
INM-1	4646	Hno111	4	00.00	00.00	00.00	48.40	27.78	21.85	27.70	20.55
INM-1	4568	E5&B1	5	00.00	00.00	00.00	56.65	33.60	27.95	32.93	25.23
INM-1	37		6	00.00	00.00	00.00	61.97	00.00	30.26	37.02	27.84
INM-1	1508	G4	6	00.00	00.00	00.00	00.00	00.00	26.43	33.43	23.56
INM-1	4557	X3	7	00.00	00.00	00.00	56.75	31.45	26.15	30.36	24.35
INM-1	7931	Hno51	7	00.00	00.00	00.00	00.00	00.00	23.85	29.50	21.85
INM-1	5047	B1	8	00.00	00.00	00.00	00.00	00.00	23.20	29.40	21.51
INM-1	4384	C5	8	00.00	00.00	00.00	50.30	28.25	25.08	28.77	21.19
INM-1	4667	D4	8	00.00	00.00	00.00	50.25	28.60	23.95	28.00	20.75
INM-1	859	D5	8	00.00	00.00	00.00	45.94	27.03	21.47	26.43	21.39
INM-1	5128	D7	8	00.00	00.00	00.00	48.80	28.30	22.30	28.10	21.75
INM-1	4616	E7	8	00.00	00.00	00.00	00.00	00.00	24.50	28.46	21.0
INM-1	3897	B3&C3	8	00.00	00.00	00.00	54.65	31.08	25.29	30.70	24.00
INM-1	732	Hno7073	8	00.00	00.00	00.00	48.34	28.47	26.88	22.85	21.08
INM-1	4891	Hno71	8	00.00	40.58	41.15	00.00	00.00	00.00	00.00	00.00
INM-1	4894	Hno71	8	00.00	00.00	00.00	00.00	00.00	23.60	28.27	20.85
INM-1	711	B1	9	00.00	00.00	00.00	54.39	30.31	24.48	30.30	21.27
INM-1	4594	E3	9	00.00	46.30	44.50	00.00	00.00	00.00	00.00	00.00
INM-1	601	C2	10	00.00	00.00	00.00	47.12	28.83	23.43	28.82	23.43
INM-1	6060	D7	11	00.00	00.00	00.00	00.00	00.00	28.25	30.15	24.05
INM-1	5529	D8	11	00.00	00.00	00.00	58.77	34.15	25.70	33.67	24.60
INM-1	6018	E3	11	00.00	00.00	00.00	00.00	00.00	22.20	27.00	18.75
INM-1	4180	B1	12	00.00	00.00	00.00	48.75	29.52	22.10	28.80	19.70
INM-1	810	C1	12	00.00	00.00	00.00	55.26	30.37	25.93	30.08	20.10
INM-1	5844	C6	12	00.00	46.85	44.00	00.00	00.00	00.00	00.00	00.00
INM-1	5845	C6	12	00.00	00.00	00.00	54.25	32.00	25.90	31.65	24.75
INM-1	6156	Hno75	12	00.00	00.00	00.00	50.45	28.85	23.10	28.65	21.35
INM-1	5671	A4	14	00.00	42.22	38.90	00.00	00.00	00.00	00.00	00.00
INM-1	6286	B3	14	00.00	00.00	00.00	00.00	00.00	23.95	29.45	22.50

INM-1	6287	B3	14	00.00	00.00	00.00	00.00	00.00	29.65	30.30	22.50
INM-1	5489	C1	14	00.00	00.00	00.00	48.65	27.69	22.98	27.66	21.36
INM-1	5645	C6	14	00.00	00.00	00.00	49.65	20.20	21.75	28.02	19.40
INM-1	5379	C3	15	00.00	00.00	00.00	57.50	32.95	27.07	32.65	22.90
INM-1	5581	E3	15	26.35	44.70	44.95	00.00	00.00	00.00	00.00	00.00
INM-1	5582	E3	15	00.00	49.65	49.60	00.00	00.00	00.00	00.00	00.00
INM-1	5583	E3	15	00.00	00.00	00.00	58.20	31.30	26.85	30.85	22.90
INM-1	5584	E3	15	00.00	00.00	00.00	00.00	00.00	26.30	34.25	25.15

Table A.2.12. Measurements of Astragalus of *Bos indicus*

Site	Reg.No.	Trench	Layer	GL	GLL	GLM	TL	TM	BD
INM-1	1159	EMBANK	0	71.46	69.56	64.08	39.39	38.83	40.29
INM-1	1163	EMBANK	0	00.00	00.00	56.97	00.00	00.00	00.00
INM-1	8179	I11	1	00.00	62.90	58.30	40.90	31.83	34.10
INM-1	8180	I11	1	00.00	69.80	63.20	42.50	36.50	35.50
INM-1	1120	G4	1	75.20	73.72	66.84	40.46	42.43	00.00
INM-1	2036	G7	1	68.43	67.48	64.17	37.13	38.91	45.13
INM-1	1002	E10	1	00.00	00.00	50.50	00.00	31.11	00.00
INM-1	3322	F11	1	00.00	00.00	48.65	00.00	28.58	00.00
INM-1	3323	F11	1	64.92	64.54	59.04	34.75	37.35	39.00
INM-1	7387	F26	1	70.00	69.55	64.40	36.90	37.60	43.75
INM-1	1173	G10	1	60.10	58.50	54.08	30.32	33.73	39.11
INM-1	1657	G11	1	63.97	63.89	59.99	32.78	35.97	38.41
INM-1	2141	Hno132	1	69.23	67.25	60.89	34.76	35.21	40.69
INM-1	2336		2	62.58	62.54	58.03	34.24	34.73	40.82
INM-1	2338		2	70.44	68.09	63.32	38.18	39.34	42.68
INM-1	2339		2	00.00	00.00	00.00	00.00	00.00	33.74
INM-1	1569	H5	2	71.54	70.41	62.44	36.60	37.53	42.44
INM-1	1093	H7	2	00.00	00.00	00.00	28.45	28.92	00.00
INM-1	1989	D10	2	68.89	68.44	61.37	35.21	37.35	36.45
INM-1	2820	D10	2	59.80	59.22	55.08	32.75	33.92	39.54
INM-1	2821	D10	2	64.95	63.67	57.90	32.82	33.83	37.33
INM-1	2822	D10	2	65.38	64.15	61.41	35.72	36.78	43.33
INM-1	1151	E10	2	00.00	67.12	00.00	34.08	00.00	00.00
INM-1	7504	F24	2	65.06	64.98	58.45	34.75	35.70	41.27
INM-1	2761	H11	2	63.30	62.18	57.44	34.38	34.32	37.59
INM-1	7910	Hno108	2	70.60	69.55	62.10	00.00	00.00	00.00
INM-1	622	Hno125	2	00.00	67.51	00.00	38.47	00.00	00.00
INM-1	4475	Hno129	2	67.00	66.69	61.00	34.85	00.00	39.55
INM-1	8182	D8	3	00.00	63.00	58.42	39.55	34.43	33.05
INM-1	8181	A3	3	00.00	60.00	54.90	36.82	30.00	29.80
INM-1	3451	H7	3	00.00	00.00	62.75	36.12	35.46	40.58
INM-1	4084	D10	3	65.38	64.85	60.25	36.15	37.05	43.40
INM-1	4085	D10	3	65.82	64.52	59.00	35.35	35.75	38.60
INM-1	4086	D10	3	62.85	61.27	57.65	33.20	35.45	39.00
INM-1	2011	D24	3	68.24	68.14	61.91	37.08	37.96	41.55
INM-1	2012	D24	3	53.85	53.54	50.08	29.32	29.85	34.19
INM-1	2312	E10	3	69.13	00.00	60.64	00.00	37.86	00.00
INM-1	2313	E10	3	00.00	00.00	00.00	32.95	00.00	00.00
INM-1	6569	F25	3	64.55	64.45	60.33	33.45	34.40	36.70
INM-1	2441	F25-26	3	55.59	55.37	52.32	31.25	30.56	36.14
INM-1	2221	G10	3	00.00	00.00	61.14	00.00	36.53	39.08
INM-1	1294	G11	3	61.41	61.07	57.54	32.39	31.61	38.64
INM-1	1295	G11	3	65.59	64.64	60.59	34.06	35.03	38.87
INM-1	1309	G11	3	00.00	61.67	00.00	31.77	00.00	00.00
INM-1	933	G12	3	65.33	64.91	58.63	34.84	35.58	38.26
INM-1	934	G12	3	58.42	58.20	53.56	30.91	32.90	34.78
INM-1	935	G12	3	54.39	53.61	50.64	31.68	31.65	34.28
INM-1	7726	G26	3	74.55	73.95	65.10	37.90	38.77	44.55
INM-1	7897	G27	3	00.00	00.00	62.35	36.88	38.10	00.00
INM-1	2643	H11	3	00.00	00.00	62.58	00.00	35.62	40.86

INM-1	3891	B3&C3	3	66.89	66.69	62.25	37.57	38.49	44.45
INM-1	1809	H5	4	59.19	57.92	57.62	32.75	31.37	37.34
INM-1	3049	H7	4	61.08	60.92	56.40	34.37	00.00	36.58
INM-1	8001	F24	4	60.00	59.85	57.17	32.35	35.20	36.63
INM-1	7676	G25	4	65.53	65.15	58.25	34.55	36.20	40.48
INM-1	7677	G25	4	65.75	65.65	61.10	35.65	36.49	40.25
INM-1	7678	G25	4	66.25	65.45	61.75	35.60	38.20	41.95
INM-1	8183	C8	5	00.00	64.40	58.90	40.20	31.65	32.27
INM-1	8184	F8	5	00.00	62.20	54.75	37.95	31.10	31.40
INM-1	8203	B6	6	00.00	58.00	52.40	35.20	28.50	28.30
INM-1	8201	E3	6	00.00	59.40	54.30	35.00	31.10	29.60
INM-1	8197	B3	7	00.00	64.30	58.80	37.70	32.10	30.95
INM-1	8200	B3	7	00.00	67.60	61.65	40.70	36.10	35.30
INM-1	4136	C3	7	71.86	71.11	63.47	39.18	42.98	46.83
INM-1	4364	E1	7	65.67	65.25	59.44	35.09	34.23	38.87
INM-1	4298	E4	7	00.00	00.00	58.95	35.70	37.35	41.18
INM-1	8195	E5	7	00.00	70.40	62.30	43.80	34.30	35.50
INM-1	8193	A5	8	00.00	68.14	61.90	40.80	35.80	34.90
INM-1	4260	B2	8	64.51	64.40	58.71	33.30	33.62	39.10
INM-1	4261	B2	8	63.34	63.29	57.70	34.73	35.78	39.04
INM-1	4162	B2	8	71.70	71.09	66.44	37.45	39.47	42.85
INM-1	8192	B3	8	00.00	62.10	55.90	37.95	33.13	34.50
INM-1	636	B5	8	60.87	60.83	54.64	28.05	00.00	34.93
INM-1	637	B5	8	00.00	00.00	57.10	30.94	32.97	00.00
INM-1	8188	B6	8	00.00	60.52	56.30	38.10	30.90	31.20
INM-1	8207	B6	8	00.00	65.90	60.05	43.30	34.60	35.40
INM-1	4734	C1	8	00.00	00.00	57.90	00.00	33.65	38.72
INM-1	4735	C1	8	63.90	63.67	59.30	34.35	35.45	39.10
INM-1	4736	C1	8	00.00	70.55	00.00	39.50	00.00	00.00
INM-1	8204	C4	8	00.00	63.80	54.30	38.95	31.00	32.00
INM-1	8209	C4	8	00.00	61.50	57.80	37.90	30.50	30.90
INM-1	8198	D1	8	00.00	63.65	56.80	37.90	32.10	31.60
INM-1	8194	D3	8	00.00	65.13	59.35	42.38	34.50	34.10
INM-1	871	D5	8	00.00	00.00	00.00	34.42	34.05	00.00
INM-1	5177	D6	8	65.37	64.85	58.40	35.65	37.79	45.00
INM-1	5178	D6	8	00.00	64.35	00.00	36.20	00.00	00.00
INM-1	8189	D6	8	00.00	63.60	58.00	41.15	32.70	31.50
INM-1	8202	D6	8	00.00	66.40	58.50	38.50	33.80	34.40
INM-1	8206	D6	8	00.00	64.80	58.50	40.73	33.80	33.50
INM-1	8190	D8	8	00.00	66.80	61.80	43.80	34.20	35.30
INM-1	8191	D9	8	00.00	67.40	60.00	42.70	35.90	3636
INM-1	968	E2	8	59.26	59.93	52.62	30.02	31.37	37.07
INM-1	5062	E5	8	71.35	71.05	63.25	37.90	40.50	43.25
INM-1	5078	E5	8	00.00	00.00	58.05	00.00	36.45	00.00
INM-1	4947	E6	8	67.60	65.89	60.15	36.57	38.05	43.05
INM-1	8187	E7	8	00.00	66.90	61.37	43.30	34.15	35.80
INM-1	8185	F3	8	00.00	61.90	56.45	38.65	33.10	32.00
INM-1	8199	G3	8	00.00	60.85	56.40	40.20	33.30	32.70
INM-1	8196	C1	8	00.00	63.50	58.44	40.60	32.11	33.50
INM-1	4710	Hno61	8	00.00	00.00	54.20	30.18	30.26	35.75
INM-1	8205	B5	9	00.00	63.50	60.10	39.80	32.30	32.20
INM-1	3062	C4	9	63.48	63.11	59.62	33.43	33.83	41.16
INM-1	3063	C4	9	61.59	61.17	56.97	33.07	33.10	00.00
INM-1	8208	C4	9	00.00	63.50	60.60	41.30	32.00	33.40
INM-1	797	C5	9	67.49	66.34	00.00	35.38	00.00	00.00
INM-1	4502	D2	9	70.60	69.95	66.11	36.45	37.10	00.00
INM-1	2880	D5	9	63.41	63.32	57.15	32.60	33.40	36.17
INM-1	2881	D5	9	60.25	60.14	56.26	31.59	3.25	35.66
INM-1	2882	D5	9	66.49	66.32	61.13	35.62	34.39	00.00
INM-1	4241	E4	9	00.00	00.00	60.89	34.57	00.00	38.46
INM-1	4233	C1	10	59.75	58.45	00.00	34.74	37.15	00.00
INM-1	681	A1	11	64.64	64.38	00.00	34.37	00.00	00.00
INM-1	5104	C1	11	00.00	00.00	55.10	00.00	31.90	39.40

INM-1	6057	C3	11	00.00	00.00	59.95	00.00	34.90	40.40
INM-1	5678	C8	11	61.55	60.95	56.85	32.95	33.45	39.35
INM-1	5551	E9	11	49.80	49.80	45.83	27.09	00.00	31.35
INM-1	5552	E9	11	63.70	61.04	57.03	33.15	33.72	38.80
INM-1	8173		12	00.00	67.80	62.60	44.30	34.00	35.05
INM-1	8176		12	00.00	00.00	57.00	37.50	29.95	31.30
INM-1	6212	B6	12	71.94	71.70	66.76	39.85	43.70	44.80
INM-1	4292	C2	12	00.00	00.00	00.00	32.58	37.87	00.00
INM-1	8164	C3	12	00.00	61.15	56.70	40.80	32.80	31.80
INM-1	5847	C6	12	00.00	00.00	64.35	36.25	37.25	41.60
INM-1	5848	C6	12	00.00	68.20	00.00	38.70	00.00	00.00
INM-1	5849	C6	12	00.00	00.00	00.00	35.35	35.65	39.50
INM-1	5861	C6	12	00.00	00.00	00.00	32.25	34.20	38.85
INM-1	3816	C1	12	60.40	60.75	55.15	33.77	36.18	38.15
INM-1	4819	B0	12	69.83	69.38	65.75	38.05	40.00	47.45
INM-1	8168	B0	12	00.00	69.80	65.70	47.60	36.60	38.00
INM-1	6157	Hno75	12	65.10	64.75	60.40	34.30	35.10	40.30
INM-1	6158	Hno75	12	70.80	68.05	62.90	39.15	41.22	48.90
INM-1	5504	Hno85	12	63.58	63.02	57.10	32.80	33.89	37.45
INM-1	5505	Hno85	12	00.00	56.77	00.00	33.55	00.00	00.00
INM-1	5547		13	60.92	60.83	56.64	32.35	34.70	00.00
INM-1	8169		13	00.00	72.95	68.80	48.40	37.40	38.40
INM-1	8174		13	00.00	00.00	62.80	44.90	35.50	34.10
INM-1	8175		13	00.00	00.00	59.75	38.60	32.53	34.00
INM-1	8163	B2	13	00.00	63.40	59.50	41.70	35.41	33.90
INM-1	8167	B4	13	00.00	71.50	63.60	47.20	36.30	38.10
INM-1	5818	C4	13	00.00	00.00	60.75	33.76	35.20	39.05
INM-1	5817	E4	13	00.00	00.00	59.45	00.00	33.60	00.00
INM-1	8170		14	00.00	68.50	63.50	46.30	34.70	33.95
INM-1	8172		14	00.00	69.40	65.10	48.10	37.20	38.50
INM-1	8177		14	00.00	00.00	61.70	41.62	33.10	33.50
INM-1	6310	A5	14	63.55	62.80	58.10	31.90	34.06	30.76
INM-1	6311	A5	14	62.32	61.95	55.70	33.10	35.30	00.00
INM-1	5911	B3	14	69.16	68.50	62.55	35.45	36.36	44.55
INM-1	5487	C1	14	61.12	60.56	55.80	32.70	33.45	37.70
INM-1	5488	C1	14	56.30	56.15	53.30	31.30	33.62	36.57
INM-1	5646	C6	14	53.96	53.30	51.30	28.20	29.95	35.75
INM-1	6090	D7	14	66.40	65.25	59.40	33.84	33.94	39.20
INM-1	5315	E3	14	00.00	00.00	57.22	00.00	32.95	37.85
INM-1	6337	E4	14	59.46	59.24	56.37	32.13	32.28	32.75
INM-1	6260	E9	14	64.70	64.40	58.23	33.55	34.60	38.52
INM-1	6261	E9	14	00.00	69.40	00.00	37.50	00.00	00.00
INM-1	8165	B0	14	00.00	62.80	58.20	38.65	30.70	31.80
INM-1	8166	B0	14	00.00	68.00	64.30	43.10	32.85	33.80
INM-1	8178	B0	14	00.00	66.31	61.80	41.15	31.47	31.80
INM-1	8171		15	00.00	63.80	57.10	38.10	31.30	31.20
INM-1	5619	A4	15	68.33	68.08	63.72	37.60	37.17	43.05
INM-1	5373	C3	15	60.45	59.57	55.25	33.30	35.19	40.38
INM-1	5377	C3	15	00.00	64.77	00.00	34.12	34.35	00.00
INM-1	5565	D4	15	61.60	61.33	56.75	31.65	33.60	38.45
INM-1	5428	E4	15	64.65	64.25	60.27	34.64	34.95	38.45
INM-1	5695	E9	15	64.75	63.25	59.10	32.15	00.00	38.80
INM-1	5696	E9	15	64.10	62.35	57.35	32.08	32.80	36.20
INM-1	5697	E9	15	59.85	58.10	5.70	30.60	31.80	33.87
INM-1	5698	E9	15	46.45	46.20	43.55	00.00	00.00	28.50
INM-1	5357	Hno95	16	69.32	67.46	62.45	35.20	00.00	39.12
INM-1	5365	Hno95	16	64.40	62.10	9.42	33.50	33.88	38.65

Table A.2.13. Measurements of Calcaneum of *Bos indicus*

Site	Reg.No.	Trench	Layer	GL	GB
INM-1	106	LJP	0	92.87	28.61
INM-1	2034	G7	1	00.00	42.70
INM-1	3327	F11	1	128.66	44.59
INM-1	1926	H11	1	00.00	53.70
INM-1	2135	Hno132	1	126.98	46.02
INM-1	3955	G4	2	00.00	50.65
INM-1	3956	G4	2	122.78	40.00
INM-1	1631	H5	2	122.14	53.69
INM-1	3386	H6	2	124.56	55.50
INM-1	3387	H6	2	107.20	37.95
INM-1	1433	D25	2	113.94	51.45
INM-1	2548	F11	2	00.00	36.06
INM-1	2813	H11	2	118.68	00.00
INM-1	2816	H11	2	123.13	00.00
INM-1	4981	C2	3	11.51	43.05
INM-1	1894	E10	3	00.00	37.68
INM-1	6568	F25	3	00.00	45.55
INM-1	7748	G26	3	00.00	42.90
INM-1	4542	H12	3	121.90	45.80
INM-1	8002	F24	4	124.45	46.95
INM-1	5984	F25	4	127.87	44.65
INM-1	3812	E8	5	00.00	47.40
INM-1	3877	B6	6	118.69	41.90
INM-1	4137	C3	7	119.03	42.58
INM-1	4856	C0	7	131.45	55.45
INM-1	4733	C1	8	00.00	51.65
INM-1	4551	C4	8	00.00	43.90
INM-1	4383	C5	8	118.80	45.10
INM-1	4583	D6	8	00.00	53.05
INM-1	4936	E6	8	00.00	44.90
INM-1	4458	E7	8	00.00	46.30
INM-1	3064	C4	9	55.98	00.00
INM-1	4570	E3	9	41.05	00.00
INM-1	4838	E1	10	106.75	40.25
INM-1	6322	B5	12	00.00	46.45
INM-1	6213	B6	12	121.90	44.10
INM-1	6214	B6	12	00.00	45.40
INM-1	4287	C2	12	00.00	36.94
INM-1	6272	C8	12	00.00	40.43
INM-1	6155	Hno75	12	00.00	38.20
INM-1	5822	D3	13	00.00	47.30
INM-1	5840	D3	13	00.00	40.00
INM-1	5458	B1	14	134.45	41.40
INM-1	6283	B3	14	125.25	39.70
INM-1	5494	C1	14	00.00	39.83
INM-1	6262	E9	14	00.00	38.70
INM-1	5620	A4	15	125.80	44.85
INM-1	5660	A5	15	122.50	38.80
INM-1	5699	E9	15	00.00	38.45

Table A.2.14. Measurements of Centrotarsal of *Bos indicus*

Site	Reg.No.	Trench	Layer	GB
INM-1	335	LJP	0	50.53
INM-1	336	LJP	0	5.64
INM-1	337	LJP	0	58.78
INM-1	338	LJP	0	51.47
INM-1	2037	G7	1	52.68
INM-1	3534	H5	1	53.80

INM-1	1003	E10	1	55.46
INM-1	1934	H11	1	54.46
INM-1	4008	G4	2	48.00
INM-1	2823	D10	2	58.58
INM-1	842	F12	2	45.43
INM-1	4083	D10	3	49.35
INM-1	2248	F11	3	44.05
INM-1	6570	F25	3	43.75
INM-1	6571	F25	3	56.45
INM-1	2902	G10	3	47.60
INM-1	2903	G10	3	56.10
INM-1	932	G12	3	53.07
INM-1	7852	G25	3	44.35
INM-1	2641	H11	3	61.15
INM-1	3892	B3&C3	3	52.15
INM-1	980	C1	4	52.73
INM-1	2959	D10	4	59.00
INM-1	8000	F24	4	41.65
INM-1	5963	F25	4	48.30
INM-1	7137	G24	4	48.90
INM-1	7696	G25	4	45.50
INM-1	4417		6	50.90
INM-1	4115	C3	7	51.30
INM-1	4370	E5	7	48.10
INM-1	4101	A1&A2	7	50.30
INM-1	4552	C4	8	48.70
INM-1	4381	C5	8	52.24
INM-1	4382	C5	8	51.35
INM-1	4409	C7	8	49.86
INM-1	4584	D6	8	52.08
INM-1	5063	E5	8	46.85
INM-1	4447	E6	8	44.70
INM-1	3738	Hno69	8	53.00
INM-1	703	B1	9	57.63
INM-1	3066	C4	9	51.68
INM-1	795	C5	9	50.53
INM-1	796	C5	9	57.30
INM-1	5714	B6	11	47.30
INM-1	6007	B7	11	55.57
INM-1	5335	D5	11	49.70
INM-1	5553	E9	11	47.85
INM-1	6321	B5	12	58.80
INM-1	5722	Hno78	12	50.77
INM-1	6300	Hno81	12	50.35
INM-1	5601	Hno84	12	50.75
INM-1	5506	Hno85	12	46.70
INM-1	5774	B5	13	50.75
INM-1	5459	B1	14	48.45
INM-1	5460	B1	14	46.91
INM-1	6076	B5	14	48.38
INM-1	5996	D6	14	45.70
INM-1	5305	E3	14	47.35
INM-1	4796	B0	14	52.92
INM-1	6363	B2	15	54.54
INM-1	5474	D3	15	47.75
INM-1	5475	D3	15	47.30
INM-1	6240	E5	15	49.90
INM-1	6029	Hno87	16	48.19
INM-1	5358	Hno95	16	48.03
INM-1	2505	D3	16	51.52
INM-1	2506	D3	16	56.53
INM-1	2507	D3	16	57.40
INM-1	4150	D3	16	49.90

Table A.2.15. Measurements of First Phalanx of *Bos indicus*

Site	Reg.No.	Trench	Layer	GL	BP	TP	BD	TD
INM-1	39	LJP	0	63.86	24.75	29.77	23.04	18.27
INM-1	94	LJP	0	63.72	26.35	30.29	24.69	18.28
INM-1	116	LJP	0	58.80	22.70	28.84	21.75	17.73
INM-1	118	LJP	0	43.91	14.78	28.07	12.45	18.52
INM-1	198	LJP	0	70.02	26.37	30.64	26.13	20.47
INM-1	257	LJP	0	56.80	26.77	30.31	23.27	18.05
INM-1	258	LJP	0	00.00	00.00	00.00	23.23	00.00
INM-1	345	LJP	0	55.19	27.81	28.88	00.00	00.00
INM-1	6348	A3&B3	0	61.80	26.65	30.80	25.80	19.80
INM-1	6349	A3&B3	0	00.00	23.80	28.50	00.00	00.00
INM-1	1110	G4	1	59.30	26.21	31.66	25.81	19.81
INM-1	3274	D10	1	59.30	00.00	00.00	25.75	18.97
INM-1	3576	B0	1	58.65	26.22	30.15	22.55	18.08
INM-1	4902	F13	1	60.90	26.72	34.05	27.45	20.78
INM-1	8068	F27	1	64.15	28.20	33.65	28.10	22.16
INM-1	8069	F27	1	00.00	26.60	29.05	00.00	00.00
INM-1	8070	F27	1	00.00	27.45	28.75	00.00	00.00
INM-1	8071	F27	1	00.00	00.00	00.00	28.80	23.21
INM-1	1169	G10	1	64.32	27.27	32.20	25.57	18.79
INM-1	1635	G11	1	67.56	33.47	35.62	29.98	22.67
INM-1	1636	G11	1	55.60	29.15	27.93	26.22	20.69
INM-1	7533	G24	1	59.35	29.50	29.85	26.25	19.60
INM-1	3760	I11	1	00.00	00.00	00.00	26.17	20.20
INM-1	2328		2	56.35	24.18	29.74	00.00	00.00
INM-1	2329		2	00.00	28.01	30.59	00.00	00.00
INM-1	3115		2	65.55	28.21	30.85	26.25	19.43
INM-1	4451	B7	2	00.00	25.78	27.65	00.00	00.00
INM-1	2305	E1	2	00.00	32.37	36.38	00.00	00.00
INM-1	3069	G3	2	00.00	00.00	00.00	26.41	19.57
INM-1	3929	G4	2	59.55	25.55	30.95	24.55	18.70
INM-1	2066	G6	2	00.00	00.00	00.00	23.99	19.19
INM-1	1526	H5	2	65.13	28.45	33.20	29.18	21.32
INM-1	1527	H5	2	68.79	29.54	34.90	28.34	22.34
INM-1	1530	H5	2	00.00	21.58	29.53	00.00	00.00
INM-1	3577	B0	2	00.00	26.65	32.48	00.00	00.00
INM-1	4068	C0	2	51.15	25.15	26.95	25.15	18.93
INM-1	4069	C0	2	58.60	28.10	32.05	00.00	00.00
INM-1	1143	E10	2	61.02	25.25	31.08	00.00	00.00
INM-1	7325	E24	2	00.00	23.90	28.45	00.00	00.00
INM-1	7326	E24	2	00.00	25.05	27.62	00.00	00.00
INM-1	7327	E24	2	00.00	00.00	00.00	20.55	17.05
INM-1	6372	E25	2	69.30	36.80	33.95	32.25	23.58
INM-1	6373	E25	2	60.80	27.25	28.65	18.50	00.00
INM-1	6374	E25	2	00.00	00.00	00.00	24.39	17.95
INM-1	2985	F10	2	57.62	27.21	31.91	25.90	20.64
INM-1	845	F12	2	00.00	00.00	00.00	27.25	29.36
INM-1	6646	F24	2	00.00	00.00	00.00	24.85	19.70
INM-1	6647	F24	2	00.00	28.15	27.65	00.00	00.00
INM-1	3121	G11	2	56.23	26.46	27.61	24.69	18.46
INM-1	7040	G25	2	62.05	27.20	28.95	25.67	18.95
INM-1	7041	G25	2	56.45	25.75	28.12	25.30	19.45
INM-1	7190	G25	2	58.55	25.10	28.80	26.25	18.00
INM-1	7191	G25	2	00.00	00.00	00.00	24.00	17.45
INM-1	7192	G25	2	00.00	00.00	00.00	29.35	23.60
INM-1	3564	I10	2	00.00	00.00	00.00	24.60	18.50
INM-1	7905	Hno108	2	56.95	21.95	25.65	21.55	15.55
INM-1	7906	Hno108	2	58.30	23.95	29.97	24.07	18.20
INM-1	7907	Hno108	2	00.00	32.78	34.05	00.00	00.00
INM-1	624	Hno125	2	00.00	00.00	00.00	18.74	14.45

INM-1	2659	G7	3	54.61	25.43	27.03	22.77	18.86
INM-1	3243	D10	3	00.00	00.00	00.00	24.30	18.65
INM-1	3244	D10	3	61.90	26.10	30.60	25.70	19.55
INM-1	3245	D10	3	61.08	25.35	28.75	24.25	18.75
INM-1	3246	D10	3	00.00	00.00	00.00	24.55	18.75
INM-1	4077	D10	3	57.40	30.00	31.95	28.36	20.03
INM-1	4078	D10	3	60.95	24.35	30.55	23.15	18.75
INM-1	4079	D10	3	00.00	24.00	30.00	00.00	00.00
INM-1	4080	D10	3	61.75	26.85	31.58	25.45	18.36
INM-1	4081	D10	3	00.00	25.95	28.57	00.00	00.00
INM-1	1876	E10	3	55.14	28.55	28.84	25.72	19.60
INM-1	1877	E10	3	00.00	00.00	00.00	16.29	13.14
INM-1	6765	E23	3	58.45	29.85	34.30	27.25	21.75
INM-1	6956	E26	3	69.70	31.45	35.25	30.95	21.85
INM-1	6957	E26	3	53.25	25.65	26.00	24.55	18.25
INM-1	6958	E26	3	66.95	32.00	34.15	30.70	22.95
INM-1	6959	E26	3	58.35	26.15	29.60	24.55	00.00
INM-1	2241	F11	3	62.96	27.93	32.62	26.57	19.63
INM-1	6562	F25	3	54.90	26.10	28.35	25.10	19.70
INM-1	6563	F25	3	59.70	30.10	31.85	30.45	22.10
INM-1	2178	G10	3	62.23	28.90	32.60	00.00	23.18
INM-1	2179	G10	3	60.46	26.31	31.35	25.21	19.29
INM-1	878	G12	3	56.51	2628	28.08	26.27	19.63
INM-1	7850	G25	3	61.70	32.40	31.60	29.00	22.85
INM-1	7895	G27	3	57.60	29.45	30.55	27.68	21.45
INM-1	4540	H12	3	55.30	27.06	28.85	23.78	17.25
INM-1	6171	Hno106	3	00.00	00.00	00.00	29.75	22.15
INM-1	3653	Hno127	3	57.62	26.68	29.75	23.63	18.70
INM-1	2956	D10	4	58.57	24.70	27.35	24.39	18.56
INM-1	8027	E24	4	00.00	00.00	00.00	25.40	18.20
INM-1	5846	E27	4	00.00	25.85	30.25	00.00	00.00
INM-1	5847	F27	4	00.00	00.00	00.00	23.75	18.41
INM-1	7994	F24	4	00.00	16.42	20.35	00.00	00.00
INM-1	7995	F24	4	62.80	30.35	31.25	29.10	23.35
INM-1	8007	F24	4	00.00	00.00	00.00	24.50	18.60
INM-1	8009	F24	4	00.00	11.55	15.80	00.00	00.00
INM-1	5945	F25	4	59.20	25.20	27.90	22.60	17.37
INM-1	5946	F25	4	56.56	25.65	29.00	24.56	17.75
INM-1	5947	F25	4	00.00	23.06	28.45	00.00	00.00
INM-1	7131	G24	4	61.00	25.40	31.95	25.00	18.91
INM-1	7132	G24	4	61.50	26.25	31.35	24.35	19.00
INM-1	7667	G25	4	55.15	27.40	29.15	24.55	18.60
INM-1	7668	G25	4	61.20	22.65	29.30	23.15	17.40
INM-1	7669	G25	4	61.27	26.85	30.70	24.54	17.95
INM-1	6832	G26	4	64.40	30.90	34.75	33.96	23.30
INM-1	6833	G26	4	00.00	27.00	29.25	00.00	00.00
INM-1	6834	G26	4	00.00	29.15	29.30	00.00	00.00
INM-1	6835	G26	4	00.00	27.45	29.05	00.00	00.00
INM-1	6836	G26	4	00.00	00.00	00.00	24.20	18.00
INM-1	4773	B0	4	00.00	25.55	28.40	00.00	00.00
INM-1	4774	B0	4	00.00	24.32	28.00	00.00	00.00
INM-1	7765	E25&D24	4	57.13	00.00	27.45	22.45	16.65
INM-1	7766	E25&D24	4	56.35	25.55	27.00	24.20	18.75
INM-1	7767	E25&D24	4	62.60	26.95	32.55	26.15	19.70
INM-1	776	E25&D24	4	00.00	27.60	28.75	00.00	00.00
INM-1	779	E25&D24	4	00.00	25.82	28.40	00.00	00.00
INM-1	7248	Hno104	4	57.80	24.80	26.27	21.90	16.80
INM-1	7249	Hno104	4	53.20	23.80	25.75	22.35	17.80
INM-1	7250	Hno104	4	00.00	00.00	00.00	27.37	19.75
INM-1	648	Hno110	4	59.28	23.99	28.62	24.52	18.37
INM-1	649	Hno110	4	61.70	25.28	29.82	24.59	19.07
INM-1	650	Hno110	4	00.00	00.00	00.00	26.13	20.68
INM-1	652	Hno110	4	00.00	26.27	28.72	00.00	00.00

INM-1	659	Hno110	4	00.00	00.00	00.00	25.56	00.00
INM-1	4631	Hno111	4	00.00	00.00	00.00	17.85	16.00
INM-1	4632	Hno111	4	00.00	26.50	32.52	00.00	00.00
INM-1	4633	Hno111	4	00.00	24.65	29.45	00.00	00.00
INM-1	4634	Hno111	4	00.00	00.00	00.00	27.75	21.60
INM-1	4628	Hno111	4	00.00	19.80	23.05	00.00	00.00
INM-1	4629	Hno111	4	00.00	00.00	00.00	18.15	16.10
INM-1	4630	Hno111	4	00.00	00.00	00.00	18.80	16.67
INM-1	787	Hno113	4	65.00	29.77	34.56	31.66	22.06
INM-1	5025	Hno119	4	57.30	25.30	30.90	24.15	17.50
INM-1	5038	Hno119	4	56.75	28.64	29.25	27.10	00.00
INM-1	5016	Hno122	4	57.30	29.95	27.85	26.15	20.15
INM-1	784	Hno124	4	58.53	28.42	28.77	26.61	20.72
INM-1	785	Hno124	4	00.00	00.00	29.26	00.00	00.00
INM-1	2970	B3	5	57.88	29.49	30.13	26.87	19.70
INM-1	2985	B3	5	00.00	27.80	23.67	26.15	00.00
INM-1	4724	B4	5	00.00	24.30	28.50	00.00	00.00
INM-1	4101	C3	5	00.00	20.50	26.35	00.00	00.00
INM-1	3804	E8	5	00.00	24.66	25.95	00.00	00.00
INM-1	4308	F8	5	59.05	31.25	35.05	29.00	22.10
INM-1	4415		6	62.65	25.67	32.00	25.92	20.15
INM-1	3703	A3	6	00.00	00.00	00.00	24.10	17.95
INM-1	5202	B3	6	00.00	26.80	28.30	00.00	00.00
INM-1	5203	B3	6	61.77	27.76	29.66	25.50	18.46
INM-1	5204	B3	6	59.70	31.45	31.86	31.66	20.00
INM-1	5218	B3	6	00.00	22.45	26.75	00.00	00.00
INM-1	4206	E1	6	42.55	30.35	32.80	26.25	28.29
INM-1	1490	G4	6	59.17	25.13	27.43	23.80	17.41
INM-1	1491	G4	6	57.21	25.90	30.94	27.78	21.91
INM-1	2258	G5	6	59.52	2270	27.92	23.50	18.27
INM-1	3868	D3	6	00.00	00.00	00.00	25.90	20.00
INM-1	4477	X4	6	60.06	31.60	32.40	30.60	22.65
INM-1	4478	X4	6	59.55	30.77	32.75	29.90	21.92
INM-1	4558		7	67.20	31.17	36.50	29.35	22.30
INM-1	4559		7	65.85	29.75	34.90	28.00	21.25
INM-1	4696	E2	7	00.00	00.00	00.00	28.20	20.25
INM-1	4697	E2	7	61.60	23.60	29.45	23.50	17.90
INM-1	4698	E2	7	41.42	00.00	00.00	25.50	26.35
INM-1	4699	E2	7	40.85	28.65	30.70	25.13	27.45
INM-1	4296	E4	7	61.80	25.35	30.60	24.85	18.35
INM-1	4191	E8	7	00.00	00.00	00.00	18.80	14.10
INM-1	1827	G4	7	59.93	26.88	28.08	25.60	19.98
INM-1	6359	Hno51A	7	65.85	28.90	33.05	27.20	20.05
INM-1	6360	Hno51A	7	65.30	00.00	35.00	28.30	20.63
INM-1	6117	Hno51	7	58.10	23.80	27.70	23.10	17.50
INM-1	7698	Hno51	7	00.00	27.95	29.05	00.00	00.00
INM-1	5046	B1	8	58.55	25.30	28.30	21.75	23.05
INM-1	4257	B2	8	00.00	26.93	26.26	00.00	00.00
INM-1	4258	B2	8	64.35	27.90	33.15	25.60	19.60
INM-1	4964	B3	8	57.95	31.37	29.02	29.00	20.75
INM-1	631	B5	8	00.00	00.00	00.00	23.75	00.00
INM-1	4389	B6	8	52.83	25.00	00.00	26.25	18.45
INM-1	4729	C1	8	60.23	23.90	30.20	23.87	18.90
INM-1	4730	C1	8	00.00	00.00	00.00	25.90	18.60
INM-1	4972	C2	8	54.40	26.35	30.95	25.20	18.90
INM-1	4973	C2	8	62.65	32.67	33.60	00.00	00.00
INM-1	4974	C2	8	00.00	26.82	31.12	00.00	00.00
INM-1	4975	C2	8	00.00	00.00	00.00	25.65	18.88
INM-1	4546	C4	8	00.00	25.78	29.25	00.00	00.00
INM-1	4547	C4	8	00.00	00.00	00.00	30.05	20.40
INM-1	4377	C5	8	62.30	32.25	34.28	29.80	22.16
INM-1	4378	C5	8	00.00	00.00	00.00	22.90	17.25
INM-1	4405	C7	8	58.52	26.62	00.00	25.80	22.30

INM-1	4407	C7	8	55.30	25.78	29.65	23.60	18.65
INM-1	4497	D2	8	00.00	00.00	00.00	27.86	21.69
INM-1	864	D5	8	66.05	27.83	34.54	27.01	20.82
INM-1	866	D5	8	00.00	00.00	00.00	25.97	19.96
INM-1	867	D5	8	00.00	27.51	28.62	00.00	00.00
INM-1	4577	D6	8	00.00	00.00	00.00	26.75	18.85
INM-1	5161	D6	8	00.00	00.00	00.00	27.00	19.35
INM-1	5119	D7	8	00.00	24.15	27.60	00.00	00.00
INM-1	5120	D7	8	56.85	27.80	28.95	28.15	19.60
INM-1	4845	E4	8	00.00	00.00	00.00	23.50	17.55
INM-1	5052	E5	8	61.05	28.10	31.75	30.15	21.65
INM-1	5053	E5	8	62.52	31.15	32.35	30.35	20.50
INM-1	5054	E5	8	00.00	26.40	30.20	00.00	00.00
INM-1	5055	E5	8	00.00	26.95	30.30	00.00	00.00
INM-1	5056	E5	8	00.00	24.86	18.59	00.00	00.00
INM-1	5076	E5	8	00.00	00.00	00.00	25.45	18.33
INM-1	4438	E6	8	55.75	25.25	26.40	25.95	19.08
INM-1	4439	E6	8	63.15	30.65	36.00	30.40	22.35
INM-1	4604	E7	8	00.00	27.60	26.20	00.00	00.00
INM-1	4605	E7	8	57.25	25.62	25.76	23.60	19.50
INM-1	4243	E8	8	65.65	29.15	36.45	29.95	21.15
INM-1	4244	E8	8	00.00	24.80	29.03	00.00	00.00
INM-1	4928	E26	8	61.35	31.10	34.40	27.90	22.00
INM-1	4929	E26	8	64.70	27.36	32.17	26.08	00.00
INM-1	4930	E26	8	60.50	27.18	27.70	25.25	20.80
INM-1	4931	E26	8	00.00	00.00	00.00	27.05	20.33
INM-1	4932	E26	8	57.50	30.40	31.60	00.00	00.00
INM-1	3882	B3&C3	8	59.70	22.45	28.20	21.85	17.17
INM-1	3883	B3&C3	8	60.46	23.86	28.54	24.50	17.55
INM-1	3884	B3&C3	8	53.46	25.52	26.09	22.87	18.67
INM-1	3885	B3&C3	8	54.68	27.17	28.40	25.04	19.79
INM-1	3887	B3&C3	8	60.15	25.75	31.13	25.98	18.48
INM-1	3888	B3&C3	8	60.60	31.45	34.40	27.28	20.20
INM-1	3889	B3&C3	8	72.65	32.65	39.46	30.60	22.50
INM-1	4126	B3&C3	8	62.34	28.88	30.97	27.66	18.8
INM-1	4127	X3&B2	8	00.00	23.45	26.68	00.00	00.00
INM-1	4128	X3&B2	8	00.00	00.00	00.00	26.96	21.98
INM-1	4491	Hno49	8	00.00	27.15	38.35	00.00	00.00
INM-1	4492	Hno49	8	00.00	26.98	00.00	00.00	00.00
INM-1	4598	Hno65	8	00.00	00.00	00.00	26.46	00.00
INM-1	4599	Hno65	8	00.00	00.00	00.00	15.60	13.85
INM-1	3730	Hno68	8	60.58	27.50	29.40	26.25	18.65
INM-1	3731	Hno68	8	00.00	00.00	00.00	27.18	18.18
INM-1	3732	Hno68	8	00.00	00.00	00.00	25.55	17.00
INM-1	4882	Hno71	8	58.42	24.92	29.60	23.45	17.65
INM-1	4144	Hno73	8	70.70	29.80	37.00	27.85	21.60
INM-1	4145	Hno73	8	00.00	00.00	00.00	28.80	22.45
INM-1	734	Hno70&73	8	00.00	00.00	00.00	24.98	18.57
INM-1	1743	A1	9	00.00	27.00	38.15	00.00	00.00
INM-1	701	B1	9	00.00	33.69	38.68	00.00	00.00
INM-1	702	B1	9	65.60	26.71	33.03	27.67	19.44
INM-1	4347	C1	9	70.68	30.76	34.25	31.60	21.80
INM-1	4348	C1	9	58.25	25.37	27.40	23.86	17.55
INM-1	802	C5	9	53.68	27.46	30.98	25.02	20.95
INM-1	5113	C5	9	00.00	00.00	00.00	24.30	18.20
INM-1	2873	D5	9	00.00	00.00	00.00	24.46	18.89
INM-1	722	D6	9	00.00	26.31	00.00	22.51	24.73
INM-1	4574	E3	9	51.55	24.22	26.26	23.90	19.15
INM-1	5261	A5	10	60.15	00.00	28.85	26.40	19.10
INM-1	5011	B1	10	57.20	22.75	27.70	22.15	17.15
INM-1	4229	C1	10	00.00	00.00	00.00	24.20	18.39
INM-1	600	C2	10	00.00	27.26	32.01	00.00	00.00
INM-1	607	C2	10	60.32	28.79	33.04	26.27	21.44

INM-1	604	C3	10	00.00	26.75	30.22	00.00	00.00
INM-1	612	C3	10	00.00	00.00	00.00	22.28	00.00
INM-1	4832	E1	10	69.27	31.45	36.60	31.40	22.77
INM-1	4810	B0	10	00.00	39.73	24.06	00.00	00.00
INM-1	4765	C0	10	50.45	00.00	00.00	19.75	16.15
INM-1	5241	B1	11	00.00	00.00	00.00	18.00	14.85
INM-1	5146	B2	11	00.00	00.00	00.00	14.86	13.10
INM-1	5147	B2	11	59.30	24.75	28.05	25.75	17.86
INM-1	5148	B2	11	54.70	25.90	29.25	24.64	19.10
INM-1	6003	B7	11	00.00	25.40	28.90	00.00	00.00
INM-1	6004	B7	11	00.00	25.80	27.55	00.00	00.00
INM-1	6005	B7	11	00.00	00.00	00.00	16.94	15.00
INM-1	5097	C1	11	61.25	26.00	30.10	27.95	20.27
INM-1	6092	C7	11	00.00	00.00	00.00	20.01	17.40
INM-1	6127	D4	11	00.00	00.00	00.00	15.70	13.15
INM-1	5333	D5	11	00.00	25.78	29.45	00.00	00.00
INM-1	5518	D8	11	00.00	32.40	33.88	00.00	00.00
INM-1	5519	D8	11	00.00	22.90	30.75	00.00	00.00
INM-1	5221	E1	11	63.50	32.40	36.00	30.20	24.10
INM-1	5764	E4	11	56.15	25.95	26.80	25.20	18.95
INM-1	0	B0	11	00.00	00.00	00.00	20.50	13.40
INM-1	3686	B1	12	58.70	28.03	32.30	26.76	19.55
INM-1	4172	B1	12	64.85	29.62	33.80	00.00	00.00
INM-1	5	B2	12	00.00	00.00	00.00	26.03	19.28
INM-1	5934	B3	12	00.00	25.25	28.57	00.00	00.00
INM-1	5935	B3	12	00.00	26.10	26.85	00.00	00.00
INM-1	6317	B5	12	00.00	00.00	00.00	16.75	14.35
INM-1	6318	B5	12	00.00	00.00	00.00	28.90	22.30
INM-1	6319	B5	12	64.05	25.80	31.14	26.45	18.90
INM-1	6101	B6	12	00.00	00.00	00.00	24.00	16.55
INM-1	6102	B6	12	60.50	29.10	31.20	27.85	19.63
INM-1	6103	B6	12	62.15	30.82	31.60	30.05	21.90
INM-1	6104	B6	12	69.60	30.60	35.50	28.15	22.50
INM-1	5603	C3	12	63.72	32.65	33.20	31.75	23.20
INM-1	5766	C5	12	64.50	30.15	34.00	30.00	21.20
INM-1	5837A	C6	12	63.36	34.40	34.90	31.30	24.40
INM-1	5838A	C6	12	59.35	26.36	30.65	25.30	18.65
INM-1	5733	E3	12	00.00	00.00	00.00	32.90	28.48
INM-1	5734	E3	12	00.00	00.00	00.00	26.85	15.35
INM-1	5735	E3	12	00.00	00.00	00.00	13.20	11.95
INM-1	522	Hno50	12	00.00	26.04	33.17	00.00	00.00
INM-1	6145	Hno75	12	56.30	24.70	27.40	23.20	17.90
INM-1	6146	Hno75	12	59.60	24.60	28.50	24.75	18.25
INM-1	6147	Hno75	12	58.80	23.70	29.14	24.90	17.92
INM-1	6148	Hno75	12	56.67	00.00	00.00	24.65	19.35
INM-1	6149	Hno75	12	00.00	26.70	30.0	00.00	00.00
INM-1	5723	Hno77	12	59.45	26.45	30.40	23.20	17.50
INM-1	5720	Hno78	12	64.85	28.90	31.90	28.40	20.40
INM-1	5316	Hno79	12	53.95	25.65	25.68	24.50	20.65
INM-1	5317	Hno79	12	58.05	27.75	29.65	26.35	20.00
INM-1	6365	Hno80	12	00.00	00.00	00.00	25.20	18.74
INM-1	6366	Hno80	12	00.00	20.95	22.65	00.00	00.00
INM-1	5592	Hno84	12	58.75	24.20	27.10	23.60	17.50
INM-1	5593	Hno84	12	58.73	26.30	29.70	23.55	18.15
INM-1	5594	Hno84	12	00.00	00.00	00.00	19.43	16.00
INM-1	5598	Hno84	12	61.85	26.95	28.45	26.03	18.75
INM-1	6294	Hno84	12	62.80	30.90	33.10	28.95	22.35
INM-1	6295	Hno84	12	55.90	24.05	26.20	21.70	17.25
INM-1	5496	Hno85	12	68.00	29.65	35.65	29.36	22.00
INM-1	5497	Hno85	12	53.40	24.60	25.10	23.15	18.60
INM-1	5498	Hno85	12	00.00	00.00	00.00	21.20	17.27
INM-1	5541		13	62.40	27.06	31.33	27.65	20.85
INM-1	5542		13	00.00	00.00	00.00	22.25	17.70

INM-1	5543		13	00.00	00.00	00.00	28.02	21.30
INM-1	5082	B2	13	58.80	24.58	28.60	24.96	19.60
INM-1	5083	B2	13	56.85	28.00	27.50	25.80	19.55
INM-1	5084	B2	13	60.85	26.77	28.90	24.40	19.15
INM-1	5832	B3	13	00.00	00.00	00.00	17.60	14.20
INM-1	5728	C3	13	57.75	29.30	30.70	28.00	21.90
INM-1	5729	C3	13	00.00	00.00	00.00	19.10	15.24
INM-1	5808	D3	13	53.00	25.85	25.52	00.00	00.00
INM-1	5809	E5	13	00.00	26.90	19.25	00.00	00.00
INM-1	4673	B0	13	00.00	00.00	00.00	30.85	21.10
INM-1	8159	A1	14	54.85	23.25	26.45	23.63	17.25
INM-1	5668	A4	14	54.87	25.75	28.44	23.90	18.30
INM-1	6308	A5	14	60.62	25.85	28.45	24.68	18.60
INM-1	5445	B1	14	62.10	29.95	33.22	28.15	20.12
INM-1	5446	B1	14	56.64	25.95	27.35	24.10	19.60
INM-1	5447	B1	14	55.85	24.48	28.05	21.55	17.78
INM-1	5450	B1	14	00.00	25.80	29.65	00.00	00.00
INM-1	5899	B3	14	00.00	24.45	27.30	00.00	00.00
INM-1	5900	B3	14	00.00	28.95	30.60	00.00	00.00
INM-1	6279	B3	14	00.00	31.90	31.40	00.00	00.00
INM-1	5792	B5	14	68.75	31.15	33.60	29.20	20.05
INM-1	5793	B5	14	62.05	23.86	00.00	26.50	18.76
INM-1	5991	D6	14	00.00	00.00	00.00	21.70	16.45
INM-1	5296	E3	14	00.00	27.15	00.00	00.00	00.00
INM-1	5297	E3	14	58.00	27.50	29.15	26.00	18.55
INM-1	6331	E4	14	62.55	00.00	29.60	24.35	17.95
INM-1	6332	E4	14	00.00	00.00	00.00	24.50	18.50
INM-1	6251	E9	14	61.35	30.50	31.85	28.25	22.15
INM-1	6252	E9	14	00.00	27.50	29.55	00.00	00.00
INM-1	6253	E9	14	00.00	27.00	29.15	00.00	00.00
INM-1	4787	B0	14	63.28	31.35	33.77	29.30	21.23
INM-1	4788	B0	14	60.25	24.00	30.30	23.77	19.25
INM-1	5613	A4	15	00.00	24.45	27.50	00.00	00.00
INM-1	5614	A4	15	58.40	24.37	28.00	24.65	18.35
INM-1	5615	A4	15	00.00	00.00	00.00	25.05	19.10
INM-1	5616	A4	15	00.00	30.02	34.45	00.00	00.00
INM-1	5650	A5	15	62.00	26.30	30.30	27.05	18.95
INM-1	5651	A5	15	52.75	26.55	29.90	21.40	17.70
INM-1	6101	B3	15	00.00	26.80	32.50	00.00	00.00
INM-1	6231	C4	15	51.65	26.30	27.70	21.60	17.00
INM-1	5633	C5	15	59.35	25.07	27.30	25.00	17.36
INM-1	5634	C5	15	00.00	00.00	00.00	23.10	16.55
INM-1	5638	C5	15	60.05	24.80	29.35	23.68	17.25
INM-1	5640	C5	15	00.00	00.00	00.00	21.00	16.80
INM-1	5467	D3	15	57.77	27.76	30.14	25.05	21.95
INM-1	5468	D3	15	54.87	24.75	24.95	23.90	18.95
INM-1	5469	D3	15	00.00	00.00	00.00	23.90	17.90
INM-1	4812	B0	15	55.87	25.71	26.32	23.40	18.60
INM-1	4819	B0	15	00.00	27.30	28.85	00.00	00.00
INM-1	6022	Hno87	16	63.50	26.95	34.50	28.34	21.80
INM-1	6023	Hno87	16	00.00	00.00	00.00	18.47	15.25
INM-1	5354	Hno95	16	00.00	00.00	00.00	17.95	15.40
INM-1	6000	A5	16	00.00	30.30	33.15	00.00	00.00
INM-1	6000	A5	16	59.80	24.45	29.70	23.55	18.35
INM-1	2495	D3	16	62.51	26.83	34.41	25.76	20.21

Table A.2.16. Measurements of Second Phalanx of *Bos indicus*

Site	Reg.No.	Trench	Layer	GL	BP	TP	BD	TD
INM-1	113	LJP	0	43.18	26.70	30.58	24.08	25.53
INM-1	117	LJP	0	41.42	28.05	31.32	25.14	27.83
INM-1	229	LJP	0	44.02	28.13	30.93	24.75	27.52
INM-1	230	LJP	0	39.33	24.28	28.48	21.79	24.61
INM-1	232	LJP	0	37.70	24.73	28.97	21.39	25.29
INM-1	233	LJP	0	42.28	26.50	00.00	21.55	00.00
INM-1	379	LJP	0	41.61	31.87	33.55	26.10	31.92
INM-1	5271	LJP	0	44.33	29.70	31.86	25.20	24.95
INM-1	5272	LJP	0	40.62	27.33	28.26	24.25	25.60
INM-1	6350	A3&B3	0	00.00	27.00	00.00	23.30	24.40
INM-1	6351	A3&B3	0	38.10	26.30	27.39	23.08	00.00
INM-1	3523	H5	1	45.85	30.00	32.50	26.33	27.80
INM-1	3524	H5	1	40.95	26.94	28.68	22.33	24.90
INM-1	3275	D10	1	41.40	26.80	30.05	24.95	24.50
INM-1	3276	D10	1	40.55	24.99	26.99	21.28	23.68
INM-1	3578	B0	1	00.00	00.00	00.00	25.15	27.57
INM-1	987	E10	1	41.91	25.66	28.30	22.05	24.75
INM-1	4907	F13	1	43.88	29.52	32.12	25.35	27.42
INM-1	8073	F24	1	35.20	23.68	26.30	18.95	22.48
INM-1	8074	F24	1	00.00	30.75	33.75	00.00	00.00
INM-1	8075	F24	1	00.00	00.00	00.00	24.48	26.90
INM-1	8076	F24	1	00.00	00.00	00.00	21.70	22.35
INM-1	8072	F27	1	44.10	30.45	30.52	24.65	26.39
INM-1	1649	G11	1	42.25	28.73	30.81	24.16	25.76
INM-1	3111	E2	2	24.32	10.15	13.80	8.80	11.48
INM-1	3112	E2	2	23.67	9.83	14.05	9.23	10.87
INM-1	3948	G4	2	40.10	27.60	29.20	24.80	28.20
INM-1	3949	G4	2	41.31	26.20	27.45	21.90	26.86
INM-1	3950	G4	2	37.05	24.31	25.40	22.48	25.70
INM-1	3626	G7	2	42.70	29.95	32.95	27.93	32.67
INM-1	1549	H5	2	42.05	25.80	00.00	21.61	24.71
INM-1	3379	H6	2	00.00	25.70	00.00	23.00	25.87
INM-1	1065	H7	2	39.16	29.65	22.72	28.56	00.00
INM-1	1992	D10	2	37.08	24.31	00.00	19.98	25.53
INM-1	2817	D10	2	40.12	25.62	28.38	28.68	25.36
INM-1	1419	D25	2	39.79	23.20	26.35	21.12	21.83
INM-1	4070	C0	2	36.20	23.85	37.15	22.20	26.05
INM-1	4071	C0	2	38.60	24.40	28.10	21.75	24.05
INM-1	7328	E24	2	00.00	00.00	00.00	20.95	22.65
INM-1	7329	E24	2	00.00	26.20	27.70	00.00	00.00
INM-1	6386	E25	2	40.90	25.15	27.25	21.95	23.00
INM-1	6387	E25	2	44.05	30.40	32.70	24.50	24.35
INM-1	7452	E25	2	41.50	25.49	27.20	22.52	23.28
INM-1	2989	F10	2	40.81	24.91	27.88	20.97	23.08
INM-1	2990	F10	2	41.79	24.52	28.91	20.66	24.42
INM-1	2991	F10	2	38.82	25.30	29.55	24.24	28.36
INM-1	2992	F10	2	00.00	24.97	27.45	00.00	00.00
INM-1	2528	F11	2	46.95	30.30	34.11	25.92	28.73
INM-1	2529	F11	2	46.48	26.13	28.43	21.24	24.12
INM-1	6648	F24	2	44.10	30.10	33.35	25.30	26.25
INM-1	7580	F26	2	48.90	33.95	35.65	29.15	28.64
INM-1	7581	F26	2	44.00	28.90	29.90	24.87	27.45
INM-1	1330	G11	2	00.00	26.80	29.90	00.00	00.00
INM-1	3147	G11	2	43.68	31.52	32.09	28.59	32.29
INM-1	3148	G11	2	41.49	25.03	28.11	20.98	23.43
INM-1	3149	G11	2	40.66	27.23	30.10	23.70	24.30
INM-1	7042	G24	2	39.30	00.00	28.00	21.80	21.85
INM-1	7194	G25	2	42.50	26.36	30.08	20.75	23.40
INM-1	7195	G25	2	00.00	25.95	28.50	00.00	00.00
INM-1	760	I12	2	00.00	00.00	00.00	25.43	00.00

INM-1	7908	Hno108	2	44.90	29.05	28.26	21.60	24.35
INM-1	7909	Hno108	2	38.46	23.80	28.26	21.60	24.35
INM-1	4467	Hno129	2	45.45	28.60	31.85	23.70	28.25
INM-1	2662	G7	3	00.00	28.44	31.38	00.00	00.00
INM-1	3431	H7	3	45.70	31.60	33.75	24.65	25.35
INM-1	3432	H7	3	00.00	27.35	00.00	26.10	26.54
INM-1	3433	H7	3	40.50	25.60	27.62	21.25	24.05
INM-1	3594	I6	3	38.80	00.00	00.00	20.66	23.63
INM-1	3247	D10	3	39.00	24.85	28.72	20.42	24.05
INM-1	3249	D10	3	39.65	28.79	00.00	25.83	26.12
INM-1	3250	D10	3	38.40	25.81	28.85	23.67	27.64
INM-1	4035	C0	3	38.85	24.68	27.75	20.13	20.90
INM-1	1880	E10	3	40.21	22.99	26.84	00.00	00.00
INM-1	2308	E10	3	00.00	27.26	31.97	00.00	00.00
INM-1	6766	E23	3	42.45	27.50	30.65	23.95	25.80
INM-1	6767	E23	3	37.70	26.40	28.65	21.37	23.67
INM-1	6768	E23	3	40.05	25.10	27.60	21.62	22.65
INM-1	6769	E23	3	40.44	24.65	29.30	23.50	26.20
INM-1	6787	E23	3	00.00	24.32	28.70	00.00	00.00
INM-1	6788	E23	3	00.00	18.50	22.50	00.00	00.00
INM-1	1819	F10	3	39.66	26.15	29.12	23.19	26.09
INM-1	3263	F11	3	00.00	24.18	00.00	21.40	23.65
INM-1	6564	F25	3	43.78	30.05	00.00	27.00	28.00
INM-1	6565	F25	3	43.16	29.45	00.00	25.20	26.45
INM-1	6566	F25	3	38.20	24.53	28.04	21.66	23.40
INM-1	6567	F25	3	41.50	25.25	29.80	21.96	24.50
INM-1	2376	F25&F26	3	38.41	25.74	27.04	21.17	23.64
INM-1	2377	F25&F26	3	00.00	24.41	30.17	00.00	00.00
INM-1	2378	F25&F26	3	39.32	23.06	26.24	19.90	22.36
INM-1	2379	F25&F26	3	39.42	26.22	30.09	22.69	26.25
INM-1	2380	F25&F26	3	37.66	23.06	28.26	19.55	21.90
INM-1	2381	F25&F26	3	37.73	22.91	24.65	19.36	21.82
INM-1	2382	F25&F26	3	38.59	25.59	28.21	21.50	23.36
INM-1	2383	F25&F26	3	00.00	00.00	00.00	23.27	24.90
INM-1	2384	F25&F26	3	38.51	23.54	26.24	20.42	21.65
INM-1	2385	F25&F26	3	41.16	28.93	32.12	24.01	25.15
INM-1	2386	F25&F26	3	40.29	24.85	28.76	22.13	22.21
INM-1	2387	F25&F26	3	38.91	25.34	27.37	20.75	24.24
INM-1	2388	F25&F26	3	37.24	24.27	00.00	20.15	25.50
INM-1	2389	G10	3	00.00	00.00	00.00	16.00	22.10
INM-1	2893	G10	3	40.76	25.78	30.50	23.38	28.12
INM-1	2894	G10	3	40.91	26.30	30.33	22.87	25.02
INM-1	7851	G25	3	45.35	31.15	33.85	25.60	28.10
INM-1	4541	H11	3	39.64	24.15	28.85	23.78	22.33
INM-1	7926	Hno103	3	45.50	31.30	36.00	28.08	30.05
INM-1	7927	Hno103	3	00.00	00.00	00.00	26.65	27.85
INM-1	6173	Hno106	3	00.00	26.18	00.00	21.90	00.00
INM-1	6174	Hno106	3	00.00	23.55	00.00	17.95	22.75
INM-1	3028	B3	4	41.11	25.74	27.83	22.38	24.26
INM-1	3029	B3	4	40.23	26.62	30.38	22.49	26.07
INM-1	975	C1	4	42.39	26.96	30.15	25.14	28.89
INM-1	976	C1	4	43.31	30.83	32.91	28.35	29.85
INM-1	2971	D10	4	00.00	28.39	33.23	00.00	00.00
INM-1	2972	D10	4	37.76	23.95	27.80	20.95	22.69
INM-1	2973	D10	4	38.36	22.83	25.00	20.78	21.79
INM-1	5867A	E25	4	34.65	22.60	00.00	19.39	22.19
INM-1	5853	E27	4	41.06	24.86	27.25	22.91	22.55
INM-1	7996	F24	4	38.52	23.05	24.55	20.85	22.56
INM-1	7997	F24	4	40.30	26.51	27.50	22.95	25.00
INM-1	5950	F25	4	00.00	24.70	00.00	21.70	22.75
INM-1	1856	G11	4	39.80	25.18	28.61	21.23	23.61
INM-1	7134	G24	4	40.07	26.50	28.90	21.65	23.00
INM-1	7671	G25	4	38.65	24.20	26.90	19.53	21.65

INM-1	7673	G25	4	39.80	25.00	28.23	21.40	23.18
INM-1	7674	G25	4	39.87	25.40	00.00	20.85	23.55
INM-1	7675	G25	4	40.20	25.50	00.00	21.00	22.00
INM-1	6837	G26	4	39.50	00.00	29.15	22.60	23.50
INM-1	6838	G26	4	39.35	23.70	27.60	20.90	21.30
INM-1	6839	G26	4	43.65	28.85	32.95	27.90	27.30
INM-1	4775	B0	4	39.55	24.60	26.88	21.80	23.61
INM-1	4776	B0	4	00.00	27.10	00.00	22.70	26.25
INM-1	4777	B0	4	00.00	25.73	00.00	23.57	23.63
INM-1	7770	D24&E25	4	44.95	30.32	34.37	24.45	30.45
INM-1	7771	D24&E25	4	37.90	00.00	26.45	19.85	22.90
INM-1	656	Hno110	4	00.00	00.00	00.00	26.91	32.57
INM-1	657	Hno110	4	40.15	26.51	27.71	22.06	24.59
INM-1	4637	Hno111	4	00.00	00.00	00.00	16.10	21.50
INM-1	4638	Hno111	4	00.00	00.00	00.00	15.20	20.32
INM-1	4639	Hno111	4	00.00	19.60	24.20	00.00	00.00
INM-1	4053	Hno115	4	43.10	28.40	33.95	22.30	28.25
INM-1	5027	Hn0119	4	38.60	23.06	00.00	21.00	22.65
INM-1	5017	Hno122	4	00.00	23.75	00.00	20.80	24.20
INM-1	4225	B6	5	00.00	24.00	27.00	00.00	00.00
INM-1	4226	B6	5	39.45	23.85	28.13	20.84	22.30
INM-1	5247	C1	5	42.15	27.15	30.60	24.15	24.56
INM-1	5248	C1	5	00.00	24.30	26.52	00.00	00.00
INM-1	4112	C3	5	45.25	30.55	34.40	00.00	00.00
INM-1	4113	C3	5	41.15	28.30	00.00	25.05	27.05
INM-1	4114	C3	5	42.20	26.30	27.55	22.80	25.92
INM-1	3805	E8	5	39.85	24.85	24.78	20.47	23.70
INM-1	3806	E8	5	00.00	25.50	28.06	00.00	00.00
INM-1	4317	F8	5	40.96	28.45	32.05	24.80	26.70
INM-1	4562	E5&B0	5	43.81	27.09	28.89	00.00	24.45
INM-1	4416		6	00.00	25.00	00.00	21.30	26.05
INM-1	4216	A3	6	00.00	30.55	00.00	25.60	28.95
INM-1	4217	A3	6	00.00	27.75	00.00	23.75	26.40
INM-1	5205	B3	6	34.85	23.80	25.35	19.90	23.75
INM-1	5206	B3	6	40.55	27.20	30.43	23.67	24.83
INM-1	5207	B3	6	39.45	24.02	28.00	22.69	23.55
INM-1	5208	B3	6	43.77	30.83	30.72	25.09	26.00
INM-1	3874	B6	6	40.85	26.23	00.00	21.70	25.30
INM-1	3875	B6	6	00.00	00.00	00.00	21.10	22.65
INM-1	2009	G3	6	00.00	00.00	00.00	23.00	24.76
INM-1	1492	G4	6	41.79	00.00	28.28	00.00	23.13
INM-1	1493	G4	6	40.54	23.10	27.50	20.37	23.67
INM-1	2262	G5	6	40.71	25.76	21.13	23.54	00.00
INM-1	3869	C0	7	44.60	30.45	32.25	26.72	29.80
INM-1	3869A	C0	7	42.55	26.05	29.90	22.95	23.20
INM-1	3784	H12	7	37.88	26.00	29.54	24.20	26.15
INM-1	4479	X4	7	44.74	29.04	33.58	23.49	27.43
INM-1	3865		7	44.60	30.45	32.25	26.72	29.80
INM-1	4768	A5	7	40.80	26.00	31.03	21.93	28.79
INM-1	4769	A5	7	42.90	28.05	00.00	23.92	24.65
INM-1	4770	A5	7	57.5	20.40	39.80	00.00	00.00
INM-1	3607	B6	7	41.30	28.93	00.00	23.28	27.20
INM-1	4360	E1	7	40.18	26.60	31.55	23.75	23.99
INM-1	580	E6	7	42.84	31.42	35.05	00.00	31.49
INM-1	4425	E7	7	40.70	23.88	28.50	23.25	25.95
INM-1	1829	G4	7	44.59	30.02	32.75	00.00	25.38
INM-1	3658	C0	7	40.50	23.52	27.50	20.50	23.90
INM-1	4852	C0	7	40.17	25.57	27.95	20.80	00.00
INM-1	4854	C0	7	00.00	26.57	00.00	23.30	27.15
INM-1	4096	A1&A2	7	47.30	28.45	31.80	23.25	26.35
INM-1	4097	A1&A2	7	41.85	25.95	27.25	22.65	22.30
INM-1	4098	A1&A2	7	39.50	25.40	29.85	21.63	25.25
INM-1	6118	Hno51	7	44.96	29.06	33.70	26.09	29.70

INM-1	6361	Hno51A	7	00.00	27.68	00.00	22.90	24.35
INM-1	633	B5	8	41.79	27.85	28.79	24.36	25.45
INM-1	635	B5	8	40.91	29.89	31.03	26.97	27.64
INM-1	4847	B5	8	40.10	25.15	29.72	21.48	26.50
INM-1	4391	B6	8	00.00	23.95	00.00	21.05	22.60
INM-1	4731	C1	8	37.30	25.08	29.40	21.67	23.05
INM-1	4732	C1	8	00.00	24.25	00.00	21.52	23.38
INM-1	4976	C2	8	38.30	26.50	26.80	22.00	25.10
INM-1	4977	C2	8	37.85	24.35	26.71	20.50	22.65
INM-1	4978	C2	8	37.80	25.35	26.59	21.40	25.35
INM-1	4549	C4	8	36.90	24.28	28.45	20.86	24.80
INM-1	4380	C5	8	40.00	26.50	00.00	23.00	22.09
INM-1	3671	C6	8	46.30	30.00	30.60	24.90	26.30
INM-1	3674	C6	8	46.30	30.57	32.13	24.70	26.42
INM-1	3675	C6	8	00.00	25.33	00.00	21.59	24.80
INM-1	3676	C6	8	00.00	00.00	00.00	25.04	29.15
INM-1	3677	C6	8	00.00	00.00	00.00	20.10	24.02
INM-1	4407	C7	8	00.00	20.90	00.00	19.65	21.40
INM-1	4824	D3	8	00.00	00.00	00.00	23.25	25.25
INM-1	865	D5	8	38.14	23.23	27.57	20.43	22.39
INM-1	868	D5	8	00.00	00.00	00.00	21.7	23.84
INM-1	5163	D6	8	00.00	00.00	00.00	15.45	19.05
INM-1	5164	D6	8	38.50	24.57	26.87	20.75	23.03
INM-1	5165	D6	8	40.80	24.90	26.76	22.35	23.40
INM-1	5166	D6	8	43.65	25.58	32.20	24.35	27.10
INM-1	5167	D6	8	39.39	27.31	29.60	23.65	24.35
INM-1	5168	D6	8	40.55	00.00	28.35	21.38	23.10
INM-1	5169	D6	8	38.45	25.55	28.10	20.64	25.78
INM-1	5170	D6	8	35.34	22.81	26.20	20.74	23.20
INM-1	5121	D7	8	00.00	29.75	00.00	25.60	27.45
INM-1	5122	D7	8	40.62	25.40	30.10	22.20	25.25
INM-1	5123	D7	8	00.00	31.10	00.00	25.95	00.00
INM-1	4245	E1	8	00.00	28.60	00.00	25.55	25.70
INM-1	4844	E4	8	00.00	28.30	30.25	00.00	00.00
INM-1	5058	E5	8	46.50	28.25	33.05	23.07	28.00
INM-1	5065	E5	8	42.95	30.60	32.80	27.30	28.55
INM-1	4440	E6	8	00.00	26.20	31.40	00.00	00.00
INM-1	4441	E6	8	36.50	24.50	26.70	21.33	23.45
INM-1	4442	E6	8	40.00	24.33	27.83	20.18	25.74
INM-1	4443	E6	8	00.00	22.13	24.34	00.00	00.00
INM-1	4933	E6	8	43.15	30.75	00.00	29.50	31.32
INM-1	4455	E7	8	43.26	28.65	29.90	23.65	25.95
INM-1	4608	E7	8	00.00	25.17	00.00	20.90	24.40
INM-1	4609	E7	8	36.69	24.83	25.45	20.75	22.10
INM-1	4610	E7	8	00.00	28.35	00.00	24.32	27.80
INM-1	4611	E7	8	00.00	28.46	00.00	24.09	25.96
INM-1	4620	E7	8	00.00	19.86	22.25	00.00	00.00
INM-1	3610	C0	8	43.66	29.83	00.00	26.00	26.62
INM-1	4716	B0	8	00.00	24.00	00.00	19.67	16.65
INM-1	4493	Hno49	8	45.15	30.96	32.17	27.15	28.12
INM-1	4707	Hno61	8	00.00	24.85	28.10	00.00	00.00
INM-1	3734	Hno69	8	39.49	24.00	27.67	21.03	22.15
INM-1	4884	Hno71	8	45.20	29.70	35.55	28.25	33.20
INM-1	4885	Hno71	8	40.60	27.10	28.85	24.20	24.40
INM-1	4886	Hno71	8	38.85	25.05	27.45	21.65	22.65
INM-1	4887	Hno71	8	00.00	00.00	00.00	20.15	24.45
INM-1	4888	Hno71	8	00.00	00.00	00.00	15.77	18.85
INM-1	4158	Hno72	8	00.00	31.00	00.00	29.00	30.56
INM-1	4159	Hno72	8	00.00	29.27	00.00	25.80	28.28
INM-1	4146	Hno73	8	40.45	27.30	30.82	22.80	25.55
INM-1	1742	A1	9	46.06	26.72	00.00	24.23	25.27
INM-1	704	B1	9	41.61	26.75	30.04	25.30	24.87
INM-1	706	B1	9	39.69	26.08	29.88	20.58	23.22

INM-1	709	B1	9	39.18	25.12	27.42	20.02	21.00
INM-1	4351	C1	9	39.25	24.60	27.10	19.95	21.85
INM-1	3060	C4	9	40.21	26.49	28.66	22.97	24.65
INM-1	3061	C4	9	43.64	28.65	31.18	25.88	25.90
INM-1	800	C5	9	40.29	25.11	29.58	23.13	28.40
INM-1	801	C5	9	40.68	28.64	30.54	26.56	30.09
INM-1	5114	C5	9	00.00	00.00	00.00	24.30	25.95
INM-1	5115	C5	9	00.00	00.00	00.00	12.68	15.33
INM-1	5263	D4	9	00.00	30.25	00.00	26.45	27.75
INM-1	2875	D5	9	44.69	30.55	32.20	28.79	28.07
INM-1	723	D6	9	00.00	26.31	00.00	22.51	24.73
INM-1	4575	E3	9	00.00	29.65	00.00	26.70	29.00
INM-1	4231	C1	10	00.00	28.50	00.00	24.60	24.85
INM-1	599	C3	10	00.00	23.49	26.89	22.55	25.60
INM-1	602	C3	10	00.00	00.00	00.00	17.93	23.08
INM-1	682	A1	11	42.62	29.11	00.00	25.39	27.84
INM-1	683	A1	11	38.72	24.57	27.62	20.06	24.69
INM-1	5250	B1	11	00.00	24.20	27.95	00.00	00.00
INM-1	5251	B1	11	00.00	00.00	00.00	15.10	19.45
INM-1	5252	B1	11	00.00	00.00	00.00	21.88	25.25
INM-1	5713	B6	11	42.60	27.40	32.60	25.45	27.40
INM-1	6006	B7	11	44.90	31.25	33.95	28.00	28.35
INM-1	5232	C3	11	00.00	00.00	00.00	23.90	24.25
INM-1	5707	C5	11	41.00	26.65	28.90	23.90	23.80
INM-1	6093	C7	11	40.00	25.56	27.00	00.00	22.78
INM-1	6094	C7	11	39.80	25.80	28.52	21.84	22.75
INM-1	6128	D4	11	43.06	00.00	31.05	23.05	25.30
INM-1	6129	D4	11	00.00	00.00	00.00	24.75	27.78
INM-1	5744	D6	11	48.00	29.80	36.00	27.00	28.45
INM-1	5549	E9	11	39.45	26.55	28.90	21.71	23.00
INM-1	3689	B1	12	45.75	31.00	35.05	30.05	31.50
INM-1	4174	B1	12	42.40	29.40	28.90	24.15	26.67
INM-1	4175	B1	12	36.80	22.35	24.05	18.05	23.07
INM-1	6205	B6	12	45.96	32.50	35.60	29.55	29.00
INM-1	6206	B6	12	43.55	35.30	34.00	27.60	29.82
INM-1	6207	B6	12	44.55	31.23	36.45	26.95	31.45
INM-1	6208	B6	12	45.75	30.30	33.05	25.08	28.70
INM-1	5675	B7	12	48.02	32.62	35.55	28.03	29.45
INM-1	811	C1	12	42.04	00.00	00.00	21.87	24.48
INM-1	4275	C2	12	40.40	27.30	29.55	23.85	26.25
INM-1	5604	C3	12	45.77	28.75	30.82	23.35	25.33
INM-1	5757	C5	12	48.65	33.65	36.50	28.30	30.50
INM-1	5758	C5	12	00.00	25.75	00.00	22.00	22.60
INM-1	5840A	C6	12	41.00	28.30	30.90	24.75	25.35
INM-1	5841A	C6	12	40.95	26.30	28.55	22.30	24.85
INM-1	6291	C7	12	00.00	13.45	17.30	00.00	00.00
INM-1	6268	C8	12	40.57	25.75	28.61	22.55	24.35
INM-1	6269	C8	12	43.40	30.05	31.55	28.50	27.10
INM-1	6270	C8	12	00.00	27.45	32.25	23.36	22.85
INM-1	3817	C0	12	00.00	24.88	27.30	00.00	00.00
INM-1	6150	Hno75	12	4495	27.20	29.05	23.35	24.60
INM-1	6151	Hno75	12	00.00	29.75	00.00	25.70	29.30
INM-1	6152	Hno75	12	00.00	26.00	27.75	00.00	00.00
INM-1	6153	Hno75	12	00.00	31.00	35.75	00.00	00.00
INM-1	6154	Hno75	12	00.00	26.80	00.00	00.00	00.00
INM-1	5724	Hno77	12	39.52	24.60	27.75	21.55	22.85
INM-1	5721	Hno78	12	39.50	24.45	27.25	20.90	21.00
INM-1	6138	Hno83	12	42.30	27.93	33.78	24.17	26.02
INM-1	5499	Hno85	12	40.05	22.95	26.40	20.70	21.75
INM-1	5502	Hno85	12	00.00	00.00	00.00	18.95	22.65
INM-1	5544		13	41.85	26.07	30.60	22.50	26.05
INM-1	5641	A1	13	42.05	27.15	30.05	23.85	26.20
INM-1	5085	B2	13	37.90	23.23	26.33	19.85	21.02

INM-1	5086	B2	13	38.32	24.95	28.26	22.50	24.40
INM-1	5087	B2	13	40.88	30.00	31.45	25.60	27.15
INM-1	5833	B3	13	45.95	30.50	34.75	25.10	30.70
INM-1	5835	B3	13	36.63	23.60	00.00	20.40	22.80
INM-1	5836	B3	13	00.00	00.00	00.00	22.10	22.05
INM-1	5769	B5	13	44.75	28.95	31.46	00.00	00.00
INM-1	6063	C6	13	00.00	29.75	00.00	26.70	26.00
INM-1	5815	E4	13	00.00	00.00	00.00	19.50	23.10
INM-1	4674	B0	13	44.70	27.65	30.75	23.78	27.10
INM-1	4675	B0	13	38.75	24.97	28.70	2.63	25.70
INM-1	8158	A2	14	00.00	30.00	33.00	00.00	00.00
INM-1	5669	A4	14	00.00	24.20	00.00	20.55	25.62
INM-1	6309	A5	14	40.96	25.30	27.30	21.76	00.00
INM-1	5454	B1	14	38.95	24.73	27.50	20.87	25.80
INM-1	5902	B3	14	37.90	25.65	28.85	21.50	24.45
INM-1	5903	B3	14	35.21	22.54	00.00	18.93	21.27
INM-1	5904	B3	14	00.00	19.43	00.00	17.50	20.70
INM-1	5905	B3	14	36.34	24.25	26.65	21.70	23.48
INM-1	5906	B3	14	39.92	28.49	00.00	23.60	24.78
INM-1	5907	B3	14	47.20	30.00	32.18	27.85	26.86
INM-1	5909	B3	14	42.65	26.25	27.90	22.10	22.00
INM-1	5930	B3	14	00.00	22.75	26.10	00.00	00.00
INM-1	5485	C1	14	41.06	25.78	27.90	21.02	23.09
INM-1	5644	C6	14	40.90	2915	28.68	24.34	24.60
INM-1	5992	D6	14	41.68	27.85	29.47	23.80	24.00
INM-1	5994	D6	14	31.10	18.75	22.75	15.45	20.95
INM-1	5765	D8	14	47.50	31.40	33.80	24.90	27.40
INM-1	5298	E3	14	46.75	30.00	35.35	25.05	26.69
INM-1	5299	E3	14	39.80	25.78	00.00	21.65	23.90
INM-1	5300	E3	14	43.17	27.04	28.70	23.06	25.09
INM-1	5301	E3	14	38.70	25.63	28.38	23.25	25.45
INM-1	5302	E3	14	39.00	24.65	29.05	21.25	25.65
INM-1	5303	E3	14	44.33	28.45	33.45	25.45	30.72
INM-1	6333	E4	14	00.00	23.34	00.00	20.25	23.05
INM-1	6254	E9	14	46.60	29.40	33.95	00.00	00.00
INM-1	6255	E9	14	42.65	28.60	32.65	25.25	27.60
INM-1	6256	E9	14	44.40	28.89	32.55	26.15	26.90
INM-1	4789	B0	14	39.46	25.80	29.09	21.50	25.50
INM-1	4790	B0	14	00.00	28.38	00.00	27.25	28.35
INM-1	5617	A4	15	44.15	29.82	29.75	23.71	26.60
INM-1	5618	A4	15	39.80	24.70	28.10	20.88	21.85
INM-1	5653	A5	15	40.27	26.50	28.56	23.00	25.04
INM-1	5654	A5	15	40.65	27.85	30.17	23.65	24.70
INM-1	6362	B2	15	00.00	26.17	00.00	21.60	24.75
INM-1	6067	B5	15	00.00	28.11	31.25	00.00	00.00
INM-1	6068	B5	15	42.60	28.17	30.17	24.15	00.00
INM-1	5367	C3	15	39.80	25.75	28.00	23.75	25.96
INM-1	5368	C3	15	41.55	28.15	31.20	24.15	25.87
INM-1	5563	D4	15	00.00	00.00	00.00	18.90	21.70
INM-1	5564	D4	15	00.00	00.00	00.00	15.75	20.10
INM-1	5424	E4	15	39.55	26.35	26.92	21.30	24.10
INM-1	5425	E4	15	40.67	25.73	30.60	23.19	26.13
INM-1	5426	E4	15	37.95	26.90	28.60	23.50	24.00
INM-1	6236	E5	15	37.65	23.80	26.40	19.25	22.05
INM-1	6024	Hno87	15	43.13	29.10	32.60	27.10	26.65
INM-1	6025	Hno87	15	00.00	00.00	00.00	18.54	24.25
INM-1	6026	Hno87	15	40.60	26.72	00.00	24.18	23.38
INM-1	2496	D3	16	00.00	26.95	00.00	22.99	26.33
INM-1	2497	D3	16	42.43	26.61	30.97	22.45	25.55

Table A.2.17. Measurements of Third Phalanx of *Bos indicus*

Site	Reg.No.	Trench	Layer	GL	DL	BP	TP
INM-1	58	LJP	0	63.49	53.80	25.56	00.00
INM-1	121	LJP	0	58.38	47.17	22.63	00.00
INM-1	129	LJP	0	00.00	00.00	13.74	00.00
INM-1	130	LJP	0	00.00	00.00	12.41	00.00
INM-1	207	LJP	0	52.84	44.16	20.22	00.00
INM-1	238	LJP	0	50.57	38.89	17.52	00.00
INM-1	241	LJP	0	60.60	52.03	22.91	00.00
INM-1	243	LJP	0	58.03	46.37	21.86	00.00
INM-1	244	LJP	0	56.28	46.46	22.16	00.00
INM-1	370	LJP	0	66.00	53..	24.06	00.00
INM-1	5273	LJP	0	00.00	00.00	21.70	37.65
INM-1	1114	G4	1	54.05	40.43	20.55	38.67
INM-1	1115	G4	1	59.77	47.42	21.84	42.11
INM-1	3528	H5	1	55.96	44.50	20.17	38.75
INM-1	3277	D10	1	58.00	46.71	24.37	42.87
INM-1	3579	B0	1	64.70	55.45	27.27	47.10
INM-1	3317	F11	1	54.45	45.40	20.00	37.70
INM-1	1652	G11	1	55.76	45.48	20.83	40.28
INM-1	1653	G11	1	54.75	42.23	23.14	41.25
INM-1	1654	G11	1	55.85	46.20	21.19	40.29
INM-1	2332		2	00.00	00.00	19.56	33.67
INM-1	2333		2	00.00	00.00	22.39	44.51
INM-1	2272	E1	2	00.00	00.00	21.34	36.45
INM-1	3077	G3	2	61.98	48.53	23.85	41.85
INM-1	3952	G4	2	54.80	45.00	18.37	32.85
INM-1	3630	G7	2	51.65	45.50	19.50	35.85
INM-1	1555	H5	2	58.57	50.08	21.50	38.33
INM-1	1561	H5	2	57.42	46.91	20.37	36.99
INM-1	2818	D10	2	61.97	49.85	24.85	40.88
INM-1	4302	D1	2	00.00	00.00	27.06	00.00
INM-1	7331	E24	2	57.00	46.30	21.45	39.00
INM-1	7332	E24	2	53.60	46.40	20.10	37.30
INM-1	6396	E25	2	68.57	52.31	21.60	45.45
INM-1	6397	E25	2	59.74	49.55	21.75	45.55
INM-1	2539	F11	2	57.82	47.85	22.78	41.58
INM-1	6649	F24	2	64.10	48.90	23.90	42.00
INM-1	6650	F24	2	71.40	54.85	25.00	44.80
INM-1	7502	F24	2	56.40	47.58	18.85	40.03
INM-1	7503	F24	2	60.46	50.85	20.65	40.40
INM-1	7043	G24	2	59.35	48.93	20.80	43.35
INM-1	7196	G25	2	51.40	42.55	19.00	36.96
INM-1	777	I12	2	00.00	00.00	22.44	00.00
INM-1	2845	G3	3	58.28	00.00	21.90	00.00
INM-1	2664	G7	3	51.00	37.15	21.89	00.00
INM-1	2665	G7	3	56.96	48.63	21.30	00.00
INM-1	3444	H7	3	66.56	53.44	24.98	45.53
INM-1	3251	D10	3	61.75	47.00	23.70	43.65
INM-1	3252	D10	3	49.92	42.60	18.85	32.61
INM-1	4036	C1	3	66.15	51.33	23.05	41.30
INM-1	1883	E10	3	58.36	48.23	20.96	38.98
INM-1	1884	E10	3	57.04	00.00	18.99	00.00
INM-1	1885	E10	3	00.00	00.00	20.83	00.00
INM-1	6961	E26	3	58.85	48.55	19.85	37.70
INM-1	3264	F11	3	52.60	43.75	19.70	39.10
INM-1	6591	F25	3	00.00	00.00	20.50	40.80
INM-1	2194	G10	3	64.23	50.12	24.00	42.37
INM-1	2223	G10	3	62.66	52.04	25.27	45.47
INM-1	1286	G11	3	56.84	45.55	21.82	38.23
INM-1	1766	H11	3	46.57	38.34	17.38	33.78
INM-1	1767	H11	3	64.85	55.28	24.31	42.17

INM-1	1767A	H11	3	00.00	00.00	22.22	40.67
INM-1	3654	Hno127	3	70.51	53.25	28.44	50.89
INM-1	977	C1	4	49.95	00.00	24.12	00.00
INM-1	978	C1	4	00.00	00.00	20.55	29.94
INM-1	2957	D10	4	53.07	44.02	20.57	36.53
INM-1	8028	E24	4	00.00	00.00	17.05	31.65
INM-1	8138	E24	4	65.55	48.50	21.50	43.85
INM-1	8139	E24	4	51.85	46.57	19.40	36.07
INM-1	8140	E24	4	00.00	00.00	17.15	33.15
INM-1	5854	E27	4	00.00	00.00	22.25	37.90
INM-1	5855	E27	4	00.00	00.00	22.90	42.85
INM-1	7998	F24	4	61.05	46.93	21.75	43.65
INM-1	7999	F24	4	00.00	00.00	19.25	38.75
INM-1	5956	F25	4	53.15	42.60	19.10	35.00
INM-1	5958	F25	4	59.70	45.90	20.22	43.48
INM-1	7135	G24	4	41.00	30.40	13.50	23.63
INM-1	7136	G24	4	51.80	40.00	14.05	28.00
INM-1	7900	G27	4	78.45	62.15	25.78	48.20
INM-1	4778	B0	4	51.86	44.92	21.60	37.65
INM-1	4779	B0	4	52.20	40.65	20.00	35.93
INM-1	7772	D24&E25	4	56.62	45.25	21.55	40.90
INM-1	660	Hno110	4	00.00	00.00	25.23	00.00
INM-1	4641	Hno111	4	40.40	32.00	13.85	27.10
INM-1	4752	Hno117	4	61.57	50.25	23.52	44.79
INM-1	5028	Hno119	4	00.00	00.00	21.70	44.05
INM-1	5029	Hno119	4	00.00	00.00	20.45	41.00
INM-1	4227	B6	5	68.30	54.80	27.65	46.69
INM-1	4322	F8	5	00.00	00.00	22.45	38.93
INM-1	4563	E5&B0	5	63.00	48.15	22.55	41.65
INM-1	3705	A3	6	67.25	56.02	25.65	47.50
INM-1	4167	B1	6	57.73	47.60	20.53	40.75
INM-1	4201	B2	6	63.30	48.30	25.40	41.34
INM-1	5209	B3	6	51.90	00.00	20.75	42.55
INM-1	3876	B6	6	00.00	00.00	37.80	20.40
INM-1	1496	G4	6	50.21	00.00	20.45	00.00
INM-1	1498	G4	6	00.00	00.00	19.80	39.23
INM-1	3870	B1	6	55.50	45.70	21.25	40.05
INM-1	4361	E1	7	00.00	00.00	17.10	31.36
INM-1	4363	E1	7	44.57	34.04	19.75	32.48
INM-1	4296	E4	7	40.45	00.00	24.80	27.85
INM-1	581	E6	7	57.60	45.66	22.07	00.00
INM-1	582	E6	7	48.56	42.23	18.84	00.00
INM-1	4193	E8	7	45.85	35.00	14.75	27.65
INM-1	4194	E8	7	61.15	47.48	22.20	41.90
INM-1	1833	G4	7	00.00	00.00	21.04	36.04
INM-1	4855	C0	7	56.45	42.15	21.40	38.70
INM-1	4100	A1&A2	7	66.30	53.05	21.60	39.45
INM-1	6119	Hno51	7	71.18	52.50	23.85	47.10
INM-1	4259	B2	8	56.60	44.40	21.10	34.95
INM-1	4965	B3	8	00.00	00.00	19.90	40.55
INM-1	634	B5	8	63/61	50.42	25.73	00.00
INM-1	4979	C2	8	60.00	48.80	20.50	39.52
INM-1	4980	C2	8	59.10	51.05	20.80	45.49
INM-1	4498	D2	8	00.00	00.00	26.35	47.60
INM-1	869	D5	8	38.86	29.28	13.68	26.20
INM-1	5174	D6	8	48.90	41.20	19.90	35.90
INM-1	5175	D6	8	00.00	00.00	20.52	40.55
INM-1	5176	D6	8	00.00	00.00	24.25	38.91
INM-1	5124	D7	8	64.05	51.70	22.55	42.70
INM-1	5125	D7	8	00.00	00.00	20.30	36.95
INM-1	5126	D7	8	00.00	00.00	23.26	43.70
INM-1	967	E2	8	65.13	50.97	19.21	43.40
INM-1	5060	E5	8	00.00	00.00	18.50	34.75

INM-1	5061	E5	8	00.00	00.00	22.65	41.95
INM-1	5073	E5	8	00.00	00.00	26.15	43.75
INM-1	5077	E5	8	44.57	00.00	00.00	00.00
INM-1	4934	E6	8	54.05	44.00	21.05	37.98
INM-1	4935	E6	8	61.80	49.58	24.10	45.20
INM-1	4614	E7	8	58.80	49.00	24.00	42.17
INM-1	3665	E8	8	58.45	46.98	22.25	40.69
INM-1	3669	E8	8	00.00	00.00	23.00	40.44
INM-1	4717	B0	8	54.46	44.85	20.85	38.75
INM-1	4718	B0	8	59.08	47.49	21.47	40.22
INM-1	3890	B3&C3	8	66.95	48.32	24.82	43.13
INM-1	4708	Hno61	8	56.99	44.65	21.20	39.85
INM-1	4481	Hno64	8	57.40	40.30	22.35	42.56
INM-1	4600	Hno65	8	67.00	51.90	23.56	43.80
INM-1	3733	Hno69	8	53.65	46.05	19.00	38.15
INM-1	4889	Hno71	8	54.11	44.65	19.65	38.20
INM-1	4132	X3&B2	8	65.88	55.52	26.45	45.06
INM-1	4133	X3&B2	8	49.25	45.56	17.28	34.90
INM-1	4959	KILN	8	58.67	50.15	21.60	41.21
INM-1	1744	A1	9	00.00	00.00	20.28	38.15
INM-1	1745	A1	9	00.00	00.00	19.77	37.58
INM-1	705	B1	9	56.80	47.63	20.52	00.00
INM-1	4684	B6	9	52.65	44.20	19.15	36.68
INM-1	4685	B6	9	55.45	47.65	19.55	35.84
INM-1	4352	C1	9	66.75	53.39	24.00	46.40
INM-1	798	C5	9	00.00	49.11	21.02	00.00
INM-1	5116	C5	9	63.55	52.37	24.65	00.00
INM-1	2876	D5	9	00.00	00.00	26.19	44.57
INM-1	2877	D5	9	00.00	00.00	24.18	45.18
INM-1	2878	D5	9	65.03	52.33	23.70	46.93
INM-1	2879	D5	9	00.00	37.50	19.37	00.00
INM-1	721	D6	9	00.00	00.00	22.62	00.00
INM-1	4236	E4	9	60.65	47.85	22.65	38.27
INM-1	4237	E4	9	59.56	49.40	23.88	45.08
INM-1	4238	E4	9	69.05	55.75	26.35	47.10
INM-1	4230	C1	10	74.45	60.99	27.89	49.85
INM-1	4835	E1	10	41.55	00.00	28.25	31.46
INM-1	4836	E1	10	39.65	00.00	25.65	27.87
INM-1	4837	E1	10	55.61	40.36	22.00	37.19
INM-1	686	A1	11	63.71	50.56	22.35	00.00
INM-1	687	A1	11	55.40	45.15	23.09	00.00
INM-1	5245	B1	11	00.00	00.00	20.05	39.07
INM-1	5149	B2	11	00.00	00.00	14.75	29.33
INM-1	6100	B4	11	52.20	43.45	18.40	38.50
INM-1	6009	B7	11	00.00	00.00	26.70	48.80
INM-1	5233	C3	11	00.00	00.00	22.90	42.82
INM-1	5334	D5	11	54.10	45.20	20.30	38.55
INM-1	5523	D8	11	54.20	45.25	19.50	35.30
INM-1	3691	B1	12	49.10	43.27	18.88	34.40
INM-1	3692	B1	12	33.93	28.30	14.19	24.95
INM-1	4177	B1	12	61.05	50.45	22.95	40.80
INM-1	4178	B1	12	47.45	40.75	16.10	29.45
INM-1	6320	B5	12	80.80	62.70	29.20	47.65
INM-1	6209	B6	12	63.73	52.01	22.65	44.11
INM-1	6210	B6	12	00.00	00.00	24.08	49.45
INM-1	4276	C2	12	33.80	28.65	11.90	25.50
INM-1	4277	C2	12	49.88	41.40	19.15	36.95
INM-1	4278	C2	12	48.00	40.30	19.67	38.65
INM-1	4279	C2	12	57.00	47.65	20.65	39.85
INM-1	5842A	C6	12	50.40	41.05	21.40	36.53
INM-1	5843A	C6	12	51.80	44.72	19.85	00.00
INM-1	6271	C8	12	00.00	00.00	19.67	38.08
INM-1	5736	E3	12	62.60	48.95	21.08	40.76

INM-1	5725	Hno77	12	55.30	46.70	21.45	38.65
INM-1	5318	Hno79	12	53.60	43.95	22.25	40.50
INM-1	6136	Hno83	12	41.05	31.10	15.60	30.00
INM-1	6137	Hno83	12	43.15	34.15	15.95	32.70
INM-1	5503	Hno85	12	56.82	47.75	20.40	39.35
INM-1	5837	B3	13	55.75	44.72	22.75	39.75
INM-1	5838	B3	13	73.05	63.30	26.64	46.70
INM-1	5839	B3	13	62.58	48.00	26.84	41.50
INM-1	5441	B4	13	54.85	46.60	22.30	42.80
INM-1	5770	B5	13	53.67	45.30	17.85	33.65
INM-1	5771	B5	13	63.80	49.85	22.85	44.60
INM-1	5772	B5	13	63.55	51.50	24.25	43.90
INM-1	5773	B5	13	00.00	00.00	19.30	36.85
INM-1	5816	E4	13	48.15	41.75	19.55	35.55
INM-1	4677	B0	13	63.55	47.09	22.10	43.75
INM-1	4678	B0	13	72.89	59.75	24.15	49.50
INM-1	5455	B1	14	51.85	41.85	20.08	36.88
INM-1	5456	B1	14	51.88	43.10	20.02	36.35
INM-1	5910	B3	14	54.80	43.95	20.38	40.05
INM-1	6280	B3	14	50.45	38.70	19.60	33.70
INM-1	6281	B3	14	55.35	44.00	19.95	40.15
INM-1	6075	B5	14	51.95	43.20	20.35	37.76
INM-1	5486	C1	14	68.85	53.40	27.85	47.11
INM-1	6088	D4	14	59.45	46.20	19.20	37.87
INM-1	6089	D4	14	54.20	40.30	21.00	38.07
INM-1	6334	E4	14	57.50	47.75	21.15	40.05
INM-1	6257	E9	14	71.51	57.05	27.10	44.75
INM-1	6258	E9	14	61.10	49.10	21.65	40.10
INM-1	6259	E9	14	62.35	00.00	20.75	00.00
INM-1	4793	B0	14	54.15	45.45	21.65	36.80
INM-1	4794	B0	14	62.10	48.75	22.05	37.48
INM-1	5658	A5	15	60.42	49.68	23.75	42.80
INM-1	5659	A5	15	61.20	46.00	21.60	42.80
INM-1	5635	C5	15	65.77	53.00	25.95	45.70
INM-1	5636	C5	15	67.68	54.61	27.65	45.45
INM-1	5685	D7	15	68.40	50.65	24.10	45.15
INM-1	5578	E3	15	58.83	49.00	20.20	44.45
INM-1	5579	E3	15	52.05	43.97	19.70	37.55
INM-1	5580	E3	15	00.00	00.00	20.55	41.20
INM-1	5427	E4	15	65.15	54.35	27.35	46.21
INM-1	6237	E5	15	39.70	32.30	14.00	28.06
INM-1	5701	E9	15	66.80	59.50	25.15	47.05
INM-1	6028	Hno87	16	41.55	33.75	15.13	27.45
INM-1	5356	Hno95	16	52.65	41.38	21.85	39.20
INM-1	240	C4	16	63.11	48.63	22.20	00.00

Table A.2.18. Measurements of Maxillary first/second premolar of *Bos indicus*

Site	Reg.No.	Trench	Layer	GL	GB
INM-1	2299	E1	2	17.89	17.45
INM-1	4028	G4	2	17.17	15.80
INM-1	2835	D10	2	19.20	18.86
INM-1	3234	G11	2	16.45	11.11
INM-1	3599	I6	3	16.49	12.85
INM-1	6589	F25	3	14.90	17.20
INM-1	2911	G10	3	17.43	13.59
INM-1	5878	E27	4	17.30	10.65
INM-1	5877	E27	4	15.55	16.60
INM-1	1869	G11	4	18.65	12.93
INM-1	3814	F8	6	15.33	13.72
INM-1	3871	F8	6	15.45	18.45
INM-1	4210	D8&E7	6	17.35	11.30
INM-1	4272	B2	8	10.07	18.06

INM-1	4999	C2	8	17.75	15.88
INM-1	5857A	C6	12	15.12	18.00
INM-1	8160	A1	14	16.35	10.95

Table A.2.19. Measurements of Maxillary fourth premolar of *Bos indicus*

Site	Reg.No.	Trench	Layer	GL	GB
INM-1	2967	D10	4	15.89	10.39
INM-1	5229	E1	11	14.75	18.10

Table A.2.20. Measurements of Maxillary first/second molar of *Bos indicus*

Site	Reg.No.	Trench	Layer	GL	GB
INM-1	5290	LJP	0	28.00	18.95
INM-1	6354	A3&B3	0	27.15	17.35
INM-1	3561	H5	1	32.52	20.90
INM-1	3361	F11	1	26.80	17.42
INM-1	1734	G11	1	28.74	16.13
INM-1	1735	G11	1	25.67	18.34
INM-1	3117	E2	2	30.07	24.96
INM-1	3102	G3	2	28.11	19.89
INM-1	3100	G3	2	22.83	18.27
INM-1	4009	G4	2	29.85	15.15
INM-1	4029	G4	2	22.45	16.63
INM-1	4010	G4	2	29/40	15.10
INM-1	3592	I6	2	29.10	18.60
INM-1	2831	D10	2	27.54	23.80
INM-1	2833	D10	2	29.16	21.39
INM-1	2834	D10	2	25.62	22.72
INM-1	2832	D10	2	28.30	17.83
INM-1	4076	C0	2	27.30	17.85
INM-1	7342	E24	2	30.00	16.90
INM-1	3018	F10	2	28.05	18.18
INM-1	2602	F11	2	31.50	17.77
INM-1	7585	F26	2	28.57	19.10
INM-1	7584	F26	2	27.68	18.70
INM-1	7583	F26	2	28.40	17.80
INM-1	2390	G26	2	29.41	15.39
INM-1	2799	H11	2	27.03	15.56
INM-1	2800	H11	2	30.82	20.14
INM-1	3501	H7	3	28.70	19.85
INM-1	3259	D10	3	28.16	18.30
INM-1	2322	E10	3	28.09	15.93
INM-1	2323	E10	3	31.10	16.60
INM-1	2250	F11	3	31.78	19.01
INM-1	2235	G10	3	26.95	15.18
INM-1	2914	G10	3	24.40	20.73
INM-1	2391	F25&F26	3	29.45	19.81
INM-1	2400	F25&F26	3	30.33	20.45
INM-1	2398	F25&F26	3	25.54	16.53
INM-1	2395	F25&F26	3	27.16	22.34
INM-1	2401	F25&F26	3	29.43	17.01
INM-1	2392	F25&F26	3	26.47	18.69
INM-1	2397	F25&F26	3	26.18	19.22
INM-1	2433	F25&F26	3	00.00	16.53
INM-1	2394	F25&F26	3	28.84	17.94
INM-1	2399	F25&F26	3	31.94	19.00
INM-1	2402	F25&F26	3	25.75	18.14
INM-1	2403	F25&F26	3	28.23	20.02
INM-1	2404	F25&F26	3	27.17	19.92
INM-1	2405	F25&F26	3	27.22	19.19
INM-1	2406	F25&F26	3	27.43	19.42
INM-1	2407	F25&F26	3	26.62	19.62

INM-1	2393	F25&F26	3	27.52	18.66
INM-1	2408	F25&F26	3	29.56	21.31
INM-1	2396	F25&F26	3	31.15	22.60
INM-1	2409	F25&F26	3	26.42	20.78
INM-1	2411	F25&F26	3	28.80	20.04
INM-1	2412	F25&F26	3	25.82	17.13
INM-1	2413	F25&F26	3	25.11	19.90
INM-1	2410	F25&F26	3	25.09	18.20
INM-1	2419	F25&F26	3	29.07	19.89
INM-1	1816	H5	4	28.96	17.75
INM-1	5982	F25	4	24.40	14.65
INM-1	5981	F25	4	25.60	14.88
INM-1	1873	G11	4	27.81	18.17
INM-1	4066	Hno115	4	22.15	19.70
INM-1	4065	Hno115	4	27.30	17.35
INM-1	2981	B3	5	30.05	19.65
INM-1	3813	F8	6	28.70	15.71
INM-1	4771	A5	7	26.25	17.80
INM-1	3608	B6	7	27.40	17.00
INM-1	4431	E7	7	27.42	15.33
INM-1	4430	E7	7	23.65	15.05
INM-1	3662	B0	8	26.50	17.33
INM-1	4401	B6	8	26.67	13.55
INM-1	4998	C2	8	29.25	17.10
INM-1	3684	C6	8	27.15	14.14
INM-1	3682	C6	8	28.35	16.40
INM-1	4500	D2	8	26.10	14.15
INM-1	5140	D7	8	26.30	15.30
INM-1	5141	D7	8	28.45	19.80
INM-1	5137	D7	8	21.45	15.00
INM-1	5138	D7	8	24.20	14.55
INM-1	5139	D7	8	24.65	15.85
INM-1	4619	E7	8	29.30	17.45
INM-1	3907	B3&C3	8	23.84	19.80
INM-1	3906	B3&C3	8	27.90	20.95
INM-1	3908	B3&C3	8	24.93	23.80
INM-1	4722	Hno68	8	30.03	17.75
INM-1	3740	Hno69	8	27.99	29.65
INM-1	4898	Hno71	8	29.30	19.40
INM-1	4899	Hno71	8	25.60	17.95
INM-1	5117	C5	9	24.27	14.15
INM-1	5118	C5	9	24.95	19.30
INM-1	5718	B6	11	30.30	18.10
INM-1	6097	C7	11	28.78	24.20
INM-1	5348	D5	11	23.53	16.95
INM-1	5345	D5	11	29.75	17.57
INM-1	5349	D5	11	29.71	17.00
INM-1	5347	D5	11	20.94	19.68
INM-1	5346	D5	11	30.27	20.45
INM-1	5745	D6	11	28.40	18.85
INM-1	5746	D6	11	28.25	18.80
INM-1	6019	E3	11	31.60	23.00
INM-1	5944	B3	12	34.57	23.35
INM-1	6223	B6	12	21.35	15.30
INM-1	5859A	C6	12	24.40	18.95
INM-1	5854A	C6	12	29.05	17.55
INM-1	5853A	C6	12	26.90	16.80
INM-1	5856A	C6	12	28.80	20.10
INM-1	5855A	C6	12	28.10	18.10
INM-1	5627	D7	12	22.22	18.20
INM-1	5626	D7	12	21.35	19.75
INM-1	6303	Hno81	12	30.31	20.00
INM-1	6304	Hno81	12	25.90	17.95

INM-1	5790	Hno74	12	25.25	14.90
INM-1	4683	B0	13	26.85	16.25
INM-1	6314	A5	14	31.60	18.55
INM-1	6080	B5	14	28.02	21.55
INM-1	4806	B0	14	29.60	20.34
INM-1	4807	B0	14	25.78	18.95
INM-1	4804	B0	14	27.40	20.30
INM-1	5589	E3	15	27.10	17.20
INM-1	5591	E3	15	25.50	14.88
INM-1	6244	E5	15	29.80	17.60
INM-1	6243	E5	15	29.40	17.75
INM-1	5705	E9	15	29.85	21.55
INM-1	5704	E9	15	27.60	20.80
INM-1	6034	Hno87	16	28.43	21.35
INM-1	6036	Hno87	16	22.35	14.30
INM-1	6035	Hno87	16	28.55	20.10
INM-1	2514	D3	16	18.17	14.03
INM-1	2513	D3	16	20.59	17.82

Table A.2.21. Measurements of Maxillary second molar of *Bos indicus*

Site	Reg.No.	Trench	Layer	GL	GB
INM-1	3118	E2	2	28.88	20.68
INM-1	675	Hno110	4	27.95	10.50

Table A.2.22. Measurements of Maxillary third molar of *Bos indicus*

Site	Reg.No.	Trench	Layer	GL	GB
INM-1	5291	LJP	0	28.25	18.45
INM-1	1132	G4	1	27.97	00.00
INM-1	1131	G4	1	32.01	16.26
INM-1	3560	H5	1	32.55	18.52
INM-1	2324	E10	3	28.78	19.28
INM-1	2418	F25&F26	3	30.56	21.86
INM-1	5980	F25	4	29.10	19.05
INM-1	4569	E5&B0	5	29.40	20.05
INM-1	3663	C0	7	30.18	17.60
INM-1	5852A	C6	12	27.08	17.10
INM-1	6292	C7	12	27.50	18.95
INM-1	6293	C7	12	27.40	18.70
INM-1	4805	B0	14	28.08	18.87
INM-1	4803	B0	14	28.42	18.75
INM-1	6242	E5	15	29.40	19.80

Table A.2.23. Measurements of Mandibular incisors of *Bos indicus*

Site	Reg.No.	Trench	Layer	GL	GB
INM-1	2742	G6	1	14.11	8.89
INM-1	4014	G4	2	15.75	9.15
INM-1	2742	D10	4	16.88	9.92
INM-1	4723	Hno68	8	11.90	6.26
INM-1	4900	Hno71	8	13.75	9.90
INM-1	5631	B4	11	15.30	9.20
INM-1	5751	D6	11	14.86	7.55
INM-1	5750	D6	11	14.33	9.70
INM-1	6305	Hno81	12	14.35	8.60

Table A.2.24. Measurements of Mandibular first/second premolar of *Bos indicus*

Site	Reg.No.	Trench	Layer	GL	GB
INM-1	2963	D10	4	18.65	12.17
INM-1	3706	A3	6	20.20	12.30
INM-1	3872	F8	6	18.80	8.65

INM-1	3873	F8	6	21.73	10.85
INM-1	4712	Hno61	8	18.16	9.20
INM-1	4829	D3	8	21.13	11.10
INM-1	3904	B3&C3	8	18.83	10.87
INM-1	6010	B7	11	24.70	11.30
INM-1	6020	E3	11	22.25	11.10
INM-1	6139	Hno83	12	21.74	12.17
INM-1	6222	B6	12	20.21	11.00
INM-1	6221	B6	12	21.94	12.60

Table A.2.25. Measurements of Mandibular second premolar of *Bos indicus*

Site	Reg.No.	Trench	Layer	GL	GB
INM-1	2292	E1	2	15.54	6.73
INM-1	10	B2	12	17.47	11.77
INM-1	4822	B0	12	13.15	9.25

Table A.2.26. Measurements of Mandibular fourth premolar of *Bos indicus*

Site	Reg.No.	Trench	Layer	GL	GB
INM-1	7586	F26	2	22.77	11.50
INM-1	3261	D10	3	19.50	11.30
INM-1	3262	D10	3	19.41	11.55
INM-1	5219	B3	6	21.13	11.20
INM-1	4199	E8	7	21.50	13.20
INM-1	4713	Hno61	8	19.60	7.42
INM-1	5200	C3	9	17.85	7.65
INM-1	5201	C3	9	33.55	10.85
INM-1	5199	C3	9	18.05	7.55

Table A.2.27. Measurements of Mandibular first/second molars of *Bos indicus*

Site	Reg.No.	Trench	Layer	GL	GB
INM-1	6353	A3&B3	0	26.25	11.55
INM-1	2948	H10	1	24.30	14.78
INM-1	1978	H11	1	29.06	11.15
INM-1	2293	E1	2	28.85	11.59
INM-1	4011	G4	2	32.62	12.10
INM-1	1626	H5	2	25.99	12.10
INM-1	2836	D10	2	30.51	12.36
INM-1	2604	F11	2	27.82	11.84
INM-1	2603	F11	2	27.82	11.84
INM-1	2801	H11	2	27.33	10.42
INM-1	2647	H11	3	26.09	10.93
INM-1	2251	F11	3	29.73	10.12
INM-1	2253	F11	3	30.56	10.38
INM-1	2252	F11	3	29.98	11.21
INM-1	2215	G10	3	29.05	12.38
INM-1	2216	G10	3	31.11	11.06
INM-1	2214	G10	3	28.87	10.35
INM-1	4543	H12	3	28.00	12.20
INM-1	4544	H12	3	30.07	12.22
INM-1	2432	F25&F26	3	28.50	12.98
INM-1	2414	F25&F26	3	25.99	14.31
INM-1	2415	F25&F26	3	27.36	17.32
INM-1	2416	F25&F26	3	00.00	17.28
INM-1	2431	F25&F26	3	28.40	11.66
INM-1	2434	F25&F26	3	25.68	12.97
INM-1	2435	F25&F26	3	25.39	14.02
INM-1	2965	D10	4	26.26	13.91
INM-1	6852	G26	4	30.65	11.60
INM-1	6853	G26	4	26.25	11.55
INM-1	6854	G26	4	29.80	12.15

INM-1	4783	B0	4	28.00	14.67
INM-1	4122	C3	5	31.83	11.24
INM-1	4366	E1	7	29.75	11.14
INM-1	4373	E5	7	27.70	13.10
INM-1	4432	E7	7	24.40	11.40
INM-1	4437	E7	7	30.55	12.25
INM-1	4198	E8	7	27.05	15.25
INM-1	3664	C0	7	29.94	12.75
INM-1	4271	B2	8	25.38	13.99
INM-1	4400	B6	8	28.97	11.55
INM-1	4402	B6	8	30.25	11.72
INM-1	4403	B6	8	27.89	10.60
INM-1	4742	C1	8	26.35	11.85
INM-1	5004	C2	8	24.20	13.18
INM-1	4828	D3	8	28.51	10.23
INM-1	5188	D6	8	27.65	11.95
INM-1	5189	D6	8	30.05	11.35
INM-1	5142	D7	8	25.05	11.55
INM-1	4706	F3	8	33.09	13.25
INM-1	3905	B2&C2	8	24.86	11.60
INM-1	4143	X0	8	31.70	12.25
INM-1	3739	Hno69	8	30.02	11.15
INM-1	4896	Hno71	8	24.40	14.40
INM-1	4148	Hno73	8	27.55	11.35
INM-1	4503	D2	9	23.95	11.90
INM-1	4571	E3	9	28.20	10.75
INM-1	5015	B1	10	28.25	11.48
INM-1	4235	C1	10	27.58	13.23
INM-1	5242	B1	11	27.55	12.07
INM-1	5350	D5	11	25.50	14.83
INM-1	5748	D6	11	26.65	11.30
INM-1	5747	D6	11	29.10	12.50
INM-1	5749	D6	11	29.80	12.65
INM-1	5943	B3	12	28.70	11.47
INM-1	6217	B6	12	28.99	14.10
INM-1	6219	B6	12	25.20	12.18
INM-1	6218	B6	12	30.33	13.46
INM-1	6220	B6	12	32.30	13.90
INM-1	5608	C3	12	28.60	12.45
INM-1	5607	C3	12	25.65	12.15
INM-1	5606	C3	12	25.65	11.50
INM-1	6275	C8	12	29.30	12.70
INM-1	6276	C8	12	30.50	11.30
INM-1	5739	E3	12	28.00	11.75
INM-1	531	Hno50	12	27.81	13.34
INM-1	6368	Hno80	12	26.75	14.50
INM-1	5546		13	28.12	11.12
INM-1	5545		13	28.00	12.00
INM-1	4682	B0	13	32.80	12.50
INM-1	5674	A4	14	26.65	13.20
INM-1	6316	A5	14	31.06	12.25
INM-1	6315	A5	14	26.85	13.20
INM-1	5800	B5	14	26.10	12.50
INM-1	5491	C1	14	31.90	13.55
INM-1	5490	C1	14	28.60	12.86
INM-1	5649	C6	14	25.50	10.20
INM-1	5647	C6	14	31.40	12.26
INM-1	5997	D6	14	31.58	13.65
INM-1	5998	D6	14	29.28	12.56
INM-1	5312	E3	14	30.30	12.80
INM-1	5313	E3	14	28.90	16.50
INM-1	6266	E9	14	27.60	11.10
INM-1	4808	B0	14	26.70	11.50

INM-1	4809	B0	14	24.12	13.68
INM-1	5420	C3	15	29.0	11.75
INM-1	5588	E3	15	30.15	13.95
INM-1	5590	E3	15	26.45	15.65
INM-1	5439	E4	15	26.05	14.69
INM-1	6038	Hno87	16	31.45	12.25
INM-1	6037	Hno87	16	28.22	12.00

Table A.2.28. Measurements of Mandibular third molars of *Bos indicus*

Site	Reg.No.	Trench	Layer	GL	GB
INM-1	3099	G3	2	37.95	16.46
INM-1	4012	G4	2	33.55	10.95
INM-1	3260	D10	3	36.15	14.30
INM-1	2230	G10	3	37.45	13.74
INM-1	2437	F25&F26	3	40.08	13.52
INM-1	2436	F25&F26	3	37.38	12.45
INM-1	2962	D10	4	33.85	14.25
INM-1	2963	D10	4	35.74	12.15
INM-1	4782	B0	4	36.54	12.30
INM-1	2980	B3	5	32.61	13.32
INM-1	2982	B3	5	34.77	12.87
INM-1	3777	F8	6	34.80	13.95
INM-1	4433	7	7	37.55	13.15
INM-1	4412	C7	8	37.75	15.10
INM-1	4482	Hno64	8	36.85	14.90
INM-1	4897	Hno71	8	35.75	14.40
INM-1	5717	B6	11	35.50	12.55
INM-1	4295	C2	12	34.25	12.80
INM-1	6141	Hno83	12	37.70	14.15
INM-1	5766	D8	14	37.75	13.70
INM-1	5310	E3	14	42.10	15.00
INM-1	5311	E3	14	37.68	13.95

Table A.2.29. Measurements of Mandibles with symphysis and teeth of *Bos indicus*

MS – Mandibular Symphysis, MC- Mandibular condyle, p2 –second premolar, p3- third premolar, p4- fourth premolar, m1- first molar, m2- second molar, m3- third molar, dp2,dp3, dp4 –deciduous second, third and fourth premolars.

Site	Reg.No.	Trench	Layer	Mandible	Length	Width	Height
INM-1	1129	G4	1	MC	00.00	42.32	00.00
INM-1	1017	E10	1	MC	00.00	41.44	00.00
INM-1	1200	F10	1	p2	8.58	6.93	00.00
				p3	19.49	9.38	00.00
				p4	22.14	11.29	00.00
INM-1	1728	G11	1	p4	22.18	10.36	00.00
INM-1	1980	H11	1	MC	00.00	47.13	00.00
INM-1	1974	H11	1	m1	21.81	11.92	32.40
				m2	24.64	12.50	37.71
INM-1	4922	F13	1	p2	9.47	5.75	27.55
				p3	20.85	9.43	29.82
INM-1	8079	F24	1	MC	00.00	44.86	00.00
INM-1	3580	B0	1	MC	00.00	30.03	00.00
INM-1	5513	Hno85	1	p4	22.50	10.86	00.00
INM-1	2287	E1	2	MC	00.00	44.70	00.00
INM-1	3096	G3	2	p2	11.83	00.00	00.00
				p3	16.37	9.54	00.00
INM-1	3097	G3	2	p2	11.75	8.70	00.00
				p3	19.40	9.84	00.00

INM-1	3996	G4	2	dp2	9.40	5.90	22.75
				dp3	19.25	8.30	26.80
				dp4	31.25	10.00	30.40
				m1	26.45	10.08	36.50
INM-1	3997	G4	2	p3	17.36	7.58	00.00
				p4	17.10	9.94	31.48
				m1	00.00	00.00	34.00
INM-1	4023	G4	2	MC	00.00	44.00	00.00
INM-1	1622	H5	2	m2	25.71	16.03	00.00
INM-1	1090	H7	2	MS	00.00	00.00	32.00
				p2	00.00	00.00	32.44
INM-1	1141	E10	2	MS	00.00	00.00	31.52
INM-1	2591	F11	2	MC	00.00	21.64	00.00
INM-1	850	F12	2	MC	00.00	41.69	00.00
INM-1	7507	F24	2	MS	00.00	00.00	41.85
INM-1	1440	D25	2	MC	00.00	32.59	00.00
INM-1	4075	C0	2	p4	19.85	14.40	00.00
INM-1	2867	G3	3	MS	00.00	00.00	36.37
				p2	00.00	00.00	38.04
				p3	00.00	00.00	40.08
INM-1	1897	E10	3	MS	00.00	00.00	31.66
INM-1	2320	E10	3	MS	00.00	00.00	32.20
INM-1	2360	E10	3	m3	33.71	13.08	00.00
INM-1	2231	G10	3	MS	00.00	00.00	30.70
INM-1	2234	G10	3	p2	00.00	00.00	40.77
INM-1	2249	F11	3	MS	00.00	00.00	29.99
INM-1	3793	H12	3	p2	00.00	00.00	23.45
				p3	00.00	00.00	27.66
INM-1	4545	H12	3	MC	00.00	39.85	00.00
INM-1	2459	D24	3	dp2	00.00	00.00	31.69
				dp3	00.00	00.00	40.65
				dp4	27.89	12.31	46.06
				m1	27.81	12.09	52.72
				m2	29.87	11.98	62.05
				m3	00.00	00.00	71.26
INM-1	2460	D24	3	m1	27.23	14.83	00.00
				m2	27.75	11.00	62.73
				m3	30.91	9.65	67.86
INM-1	670	Hno110	4	MC	00.00	30.51	00.00
INM-1	671	Hno110	4	MC	00.00	44.63	00.00
INM-1	5214	B3	6	MC	00.00	32.02	00.00
INM-1	4418		6	MC	00.00	29.47	00.00
INM-1	4741	C1	8	m3	33.73	15.65	00.00
INM-1	4553	C4	8	MC	00.00	41.05	00.00
INM-1	639	B5	8	MC	00.00	46.35	00.00
INM-1	4411	C7	8	MC	00.00	43.98	00.00
INM-1	3611	C0	8	MC	00.00	41.50	00.00
INM-1	3900	B3&C3	8	m1	21.35	12.90	00.00
				m2	26.15	13.30	00.00
INM-1	3901	B3&C3	8	MS	00.00	00.00	34.85
INM-1	3903	B3&C3	8	MS	00.00	00.00	40.35
INM-1	4596	E3	9	MC	00.00	00.00	46.40
INM-1	1	B2	12	MC	00.00	41.41	00.00
INM-1	4289	C2	12	MC	00.00	21.61	00.00
INM-1	5605	C3	12	m1	25.35	14.90	00.00
INM-1	5512	Hno85	12	MC	00.00	42.30	00.00
INM-1	5568	D4	15	p2	00.00	00.00	26.00

				p3	00.00	00.00	31.95
				p4	00.00	00.00	33.80

Table A.2.30. Measurements of Maxilla with teeth of *Bos indicus*

M1- Maxillary first molar, M2- Maxillary second molar, M3-Maxillary third molar

Site	Reg.No.	Trench	Layer	Teeth	Length	Width	Height
INM-1	2361	E10	3	M1	27.68	15.05	00.00
				M2	26.75	17.26	00.00
INM-1	2362	E10	3	M3	28.21	25.92	00.00
INM-1	2213	G10	3	M2	22.50	14.43	00.00
				M3	24.42	14.87	00.00

Table A.3. Measurements of Bones of *Bubalus bubalis*

Table A.3.1. Measurements of Scapula of *Bubalus bubalis*

Site	Reg.No.	Trench	Layer	GLP	LG	BG	SLC	LA
INM-1	312	LJP	1-4	00.00	52.56	43.76	49.77	00.00

Table A.3.2. Measurements of Radius of *Bubalus bubalis*

Site	Reg.No.	Trench	Layer	DD	BP	TP
INM-1	6962	E26	3	34.40	68.45	35.00

Table A.3.3. Measurements of Metacarpal of *Bubalus bubalis*

Site	Reg.No.	Trench	Layer	BP	TP	BD	TD	WT	TT	TLT
INM-1	35	LJP	1-4	00.00	00.00	58.81	00.00	28.33	30.63	23.00
INM-1	193	LJP	1-4	50.04	29.68	00.00	00.00	00.00	00.00	00.00
INM-1	3116		2	00.00	00.00	54.10	30.60	25.63	29.92	22.35
INM-1	4304	B0	2	00.00	00.00	62.46	33.17	28.63	31.85	21.40
INM-1	6652	F24	2	00.00	00.00	54.40	30.40	25.90	31.36	23.20
INM-1	2782	H11	2	59.95	36.05	00.00	00.00	00.00	00.00	00.00
INM-1	2904	G10	3	00.00	00.00	48.77	26.38	22.38	26.14	19.59
INM-1	1280	G11	3	00.00	00.00	65.78	34.55	32.04	34.27	28.65
INM-1	1776	H11	3	00.00	00.00	62.35	34.79	29.78	34.52	25.28
INM-1	1777	H11	3	00.00	00.00	52.38	28.73	25.30	28.20	22.44
INM-1	1778	H11	3	00.00	00.00	50.92	28.02	23.45	28.14	21.13
INM-1	4992	C2	8	00.00	00.00	51.60	28.55	24.40	28.60	20.94
INM-1	4993	C2	8	00.00	00.00	52.65	28.90	24.75	28.55	20.80
INM-1	5132	D7	8	00.00	00.00	60.05	00.00	28.79	31.30	21.00
INM-1	4946	E6	8	59.05	32.10	00.00	00.00	00.00	00.00	00.00
INM-1	4618	E7	8	00.00	00.00	65.25	35.50	30.80	35.47	24.05
INM-1	4719	B0	8	00.00	00.00	50.25	27.12	23.90	26.58	21.08
INM-1	720	B1	9	45.35	25.53	00.00	00.00	00.00	00.00	00.00
INM-1	5338	D5	11	00.00	00.00	60.45	31.90	28.85	31.85	23.60
INM-1	5527	D8	11	54.75	32.90	00.00	00.00	00.00	00.00	00.00
INM-1	5528	D8	11	47.80	28.00	00.00	00.00	00.00	00.00	00.00
INM-1	5267	C0	11	00.00	00.00	62.15	32.85	28.95	32.80	22.60
INM-1	5661	A5	15	00.00	00.00	51.22	28.35	24.73	28.16	22.00
INM-1	5359	Hno95	16	00.00	00.00	62.45	33.97	29.08	33.50	25.97

Table A.3.4. Measurements of Metapodia of *Bubalus bubalis*

Site	Reg.No.	Trench	Layer	BD	TD	WT	TT	TLT
INM-1	733	Hno70&73	0	73.66	33.98	36.82	33.90	27.12
INM-1	6410	E25	2	65.85	33.75	31.00	33.05	26.50

Table A.3.5. Measurements of Tibia of *Bubalus bubalis*

Site	Reg.No.	Trench	Layer	BD	TD
INM-1	4371	E5	7	53.30	40.80

Table A.3.6. Measurements of Metatarsal of *Bubalus bubalis*

Site	Reg.No.	Trench	Layer	BP	TP	BD	TD	WT	TT	TLT
INM-1	61	LJP	1-4	00.00	00.00	67.79	00.00	33.77	34.10	25.74
INM-1	2787	H11	2	00.00	00.00	52.15	28.56	23.61	28.28	22.25
INM-1	2850	G3	3	49.70	45.86	00.00	00.00	00.00	00.00	00.00
INM-1	5179	D6	8	00.00	00.00	53.60	28.76	25.55	28.76	21.49
INM-1	3668	E8	8	00.00	00.00	67.70	31.40	30.14	30.96	22.35
INM-1	5324	Hn079	12	00.00	00.00	56.55	30.60	24.90	30.52	21.17

Table A.3.7. Measurements of Calcaneum of *Bubalus bubalis*

Site	Reg.No.	Trench	Layer	GL	DD
INM-1	3509	H7	3	00.00	40.85
INM-1	2309	E10	3	00.00	46.44
INM-1	5796	B5	14	00.00	39.65

Table A.3.8. Measurements of Centrotarsal of *Bubalus bubalis*

Site	Reg.No.	Trench	Layer	GL
INM-1	334	LJP	1-4	71.95
INM-1	4982	C2	3	57.40
INM-1	4983	C2	3	47.20
INM-1	4984	C2	3	49.95
INM-1	7679	G25	4	51.20
INM-1	7695	G25	4	52.55
INM-1	5030	HnO119	4	50.15
INM-1	5031	Hno119	4	54.15
INM-1	514	F3	6	45.80
INM-1	4966	B3	8	52.60
INM-1	4950	E6	8	55.15
INM-1	4890	Hno71	8	51.90
INM-1	5150	B2	11	53.35
INM-1	5677	C8	11	55.65
INM-1	5787	Hno74	12	53.33
INM-1	5327	Hno79	12	50.60
INM-1	5473	D3	15	45.58
INM-1	5686	D7	15	57.70

Table A.3.9. Measurements of First Phalanx of *Bubalus bubalis*

Site	Reg.No.	Trench	Layer	GL	BP	TP	BD	TD
INM-1	120	LJP	1-4	62.03	00.00	00.00	00.00	00.00
INM-1	347	LJP	1-4	65.01	33.12	35.36	30.89	24.28
INM-1	2022	G7	1	00.00	24.98	27.73	00.00	00.00
INM-1	990	E10	1	57.66	31.81	35.42	28.74	00.00
INM-1	991	E10	1	60.67	33.48	35.84	29.68	22.76
INM-1	2918	H10	1	61.56	31.07	32.13	30.16	00.00
INM-1	2327		2	67.06	33.08	34.67	30.31	23.17
INM-1	2067	G6	2	00.00	00.00	00.00	00.00	26.79

INM-1	3601	G4	3	68.05	29.80	34.36	28.50	21.40
INM-1	3242	D10	3	57.80	27.45	28.68	26.07	20.70
INM-1	1230	D25	3	58.81	27.49	29.15	27.29	21.33
INM-1	2177	G10	3	67.61	35.55	39.78	31.43	24.49
INM-1	2180	G10	3	62.81	32.27	33.92	28.04	20.95
INM-1	2888	G10	3	62.44	27.85	33.92	26.71	20.67
INM-1	1287	G11	3	62.33	29.01	28.43	28.32	00.00
INM-1	1288	G11	3	00.00	00.00	00.00	25.57	20.52
INM-1	1289	G11	3	00.00	00.00	00.00	29.13	22.93
INM-1	2635	H11	3	00.00	00.00	00.00	29.56	22.81
INM-1	651	Hno110	4	00.00	00.00	00.00	27.98	20.72
INM-1	578	E6	7	59.26	30.32	36.00	27.89	20.64
INM-1	4851	C0	7	60.27	33.50	35.16	31.65	23.70
INM-1	5026	Hno119	7	58.54	32.80	33.85	30.80	21.38
INM-1	640	B5	8	00.00	29.73	29.31	00.00	00.00
INM-1	857	Hno60&67	8	00.00	00.00	00.00	29.39	22.12
INM-1	3670	C6	8	00.00	30.44	33.20	00.00	00.00
INM-1	3672	C6	8	63.03	30.36	32.72	29.07	00.00
INM-1	4671	D4	8	58.05	30.70	31.05	30.57	27.85
INM-1	5160	D6	8	00.00	00.00	00.00	32.15	24.00
INM-1	4606	E7	8	59.05	30.30	33.90	30.90	23.00
INM-1	4607	E7	8	66.50	33.55	33.76	33.66	24.32
INM-1	3886	B3&C3	8	53.58	28.00	29.43	25.91	19.20
INM-1	2874	D5	9	00.00	27.27	29.64	00.00	00.00
INM-1	603	C3	10	00.00	00.00	00.00	24.45	18.09
INM-1	593		11	00.00	31.07	24.15	00.00	00.00

Table A.3.10. Measurements of Second Phalanx of *Bubalus bubalis*

Site	Reg.No.	Trench	Layer	GL	BP	TP	BD	TD
INM-1	132	LJP	1-4	41.48	29.24	32.32	23.59	26.71
INM-1	228	LJP	1-4	40.81	26.01	25.49	00.00	00.00
INM-1	231	LJP	1-4	37.90	24.80	29.03	22.08	25.58
INM-1	248	LJP	1-4	42.16	33.31	00.00	31.10	30.98
INM-1	3796	G6	0	45.07	34.05	33.53	26.80	31.35
INM-1	988	E10	1	45.80	30.84	36.58	29.37	31.06
INM-1	989	E10	1	43.44	31.58	33.87	27.84	29.39
INM-1	2061	G6	2	00.00	24.07	00.00	20.33	24.57
INM-1	1550	H5	2	42.99	26.22	30.06	23.16	26.62
INM-1	3380	H6	2	45.35	31.50	00.00	26.65	29.46
INM-1	7330	E24	2	41.85	28.30	30.20	26.30	27.75
INM-1	7451	E25	2	47.18	31.25	32.12	26.05	26.32
INM-1	7453	E25	2	46.20	32.28	34.50	26.78	29.05
INM-1	7044	G24	2	00.00	00.00	00.00	29.15	28.55
INM-1	7045	G24	2	63.55	22.50	42.35	00.00	00.00
INM-1	2843	G3	3	44.47	28.31	31.14	23.93	25.22
INM-1	2844	G3	3	40.96	26.36	27.84	24.10	26.76
INM-1	3248	D10	3	39.15	28.00	00.00	24.08	25.05
INM-1	1220	D25	3	46.37	31.39	31.71	25.08	28.65
INM-1	4033	C0	3	46.65	30.40	33.85	28.70	31.43
INM-1	4034	C0	3	44.30	31.18	31.38	24.85	26.30
INM-1	1881	E10	3	40.70	25.19	29.21	22.77	24.62
INM-1	1882	E10	3	42.42	27.56	29.99	24.10	27.82
INM-1	2307	E10	3	47.21	32.37	35.90	25.71	29.02
INM-1	1819A	F10	3	38.60	23.68	26.33	21.11	24.49
INM-1	1282	G11	3	45.06	28.83	33.85	00.00	31.51
INM-1	2368	F25&F26	3	41.91	25.43	30.37	22.83	25.36
INM-1	2369	F25&F26	3	40.80	25.71	28.09	20.79	23.74
INM-1	2370	F25&F26	3	40.19	26.82	29.42	23.91	25.39
INM-1	1374	E1	4	45.41	30.70	32.78	28.18	29.81
INM-1	567	F8	4	39.78	30.34	31.56	28.97	31.35
INM-1	1803	H5	4	37.23	31.00	29.77	27.37	30.00
INM-1	3039	H7	4	45.70	33.06	00.00	30.05	31.82

INM-1	1855	G11	4	43.70	29.04	31.93	25.77	26.69
INM-1	7672	G25	4	39.17	25.82	27.75	23.30	23.80
INM-1	4052	Hno115	4	00.00	26.85	00.00	00.00	00.00
INM-1	3704	A3	6	37.45	28.07	28.31	23.50	28.45
INM-1	2004	G3	6	44.57	28.25	31.98	24.89	27.09
INM-1	1494	G4	6	40.06	26.26	26.95	26.77	28.17
INM-1	579	E6	7	45.57	32.74	34.66	29.92	32.66
INM-1	4853	C0	7	46.60	35.08	36.70	29.94	30.40
INM-1	4548	C4	8	38.05	24.35	26.10	21.10	23.45
INM-1	4379	C5	8	47.65	37.20	36.13	30.59	33.90
INM-1	4578	D6	8	47.68	32.11	34.50	28.45	30.15
INM-1	4579	D6	8	45.58	31.13	32.10	26.65	27.03
INM-1	4580	D6	8	00.00	29.62	00.00	27.31	29.03
INM-1	4581	D6	8	42.84	27.76	30.38	23.44	25.65
INM-1	4582	D6	8	42.47	27.97	28.95	23.16	25.53
INM-1	5171	D6	8	43.42	30.60	34.58	27.38	29.55
INM-1	5172	D6	8	42.35	31.65	33.41	30.40	30.69
INM-1	5173	D6	8	43.45	33.30	00.00	30.70	00.00
INM-1	969	E2	8	00.00	29.00	29.77	23.97	26.28
INM-1	5057	E5	8	45.65	30.30	30.52	27.10	27.70
INM-1	5059	E5	8	43.10	31.12	28.10	28.72	29.20
INM-1	4444	E6	8	40.50	28.40	00.00	26.65	28.60
INM-1	4456	E7	8	00.00	23.65	00.00	29.20	32.60
INM-1	4612	E7	8	00.00	30.85	00.00	27.18	28.25
INM-1	4613	E7	8	47.25	31.57	36.27	30.50	35.06
INM-1	4129	X3	8	40.66	27.33	29.22	23.20	28.21
INM-1	4130	X3	8	41.65	28.07	28.09	27.04	23.68
INM-1	4131	X3	8	40.75	26.40	30.20	22.25	25.80
INM-1	707	B1	9	00.00	24.86	26.12	24.27	00.00
INM-1	708	B1	9	00.00	31.10	32.28	28.56	00.00
INM-1	4349	C1	9	40.95	28.70	32.85	28.00	28.95
INM-1	4350	C1	9	45.25	32.65	33.88	30.36	29.00
INM-1	799	C5	9	41.67	29.78	31.16	28.07	30.49
INM-1	4032	B1&B2	9	45.65	32.45	34.75	30.50	30.75
INM-1	5012	B1	10	53.63	37.85	38.60	31.00	35.40
INM-1	6095	C7	11	43.30	31.95	32.05	28.00	28.50
INM-1	6096	C7	11	44.20	30.53	35.25	29.85	29.73
INM-1	5520	D8	11	43.02	30.65	34.18	29.45	30.35
INM-1	5521	D8	11	38.54	26.97	27.90	24.49	25.25
INM-1	5522	D8	11	36.55	25.05	27.05	21.83	24.67
INM-1	5550	E9	11	00.00	30.58	00.00	26.47	27.30
INM-1	4176	B1	12	00.00	31.35	00.00	27.67	31.26
INM-1	5500	Hno85	12	42.55	24.57	28.18	21.28	22.18
INM-1	5501	Hno85	12	43.16	27.06	29.15	21.20	00.00
INM-1	5834	B3	13	45.33	29.65	32.65	28.17	27.35
INM-1	5794	B5	14	44.10	32.98	33.15	26.80	27.93
INM-1	5795	B5	14	46.95	31.00	34.35	24.38	26.95
INM-1	5484	C1	14	42.74	27.80	30.85	25.42	27.00
INM-1	5652	A5	14	45.10	30.26	31.58	27.15	28.85
INM-1	5470	D3	15	00.00	26.65	00.00	27.18	30.70
INM-1	5471	D3	15	00.00	00.00	00.00	23.85	24.86
INM-1	5562	D4	15	39.07	27.15	28.15	26.20	24.80
INM-1	5694	E9	15	43.45	28.45	32.60	26.02	29.68

Table A.3.11. Measurements of Third Phalanx of *Bubalus bubalis*

Site	Reg.No.	Trench	Layer	GL	DL	BP	TP
INM-1	2030	G7	1	52.30	47.05	00.00	00.00
INM-1	1194	F10	1	00.00	00.00	24.81	48.98
INM-1	4301	B0	3	41.30	00.00	28.54	29.60
INM-1	4303	B0	3	46.32	00.00	30.80	32.96
INM-1	1142	E10	3	61.22	00.00	00.00	00.00
INM-1	1334	G11	3	65.43	48.63	25.77	45.00

INM-1	758	I12	3	57.40	48.97	23.91	00.00
INM-1	1244	D25	3	66.59	51.57	24.64	46.62
INM-1	2240	F11	3	93.04	79.57	35.60	60.57
INM-1	2617	H10	3	61.33	00.00	22.13	36.51
INM-1	4392	B6	8	53.20	44.05	23.65	42.50
INM-1	699	B1	9	00.00	52.75	00.00	00.00
INM-1	4833	E1	10	46.00	00.00	31.25	33.20

Table A.3.12. Measurements of Maxillary first/second molar of *Bubalus bubalis*

Site	Reg.No.	Trench	Layer	GL	GB
INM-1	1022	E10	1	31.42	21.18
INM-1	2913	G10	3	18.45	20.77

Table A.3.13. Measurements of Maxillary third molar of *Bubalus bubalis*

Site	Reg.No.	Trench	Layer	GL	GB
INM-1	1130	G4	1	30.05	21.03

Table A.3.14. Measurements of Mandibles with teeth of *Bubalus bubalis*

Site	Reg.No.	Trench	Layer	Mandible	Length	Width	Height
INM-1	4923	F13	1	m2	24.75	15.67	00.00
INM-1				m3	00.00	14.95	00.00
INM-1	1091	H7	2	p2	00.00	00.00	32.96

p2 –second premolar, p3- third premolar, p4- fourth premolar, m1- first molar, m2- second molar, m3- third molar.

Table A.4. Measurements of Bones of *Bos/Bubalus*

Table A.4.1. Measurements of Humerus of *Bos/Bubalus*

Site	Reg.No.	Trench	Layer	GL	DD	BP	TP	BD	TD
INM-1	8146	E24	4	00.00	00.00	00.00	00.00	77.70	35.50

Table A.4.2. Measurements of Metacarpal of *Bos/Bubalus*

Site	Reg.No.	Trench	Layer	DD	BP	TP	BD	TD	WT	TT	TLT
INM-1	4913	F13	1	00.00	00.00	00.00	52.15	29.00	22.55	27.90	21.60
INM-1	4927	F13	1	23.80	00.00	00.00	00.00	00.00	00.00	00.00	00.00
INM-1	1959	H11	1	00.00	00.00	00.00	00.00	00.00	29.24	33.70	26.17
INM-1	1281	G11	3	00.00	00.00	00.00	00.00	00.00	24.26	32.19	22.63
INM-1	5859	E27	4	00.00	00.00	00.00	52.05	27.50	23.25	27.20	19.35
INM-1	4445	E6	8	00.00	00.00	00.00	00.00	00.00	26.10	29.28	22.50
INM-1	4955	E6	8	00.00	53.00	31.05	00.00	00.00	00.00	00.00	00.00
INM-1	4720	B0	8	00.00	00.00	00.00	00.00	00.00	24.50	27.37	21.76

Table A.4.3. Measurements of Metapodia of *Bos/Bubalus*

Site	Reg.No.	Trench	Layer	BD	TD	WT	TT	TLT
INM-1	391	LJP	1-4	00.00	00.00	22.23	29.07	26.52
INM-1	392	LJP	1-4	00.00	00.00	23.23	25.93	21.18
INM-1	1158	EMBANK	0	00.00	00.00	30.50	34.74	26.14
INM-1	4912	F13	1	46.55	30.58	19.02	30.00	21.85
INM-1	1961	H11	1	00.00	00.00	30.00	35.01	26.43
INM-1	1009	E10	1	00.00	00.00	22.37	28.25	18.61
INM-1	1137	E10	2	00.00	00.00	21.70	27.93	21.20
INM-1	7334	E24	2	00.00	00.00	30.20	31.85	26.10
INM-1	6411	E25	2	00.00	00.00	24.85	31.31	22.50
INM-1	2581	F11	2	00.00	00.00	23.78	28.39	20.44

INM-1	2582	F11	2	00.00	00.00	24.11	31.75	20.83
INM-1	2583	F11	2	00.00	00.00	23.42	29.36	20.42
INM-1	7198	G25	2	00.00	00.00	27.73	30.55	21.05
INM-1	7199	G25	2	00.00	00.00	23.17	27.70	19.75
INM-1	7197	G25	2	00.00	00.00	28.85	31.10	23.30
INM-1	2798	H11	2	00.00	00.00	28.63	34.91	26.48
INM-1	7454	E25	2	00.00	00.00	24.25	31.95	24.90
INM-1	7455	E25	2	00.00	00.00	25.91	29.07	21.78
INM-1	2688	G7	3	00.00	00.00	27.96	30.63	23.30
INM-1	1891	E10	3	00.00	00.00	25.88	30.20	22.87
INM-1	1892	E10	3	00.00	00.00	25.15	30.17	24.19
INM-1	3755	G11	3	00.00	00.00	25.85	33.40	25.40
INM-1	1786	H11	3	00.00	00.00	22.12	00.00	22.77
INM-1	3655	Hno127	3	00.00	00.00	23.67	27.07	20.78
INM-1	8156	A2	4	00.00	00.00	26.19	32.80	23.13
INM-1	8143	E24	4	00.00	00.00	25.00	30.85	23.65
INM-1	1863	G11	4	60.22	31.92	29.04	31.58	24.49
INM-1	4649	Hno111	4	00.00	00.00	19.48	26.20	17.60
INM-1	4647	Hno111	4	00.00	00.00	22.79	28.09	22.05
INM-1	4648	Hno111	4	00.00	00.00	25.93	30.87	24.60
INM-1	5018	Hno122	4	47.70	28.85	21.60	28.65	22.05
INM-1	4726	B1	5	00.00	00.00	23.90	27.70	20.83
INM-1	4623	B2	5	00.00	00.00	22.35	25.62	20.90
INM-1	4203	B2	6	00.00	00.00	23.95	30.20	23.65
INM-1	4202	B2	6	00.00	00.00	22.75	28.75	21.90
INM-1	4208	E1	6	00.00	00.00	23.50	30.40	21.60
INM-1	3779	E8	6	00.00	00.00	24.05	27.80	21.25
INM-1	3778	E8	6	00.00	00.00	24.00	28.65	19.45
INM-1	516	F3	6	00.00	00.00	24.26	00.00	00.00
INM-1	4396	B6	8	00.00	00.00	23.49	27.80	21.00
INM-1	4664	A5	8	00.00	00.00	24.35	28.80	22.49
INM-1	4665	A5	8	00.00	00.00	21.00	26.45	19.50
INM-1	4968	B3	8	00.00	00.00	22.25	29.20	21.65
INM-1	4967	B3	8	00.00	00.00	26.48	30.70	24.00
INM-1	4739	C1	8	00.00	00.00	21.47	27.35	19.75
INM-1	4738	C1	8	00.00	00.00	22.70	27.62	22.35
INM-1	862	D5	8	00.00	00.00	29.81	30.21	22.51
INM-1	4589	D6	8	00.00	00.00	22.95	27.32	21.55
INM-1	4586	D6	8	00.00	00.00	25.60	30.85	20.85
INM-1	4587	D6	8	00.00	00.00	26.10	32.90	23.08
INM-1	5183	D6	8	00.00	00.00	26.80	31.25	24.36
INM-1	5181	D6	8	00.00	00.00	27.96	30.77	22.00
INM-1	5184	D6	8	00.00	00.00	27.13	30.10	23.32
INM-1	5185	D6	8	00.00	00.00	26.25	30.30	21.35
INM-1	5182	D6	8	00.00	00.00	20.87	26.00	19.55
INM-1	5131	D7	8	00.00	00.00	21.80	26.30	21.05
INM-1	4458	E7	8	00.00	00.00	29.93	32.87	25.30
INM-1	4457	E7	8	00.00	00.00	29.00	33.10	22.35
INM-1	3735	Hno69	8	00.00	00.00	24.18	28.95	19.76
INM-1	4162	Hno72	8	00.00	00.00	24.75	29.25	19.93
INM-1	4160	Hno72	8	00.00	00.00	20.85	27.05	19.10
INM-1	4161	Hno72	8	00.00	00.00	24.65	27.94	21.05
INM-1	730	Hno70&73	8	00.00	00.00	23.02	29.50	23.33
INM-1	731	Hno70&73	8	00.00	00.00	22.65	29.34	21.55
INM-1	4962	KILNB1	8	00.00	00.00	20.95	26.55	19.32
INM-1	807	C5	9	00.00	00.00	22.15	28.10	19.71
INM-1	805	C5	9	00.00	00.00	22.55	27.93	18.89
INM-1	803	C5	9	00.00	00.00	22.99	27.72	20.98

INM-1	808	C5	9	00.00	00.00	29.35	32.97	25.67
INM-1	2887	D5	9	00.00	00.00	25.30	29.80	21.04
INM-1	2886	D5	9	00.00	00.00	21.05	27.45	19.15
INM-1	5191	C3	9	00.00	00.00	24.10	29.05	22.40
INM-1	5192	C3	9	00.00	00.00	22.50	28.85	19.75
INM-1	5193	C3	9	00.00	00.00	22.80	28.65	21.54
INM-1	5194	C3	9	00.00	00.00	25.75	30.70	22.40
INM-1	6056	C3	11	00.00	00.00	28.75	32.30	24.40
INM-1	5340	D5	11	00.00	00.00	23.60	30.43	23.25
INM-1	5341	D5	11	00.00	00.00	24.50	28.95	20.05
INM-1	5342	D5	11	00.00	00.00	24.31	28.66	21.32
INM-1	5339	D5	11	00.00	00.00	28.37	32.60	00.00
INM-1	5938	B3	12	00.00	00.00	31.82	32.25	25.35
INM-1	5937	B3	12	00.00	00.00	23.90	30.53	22.75
INM-1	5936	B3	12	00.00	00.00	22.50	29.65	20.45
INM-1	6323	B5	12	00.00	00.00	24.00	31.80	21.05
INM-1	812	C1	12	00.00	00.00	27.38	29.97	20.22
INM-1	5860	C6	12	00.00	00.00	28.49	32.15	21.05
INM-1	5789	Hno74	12	00.00	00.00	23.15	30.00	20.90
INM-1	5776	B5	13	00.00	00.00	24.28	30.20	22.35
INM-1	5775	B5	13	00.00	00.00	23.85	27.05	19.38
INM-1	5824	E5	13	00.00	00.00	24.95	30.30	21.25
INM-1	5457	B1	14	00.00	00.00	20.75	26.30	19.50
INM-1	6336	E4	14	00.00	00.00	25.00	31.65	23.05
INM-1	6335	E4	14	00.00	00.00	23.55	31.00	21.05
INM-1	3613	C0	14	00.00	00.00	23.49	29.87	18.87
INM-1	5380	C3	15	00.00	00.00	30.85	32.35	22.97
INM-1	5382	C3	15	00.00	00.00	25.80	31.07	23.75
INM-1	5384	C3	15	00.00	00.00	17.76	28.35	21.35
INM-1	5381	C3	15	00.00	00.00	25.45	31.45	21.05
INM-1	6239	E5	15	00.00	00.00	23.25	30.25	22.90
INM-1	4817	B0	15	00.00	00.00	28.35	28.75	21.50

Table A.4.4. Measurements of Femur heads of *Bos/Bubalus*

Site	Reg.No.	Trench	Layer	LC	TC
INM-1	482	LJP	1-4	00.00	41.20
INM-1	483	LJP	1-4	00.00	38.35
INM-1	484	LJP	1-4	61.40	44.95
INM-1	485	LJP	1-4	00.00	46.60
INM-1	486	LJP	1-4	00.00	40.30
INM-1	488	LJP	1-4	00.00	37.70
INM-1	489	LJP	1-4	00.00	39.80
INM-1	492	LJP	1-4	00.00	41.00
INM-1	1730	G11	1	42.17	00.00
INM-1	6430	E25	2	64.15	48.40
INM-1	4358	A5	6	67.20	45.50
INM-1	856	Hno60&67	8	56.98	40.79
INM-1	872	D5	8	00.00	44.20
INM-1	4446	E6	8	69.85	45.45
INM-1	4593	D6	8	54.25	40.00
INM-1	1747	A1	9	39.36	29.32

Table A.4.5. Measurements of Shaft of Tibia of *Bos/Bubalus*

Site	Reg.No.	Trench	Layer	DD
INM-1	598	LJP	1-4	38.65

Table A.4.6. Measurements of Metatarsal of *Bos/Bubalus*

Site	Reg.No.	Trench	Layer	DD	BP	TP	BD	TD	WT	TT	TLT
INM-1	1960	H11	1	00.00	00.00	00.00	00.00	00.00	26.33	32.88	23.31
INM-1	8152	E24	4	00.00	21.50	23.25	00.00	00.00	00.00	00.00	00.00
INM-1	4588	D6	8	00.00	00.00	00.00	00.00	00.00	22.75	28.57	21.97
INM-1	5180	D6	8	00.00	00.00	00.00	00.00	00.00	23.05	27.55	20.00
INM-1	804	C5	9	00.00	00.00	00.00	00.00	00.00	23.07	26.85	20.80
INM-1	5151	B2	11	00.00	00.00	00.00	00.00	00.00	25.20	32.75	23.00
INM-1	5940	B3	12	00.00	10.65	18.40	00.00	00.00	00.00	00.00	00.00
INM-1	5941	B3	12	9.70	00.00	00.00	00.00	00.00	00.00	00.00	00.00

Table A.4.7. Measurements of First Phalanx of *Bos/Bubalus*

Site	Reg.No.	Trench	Layer	BD	TD
INM-1	858	Hno60&67	8	28.82	21.58
INM-1	4621	E7	8	16.95	15.45

Table A.4.8. Measurements of Second Phalanx of *Bos/Bubalus*

Site	Reg.No.	Trench	Layer	BP	TP	BD	TD
INM-1	1155	EMBANK	0	29.80	32.40	25.12	29.54
INM-1	1156	EMBANK	0	28.74	33.14	24.73	00.00

Table A.4.9. Measurements of Third Phalanx of *Bos/Bubalus*

Site	Reg.No.	Trench	Layer	BP	TP
INM-1	512	F3	6	20.32	00.00

Table A.4.10. Measurements of Maxillary first/second molar of *Bos/Bubalus*

Site	Reg.No.	Trench	Layer	GL	GB
INM-1	400	LJP	1-4	30.13	20.33
INM-1	403	LJP	1-4	29.59	18.53
INM-1	410	LJP	1-4	28.61	18.05
INM-1	408	LJP	1-4	29.86	21.44
INM-1	404	LJP	1-4	29.31	18.51
INM-1	412	LJP	1-4	32.71	23.16
INM-1	413	LJP	1-4	27.73	20.47
INM-1	405	LJP	1-4	28.24	17.70
INM-1	406	LJP	1-4	29.65	18.99
INM-1	409	LJP	1-4	27.64	20.12
INM-1	414	LJP	1-4	30.15	19.93
INM-1	398	LJP	1-4	31.27	17.34
INM-1	399	LJP	1-4	29.53	19.54
INM-1	401	LJP	1-4	25.64	17.68
INM-1	411	LJP	1-4	27.09	17.45
INM-1	1162	EMBANK	0	21.68	20.96
INM-1	1161	EMBANK	0	24.55	14.80
INM-1	4305	B0	2	26.86	17.63
INM-1	1144	E10	2	29.82	18.43
INM-1	1147	E10	2	24.59	18.60
INM-1	1145	E10	2	29.58	18.37
INM-1	1148	E10	2	28.13	16.30
INM-1	1146	E10	2	00.00	19.25
INM-1	1149	E10	2	26.85	19.08
INM-1	7335	E24	2	26.73	20.35
INM-1	6447	E25	2	29.50	19.40
INM-1	849	F12	2	27.69	14.87

INM-1	7509	F24	2	29.95	22.70
INM-1	1306	G11	3	20.40	14.99
INM-1	983	C1	4	30.97	15.09
INM-1	575	F8	4	30.44	18.64
INM-1	574	F8	4	29.85	16.98
INM-1	8147	E24	4	30.95	18.45
INM-1	7682	G25	4	29.13	20.60
INM-1	546	C0	4	27.43	18.46
INM-1	5036	Hno119	4	27.21	19.55
INM-1	5037	Hno119	4	29.70	21.90
INM-1	5020	Hno122	4	29.80	21.04
INM-1	5019	Hno122	4	27.35	22.15
INM-1	4624	B2	5	26.70	19.70
INM-1	4152	D3	7	24.83	14.90
INM-1	1842	G4	7	24.42	15.09
INM-1	1841	G4	7	23.46	14.26
INM-1	5049	B1	8	25.35	20.10
INM-1	644	B5	8	29.38	17.00
INM-1	643	B5	8	28.07	19.17
INM-1	4591	D6	8	25.90	15.05
INM-1	4590	D6	8	26.86	18.15
INM-1	5143	D7	8	20.55	16.65
INM-1	5068	E5	8	28.65	21.25
INM-1	5069	E5	8	28.60	19.05
INM-1	5070	E5	8	30.00	18.90
INM-1	5071	E5	8	25.95	15.05
INM-1	5072	E5	8	28.85	17.25
INM-1	740	Hno70	8	30.85	20.59
INM-1	737	Hno70	8	26.87	15.82
INM-1	726	D6	9	31.30	15.61
INM-1	727	D6	9	27.64	18.11
INM-1	724	D6	9	28.30	18.00
INM-1	618	C2	10	28.67	00.00
INM-1	611	C3	10	24.98	14.68
INM-1	5680	C8	11	22.80	14.50
INM-1	5681	C8	11	27.30	17.05
INM-1	6326	B5	12	27.50	16.35
INM-1	822	C1	12	26.81	17.80
INM-1	823	C1	12	29.46	16.05
INM-1	5690	D3	12	14.88	16.60
INM-1	6085	D6	12	31.42	24.70
INM-1	6084	D6	12	25.50	18.80
INM-1	6165	Hno75	12	25.60	15.42
INM-1	6164	Hno75	12	26.60	14.85
INM-1	6163	Hno75	12	29.56	18.08
INM-1	6338	E4	14	21.60	18.65
INM-1	6339	E4	14	27.55	16.85
INM-1	5666	A5	15	29.10	19.92
INM-1	5482	D3	15	28.26	13.33
INM-1	5483	D3	15	23.70	15.65
INM-1	6065	D8	15	27.60	15.10
INM-1	6066	D8	15	29.87	18.20
INM-1	5437	E13	15	29.88	18.54
INM-1	5362	Hno95	16	28.05	16.45
INM-1	5361	Hno95	16	29.05	18.72

Table A.4.11. Measurements of Maxillary third molar of *Bos/Bubalus*

Site	Reg.No.	Trench	Layer	GL	GB
INM-1	947	G12	3	30.59	18.74
INM-1	8148	E24	4	31.92	22.30
INM-1	5217	B3	6	27.75	22.55
INM-1	717	B1	9	26.19	17.88
INM-1	5726	Hno77	12	28.90	19.30
INM-1	5392	C3	15	28.30	20.33
INM-1	5391	C3	15	27.83	20.86
INM-1	5364	Hno95	16	31.64	18.30
INM-1	5363	Hno95	16	28.32	19.55

Table A.4.12. Measurements of Mandibular deciduous premolar of *Bos/Bubalus*

Site	Reg.No.	Trench	Layer	GL	GB
INM-1	5684	C8	11	28.25	11.50
INM-1	4185	B1	12	29.70	11.50
INM-1	5436	E13	15	32.35	11.35

Table A.4.13. Measurements of Mandibular incisor of *Bos/Bubalus*

Site	Reg.No.	Trench	Layer	GL	GB
INM-1	5144	D7	8	14.25	6.85
INM-1	5330	Hno79	12	15.25	9.65
INM-1	4184	B1	12	16.55	9.25
INM-1	5732	C3	13	15.00	9.85
INM-1	5574	D4	15	12.65	9.65

Table A.4.14. Measurements of Mandibular first/second premolar of *Bos/Bubalus*

Site	Reg.No.	Trench	Layer	GL	GB
INM-1	5024	Hno122	4	18.55	10.55
INM-1	5107	C1	11	19.29	8.45
INM-1	5719	B6	11	16.10	8.10
INM-1	5826	E5	13	16.80	10.15
INM-1	5463	B1	14	19.95	10.20
INM-1	5623	A4	15	19.72	10.50

Table A.4.15. Measurements of Mandibular second premolar of *Bos/Bubalus*

Site	Reg.No.	Trench	Layer	GL	GB
INM-1	518	F3	6	10.57	6.81
INM-1	5080	E5	8	10.35	6.10

Table A.4.16. Mandibular third premolar of *Bos/Bubalus*

Site	Reg.No.	Trench	Layer	GL	GB
INM-1	5079	E5	8	18.25	9.70

Table A.4.17. Measurements of Mandibular fourth premolar of *Bos/Bubalus*

Site	Reg.No.	Trench	Layer	GL	GB
INM-1	576	F8	4	21.34	11.35
INM-1	536	Hno50	12	23.00	11.14

Table A.4.18. Measurements of Mandibular first/second molar of *Bos/Bubalus*

Site	Reg.No.	Trench	Layer	GL	GB
INM-1	402	LJP	1-4	31.19	12.45
INM-1	407	LJP	1-4	34.20	12.61
INM-1	1199	F10	1	29.03	11.62
INM-1	1150	E10	2	31.72	12.25
INM-1	6446	E25	2	26.40	12.10
INM-1	770	I12	2	27.41	14.98
INM-1	1265	D25	3	27.15	10.78
INM-1	540	Hno127	3	28.91	10.05
INM-1	982	C1	4	27.03	00.00
INM-1	589	E6	7	29.13	10.75
INM-1	642	B5	8	24.20	13.02
INM-1	5081	E5	8	32.05	12.40
INM-1	4592	D6	9	30.60	14.60
INM-1	725	D6	9	26.39	12.55
INM-1	615	C2	10	28.34	00.00
INM-1	5352	D5	11	21.04	14.00
INM-1	5535	D8	11	21.70	13.06
INM-1	5536	D8	11	21.60	12.40
INM-1	6324	B5	12	22.75	12.80
INM-1	5329	Hno79	12	28.63	10.42
INM-1	5516	Hno85	12	30.80	12.55
INM-1	5515	Hno85	12	25.65	16.20
INM-1	5514	Hno85	12	26.85	13.65
INM-1	6341	E4	14	30.31	11.87
INM-1	6340	E4	14	29.90	11.55
INM-1	5667	A5	15	31.61	12.28
INM-1	5393	C3	15	30.13	15.05
INM-1	5570	D4	15	32.55	12.20
INM-1	5571	D4	15	27.20	11.68
INM-1	5569	D4	15	28.61	14.35
INM-1	5438	E13	15	28.06	18.35
INM-1	5360	Hno95	16	29.30	11.50

Table A.4.19. Measurements of Mandibular third molar of *Bos/Bubalus*

Site	Reg.No.	Trench	Layer	GL	GB
INM-1	417	LJP	1-4	34.90	12.78
INM-1	416	LJP	1-4	38.13	12.80
INM-1	415	LJP	1-4	37.50	13.35
INM-1	418	LJP	1-4	36.39	13.49
INM-1	420	LJP	1-4	33.84	12.87
INM-1	421	LJP	1-4	36.62	12.41
INM-1	422	LJP	1-4	35.36	13.37
INM-1	423	LJP	1-4	35.35	13.53
INM-1	419	LJP	1-4	37.19	13.88
INM-1	425	LJP	1-4	39.28	14.74
INM-1	424	LJP	1-4	38.31	00.00
INM-1	5216	B3	6	34.30	12.25
INM-1	616	C2	10	39.34	12.13
INM-1	5351	D5	11	36.07	15.00
INM-1	5689	D3	12	36.30	15.25
INM-1	6162	Hno75	12	34.80	12.70
INM-1	5394	C3	15	37.14	14.25
INM-1	5572	D4	15	3.05	12.35

Table A.4.20. Measurements of Mandible of *Bos/Bubalus*

Site	Reg.No.	Trench	Layer	Mandible	Length	Width	Height
INM-1	1015	E10	1	MC	00.00	34.60	00.00
INM-1	1016	E10	1	MC	00.00	36.00	00.00
INM-1	4585	D6	8	MC	00.00	41.83	00.00
INM-1	4994	C2	8	MC	00.00	45.65	00.00
INM-1	4995	C2	8	MC	00.00	44.00	00.00
INM-1	4996	C2	8	MS	00.00	00.00	39.85
INM-1	5215	B3	6	p3	17.25	9.70	00.00
INM-1	5383	C3	15	MC	00.00	37.10	00.00
INM-1	5622	A4	15	MC	00.00	40.65	00.00

Table A.5. Measurements of Bones of *Capra hircus*

Table A.5. 1. Measurements of Scapula of *Capra hircus*

Site	Reg.No.	Trench	Layer	LG	BG
INM-1	4429	E7	7	20.88	18.48

Table A.5. 2. Measurements of Humerus of *Capra hircus*

Site	Reg.No.	Trench	Layer	BD	TD
INM-1	1192	F10	1	28.74	14.25
INM-1	4196	E8	7	27.95	12.65

Table A.5. 3. Measurements of Radius of *Capra hircus*

Site	Reg.No.	Trench	Layer	DD	BP	TP	BD	TD
INM-1	79	LJP	1-4	00.00	00.00	00.00	31.55	21.09
INM-1	82	LJP	1-4	00.00	43.62	35.22	00.00	00.00
INM-1	4423		6	9.48	00.00	00.00	00.00	00.00
INM-1	4247	E1	8	00.00	21.55	12.20	00.00	00.00
INM-1	4449	E6	8	00.00	25.88	13.84	00.00	00.00
INM-1	5465	B1	14	00.00	00.00	00.00	25.00	17.30

Table A.5. 4. Measurements of Metacarpal of *Capra hircus*

Site	Reg.No.	Trench	Layer	BP	TP	BD	TD	WT	TT	TLT
INM-1	331	LJP	1-4	00.00	00.00	22.44	12.89	9.60	13.00	7.88
INM-1	3087	G3	2	22.12	15.78	00.00	00.00	00.00	00.00	00.00
INM-1	889	G12	3	00.00	00.00	24.03	10.52	17.30	13.29	00.00

Table A.5. 5. Measurements of Metapodia of *Capra hircus*

Site	Reg.No.	Trench	Layer	BP	TP	BD	TD	WT	TT	TLT
INM-1	1701	G11	1	00.00	00.00	00.00	00.00	11.15	14.89	8.37
INM-1	3015	F10	2	00.00	00.00	00.00	00.00	9.99	13.49	9.34

Table A.5. 6. Measurements of Femur of *Capra hircus*

Site	Reg.No.	Trench	Layer	BD	TD	LC	TC
INM-1	290	LJP	1-4	32.10	00.00	00.00	00.00
INM-1	4253	Hno130	2	42.75	58.60	00.00	00.00
INM-1	7380	F26	1	00.00	00.00	21.05	17.92
INM-1	289	G25	2	36.64	00.00	00.00	00.00
INM-1	3258	D10	3	00.00	00.00	00.00	17.25
INM-1	5804	B5	14	00.00	00.00	18.40	18.20

Table A.5. 7. Measurements of Tibia of *Capra hircus*

Site	Reg.No.	Trench	Layer	DD	BD	TD
INM-1	143	LJP	1-4	00.00	22.13	16.96
INM-1	146	LJP	1-4	00.00	25.15	19.14
INM-1	1981	H11	1	10.65	00.00	00.00

Table A.5. 8. Measurements of Metatarsal of *Capra hircus*

Site	Reg.No.	Trench	Layer	BP	TP
INM-1	3088	G3	2	22.41	24.12

Table A.5. 9. Measurements of Astragalus of *Capra hircus*

Site	Reg.No.	Trench	Layer	GL	GLL	GLM	TL	TM	BD
INM-1	1887	E10	3	00.00	27.99	25.05	16.78	00.00	00.00
INM-1	4448	E6	8	25.09	25.05	23.50	12.65	13.55	15.15

Table A.5. 10. Measurements.of Calcaneum of *Capra hircus*

Site	Reg.No.	Trench	Layer	GL	GB
INM-1	6357	A3&B3	0	39.70	13.00

Table A.5. 11. Measurements of Centrotarsal of *Capra hircus*

Site	Reg.No.	Trench	Layer	GB
INM-1	1887	E10	3	18.38

Table A.5. 12. Measurements of First Phalanx of *Capra hircus*

Site	Reg.No.	Trench	Layer	GL	BP	TP	BD	TD
INM-1	72	LJP	1-4	45.71	11.90	16.79	11.33	10.90
INM-1	95	LJP	1-4	46.30	12.72	17.00	10.08	11.90
INM-1	96	LJP	1-4	44.47	11.74	16.30	9.67	10.42
INM-1	249	LJP	1-4	00.00	11.61	13.11	00.00	00.00
INM-1	259	LJP	1-4	00.00	12.78	16.66	00.00	00.00
INM-1	1170	G10	1	40.78	14.54	17.64	12.63	12.10
INM-1	2271	E1	2	36.62	11.22	13.26	00.00	00.00
INM-1	1540	H5	2	36.12	11.52	14.57	11.23	10.79
INM-1	5190	C3	9	33.95	10.90	13.65	10.30	10.10
INM-1	4831	E1	10	34.43	10.8	13.20	10.30	00.00
INM-1	524	Hno50	12	00.00	00.00	00.00	7.98	7.71

Table A.5. 13. Measurements of Second Phalanx of *Capra hircus*

Site	Reg.No.	Trench	Layer	GL	BP	TP	BD	TD
INM-1	77	LJP	1-4	25.27	11.76	11.57	8.92	10.52
INM-1	107	LJP	1-4	25.47	10.32	12.29	8.59	11.30
INM-1	1041	H7	2	27.54	10.60	14.70	10.07	12.46
INM-1	1042	H7	2	26.32	10.72	14.74	1060	12.65
INM-1	1044	H7	2	22.98	9.93	13.61	9.05	11.29
INM-1	1046	H7	2	24.21	00.00	00.00	9.19	12.18
INM-1	1047	H7	2	23.97	9.90	13.17	8.35	11.87
INM-1	6054	C3	11	21.35	11.00	11.46	9.35	11.05

Table A.5. 14. Measurements of Maxillary second molar of *Capra hircus*

Site	Reg.No.	Trench	Layer	GL	GB
INM-1	2300	E1	2	18.39	12.02

Table A.5. 15. Maxillary first/ second molar of *Capra hircus*

Site	Reg.No.	Trench	Layer	GL	GB
INM-1	5409	C3	15	13.58	10.95

Table A.5.16. Measurements of Mandibles with teeth of *Capra hircus*

p2 –second premolar, p3- third premolar, p4- fourth premolar, m1- first molar, m2- second molar, m3- third molar.

Site	Reg.No.	Trench	Layer	Mandible	Length	Width	Height
INM-1	3915	B3&C3	8	p2	00.00	00.00	12.80
INM-1				p3	00.0	00.0	16.05
INM-1				p4	8.77	5.70	17.50
INM-1				m1	9.65	6.95	18.68
INM-1	3916	B3&C3	8	p2	00.00	00.00	12.20
INM-1				p3	7.95	4.93	15.45
INM-1				p4	00.00	00.00	17.35
INM-1	4242	E4	9	p2	00.00	00.00	9.10
INM-1				p3	00.00	00.00	10.80
INM-1				p4	00.00	00.00	11.15
INM-1				m1	00.00	00.00	15.55
INM-1	835	F12	2	m2	14.25	7.54	27.27
INM-1				m3	20.55	7.75	29.71
INM-1	836	F12	2	p4	8.49	5.61	19.02
INM-1				m1	13.21	7.61	19.69
INM-1				m2	16.29	8.11	22.41

Table A.5.17. Measurements of Maxilla with teeth of *Capra hircus*

P2 –second premolar, P3- third premolar, P4- fourth premolar, M1- first molar, M2- second molar, M3- third molar.

Site	Reg.No.	Trench	Layer	Maxilla	Length	Width	Height
INM-1	946	G12	3	M1	14.23	10.98	00.00
INM-1				M2	15.87	10.31	00.00
INM-1				M3	15.33	9.33	00.00
INM-1	2165	Hno132	1	P3	00.00	00.00	18.49
INM-1				P4	9.79	4.94	2.21
INM-1				M1	12.61	6.91	24.82
INM-1				M2	14.01	7.60	27.40
INM-1				M3	19.40	7.00	31.47
INM-1	2167	Hno132	1	M1	11.86	6.88	00.00
INM-1				M2	13.73	7.08	00.0
INM-1	1075	H7	2	P3	00.00	00.00	16.99
INM-1				P4	10.54	4.53	21.05
INM-1				M1	13.11	6.99	22.00
INM-1				M2	14.76	7.81	26.30
INM-1				M3	00.00	00.00	30.88

Table A.6. Measurements of Bones of *Ovis aries*

Table A.6.1. Measurements of Scapula of *Ovis aries*

Site	Reg.No.	Trench	Layer	GLP	LG	BG	SLC
INM-1	374		6	00.00	28.88	23.51	17.39

Table A.6.2. Measurements of Humerus of *Ovis aries*

Site	Reg.No.	Trench	Layer	BD	TD
INM-1	275	LJP	1-4	26.02	12.82
INM-1	161	F26	3	23.21	13.36
INM-1	2976	B3	5	26.63	12.67
INM-1	4703	F3	8	29.00	12.76

Table A.6.3. Measurements of Radius of *Ovis aries*

Site	Reg.No.	Trench	Layer	DD
INM-1	839	F12	2	14.03

Table A.6.4. Measurements of Femur of *Ovis aries*

Site	Reg.No.	Trench	Layer	TC
INM-1	16	LJP	1-4	16.41

Table A.6.5. Measurements of Mandibles with teeth of *Ovis aries*

p2 –second premolar, p3- third premolar, p4- fourth premolar, m1- first molar, m2- second molar, m3- third molar.

Site	Reg.No.	Trench	Layer	Mandible	Length	Width	Height
INM-1	445	G24	2	p2	00.0	00.00	12.47
INM-1				p3	8.33	5.31	15.48
INM-1				p4	10.80	6.74	16.22
INM-1				m1	12.29	7.69	21.02
INM-1				m2	14.91	7.24	23.16
INM-1	435	LJP	0	dp4	14.19	7.09	00.00
INM-1				m1	13.96	7.28	00.00

Table A.6.6. Measurements of Maxillary with teeth of *Ovis aries*

Site	Reg.No.	Trench	Layer	Maxilla	Length	Width	Height
INM-1	2467	F25	3	P4	10.12	7.30	00.00
INM-1				M1	11.15	4.54	00.00

Table A.7. Measurements of Bones of *Capra/Ovis*

Table A.7.1.Measurements of Scapula of *Capra/Ovis*

Site	Reg.No.	Trench	Layer	GLP	LG	BG	SLC
INM-1	5156	B2	11	25.06	21.35	14.00	00.00
INM-1	5102	C1	11	26.55	21.70	17.50	00.00
INM-1	5226	E1	11	25.80	21.53	16.75	00.00

Table A.7.2. Measurements of Humerus of *Capra/Ovis*

Site	Reg.No.	Trench	Layer	BP	TP	BD	TD
INM-1	274	LJP	1-4	00.00	00.00	27.27	12.93

INM-1	276	LJP	1-4	00.00	00.00	28.82	13.26
INM-1	281	LJP	1-4	00.00	00.00	26.01	12.42
INM-1	2315	E10	3	32.80	40.23	00.00	00.00
INM-1	790	Hno113	4	00.00	00.00	28.95	13.75
INM-1	7754	D24	4	00.00	00.00	28.40	14.00
INM-1	7755	D24	4	00.00	00.00	26.75	12.20
INM-1	4414	C7	8	00.00	00.00	26.70	12.58
INM-1	5066	E5	8	00.00	00.00	29.90	13.80
INM-1	5145	D7	8	00.00	00.00	24.22	13.37
INM-1	377		9	00.00	00.00	28.63	12.81
INM-1	5825	D3	13	00.00	00.00	26.70	12.75
INM-1	5767	D8	14	00.00	00.00	24.35	12.30

Table A.7.3. Measurements of Radius of *Capra/Ovis*

Site	Reg.No.	Trench	Layer	GL	DD	BP	TP	BD	TD
INM-1	170	LJP	0	00.00	00.00	00.00	00.00	28.47	20.04
INM-1	4661	Hno111	4	00.00	00.00	28.05	16.90	00.00	00.00
INM-1	295	E24	4	00.00	12.77	22.31	11.92	00.00	00.00
INM-1	5872	E27	4	00.00	00.00	23.67	13.00	00.00	00.00
INM-1	5873	E27	4	00.00	00.00	24.45	12.10	00.00	00.00
INM-1	4169	B1	6	123.57	00.00	24.57	14.60	14.45	00.00
INM-1	4157	B3	7	00.00	00.00	27.10	13.85	00.00	00.00
INM-1	608	C2	10	00.00	00.00	23.62	12.50	00.00	00.00
INM-1	4840	E1	10	00.00	00.00	24.70	13.88	00.00	00.00
INM-1	4841	E1	10	00.00	00.00	00.00	00.00	24.27	17.80
INM-1	5712	C5	11	00.00	00.00	26.80	13.45	00.00	00.00
INM-1	5782	Hno75	12	00.00	00.00	27.57	13.80	00.00	00.00
INM-1	4183	B1	12	00.00	00.00	27.75	13.80	25.50	19.65
INM-1	6277	C8	12	00.00	00.00	29.65	17.00	00.00	00.00
INM-1	5920	B3	14	00.00	00.00	23.25	13.70	00.00	00.00
INM-1	5921	B3	14	00.00	00.00	26.35	13.85	00.00	00.00

Table A.7.4. Measurements of Metacarpal of *Capra/Ovis*

Site	Reg.No.	Trench	Layer	BP	TP	BD	TD	WT	TT	TLT
INM-1	3353	F11	1	00.00	00.00	24.13	15.82	11.40	14.68	9.30
INM-1	188	LJP	1-4	00.00	00.00	25.31	00.00	111.39	17.12	13.18
INM-1	3089	G3	2	00.00	00.00	25.45	14.75	10.62	14.85	8.71
INM-1	4	B2	12	23.38	15.95	00.00	00.00	00.00	00.00	00.00
INM-1	5823	E5	13	00.00	00.00	00.00	00.00	8.95	13.10	7.70
INM-1	4680	B0	13	22.05	00.00	00.00	00.00	00.00	00.00	00.00
INM-1	4816	B0	15	00.00	00.00	20.97	13.80	9.20	13.20	9.19

Table A.7.5. Measurements of Metapodia of *Capra/Ovis*

Site	Reg.No.	Trench	Layer	BD	TD	WT	TT	TLT
INM-1	19	LJP	1-4	24.38	00.00	10.30	16.71	12.19
INM-1	6940	E26	3	00.00	00.00	10.30	14.70	8.10
INM-1	6858	G26	4	00.00	00.00	9.80	14.90	9.60
INM-1	6826	G26	4	00.00	00.00	10.75	15.20	9.00
INM-1	4336	F8	5	00.00	00.00	10.24	14.30	10.20
INM-1	5211	B3	6	00.00	00.00	11.40	15.95	10.00
INM-1	5762	C5	12	16.30	00.00	7.35	11.20	7.95
INM-1	6278	C8	12	00.00	00.00	9.30	13.80	10.55

Table A.7.6. Measurements of Femur of *Capra/Ovis*

Site	Reg.No.	Trench	Layer	BP	TP	BD	TD
INM-1	4119	C3	5	00.00	00.00	31.05	38.25
INM-1	135	B1	11	21.69	21.65	00.00	00.00
INM-1	539	Hno126	3	31.18	00.00	00.00	00.00

Table A.7.7. Measurements of Tibia of *Capra/Ovis*

Site	Reg.No.	Trench	Layer	DD	BD	TD
INM-1	7756	D24&e26	4	00.00	27.35	21.10
INM-1	350		8	12.80	23.13	16.67
INM-1	5931	B3	14	9.25	00.00	00.00

Table A.7.8. Measurements of Metatarsal of *Capra/Ovis*

INM-1	7173	G25	2	00.00	17.05	15.35
INM-1	5001	C2	8	00.00	16.80	14.25
INM-1	5154	B2	11	9.05	17.35	16.20
INM-1	2444	F25&F26	3	00.00	16.96	15.01

Table A.7. 9. Measurements of Astragalus of *Capra/Ovis*

Site	Reg.No.	Trench	Layer	GL	GLL	GLM	TL	TM	BD
INM-1	85	LJP	1-4	29.83	29.82	17.59	15.46	15.73	17.98
INM-1	88	LJP	1-4	32.13	31.96	29.57	19.08	18.14	19.16
INM-1	221	LJP	1-4	00.00	00.00	22.93	00.00	13.62	00.00
INM-1	362	LJP	1-4	28.80	28.57	27.32	16.88	16.67	16.90
INM-1	5759	C5	12	31.05	30.95	29.30	15.80	16.80	18.00
INM-1	5692	E9	15	28.15	27.90	25.90	15.60	16.07	17.65

Table A.7. 10. Measurements of Calcaneum of *Capra/Ovis*

Site	Reg.No.	Trench	Layer	GL	GB
INM-1	5628	D7	12	38.08	12.65

Table A.7. 11. Measurements of Centrotarsal of *Capra/Ovis*

Site	Reg.No.	Trench	Layer	GB
INM-1	519	Hno50	12	21.66
INM-1	5820	D3	13	19.45

Table A.7. 12. Measurements of First Phalanx of *Capra/Ovis*

Site	Reg.No.	Trench	Layer	GL	BP	TP	BD	TD
INM-1	7651	G25	4	00.00	11.95	14.65	00.00	00.00
INM-1	609	C3	10	00.00	00.00	00.00	8.67	7.83
INM-1	5153	B2	11	00.00	00.00	00.00	8.50	7.57
INM-1	5099	C1	11	00.00	00.00	00.00	8.80	7.42
INM-1	5236	C3	11	00.00	00.00	00.00	8.90	7.75
INM-1	5779	Hno75	12	00.00	00.00	00.00	10.30	9.28
INM-1	4672	B0	13	29.45	9.70	12.10	9.45	8.92
INM-1	5451	B1	14	32.31	11.55	12.81	10.55	10.05
INM-1	5422	E4	15	00.00	9.85	12.10	00.00	00.00
INM-1	5423	E4	15	00.00	10.40	11.75	00.00	00.00

Table A.7. 13. Measurements of Second Phalanx of *Capra/Ovis*

Site	Reg.No.	Trench	Layer	GL	BP	TP	BD	TD
INM-1	359	LJP	1-4	00.00	00.00	00.00	6.80	10.57
INM-1	7711	G26	3	00.00	11.40	11.45	00.00	00.00

Table A.7. 14. Measurements of Maxillary first/second molar of *Capra/Ovis*

Site	Reg.No.	Trench	Layer	GL	GB
INM-1	6358	A3&B3	0	15.60	12.60
INM-1	7247	Hno104	4	15.95	11.15
INM-1	7246	Hno104	4	17.20	11.65
INM-1	7383	F26	1	17.10	12.60
INM-1	1625	H5	2	15.78	10.39
INM-1	837	F12	2	17.21	9.43
INM-1	5880	E27	4	15.75	11.00
INM-1	5353	D5	11	17.95	11.90
INM-1	5845	B3	13	14.35	9.70
INM-1	5406	C3	15	15.45	11.75

Table A.7. 15. Measurements of Maxillary third molar of *Capra/Ovis*

Site	Reg.No.	Trench	Layer	GL	GB
INM-1	4170	B1	6	15.80	11.15
INM-1	5492	C1	14	17.67	12.30

Table A.7. 16. Measurements of Mandibular first/second molar of *Capra/Ovis*

Site	Reg.No.	Trench	Layer	GL	GB
INM-1	4165	Hno72	8	13.65	7.55
INM-1	4164	Hno72	8	12.40	7.80
INM-1	3917	B3&C3	8	15.80	7.20
INM-1	535	Hno50	12	12.02	7.09
INM-1	5410	C3	15	12.69	7.25

Table A.7. 17. Measurements of Mandibular second molar of *Capra/Ovis*

Site	Reg.No.	Trench	Layer	GL	GB
INM-1	617	C2	10	14.31	6.74
INM-1	610	C3	10	15.25	7.81
INM-1	9	B2	12	15.85	6.74

Table A.7. 18. Measurements of Mandibular third molar of *Capra/Ovis*

Site	Reg.No.	Trench	Layer	GL	GB
INM-1	426	LJP	1-4	19.60	7.02
INM-1	2427	F25&F26	3	21.71	7.48
INM-1	2426	F25&F26	3	21.62	.52

Table A.7.19. Measurements of Mandibles with teeth of *Capra/Ovis*

p2 –second premolar, p3- third premolar, p4- fourth premolar, m1- first molar, m2- second molar, m3- third molar, MC-Mandibular condyle

Site	Reg.No.	Trench	Layer	Mandible	Length	Width	Height
INM-1	450	LJP	1-4	dp4	15.72	5.98	00.00
INM-1				m1	00.00	00.00	18.07
INM-1	453	LJP	1-4	p4	00.00	00.00	21.23
INM-1				m1	00.00	00.00	21.76
INM-1				m2	00.00	00.00	23.63
INM-1	454	LJP	1-4	p3	00.00	00.00	22.56
INM-1				p4	00.00	00.00	23.76
INM-1				m1	00.00	00.00	26.06
INM-1	1014	E10	1	p2	00.00	00.00	12.30
INM-1				p3	00.00	00.00	13.29
INM-1				p4	00.00	00.00	14.49
INM-1	1121	G4	1	dp3	9.24	5.57	00.00
INM-1				dp4	14.68	7.11	00.00
INM-1	1123	G4	1	dp2	00.00	00.00	10.84
INM-1				dp3	00.00	00.00	13.30
INM-1				dp4	15.28	5.65	16.31
INM-1				m1	00.00	00.00	20.19
INM-1				m2	14.36	6.51	00.00
INM-1	1621	H5	2	p3	7.25	4.17	00.0
INM-1				p4	9.60	4.80	00.00
INM-1				m1	12.73	7.89	00.00
INM-1	4767	C0	10	MC	00.00	19.35	00.00
INM-1	5416	C3	15	p2	00.00	00.00	13.75
INM-1				p3	00.00	00.00	16.45
INM-1				p4	00.00	00.00	18.42
INM-1	5942	B3	12	p2	00.00	00.00	13.00
INM-1				p3	00.00	00.00	15.46
INM-1				p4	00.00	00.00	16.22
INM-1				m1	00.00	00.00	18.22
INM-1	8063	F27	1	p4	7.90	5.60	00.00
INM-1				m1	10.85	7.10	18.58
INM-1				m2	13.90	8.25	21.45

Table A.7.20. Measurements of Maxilla with teeth of *Capra/Ovis*
P4- fourth premolar, M1- first molar, M2- second molar, M3- third molar,

Site	Reg.No.	Trench	Layer	Maxilla	Length	Width	Height
INM-1	1175	G10	1	P4	11.17	8.19	00.00
INM-1				M1	14.57	8.93	00.00
INM-1				M2	15.97	8.20	00.00
INM-1				M3	13.80	8.38	00.00

Table A.8. Measurements of Bones of *Canis familiaris*

Table A.8.1.Measurements of Scapula of ***Canis familiaris***

Site	Reg.No.	Trench	Layer	GLP	LG	BG	SLC
INM-1	310	LJP	1-4	29.63	23.53	16.84	26.55
INM-1	6015	E3	11	25.60	23.00	15.95	00.00

Table A.8.2. Measurements of Humerus of ***Canis familiaris***

Site	Reg.No.	Trench	Layer	BD	TD
INM-1	46	LJP	1-4	32.70	14.08
INM-1	284	LJP	1-4	27.00	10.52
INM-1	378	LJP	1-4	27.89	12.71
INM-1	1672	G11	1	28.78	12.06
INM-1	1301	G11	3	29.21	11.14
INM-1	283	LJP	4	29.75	12.35
INM-1	5778	B5	13	29.70	12.20
INM-1	5924	B3	14	31.60	12.85
INM-1	5432	E4	15	26.63	10.45
INM-1	6074	B5	15	28.35	13.30

Table A.8.3. Measurements of Radius of ***Canis familiaris***

Site	Reg.No.	Trench	Layer	GL	DD	BP	TP	BD	TD
INM-1	47	LJP	1-4	179.26	13.62	19.52	13.08	24.34	14.14
INM-1	1160	EMBANK	0	00.00	00.00	17.83	11.80	00.00	00.00
INM-1	1013	E10	1	00.00	00.00	18.17	12.15	00.00	00.00
INM-1	456	F25	3	00.00	00.00	13.07	9.63	00.00	00.00
INM-1	7092	G24	4	00.00	00.00	18.55	00.00	00.00	00.00
INM-1	6302	Hno81	12	00.00	00.00	00.00	00.00	21.95	11.40
INM-1	5922	B3	14	00.00	12.60	00.00	00.00	20.60	13.40

Table A.8.4. Measurements of Ulna of ***Canis familiaris***

Site	Reg.No.	Trench	Layer	BP	SDO
INM-1	3460	H7	3	17.75	00.00
INM-1	5803	B5	14	16.50	22.36

Table A.8.5. Measurements of Pelvis of ***Canis familiaris***

Site	Reg.No.	Trench	Layer	LA
INM-1	1787	H11	3	24.94
INM-1	5870	E27	4	22.80

Table A.8.6. Measurements of Femur of ***Canis familiaris***

Site	Reg.No.	Trench	Layer	DD	BD	TD	LC	TC
INM-1	597	LJP	1-4	13.69	00.00	00.00	00.00	00.00
INM-1	208	F25	2	00.00	00.00	00.00	16.80	18.22
INM-1	6641	F24	2	00.00	30.50	36.80	00.00	00.00
INM-1	6073	B5	15	00.00	29.90	36.05	00.00	00.00

Table A.8.7. Measurements of Tibia of ***Canis familiaris***

Site	Reg.No.	Trench	Layer	DD	BD	TD
INM-1	141	LJP	1-4	12.16	20.61	15.11
INM-1	7017	G24	2	00.00	20.80	15.85
INM-1	5891	E25	4	00.00	23.75	17.65
INM-1	8025	E24	4	00.00	20.95	16.40
INM-1	4120	C3	5	00.00	19.35	14.35

Table A.8.8. Measurements of First Phalanx of ***Canis familiaris***

Site	Reg.No.	Trench	Layer	GL	BP	TP	BD	TD
INM-1	11	LJP	1-4	26.67	7.91	00.00	6.55	00.00
INM-1	12	LJP	1-4	27.21	8.62	00.00	7.24	00.00
INM-1	13	LJP	1-4	26.63	8.65	00.00	6.74	00.00
INM-1	5007	C2	8	27.35	8.65	8.26	7.00	00.00
INM-1	5258	E2	9	28.95	8.60	7.95	6.55	5.35

Table A.8.9. Measurements of Mandibles with teeth of ***Canis familiaris***

p2 –second premolar, p3- third premolar, p4- fourth premolar, m1- first molar, m2- second molar, m3- third molar, MC-Mandibular condyle

Site	Reg.No.	Trench	Layer	Mandible	Length	Width	Height
INM-1	430	LJP	1-4	p2	8.11	3.97	00.00
INM-1				p3	10.47	5.29	00.00
INM-1	455	LJP	1-4	p2	00.00	00.00	17.01
INM-1				p3	00.00	00.00	16.40
INM-1				p4	00.00	00.00	15.40
INM-1	1379	E1	4	MC	00.00	20.03	00.00
INM-1	1380	E1	4	MC	00.00	20.01	00.00
INM-1	1724	G11	1	p2	00.00	00.00	15.46
INM-1				p3	00.00	00.00	16.59
INM-1				p4	00.00	00.00	16.75
INM-1	2359	E10	3	p3	00.00	00.00	18.62
INM-1				p4	00.00	00.00	17.73
INM-1				m1	00.00	00.00	17.80
INM-1				m2	00.00	00.00	19.66
INM-1				m3	00.00	00.00	21.99
INM-1	4004	G4	2	p2	6.10	00.00	17.75
INM-1				p3	9.40	4.45	17.45
INM-1				p4	9.60	5.10	20.15
INM-1				m1	19.70	7.75	23.25
INM-1				m2	00.00	00.00	24.70
INM-1				m3	00.00	00.00	25.65
INM-1	4538	H12	3	m2	00.00	00.00	26.08
INM-1				m3	00.00	00.00	27.63
INM-1	5000	C2	8	p1	00.00	00.00	17.70
INM-1				p2	00.00	00.00	19.15
INM-1				p3	00.00	00.00	18.95
INM-1	6230	B6	12	MS	00.00	00.00	15.00
INM-1				p1	00.00	00.00	15.05
INM-1				p2	00.00	00.00	15.50
INM-1				p3	00.00	00.00	14.90
INM-1				m2	8.00	5.85	17.50
INM-1				m3	00.00	00.00	17.60
INM-1	6245	E5	15	p2	00.00	00.00	17.60
INM-1				p3	00.00	00.00	16.30
INM-1				p4	00.00	00.00	16.70
INM-1				m1	00.00	00.00	19.15

Table A.8.10. Measurements of Maxilla with teeth of ***Canis familiaris***

P2 –second premolar, P3- third premolar, P4- fourth premolar, M1- first molar, M2- second molar, M3- third molar,

Site	Reg.No.	Trench	Layer	Mandible	Length	Width	Height
INM-1	4005	G4	2	P1	6.00	3.70	00.00

A.9. Measurements of Bones of *Sus domesticus*

Table A.9.1.Measurements of Scapula of ***Sus domesticus***

Site	Reg.No.	Trench	Layer	GLP	LG	BG	SLC
INM-1	187	LJP	1-4	25.27	20.08	17.63	19.00
INM-1	3972	G4	2	00.00	20.00	17.45	00.00
INM-1	7853	G25	3	00.00	22.20	19.10	00.00

Table A.9.2.Measurements of Humerus of ***Sus domesticus***

Site	Reg.No.	Trench	Layer	BD	TD
INM-1	156	LJP	1-4	34.39	15.50
INM-1	3976	G4	2	28.45	13.55

Table A.9.3.Measurements of Radius of ***Sus domesticus***

Site	Reg.No.	Trench	Layer	GL	DD	BP	TP	BD	TD
INM-1	204	LJP	1-4	00.00	00.00	00.00	00.00	23.94	00.00
INM-1	3977	G4	2	00.00	00.00	20.68	14.88	00.00	00.00
INM-1	3910	B&C3	8	00.00	00.00	32.15	23.05	00.00	00.00
INM-1	5225	E1	11	103.20	11.90	20.25	12.07	17.65	13.30

Table A.9.4.Measurements of Ulna of ***Sus domesticus***

Site	Reg.No.	Trench	Layer	BP
INM-1	1823	F10	3	13.26

Table A.9.5.Measurements of Pelvis of ***Sus domesticus***

Site	Reg.No.	Trench	Layer	LA
INM-1	476		4	31.60

Table A.9.6.Measurements of Tibia of ***Sus domesticus***

Site	Reg.No.	Trench	Layer	BP	TP	BD	TD
INM-1	1875	E10	3	26.38	23.48	00.00	00.00
INM-1	7654	G25	4	00.00	00.00	22.50	20.05
INM-1	380		8	00.00	00.00	24.56	22.51

Table A.9.7.Measurements of Astragalus of ***Sus domesticus***

Site	Reg.No.	Trench	Layer	GL	GLL	GLM	TL	TM	BD
INM-1	2764	H11	2	00.00	37.75	32.68	18.68	20.14	20.03
INM-1	134	G25	4	33.57	33.45	30.65	17.61	17.18	17.96
INM-1	5761	C5	12	33.95	33.93	31.40	17.55	19.00	17.15
INM-1	5472	D3	15	33.35	33.21	29.00	17.95	18.6	19.63

Table A.9.8.Measurements of Calcaneum of ***Sus domesticus***

Site	Reg.No.	Trench	Layer	GL	GB
INM-1	5612	C3	12	00.00	15.95

Table A.9.9.Measurements of First Phalanx of ***Sus domesticus***

Site	Reg.No.	Trench	Layer	GL	BP	TP	BD	TD
INM-1	2758	H11	2	19.40	13.20	12.57	12.01	11.30
INM-1	5222	E1	11	00.00	12.60	12.80	00.00	00.00

A.10. Measurements of Bones of *Equus caballus*

Table A.10. 1. Measurements of First Phalanx of ***Equus caballus***

Site	Reg.No.	Trench	Layer	GL	BP	TP	BD	TD
INM-1	119	LJP	1-4	80.82	00.00	00.00	40.33	21.83
INM-1	346	LJP	1-4	84.53	56.99	00.00	49.53	24.60

Table A.10.2. Measurements of Metapodia of ***Equus caballus***

Site	Reg.No.	Trench	Layer	BD	TD	WT	TT	TLT
INM-1	459	LJP	1-4	48.54	00.00	00.00	31.95	25.78
INM-1	3291	D10	1	47.85	37.25	00.00	00.00	00.00
INM-1	171	E26	3	46.90	32.00	00.00	00.00	00.00

A.11. Measurements of Bones of *Canis lupus*

Table A.11.1Measurements of Scapula of ***Canis lupus***

Site	Reg.No.	Trench	Layer	GLP	LG	BG	SLC
INM-1	185	LJP	1-4	27.61	20.66	17.89	24.57
INM-1	309	E25	2	00.00	20.63	15.94	20.58
INM-1	4499	D2	8	33.40	25.40	19.25	00.00

Table A.11.2. Measurements of Humerus of ***Canis lupus***

Site	Reg.No.	Trench	Layer	BD	TD
INM-1	287	LJP	1-4	34.66	13.79
INM-1	1252	D25	3	32.67	14.00
INM-1	6046	Hno87	16	34.25	13.65

Table A.11.3. Measurements of Radius of ***Canis lupus***

Site	Reg.No.	Trench	Layer	BP	TP	BD	TD
INM-1	270	LJP	1-4	00.00	00.00	23.10	13.30
INM-1	294	LJP	1-4	18.52	12.50	00.00	00.00

Table A.11.4. Measurements of Ulna of ***Canis lupus***

Site	Reg.No.	Trench	Layer	BP	SDO
INM-1	460	F27	2	18.80	00.00
INM-1	2283	E1	2	18.20	22.84

Table A.11.5. Measurements of Pelvis of ***Canis lupus***

Site	Reg.No.	Trench	Layer	LA
INM-1	474	LJP	1-4	27.52
INM-1	475	LJP	1-4	25.42

Table A.11.6. Measurements of Femur of ***Canis lupus***

Site	Reg.No.	Trench	Layer	DD	BD
INM-1	285	LJP	3	15.07	31.20

Table A.11.7. Measurements of Tibia of ***Canis lupus***

Site	Reg.No.	Trench	Layer	BP	TP	BD	TD
INM-1	1944	H11	1	00.00	00.00	23.99	18.84
INM-1	4801	B0	14	00.00	00.00	22.73	17.35
INM-1	4802	B0	14	34.80	33.50	00.00	00.00

Table A.11.8. Measurements of Metapodia of ***Canis lupus***

Site	Reg.No.	Trench	Layer	GL	DD	BP	TP	BD	TD
INM-1	6190	Hno106	3	73.70	7.23	8.05	11.47	8.40	9.15
INM-1	3414	H6	2	00.00	00.00	00.00	00.00	9.35	7.83
INM-1	5009	C2	8	00.00	00.00	9.23	12.65	00.00	00.00
INM-1	5008	C2	8	00.00	00.00	00.00	00.00	8.50	10.20
INM-1	5257	E2	9	00.00	00.00	00.00	00.00	8.15	9.45
INM-1	5253	E2	9	00.00	00.00	9.75	14.50	00.00	00.00
INM-1	5256	E2	9	00.00	00.00	00.00	00.00	8.40	9.75
INM-1	5254	E2	9	00.00	00.00	11.06	6.22	00.00	00.00
INM-1	5455	E2	9	00.00	00.00	00.00	00.00	7.50	7.45
INM-1	6229	B6	12	80.26	00.00	12.46	7.90	7.20	9.20

Table A.11.9. Measurements of Astragalus of ***Canis lupus***

Site	Reg.No.	Trench	Layer	GL	GB	GLL	GLM	TL	TM	BD
INM-1	87	F25	2	27.22	00.00	16.45	25.40	9.28	13.46	12.80
INM-1	6043	Hno87	16	29.55	00.000	28.20	18.45	15.23	11.06	14.65

Table A.11.10. Measurements of Calcaneum of ***Canis lupus***

Site	Reg.No.	Trench	Layer	GL	GB
INM-1	209	LJP	1-4	44.92	00.00
INM-1	2667	G7	3	42.20	15.66

A.12. Measurements of Bones of *Canis aureus*

Table A.12.1. Measurements of Humerus of ***Canis aureus***

Site	Reg.No.	Trench	Layer	BD	TD
INM-1	160	LJP	1-4	21.24	7.10

Table A.12.2. Measurements of Ulna of ***Canis aureus***

Site	Reg.No.	Trench	Layer	GL	BP	SDO
INM-1	6953	E26	3	00.00	11.35	15.50

Table A.12.3. Measurements of Femur of ***Canis aureus***

Site	Reg.No.	Trench	Layer	BP	TP	LC	TC
INM-1	979	C1	4	27.28	13.68	11.70	13.51

Table A.12.4. Measurements of Tibia of ***Canis aureus***

Site	Reg.No.	Trench	Layer	BD	TD
INM-1	272	LJP	1-4	16.51	11.82
INM-1	384	LJP	1-4	16.82	12.01

A.13. Measurements of Bones of *Felis chaus*

Table A.13.1. Measurements of Humerus of ***Felis chaus***

Site	Reg.No.	Trench	Layer	BD	TD
INM-1	2366	F26	2	21.39	7.46

Table A.13.2. Measurements of Femur of ***Felis chaus***

Site	Reg.No.	Trench	Layer	BD	TD
INM-1	3881	B6	6	21.10	19.20

Table A.13.3. Measurements of Calcaneum of ***Felis chaus***

Site	Reg.No.	Trench	Layer	GL	GB
INM-1	2931	H10	1	35.05	12.72

Table A.13.4. Measurements of Metapodium of ***Felis chaus***

Site	Reg.No.	Trench	Layer	BD	TD
INM-1	2367	F26	2	8.17	7.38

A.14. Measurements of Bones of *Panthera pardus*

Table A.14.1. Measurements of First Phalanx of ***Panthera pardus***

Site	Reg.No.	Trench	Layer	GL	BP	TP	BD	TD
INM-1	5238	C3	11	36.50	18.77	12.77	13.15	9.90
INM-1	5239	C3	11	33.26	16.07	11.10	12.30	8.65

Table A.14.2. Measurements of Metapodia of ***Panthera pardus***

Site	Reg.No.	Trench	Layer	GL	BP	TP	BD	TD
INM-1	4189	B1	12	71.60	14.48	14.46	13.55	12.42
INM-1	4188	B1	12	00.00	11.80	14.65	00.00	00.00

A.15. Measurements of Bones of *Sus scrofa*

Table A.15.1. Measurements of Ulna of ***Sus scrofa***

Site	Reg.No.	Trench	Layer	GL	BP	SDO
INM-1	4919	F13	1	105.96	14.35	19.50
INM-1	3982	G4	2	00.00	15.34	18.55
INM-1	3984	G4	2	00.00	15.60	18.95
INM-1	18	LJP	1-4	00.00	16.82	00.00

Table A.15.2. Measurements of Tibia of ***Sus scrofa***

Site	Reg.No.	Trench	Layer	BD	TD
INM-1	3909	B3&C3	8	30.90	29.80

Table A.15.3. Measurements of Calcaneum of ***Sus scrofa***

Site	Reg.No.	Trench	Layer	GL	GB
INM-1	7806	G25	3	57.40	16.55
INM-1	3954	G4	2	59.27	16.90

A.16. Measurements of Bones of *Antilope cervicapra*

Table A.16.1. Measurements of Scapula of ***Antilope cervicapra***

Site	Reg.No.	Trench	Layer	GLP	LG	BG	SLC
INM-1	303	F27	1	00.00	26.62	25.08	20.44
INM-1	63	LJP	1-4	33.72	25.35	23.43	18.85
INM-1	128	LJP	1-4	30.81	23.33	20.77	18.40
INM-1	179	LJP	1-4	30.70	22.90	21.14	17.28
INM-1	181	LJP	1-4	35.66	27.79	25.99	21.49
INM-1	182	LJP	1-4	36.27	28.43	25.36	21.84
INM-1	299	LJP	1-4	31.85	24.19	20.18	18.41
INM-1	302	LJP	1-4	35.55	28.98	24.81	22.29
INM-1	305	LJP	1-4	38.05	28.19	25.32	00.00
INM-1	3292	D10	1	00.00	22.60	20.84	00.00
INM-1	3849	D25	1	31.97	25.55	21.55	00.00
INM-1	3860	D25	1	00.00	25.35	22.27	00.00
INM-1	1021	E10	1	34.22	26.04	23.41	00.00
INM-1	1187	F10	1	32.20	26.10	21.11	00.00
INM-1	3328	F11	1	33.55	26.16	24.60	00.00
INM-1	125	F25	1	31.97	26.27	21.30	17.74
INM-1	7377	F26	1	32.40	27.15	23.80	00.00
INM-1	1717	G11	1	35.08	26.99	25.59	00.00
INM-1	1718	G11	1	34.24	26.76	23.79	00.00
INM-1	1719	G11	1	31.91	25.34	20.29	00.00
INM-1	1720	G11	1	00.00	24.60	19.80	00.00
INM-1	311	LJP	1	00.00	00.00	23.32	00.00
INM-1	3091	G3	2	32.32	24.64	20.49	00.00
INM-1	3973	G4	2	32.75	24.58	21.55	00.00
INM-1	1581	H5	2	30.37	26.45	21.15	00.00
INM-1	1582	H5	2	32.34	26.54	23.28	00.00
INM-1	1583	H5	2	00.00	24.07	21.02	00.00
INM-1	1584	H5	2	33.97	27.49	21.92	00.00
INM-1	1437	D25	2	30.72	23.93	22.06	00.00
INM-1	2280	E10	2	00.00	26.53	21.86	00.00
INM-1	307	E24	2	00.00	00.00	22.09	19.56
INM-1	7305	E24	2	28.45	20.20	19.20	00.00
INM-1	7306	E24	2	00.00	26.30	21/20	00.00
INM-1	127	E25	2	32.64	25.27	21.62	18.16
INM-1	6405	E25	2	32.55	26.32	21.50	00.00
INM-1	6406	E25	2	33.70	26.00	00.00	00.00
INM-1	6407	E25	2	00.00	24.95	21.60	00.00
INM-1	7448	E25	2	00.00	24.85	21.75	00.00
INM-1	3573	F10	2	31.86	24.75	21.70	00.00
INM-1	3574	F10	2	00.00	22.70	21.05	00.00
INM-1	2584	F11	2	34.81	27.69	23.30	00.00
INM-1	2585	F11	2	32.66	25.10	21.95	00.00
INM-1	126	F24	2	35.00	26.40	24.18	20.74
INM-1	7494	F24	2	00.00	24.55	20.65	00.00
INM-1	184	F25	2	00.00	00.00	20.65	17.15
INM-1	178	F26	2	33.80	25.05	24.43	22.40
INM-1	183	F26	2	00.00	00.00	20.41	19.27
INM-1	375	F26	2	23.97	23.34	20.04	17.26
INM-1	1345	G11	2	31.79	26.43	22.69	00.00
INM-1	3219	G11	2	00.00	26.25	23.91	00.00
INM-1	180	G23	2	33.68	25.26	24.32	21.83
INM-1	62	G24	2	00.00	00.00	25.58	21.35
INM-1	7031	G24	2	00.00	24.40	24.10	00.00
INM-1	301	G26	2	32.74	26.62	22.06	17.88
INM-1	2776	H11	2	32.18	24.39	21.89	00.00
INM-1	300	E26	3	33.14	25.28	23.04	19.03
INM-1	6949	E26	3	00.00	25.37	22.87	00.00
INM-1	6950	E26	3	00.00	25.20	22.60	00.00

INM-1	6951	E26	3	31.15	25.35	21.40	00.00
INM-1	6507	F25	3	00.00	26.15	22.68	00.00
INM-1	6508	F25	3	29.20	23.10	20.30	00.00
INM-1	1305	G11	3	29.94	24.13	20.20	00.00
INM-1	907	G12	3	36.13	28.59	25.56	00.00
INM-1	908	G12	3	34.29	26.43	23.02	00.00
INM-1	909	G12	3	32.87	26.43	23.10	00.00
INM-1	910	G12	3	32.09	24.96	21.54	00.00
INM-1	911	G12	3	31.87	25.75	22.73	00.00
INM-1	912	G12	3	31.81	25.39	22.09	00.00
INM-1	913	G12	3	00.00	23.71	19.16	00.00
INM-1	914	G12	3	30.92	23.91	23.27	00.00
INM-1	915	G12	3	37.16	29.36	00.00	00.00
INM-1	7833	G25	3	00.00	24.75	21.15	00.00
INM-1	7735	G26	3	00.00	24.20	20.45	00.00
INM-1	7736	G26	3	34.32	24.80	22.50	00.00
INM-1	7741	G26	3	00.00	23.75	20.65	00.00
INM-1	376	G27	3	00.00	00.00	00.00	19.81
INM-1	4522	H12	3	35.10	26.00	24.50	00.00
INM-1	1377	E1	4	29.98	25.27	21.12	00.00
INM-1	8022	E24	4	32.85	26.20	23.25	00.00
INM-1	8118	E24	4	00.00	21.35	20.95	00.00
INM-1	7983	F24	4	30.85	22.70	19.75	00.00
INM-1	7100	G24	4	32.40	24.25	21.30	00.00
INM-1	306	G25	4	00.00	25.14	20.32	00.00
INM-1	544	C0	4	00.00	00.00	19.17	00.00
INM-1	7242	Hno104	4	36.75	29.95	26.75	00.00
INM-1	4759	Hno117	4	32.00	27.25	21.08	00.00
INM-1	2264	G5	6	29.78	24.01	20.75	00.00
INM-1	1846	G4	7	00.00	24.58	20.40	00.00
INM-1	5003	C2	8	36.05	29.65	24.30	00.00
INM-1	4413	C7	8	29.10	22.95	19.00	00.00
INM-1	5155	B2	11	32.17	22.65	18.90	00.00
INM-1	5343	D5	11	35.00	25.75	24.87	00.00
INM-1	6298	Hno81	12	33.75	25.45	24.95	00.00
INM-1	5307	E3	14	26.71	21.65	17.60	00.00
INM-1	6102	B3	15	35.70	29.50	24.80	00.00
INM-1	6044	B4	15	00.00	24.75	23.60	00.00

Table A.16.2. Measurements of Humerus of ***Antilope cervicapra***

Site	Reg.No.	Trench	Layer	DD	BP	TP	BD	TD
INM-1	42	LJP	1-4	00.00	00.00	00.00	31.47	00.00
INM-1	151	LJP	1-4	00.00	00.00	00.00	30.10	16.45
INM-1	152	LJP	1-4	00.00	00.00	00.00	34.75	16.51
INM-1	153	LJP	1-4	00.00	00.00	00.00	28.46	15.03
INM-1	155	LJP	1-4	00.00	00.00	00.00	00.00	16.23
INM-1	277	LJP	1-4	00.00	00.00	00.00	33.07	16.07
INM-1	278	LJP	1-4	00.00	00.00	00.00	32.25	15.53
INM-1	279	LJP	1-4	00.00	00.00	00.00	31.57	16.01
INM-1	5285	LJP	1-4	00.00	00.00	00.00	34.25	15.38
INM-1	5286	LJP	1-4	00.00	00.00	00.00	30.00	15.60
INM-1	5295	LJP	1-4	00.00	00.00	00.00	34.90	17.33
INM-1	1011	E10	1	00.00	00.00	00.00	31.74	16.16
INM-1	1103	G4	1	00.00	00.00	00.00	31.47	15.29
INM-1	1107	G4	1	00.00	12.64	00.00	00.00	00.00
INM-1	1168	G10	1	00.00	00.00	00.00	32.68	14.92
INM-1	1189	F10	1	00.00	00.00	00.00	33.14	16.08
INM-1	1191	F10	1	00.00	00.00	00.00	31.08	15.68
INM-1	1667	G11	1	00.00	00.00	00.00	00.00	16.77
INM-1	1668	G11	1	00.00	00.00	00.00	00.00	15.25
INM-1	1669	G11	1	00.00	00.00	00.00	32.33	15.80

INM-1	1670	G11	1	00.00	00.00	00.00	30.75	15.63
INM-1	1671	G11	1	00.00	00.00	00.00	00.00	16.04
INM-1	1936	H11	1	00.00	00.00	00.00	31.95	16.16
INM-1	1937	H11	1	00.00	00.00	00.00	34.26	16.76
INM-1	1938	H11	1	00.00	00.00	00.00	29.27	14.97
INM-1	1982	H11	1	00.00	7.25	00.00	00.00	00.00
INM-1	2048	G7	1	00.00	00.00	00.00	31.27	16.07
INM-1	2612	G6	1	00.00	00.00	00.00	32.53	16.15
INM-1	2933	H10	1	00.00	00.00	00.00	33.76	16.39
INM-1	2934	H10	1	00.00	00.00	00.00	30.39	14.77
INM-1	3294	D10	1	00.00	00.00	00.00	29.35	14.80
INM-1	3295	D10	1	00.00	00.00	00.00	29.16	14.15
INM-1	3329	F11	1	00.00	00.00	00.00	34.02	16.39
INM-1	3330	F11	1	00.00	00.00	00.00	30.52	14.59
INM-1	3331	F11	1	00.00	00.00	00.00	34.27	16.86
INM-1	3332	F11	1	00.00	00.00	00.00	30.35	15.00
INM-1	3333	F11	1	00.00	00.00	00.00	31.35	00.00
INM-1	3535	H5	1	00.00	00.00	00.00	31.90	15.03
INM-1	4916	F13	1	00.00	00.00	00.00	30.45	15.95
INM-1	4917	F13	1	00.00	00.00	00.00	30.35	14.55
INM-1	7378	F26	1	00.00	00.00	00.00	30.06	15.00
INM-1	7379	F26	1	00.00	00.00	00.00	33.80	16.70
INM-1	7531	G24	1	00.00	00.00	00.00	27.85	13.60
INM-1	49	E24	2	00.00	00.00	00.00	32.26	16.28
INM-1	154	F27	2	00.00	00.00	00.00	33.88	16.49
INM-1	157	F27	2	00.00	00.00	00.00	31.10	14.96
INM-1	159	G26	2	00.00	00.00	00.00	33.53	16.36
INM-1	280	E27	2	00.00	00.00	00.00	32.63	14.63
INM-1	623	Hno125	2	00.00	00.00	00.00	35.41	17.71
INM-1	763	I12	2	00.00	00.00	00.00	29.54	15.53
INM-1	764	I12	2	00.00	00.00	00.00	32.14	15.63
INM-1	765	I12	2	00.00	00.00	00.00	00.00	14.39
INM-1	766	I12	2	00.00	00.00	00.00	28.24	00.00
INM-1	1067	H7	2	00.00	00.00	00.00	31.58	15.75
INM-1	1068	H7	2	00.00	00.00	00.00	34.21	16.72
INM-1	1069	H7	2	00.00	00.00	00.00	30.08	15.60
INM-1	1088	H7	2	00.00	37.80	00.00	00.00	00.00
INM-1	1138	E10	2	00.00	00.00	00.00	30.12	14.91
INM-1	1342	G11	2	00.00	00.00	00.00	31.03	15.57
INM-1	1470	D25	2	00.00	00.00	00.00	28.22	14.00
INM-1	1572	H5	2	00.00	00.00	00.00	31.89	16.51
INM-1	1574	H5	2	00.00	00.00	00.00	00.00	15.18
INM-1	2282	E1	2	00.00	00.00	00.00	30.69	14.78
INM-1	2549	F11	2	00.00	00.00	00.00	31.84	15.72
INM-1	2550	F11	2	00.00	00.00	00.00	29.83	14.64
INM-1	2551	F11	2	00.00	00.00	00.00	32.18	15.99
INM-1	2552	F11	2	00.00	00.00	00.00	31.03	15.14
INM-1	2553	F11	2	00.00	00.00	00.00	00.00	16.15
INM-1	2778	H11	2	00.00	00.00	00.00	32.52	15.94
INM-1	2779	H11	2	00.00	00.00	00.00	32.29	16.42
INM-1	3000	F10	2	00.00	00.00	00.00	31.53	14.74
INM-1	3092	G3	2	00.00	00.00	00.00	31.71	16.85
INM-1	3165	G11	2	00.00	00.00	00.00	28.79	14.68
INM-1	3166	G11	2	00.00	00.00	00.00	30.53	14.75
INM-1	3167	G11	2	00.00	00.00	00.00	30.81	14.87
INM-1	3168	G11	2	00.00	00.00	00.00	32.34	15.94
INM-1	3391	H6	2	00.00	00.00	00.00	32.28	16.40
INM-1	3974	G4	2	00.00	00.00	00.00	31.70	15.50
INM-1	3975	G4	2	00.00	00.00	00.00	31.45	15.68
INM-1	6427	E25	2	00.00	00.00	00.00	31.60	17.00
INM-1	6428	E25	2	00.00	00.00	00.00	31.50	16.40
INM-1	6429	E25	2	00.00	00.00	00.00	35.90	16.70
INM-1	7447	E25	2	00.00	00.00	00.00	33.60	15.05

INM-1	6636	F24	2	00.00	00.00	00.00	30.95	14.85
INM-1	7007	G24	2	00.00	00.00	00.00	27.93	13.60
INM-1	7307	E24	2	00.00	00.00	00.00	34.95	16.05
INM-1	7308	E24	2	00.00	00.00	00.00	31.00	15.90
INM-1	7309	E24	2	00.00	00.00	00.00	29.40	15.07
INM-1	7310	E24	2	00.00	00.00	00.00	31.92	15.98
INM-1	7496	F24	2	00.00	00.00	00.00	30.35	16.85
INM-1	7573	F26	2	00.00	00.00	00.00	30.40	15.40
INM-1	4472	Hno129	2	00.00	00.00	00.00	30.47	15.75
INM-1	43	F6	2	00.00	00.00	00.00	32.39	16.48
INM-1	923	G12	3	00.00	00.00	00.00	32.04	16.14
INM-1	924	G12	3	00.00	00.00	00.00	00.00	16.02
INM-1	1300	G11	3	00.00	00.00	00.00	33.07	16.71
INM-1	1775	H11	3	00.00	00.00	00.00	34.76	16.56
INM-1	2204	G10	3	00.00	00.00	00.00	34.69	17.38
INM-1	2227	G10	3	00.00	00.00	00.00	00.00	15.70
INM-1	2228	G10	3	00.00	33.36	17.39	00.00	00.00
INM-1	2316	E10	3	00.00	33.89	15.95	00.00	00.00
INM-1	2847	G3	3	00.00	00.00	00.00	31.89	15.80
INM-1	2912	G10	3	00.00	00.00	00.00	29.16	14.60
INM-1	3458	H7	3	00.00	00.00	00.00	31.35	15.27
INM-1	3459	H7	3	00.00	00.00	00.00	29.92	14.95
INM-1	4523	H12	3	00.00	00.00	00.00	33.60	17.43
INM-1	6513	F25	3	00.00	00.00	00.00	34.75	18.30
INM-1	6514	F25	3	00.00	00.00	00.00	33.75	15.80
INM-1	6782	E23	3	00.00	00.00	00.00	30.15	15.60
INM-1	6941	E26	3	00.00	00.00	00.00	30.05	15.25
INM-1	6942	E26	3	00.00	00.00	00.00	29.70	16.70
INM-1	7722	G26	3	00.00	00.00	00.00	31.87	15.65
INM-1	7860	G25	3	00.00	00.00	00.00	30.30	15.35
INM-1	50	E25	4	00.00	00.00	00.00	32.44	15.88
INM-1	273	E24	4	00.00	00.00	00.00	31.86	15.45
INM-1	5875	E27	4	00.00	00.00	00.00	26.60	13.30
INM-1	5888	E25	4	00.00	00.00	00.00	31.05	15.00
INM-1	6830	G26	4	00.00	00.00	00.00	27.25	14.80
INM-1	6831	G26	4	00.00	00.00	00.00	33.70	16.80
INM-1	7086	G24	4	00.00	00.00	00.00	31.30	15.00
INM-1	7087	G24	4	00.00	00.00	00.00	32.00	15.87
INM-1	7143	G24	4	00.00	00.00	00.00	29.50	14.50
INM-1	7639	G25	4	00.00	00.00	00.00	34.00	16.95
INM-1	7640	G25	4	00.00	00.00	00.00	31.60	15.41
INM-1	8117	E24	4	00.00	00.00	00.00	31.25	15.15
INM-1	4340	F8	5	00.00	00.00	00.00	32.95	15.78
INM-1	4341	F8	5	00.00	00.00	00.00	31.61	15.69
INM-1	4428	E7	7	00.00	00.00	00.00	30.70	14.95
INM-1	4760	Hno117	7	00.00	00.00	00.00	28.70	14.43
INM-1	3612	C0	8	00.00	00.00	00.00	27.15	12.37
INM-1	4704	F3	8	00.00	00.00	00.00	26.58	12.95
INM-1	5713	B7	12	00.00	00.00	00.00	31.20	16.00
INM-1	6297	Hno81	12	00.00	00.00	00.00	32.50	16.15
INM-1	4064	Hno115	12	00.00	00.00	00.00	30.95	15.20
INM-1	5642	A4	13	14.45	00.00	00.00	30.80	14.70
INM-1	5777	B5	13	00.00	00.00	00.00	26.95	13.10
INM-1	5308	E3	14	00.00	00.00	00.00	31.95	16.26

Table A.16.3. Measurements of Radius of ***Antilope cervicapra***

Site	Reg.No.	Trench	Layer	DD	BP	TP	BD	TD
INM-1	81	LJP	1-4	00.00	00.00	00.00	26.70	18.81
INM-1	166	LJP	1-4	00.00	00.00	00.00	28.53	19.86
INM-1	167	LJP	1-4	00.00	00.00	00.00	27.26	18.87
INM-1	168	LJP	1-4	00.00	00.00	00.00	29.42	20.75
INM-1	169	LJP	1-4	00.00	00.00	00.00	25.08	18.25
INM-1	205	LJP	1-4	00.00	28.38	00.00	00.00	00.00
INM-1	206	LJP	1-4	00.00	28.13	00.00	00.00	00.00
INM-1	296	LJP	1-4	00.00	33.02	17.68	00.00	00.00
INM-1	297	LJP	1-4	00.00	31.58	16.91	00.00	00.00
INM-1	462	LJP	1-4	00.00	00.00	00.00	28.89	22.24
INM-1	1674	G11	1	00.00	29.23	15.10	00.00	00.00
INM-1	1675	G11	1	00.00	29.83	17.35	00.00	00.00
INM-1	1676	G11	1	00.00	28.86	15.28	00.00	00.00
INM-1	1680	G11	1	00.00	00.00	00.00	27.70	20.05
INM-1	1939	H11	1	00.00	29.44	16.07	00.00	00.00
INM-1	1941	H11	1	00.00	00.00	00.00	27.30	22.37
INM-1	1942	H11	1	00.00	00.00	00.00	27.29	19.56
INM-1	2609	G6	1	00.00	00.00	00.00	25.75	18.00
INM-1	3334	F11	1	00.00	32.05	18.05	00.00	00.00
INM-1	3536	H5	1	00.00	31.48	17.25	00.00	00.00
INM-1	3537	H5	1	00.00	28.90	15.36	00.00	00.00
INM-1	3538	H5	1	00.00	32.50	16.73	00.00	00.00
INM-1	3540	H5	1	00.00	00.00	00.00	30.30	21.85
INM-1	3850	D25	1	00.00	32.10	17.50	00.00	00.00
INM-1	3851	D25	1	00.00	00.00	00.00	26.10	18.83
INM-1	3857	D25	1	00.00	00.00	16.50	00.00	00.00
INM-1	4918	F13	1	00.00	29.96	15.90	00.00	00.00
INM-1	7381	F26	1	00.00	32.33	17.55	00.00	00.00
INM-1	7532	F24	1	00.00	30.73	16.05	00.00	00.00
INM-1	8059	C25	1	00.00	32.10	16.35	00.00	00.00
INM-1	2159	Hno132	2	00.00	30.95	16.85	00.00	00.00
INM-1	2160	Hno132	2	00.00	00.00	00.00	27.49	20.34
INM-1	67	LJP	2	00.00	29.83	15.79	00.00	00.00
INM-1	200	C25	2	00.00	28.08	00.00	00.00	00.00
INM-1	201	LJP	2	00.00	28.73	15.75	00.00	00.00
INM-1	202	G25	2	00.00	30.98	17.40	00.00	00.00
INM-1	298	G25	2	00.00	28.05	15.45	00.00	00.00
INM-1	461	F24	2	00.00	29.84	16.23	00.00	00.00
INM-1	762	I12	2	00.00	29.51	15.00	00.00	00.00
INM-1	1066	H7	2	00.00	00.00	00.00	28.79	20.54
INM-1	1073	H7	2	00.00	30.05	16.45	00.00	00.00
INM-1	1469	D25	2	00.00	30.39	17.73	00.00	00.00
INM-1	1580	H5	2	00.00	00.00	00.00	29.32	20.26
INM-1	1996	D10	2	00.00	32.35	17.48	00.00	00.00
INM-1	1997	D10	2	00.00	30.62	16.91	00.00	00.00
INM-1	2345		2	00.00	00.00	00.00	29.18	21.64
INM-1	2346		2	00.00	31.39	16.29	00.00	00.00
INM-1	2554	F11	2	00.00	30.08	16.20	00.00	00.00
INM-1	2555	F11	2	00.00	29.77	16.50	00.00	00.00
INM-1	2781	H11	2	00.00	00.00	00.00	28.11	19.09
INM-1	3001	F10	2	00.00	29.59	17.65	00.00	00.00
INM-1	3002	F10	2	00.00	00.00	00.00	30.45	21.59
INM-1	3093	G3	2	00.00	33.10	17.70	00.00	00.00
INM-1	3095	G3	2	00.00	00.00	00.00	26.73	19.89
INM-1	3196	G11	2	00.00	29.10	15.93	00.00	00.00
INM-1	3170	G11	2	00.00	28.26	15.37	00.00	00.00
INM-1	3171	G11	2	00.00	29.26	16.26	00.00	00.00
INM-1	3172	G11	2	00.00	31.89	18.26	00.00	00.00
INM-1	3176	G11	2	00.00	00.00	00.00	28.65	21.30

INM-1	3177	G11	2	00.00	00.00	00.00	29.33	22.10
INM-1	3392	H6	2	00.00	00.00	16.40	00.00	00.00
INM-1	3394	H6	2	00.00	00.00	00.00	30.30	21.66
INM-1	3395	H6	2	00.00	00.00	00.00	29.56	21.75
INM-1	3978	G4	2	00.00	31.70	17.15	00.00	00.00
INM-1	6437	E25	2	00.00	29.40	15.65	00.00	00.00
INM-1	7444	E25	2	00.00	00.00	00.00	27.45	19.57
INM-1	7445	E25	2	00.00	00.00	00.00	29.95	20.60
INM-1	7010	G24	2	00.00	33.40	17.10	00.00	00.00
INM-1	7011	G24	2	00.00	28.80	14.50	00.00	00.00
INM-1	7012	G24	2	00.00	30.30	16.45	00.00	00.00
INM-1	7015	G24	2	00.00	00.00	00.00	26.00	18.50
INM-1	7181	G25	2	00.00	00.00	00.00	28.70	19.90
INM-1	7182	G25	2	00.00	00.00	00.00	27.90	18.30
INM-1	7183	G25	2	00.00	00.00	00.00	30.60	22.30
INM-1	7311	E24	2	00.00	00.00	00.00	27.00	18.70
INM-1	7571	F26	2	00.00	29.95	15.80	00.00	00.00
INM-1	7572	F26	2	00.00	28.75	15.50	00.00	00.00
INM-1	6186	Hno106	3	13.45	00.00	00.00	00.00	00.00
INM-1	927	G12	3	00.00	34.05	17.95	00.00	00.00
INM-1	928	G12	3	00.00	32.20	17.54	00.00	00.00
INM-1	929	G12	3	00.00	31.16	16.98	00.00	00.00
INM-1	930	G12	3	00.00	32.64	17.18	00.00	00.00
INM-1	1304	G11	3	00.00	33.78	18.70	00.00	00.00
INM-1	1772	H11	3	00.00	29.37	16.14	00.00	00.00
INM-1	2205	G10	3	00.00	26.13	14.81	00.00	00.00
INM-1	2848	G3	3	00.00	30.16	16.92	00.00	00.00
INM-1	2849	G3	3	00.00	00.00	00.00	28.41	19.56
INM-1	3462	H7	3	00.00	00.00	00.00	27.07	18.35
INM-1	3598	I6	3	00.00	32.25	17.55	00.00	00.00
INM-1	4524	H12	3	00.00	00.00	00.00	28.60	21.90
INM-1	4525	H12	3	00.00	00.00	00.00	27.60	20.53
INM-1	6515	F25	3	00.00	29.55	15.70	00.00	00.00
INM-1	6516	F25	3	00.00	00.00	00.00	26.50	17.45
INM-1	6517	F25	3	00.00	00.00	00.00	26.45	18.10
INM-1	6730	E23	3	00.00	30.35	15.20	00.00	00.00
INM-1	6943	E26	3	00.00	29.85	15.45	00.00	00.00
INM-1	6944	E26	3	00.00	00.00	00.00	26.35	18.85
INM-1	7837	G25	3	00.00	29.88	15.35	00.00	00.00
INM-1	7838	G25	3	00.00	00.00	00.00	29.25	21.40
INM-1	7857	G25	3	00.00	29.83	16.80	00.00	00.00
INM-1	4656	Hno111	4	00.00	00.00	17.40	00.00	00.00
INM-1	7243	Hno104	4	00.00	00.00	00.00	25.65	18.30
INM-1	7244	Hno104	4	00.00	00.00	00.00	28.80	20.50
INM-1	7245	Hno104	4	00.00	00.00	00.00	22.55	18.20
INM-1	51	E25	4	00.00	00.00	00.00	21.59	19.21
INM-1	1865	G11	4	00.00	25.53	16.33	00.00	00.00
INM-1	5871	E27	4	00.00	33.05	17.05	00.00	00.00
INM-1	5874	E27	4	00.00	00.00	00.00	27.50	19.90
INM-1	5876	F25	4	00.00	00.00	00.00	29.05	21.20
INM-1	6860	G26	4	00.00	34.75	18.50	00.00	00.00
INM-1	7088	G24	4	00.00	30.90	17.75	00.00	00.00
INM-1	7089	G24	4	00.00	30.25	16.05	00.00	00.00
INM-1	7090	G24	4	00.00	31.20	16.56	00.00	00.00
INM-1	7091	G24	4	00.00	30.65	16.65	00.00	00.00
INM-1	7641	G25	4	00.00	31.60	15.41	00.00	00.00
INM-1	7642	G25	4	00.00	00.00	00.00	26.14	17.75
INM-1	7982	F24	4	00.00	30.75	18.02	00.00	00.00
INM-1	7984	F24	4	00.00	00.00	00.00	28.05	22.30
INM-1	8116	E24	4	00.00	00.00	00.00	26.40	19.60
INM-1	3810	E8	5	00.00	00.00	00.00	30.00	21.35
INM-1	710	B1	9	00.00	00.00	00.00	27.89	20.07
INM-1	4215	B4	9	00.00	28.15	14.85	00.00	00.00

INM-1	5224	E1	11	00.00	28.92	14.15	00.00	00.00
INM-1	136	B2	12	00.00	29.72	16.03	00.00	00.00
INM-1	3694	B1	12	00.00	32.15	17.53	00.00	00.00
INM-1	5444	B4	13	00.00	00.00	00.00	24.35	17.75
INM-1	5402	C3	15	00.00	00.00	00.00	26.65	20.40
INM-1	5479	D3	15	00.00	26.47	15.40	00.00	00.00
INM-1	5566	D4	15	00.00	28.00	14.75	00.00	00.00
INM-1	6048	Hno87	16	00.00	00.00	00.00	25.80	16.71

Table A.16.4. Measurements of Ulna of ***Antilope cervicapra***

Site	Reg.No.	Trench	Layer	BP	SDO
INM-1	1202	F10	1	16.20	00.00
INM-1	1677	G11	1	16.71	00.00
INM-1	1731	G11	1	10.94	00.00
INM-1	1940	H11	1	18.45	00.00
INM-1	2050	G7	1	17.74	20.50
INM-1	2610	G6	1	14.26	19.04
INM-1	2611	G6	1	16.66	20.43
INM-1	2936	H10	1	14.76	00.00
INM-1	3335	F11	1	15.05	20.52
INM-1	3336	F11	1	17.77	00.00
INM-1	3539	H5	1	17.35	22.70
INM-1	3762	I11	1	15.95	00.00
INM-1	761	I12	2	00.00	19.78
INM-1	1072	H7	2	15.42	00.00
INM-1	1579	H5	2	19.28	22.87
INM-1	2100	G6	2	16.48	00.00
INM-1	2556	F11	2	16.25	20.26
INM-1	3094	G3	2	16.16	00.00
INM-1	3173	G11	2	16.00	19.51
INM-1	3174	G11	2	16.69	21.35
INM-1	3175	G11	2	15.15	00.00
INM-1	3983	G4	2	16.55	20.56
INM-1	6431	E25	2	16.60	00.00
INM-1	7013	G24	2	15.85	21.50
INM-1	7014	G24	2	14.90	19.50
INM-1	2206	G10	3	14.75	20.35
INM-1	2317	E10	3	12.94	00.00
INM-1	2676	G7	3	17.95	00.00
INM-1	3461	H7	3	16.20	00.00
INM-1	3758	G11	3	18.09	00.00
INM-1	7858	G25	3	16.40	00.00
INM-1	5889	E25	4	15.20	19.65
INM-1	4655	Hno111	4	15.95	00.00
INM-1	3269	F11	5	14.20	00.00
INM-1	3880	B6	6	15.80	00.00
INM-1	5050	B1	7	15.25	00.00
INM-1	6123	Hno51	7	17.15	00.00
INM-1	5002	C2	8	15.50	16.50
INM-1	685	A1	11	14.21	00.00
INM-1	3695	B1	12	17.15	00.00
INM-1	6327	B5	12	15.60	00.00
INM-1	5480	D3	15	14.65	00.00

Table A.16.5. Measurements of Radius and Ulna of ***Antilope cervicapra***

Site	Reg.No.	Trench	Layer	BP	TP	BD	TD	SDO
INM-1	2161	Hno132	1	00.00	00.00	29.11	22.84	00.00
INM-1	3979	G4	2	24.95	14.05	00.00	00.00	00.00
INM-1	3980	G4	2	30.65	16.55	00.00	00.00	00.00
INM-1	3981	G4	2	00.00	00.00	25.38	20.90	00.00
INM-1	677	Hno110	4	00.00	00.00	00.00	00.00	17.25

Table A.16.6. Measurements of Metacarpal of ***Antilope cervicapra***

Site	Reg.No.	Trench	Layer	DD	BP	TP	BD	TD	WT	TT	TLT
INM-1	190	LJP	1-4	00.00	00.00	00.00	22.24	00.00	10.23	16.27	12.17
INM-1	192	LJP	1-4	00.00	00.00	00.00	22.51	00.00	10.41	16.75	12.31
INM-1	327	LJP	1-4	00.00	00.00	00.00	23.19	00.00	10.67	16.65	12.10
INM-1	329	LJP	1-4	00.00	00.00	00.00	20.65	00.00	9.40	15.70	12.10
INM-1	3800	PITG6	1-4	00.00	25.90	19.05	00.00	00.00	00.00	00.00	00.00
INM-1	3801	PITG6	1-4	00.00	00.00	00.00	21.48	16.68	9.56	16.52	12.80
INM-1	3802	PITG6	1-4	00.00	00.00	00.00	24.10	18.10	10.55	17.00	12.60
INM-1	3803	PITG6	1-4	00.00	00.00	00.00	21.75	16.25	9.86	16.08	12.10
INM-1	1098	G4	1	00.00	00.00	00.00	23.03	17.49	10.42	17.13	12.91
INM-1	2718	G6	1	00.00	22.66	17.20	00.00	00.00	00.00	00.00	00.00
INM-1	2719	G6	1	00.00	21.89	16.33	00.00	00.00	00.00	00.00	00.00
INM-1	2720	G6	1	00.00	25.69	19.69	00.00	00.00	00.00	00.00	00.00
INM-1	2722	G6	1	00.00	00.00	00.00	23.46	17.80	10.45	17.44	13.74
INM-1	2723	G6	1	00.00	00.00	00.00	24.46	18.69	10.69	18.50	14.35
INM-1	2044	G7	1	00.00	00.00	00.00	23.60	17.59	10.28	16.83	13.32
INM-1	2045	G7	1	00.00	00.00	00.00	24.09	17.59	10.57	17.07	13.59
INM-1	2046	G7	1	00.00	00.00	00.00	22.89	16.92	10.22	16.53	12.97
INM-1	2047	G7	1	00.00	00.00	00.00	00.00	00.00	10.88	17.37	13.29
INM-1	3287	D10	1	00.00	00.00	00.00	25.33	17.62	10.55	17.45	12.75
INM-1	3288	D10	1	00.00	00.00	00.00	26.15	17.70	11.05	17.55	13.40
INM-1	3840	D25	1	00.00	25.85	19.80	00.00	00.00	00.00	00.00	00.00
INM-1	3841	D25	1	00.00	00.00	00.00	24.50	18.50	10.88	18.33	13.85
INM-1	1005	E10	1	00.00	25.00	19.72	00.00	00.00	00.00	00.00	00.00
INM-1	1007	E10	1	00.00	00.00	00.00	22.46	16.86	9.48	16.90	13.17
INM-1	7051	E23	1	00.00	00.00	00.00	00.00	00.00	10.20	16.85	12.96
INM-1	7052	E23	1	00.00	00.00	00.00	00.00	00.00	10.50	17.70	13.15
INM-1	7053	E23	1	00.00	00.00	00.00	00.00	00.00	10.85	17.30	13.45
INM-1	3341	F11	1	00.00	23.17	17.76	00.00	00.00	00.00	00.00	00.00
INM-1	3342	F11	1	00.00	23.88	17.84	00.00	00.00	00.00	00.00	00.00
INM-1	3350	F11	1	00.00	00.00	00.00	25.40	19.05	11.52	18.55	13.60
INM-1	3351	F11	1	00.00	00.00	00.00	21.10	16.17	9.35	15.75	12.55
INM-1	3352	F11	1	00.00	00.00	00.00	23.90	17.22	9.65	16.85	12.34
INM-1	4911	F13	1	00.00	00.00	00.00	25.35	17.55	10.98	17.40	13.70
INM-1	8048	F24	1	00.00	25.03	17.75	00.00	00.00	00.00	00.00	00.00
INM-1	8049	F24	1	00.00	00.00	00.00	24.40	17.85	10.85	17.80	13.75
INM-1	8050	F24	1	00.00	00.00	00.00	21.60	16.25	10.15	15.45	11.65
INM-1	7374	F26	1	00.00	23.35	18.65	00.00	00.00	00.00	00.00	00.00
INM-1	7375	F26	1	00.00	00.00	00.00	22.15	15.75	10.05	15.55	11.75
INM-1	387	F27	1	15.45	00.00	00.00	22.89	00.00	10.48	17.65	14.08
INM-1	8047	F27	1	00.00	21.75	15.05	00.00	00.00	00.00	00.00	00.00
INM-1	8091	F27	1	00.00	22.50	17.15	00.00	00.00	00.00	00.00	00.00
INM-1	1166	G10	1	00.00	00.00	00.00	24.60	18.63	11.38	18.15	13.34
INM-1	1682	G11	1	00.00	26.55	20.02	00.00	00.00	00.00	00.00	00.00
INM-1	1683	G11	1	00.00	26.13	19.25	00.00	00.00	00.00	00.00	00.00
INM-1	1684	G11	1	00.00	23.85	17.32	00.00	00.00	00.00	00.00	00.00
INM-1	1685	G11	1	00.00	23.88	18.31	00.00	00.00	00.00	00.00	00.00
INM-1	1690	G11	1	00.00	00.00	00.00	22.28	17.36	10.26	16.55	13.00
INM-1	1691	G11	1	00.00	00.00	00.00	23.87	17.39	11.08	16.90	13.48
INM-1	1692	G11	1	00.00	00.00	00.00	22.64	17.19	10.46	17.19	12.78
INM-1	1693	G11	1	00.00	00.00	00.00	21.11	16.27	10.17	15.85	12.27
INM-1	1694	G11	1	00.00	00.00	00.00	21.73	16.90	10.15	16.79	12.63

INM-1	1695	G11	1	00.00	00.00	00.00	22.92	17.38	10.54	16.88	12.72
INM-1	1696	G11	1	00.00	00.00	00.00	25.61	19.40	11.16	18.01	13.26
INM-1	2937	H10	1	00.00	21.17	16.15	00.00	00.00	00.00	00.00	00.00
INM-1	2938	H10	1	00.00	25.87	18.48	00.00	00.00	00.00	00.00	00.00
INM-1	2939	H10	1	00.00	00.00	00.00	23.02	17.73	10.40	17.50	13.75
INM-1	2940	H10	1	00.00	00.00	00.00	22.95	16.78	10.37	16.34	12.20
INM-1	2941	H10	1	00.00	00.00	00.00	22.36	16.93	9.65	16.51	12.52
INM-1	2942	H10	1	00.00	00.00	00.00	00.00	00.00	12.53	17.81	13.43
INM-1	2943	H10	1	00.00	00.00	00.00	00.00	00.00	11.52	18.42	14.24
INM-1	1946	H11	1	00.00	23.63	17.79	00.00	00.00	00.00	00.00	00.00
INM-1	1947	H11	1	00.00	24.61	17.77	00.00	00.00	00.00	00.00	00.00
INM-1	1948	H11	1	00.00	25.01	18.24	00.00	00.00	00.00	00.00	00.00
INM-1	1949	H11	1	00.00	23.32	18.84	00.00	00.00	00.00	00.00	00.00
INM-1	1950	H11	1	00.00	24.99	18.94	00.00	00.00	00.00	00.00	00.00
INM-1	1951	H11	1	00.00	24.15	18.25	00.00	00.00	00.00	00.00	00.00
INM-1	1954	H11	1	00.00	00.00	00.00	23.64	16.88	10.53	17.17	13.62
INM-1	1955	H11	1	00.00	00.00	00.00	22.27	16.66	10.82	16.56	12.54
INM-1	1956	H11	1	00.00	00.00	00.00	24.88	17.50	10.71	16.78	13.35
INM-1	1957	H11	1	00.00	00.00	00.00	21.82	15.95	9.92	15.91	12.11
INM-1	1962	H11	1	00.00	00.00	00.00	00.00	00.00	10.41	17.21	11.85
INM-1	1963	H11	1	00.00	00.00	00.00	00.00	00.00	10.22	16.40	12.53
INM-1	1965	H11	1	00.00	00.00	00.00	00.00	00.00	10.95	17.32	12.70
INM-1	2142	Hno132	1	00.00	22.35	16.31	00.00	00.00	00.00	00.00	00.00
INM-1	2143	Hno132	1	00.00	24.89	17.84	00.00	00.00	00.00	00.00	00.00
INM-1	2144	Hno132	1	00.00	23.72	18.36	00.00	00.00	00.00	00.00	00.00
INM-1	2145	Hno132	1	00.00	22.90	18.20	00.00	00.00	00.00	00.00	00.00
INM-1	2146	Hno132	1	00.00	23.17	17.61	00.00	00.00	00.00	00.00	00.00
INM-1	2147	Hno132	1	00.00	24.35	18.21	00.00	00.00	00.00	00.00	00.00
INM-1	2148	Hno132	1	00.00	21.73	16.07	00.00	00.00	00.00	00.00	00.00
INM-1	2157	Hno132	1	00.00	00.00	00.00	00.00	00.00	11.43	17.42	13.46
INM-1	2158	Hno132	1	00.00	00.00	00.00	00.00	00.00	11.09	17.35	12.55
INM-1	2352		2	00.00	00.00	00.00	25.11	18.02	11.02	17.82	13.38
INM-1	2353		2	00.00	00.00	00.00	23.90	15.58	10.90	15.42	12.34
INM-1	2354		2	00.00	00.00	00.00	24.52	17.71	10.57	17.59	14.21
INM-1	2355		2	00.00	00.00	00.00	21.71	16.43	9.68	15.95	12.10
INM-1	2356		2	00.00	24.04	18.18	00.00	00.00	00.00	00.00	00.00
INM-1	2357		2	00.00	21.79	17.47	00.00	00.00	00.00	00.00	00.00
INM-1	2275	E1	2	00.00	24.45	19.87	00.00	00.00	00.00	00.00	00.00
INM-1	3084	G3	2	00.00	22.47	16.04	00.00	00.00	00.00	00.00	00.00
INM-1	3085	G3	2	00.00	21.86	17.25	00.00	00.00	00.00	00.00	00.00
INM-1	3086	G3	2	00.00	23.78	18.70	00.00	00.00	00.00	00.00	00.00
INM-1	3957	G4	2	00.00	23.00	15.65	00.00	00.00	00.00	00.00	00.00
INM-1	3958	G4	2	00.00	21.60	18.30	00.00	00.00	00.00	00.00	00.00
INM-1	3959	G4	2	00.00	23.55	18.40	00.00	00.00	00.00	00.00	00.00
INM-1	3963	G4	2	00.00	00.00	00.00	27.00	18.02	11.90	17.82	13.42
INM-1	3964	G4	2	00.00	00.00	00.00	22.83	16.52	10.04	16.26	12.09
INM-1	3965	G4	2	00.00	00.00	00.00	26.97	18.33	11.05	18.25	13.97
INM-1	3966	G4	2	00.00	00.00	00.00	26.10	17.83	10.72	17.73	13.25
INM-1	3967	G4	2	00.00	00.00	00.00	00.00	00.00	11.95	19.20	14.30
INM-1	3968	G4	2	00.00	00.00	00.00	00.00	00.00	10.35	16.40	12.32
INM-1	3106	G5	2	00.00	22.99	17.01	00.00	00.00	00.00	00.00	00.00
INM-1	2085	G6	2	00.00	25.16	18.21	00.00	00.00	00.00	00.00	00.00
INM-1	2086	G6	2	00.00	23.55	18.08	00.00	00.00	00.00	00.00	00.00
INM-1	2087	G6	2	00.00	25.34	19.31	00.00	00.00	00.00	00.00	00.00
INM-1	2088	G6	2	00.00	22.57	16.88	00.00	00.00	00.00	00.00	00.00
INM-1	2089	G6	2	00.00	20.36	15.02	00.00	00.00	00.00	00.00	00.00
INM-1	2097	G6	2	00.00	00.00	00.00	00.00	00.00	11.12	17.87	13.54
INM-1	3638	G7	2	00.00	23.15	16.21	00.00	00.00	00.00	00.00	00.00
INM-1	3639	G7	2	00.00	21.68	16.08	00.00	00.00	00.00	00.00	00.00
INM-1	3641	G7	2	00.00	00.00	00.00	21.54	16.50	9.83	15.55	11.90
INM-1	3642	G7	2	00.00	00.00	00.00	23.66	17.35	10.65	16.90	13.15
INM-1	3643	G7	2	00.00	00.00	00.00	25.52	19.15	12.23	18.80	14.00
INM-1	1587	H5	2	00.00	25.52	19.16	00.00	00.00	00.00	00.00	00.00

INM-1	1588	H5	2	00.00	21.69	17.37	00.00	00.00	00.00	00.00	00.00
INM-1	1589	H5	2	00.00	21.84	15.61	00.00	00.00	00.00	00.00	00.00
INM-1	1590	H5	2	00.00	23.70	18.14	00.00	00.00	00.00	00.00	00.00
INM-1	1591	H5	2	00.00	25.70	19.09	00.00	00.00	00.00	00.00	00.00
INM-1	1592	H5	2	00.00	26.05	21.02	00.00	00.00	00.00	00.00	00.00
INM-1	1593	H5	2	00.00	21.33	16.60	00.00	00.00	00.00	00.00	00.00
INM-1	1594	H5	2	00.00	24.49	17.34	00.00	00.00	00.00	00.00	00.00
INM-1	1598	H5	2	00.00	00.00	00.00	21.72	16.48	10.00	16.05	12.22
INM-1	1599	H5	2	00.00	00.00	00.00	21.43	16.35	9.72	16.18	12.34
INM-1	1600	H5	2	00.00	00.00	00.00	21.75	16.26	9.83	15.74	11.98
INM-1	1601	H5	2	00.00	00.00	00.00	22.23	16.49	10.07	15.98	12.15
INM-1	1602	H5	2	00.00	00.00	00.00	22.28	16.77	10.16	16.76	13.24
INM-1	1603	H5	2	00.00	00.00	00.00	24.65	17.94	10.88	17.76	13.49
INM-1	1604	H5	2	00.00	00.00	00.00	24.81	18.44	11.36	17.94	14.09
INM-1	1633	H5	2	00.00	24.56	18.54	00.00	00.00	00.00	00.00	00.00
INM-1	3398	H6	2	00.00	24.63	18.29	00.00	00.00	00.00	00.00	00.00
INM-1	3399	H6	2	00.00	23.50	18.35	00.00	00.00	00.00	00.00	00.00
INM-1	3400	H6	2	00.00	22.33	17.85	00.00	00.00	00.00	00.00	00.00
INM-1	3404	H6	2	00.00	00.00	00.00	25.25	18.07	11.15	17.75	14.11
INM-1	3405	H6	2	00.00	00.00	00.00	23.78	18.12	11.00	17.98	14.33
INM-1	1052	H7	2	00.00	00.00	00.00	23.30	17.58	10.64	17.40	13.33
INM-1	1053	H7	2	00.00	00.00	00.00	23.83	17.48	10.63	17.11	13.09
INM-1	1057	H7	2	00.00	24.13	18.47	00.00	00.00	00.00	00.00	00.00
INM-1	1058	H7	2	00.00	22.63	16.10	00.00	00.00	00.00	00.00	00.00
INM-1	1059	H7	2	00.00	22.96	18.32	00.00	00.00	00.00	00.00	00.00
INM-1	1060	H7	2	00.00	24.78	18.73	00.00	00.00	00.00	00.00	00.00
INM-1	1061	H7	2	00.00	23.44	18.67	00.00	00.00	00.00	00.00	00.00
INM-1	1062	H7	2	00.00	24.85	18.56	00.00	00.00	00.00	00.00	00.00
INM-1	1063	H7	2	00.00	21.00	23.17	00.00	00.00	00.00	00.00	00.00
INM-1	1064	H7	2	00.00	21.22	23.11	00.00	00.00	00.00	00.00	00.00
INM-1	3586	I6	2	00.00	22.59	18.22	00.00	00.00	00.00	00.00	00.00
INM-1	1449	D25	2	00.00	00.00	00.00	23.83	17.28	10.64	17.02	12.88
INM-1	1450	D25	2	00.00	00.00	00.00	24.06	17.21	10.92	16.90	12.50
INM-1	1452	D25	2	00.00	00.00	00.00	00.00	00.00	11.58	18.58	14.29
INM-1	1454	D25	2	00.00	00.00	00.00	00.00	00.00	00.00	17.57	00.00
INM-1	1455	D25	2	00.00	00.00	00.00	00.00	00.00	10.50	17.46	12.29
INM-1	1456	D25	2	00.00	00.00	00.00	00.00	00.00	10.97	17.56	13.67
INM-1	1459	D25	2	00.00	00.00	00.00	00.00	00.00	11.05	17.26	12.77
INM-1	1460	D25	2	00.00	00.00	00.00	00.00	00.00	11.92	18.26	13.82
INM-1	1461	D25	2	00.00	00.00	00.00	00.00	00.00	10.86	17.26	12.33
INM-1	1462	D25	2	00.00	00.00	00.00	00.00	00.00	11.47	18.20	13.37
INM-1	1464	D25	2	00.00	00.00	00.00	00.00	00.00	10.83	16.43	12.54
INM-1	7284	E24	2	00.00	23.75	18.10	00.00	00.00	00.00	00.00	00.00
INM-1	7285	E24	2	00.00	21.55	17.20	00.00	00.00	00.00	00.00	00.00
INM-1	7286	E24	2	00.00	23.35	16.45	00.00	00.00	00.00	00.00	00.00
INM-1	7287	E24	2	00.00	00.00	00.00	21.75	16.10	9.82	15.90	12.70
INM-1	7288	E24	2	00.00	00.00	00.00	24.35	17.27	11.20	17.20	13.45
INM-1	7289	E24	2	00.00	00.00	00.00	24.30	18.80	11.27	17.83	13.90
INM-1	7290	E24	2	00.00	00.00	00.00	22.67	17.75	10.30	17.48	13.45
INM-1	7291	E24	2	00.00	00.00	00.00	22.45	16.77	10.05	16.15	12.30
INM-1	7292	E24	2	00.00	00.00	00.00	22.70	16.50	10.87	15.92	11.80
INM-1	6408	E25	2	00.00	22.85	17.30	00.00	00.00	00.00	00.00	00.00
INM-1	6409	E25	2	00.00	24.75	19.75	00.00	00.00	00.00	00.00	00.00
INM-1	6412	E25	2	00.00	00.00	00.00	24.10	17.55	10.85	16.95	13.00
INM-1	6413	E25	2	00.00	00.00	00.00	23.50	16.65	10.65	16.50	12.75
INM-1	6414	E25	2	00.00	00.00	00.00	21.05	15.65	9.40	15.56	12.27
INM-1	7422	E25	2	00.00	21.80	14.90	00.00	00.00	00.00	00.00	00.00
INM-1	7423	E25	2	00.00	00.00	00.00	22.80	17.50	17.45	13.93	00.00
INM-1	7424	E25	2	00.00	00.00	00.00	22.60	16.74	10.30	16.30	11.90
INM-1	7425	E25	2	00.00	00.00	00.00	23.25	16.62	10.15	16.10	12.30
INM-1	3009	F10	2	00.00	00.00	00.00	22.06	15.82	10.16	15.17	11.77
INM-1	3013	F10	2	00.00	00.00	00.00	00.00	00.00	10.08	16.40	12.50
INM-1	3014	F10	2	00.00	00.00	00.00	00.00	00.00	11.07	16.77	13.04

INM-1	2561	F11	2	00.00	24.73	19.41	00.00	00.00	00.00	00.00	00.00
INM-1	2562	F11	2	00.00	22.74	17.94	00.00	00.00	00.00	00.00	00.00
INM-1	2563	F11	2	00.00	24.10	17.35	00.00	00.00	00.00	00.00	00.00
INM-1	2564	F11	2	00.00	25.02	19.24	00.00	00.00	00.00	00.00	00.00
INM-1	2565	F11	2	00.00	22.85	16.32	00.00	00.00	00.00	00.00	00.00
INM-1	2569	F11	2	00.00	00.00	00.00	24.74	18.09	11.20	17.49	13.78
INM-1	2573	F11	2	00.00	00.00	00.00	23.68	16.91	10.35	16.43	12.37
INM-1	2575	F11	2	00.00	00.00	00.00	00.00	00.00	11.30	17.14	13.75
INM-1	2576	F11	2	00.00	00.00	00.00	00.00	00.00	10.64	18.08	12.77
INM-1	2577	F11	2	00.00	00.00	00.00	00.00	00.00	11.10	17.30	12.84
INM-1	2578	F11	2	00.00	00.00	00.00	00.00	00.00	11.41	18.01	13.73
INM-1	2579	F11	2	00.00	00.00	00.00	00.00	00.00	11.06	18.25	13.36
INM-1	2580	F11	2	00.00	00.00	00.00	00.00	00.00	10.68	16.55	13.33
INM-1	827	F12	2	00.00	00.00	00.00	23.91	17.61	11.35	17.31	13.87
INM-1	829	F12	2	00.00	00.00	00.00	21.95	15.85	9.94	15.87	12.13
INM-1	831	F12	2	00.00	00.00	00.00	22.29	17.28	10.20	17.23	13.07
INM-1	838	F12	2	00.00	23.32	17.83	00.00	00.00	00.00	00.00	00.00
INM-1	6616	F24	2	00.00	23.65	16.80	00.00	00.00	00.00	00.00	00.00
INM-1	6617	F24	2	00.00	22.85	17.15	00.00	00.00	00.00	00.00	00.00
INM-1	6618	F24	2	00.00	22.20	16.80	00.00	00.00	00.00	00.00	00.00
INM-1	6619	F24	2	00.00	22.85	17.47	00.00	00.00	00.00	00.00	00.00
INM-1	6620	F24	2	00.00	23.88	18.63	00.00	00.00	00.00	00.00	00.00
INM-1	6622	F24	2	00.00	00.00	00.00	23.40	17.93	10.80	17.65	13.95
INM-1	6623	F24	2	00.00	00.00	00.00	26.30	17.60	10.95	17.35	13.65
INM-1	6624	F24	2	00.00	00.00	00.00	21.80	15.80	9.57	15.65	11.60
INM-1	6625	F24	2	00.00	00.00	00.00	21.30	16.25	9.65	15.75	12.05
INM-1	6626	F24	2	00.00	00.00	00.00	21.40	16.25	10.05	16.15	12.83
INM-1	7486	F24	2	00.00	23.40	17.20	00.00	00.00	00.00	00.00	00.00
INM-1	7487	F24	2	00.00	00.00	00.00	24.05	17.15	11.10	16.80	13.25
INM-1	7488	F24	2	00.00	00.00	00.00	22.50	15.80	10.20	15.65	12.05
INM-1	7489	F24	2	00.00	00.00	00.00	23.50	17.15	10.65	16.90	12.80
INM-1	7490	F24	2	00.00	00.00	00.00	23.05	17.65	10.50	17.45	14.00
INM-1	7561	F26	2	00.00	24.40	18.50	00.00	00.00	00.00	00.00	00.00
INM-1	7562	F26	2	00.00	22.55	17.66	00.00	00.00	00.00	00.00	00.00
INM-1	7563	F26	2	00.00	00.00	00.00	22.05	16.50	9.90	16.47	12.45
INM-1	7564	F26	2	00.00	00.00	00.00	23.25	16.65	10.50	16.15	12.45
INM-1	1351	G11	2	00.00	00.00	00.00	24.47	18.02	10.80	17.40	13.55
INM-1	1353	G11	2	00.00	00.00	00.00	22.08	16.93	10.34	16.77	12.83
INM-1	1354	G11	2	00.00	00.00	00.00	24.74	17.51	11.34	17.18	13.42
INM-1	1359	G11	2	00.00	00.00	00.00	00.00	00.00	10.34	17.72	13.38
INM-1	1360	G11	2	00.00	00.00	00.00	00.00	00.00	10.83	17.50	13.17
INM-1	1361	G11	2	00.00	00.00	00.00	00.00	00.00	11.03	00.00	13.83
INM-1	1362	G11	2	00.00	24.26	17.70	00.00	00.00	00.00	00.00	00.00
INM-1	1363	G11	2	00.00	22.26	16.59	00.00	00.00	00.00	00.00	00.00
INM-1	1367	G11	2	00.00	21.92	15.56	00.00	00.00	00.00	00.00	00.00
INM-1	3183	G11	2	00.00	22.70	17.35	00.00	00.00	00.00	00.00	00.00
INM-1	3184	G11	2	00.00	22.37	18.11	00.00	00.00	00.00	00.00	00.00
INM-1	3185	G11	2	00.00	24.01	18.44	00.00	00.00	00.00	00.00	00.00
INM-1	3186	G11	2	00.00	22.99	17.23	00.00	00.00	00.00	00.00	00.00
INM-1	3187	G11	2	00.00	24.95	17.50	00.00	00.00	00.00	00.00	00.00
INM-1	3188	G11	2	00.00	23.32	16.69	00.00	00.00	00.00	00.00	00.00
INM-1	3189	G11	2	00.00	22.44	17.55	00.00	00.00	00.00	00.00	00.00
INM-1	3190	G11	2	00.00	21.57	16.12	00.00	00.00	00.00	00.00	00.00
INM-1	3191	G11	2	00.00	21.97	16.47	00.00	00.00	00.00	00.00	00.00
INM-1	3192	G11	2	00.00	22.62	17.77	00.00	00.00	00.00	00.00	00.00
INM-1	3193	G11	2	00.00	23.35	18.45	00.00	00.00	00.00	00.00	00.00
INM-1	3194	G11	2	00.00	21.83	15.59	00.00	00.00	00.00	00.00	00.00
INM-1	3204	G11	2	00.00	00.00	00.00	23.00	00.00	9.72	16.24	12.64
INM-1	3205	G11	2	00.00	00.00	00.00	23.74	16.82	10.89	16.36	12.76
INM-1	3206	G11	2	00.00	00.00	00.00	24.75	17.76	10.60	17.33	13.52
INM-1	3211	G11	2	00.00	00.00	00.00	23.55	16.05	9.72	15.74	11.34
INM-1	3213	G11	2	00.00	00.00	00.00	00.00	00.00	10.14	17.10	13.15
INM-1	3214	G11	2	00.00	00.00	00.00	00.00	00.00	11.21	17.60	13.41

INM-1	7018	G24	2	00.00	22.10	16.15	00.00	00.00	00.00	00.00	00.00
INM-1	7019	G24	2	00.00	21.85	17.20	00.00	00.00	00.00	00.00	00.00
INM-1	7020	G24	2	00.00	21.05	16.65	00.00	00.00	00.00	00.00	00.00
INM-1	7021	G24	2	00.00	26.70	19.00	00.00	00.00	00.00	00.00	00.00
INM-1	7022	G24	2	00.00	00.00	00.00	25.50	17.14	10.90	16.95	13.40
INM-1	7023	G24	2	00.00	00.00	00.00	26.10	19.00	11.55	18.90	14.45
INM-1	7024	G24	2	00.00	00.00	00.00	23.70	17.75	10.85	17.45	13.25
INM-1	7172	G25	2	00.00	24.45	17.75	00.00	00.00	00.00	00.00	00.00
INM-1	7174	G25	2	00.00	00.00	00.00	22.48	17.02	10.05	16.55	12.83
INM-1	2783	H11	2	00.00	25.52	18.08	00.00	00.00	00.00	00.00	00.00
INM-1	2788	H11	2	00.00	00.00	00.00	24.56	18.16	11.00	17.47	13.77
INM-1	2789	H11	2	00.00	00.00	00.00	22.96	17.55	10.14	17.26	13.63
INM-1	2790	H11	2	00.00	00.00	00.00	22.62	17.29	10.27	17.34	13.53
INM-1	3569	I10	2	00.00	24.27	18.85	00.00	00.00	00.00	00.00	00.00
INM-1	3570	I10	2	00.00	21.87	17.80	00.00	00.00	00.00	00.00	00.00
INM-1	3571	I10	2	00.00	00.00	00.00	23.65	17.65	10.49	17.35	12.85
INM-1	746	I12	2	9.75	00.00	00.00	00.00	00.00	00.00	00.00	00.00
INM-1	747	I12	2	14.10	00.00	00.00	23.11	16.63	10.07	15.75	11.42
INM-1	4468	Hno129	2	00.00	22.23	17.15	00.00	00.00	00.00	00.00	00.00
INM-1	4469	Hno129	2	00.00	23.95	18.80	00.00	00.00	00.00	00.00	00.00
INM-1	2854	G3	3	00.00	00.00	00.00	22.48	16.92	10.26	16.85	12.74
INM-1	2856	G3	3	00.00	00.00	00.00	26.25	17.70	11.71	17.28	13.96
INM-1	2857	G3	3	00.00	00.00	00.00	25.56	17.87	11.34	17.73	13.90
INM-1	2858	G3	3	00.00	00.00	00.00	25.65	18.84	11.35	17.75	13.63
INM-1	2859	G3	3	00.00	00.00	00.00	00.00	00.00	10.39	17.00	12.77
INM-1	2677	G7	3	00.00	00.00	00.00	25.58	18.22	11.27	17.88	13.32
INM-1	2678	G7	3	00.00	00.00	00.00	00.00	00.00	12.05	19.35	13.34
INM-1	2679	G7	3	00.00	00.00	00.00	00.00	00.00	11.34	17.13	12.53
INM-1	2680	G7	3	00.00	00.00	00.00	00.00	00.00	11.13	17.85	13.04
INM-1	2683	G7	3	00.00	00.00	00.00	00.00	00.00	11.17	17.97	13.02
INM-1	2686	G7	3	00.00	00.00	00.00	00.00	00.00	12.02	18.55	13.74
INM-1	2687	G7	3	00.00	00.00	00.00	00.00	00.00	00.00	17.11	12.74
INM-1	2698	G7	3	00.00	23.14	15.66	00.00	00.00	00.00	00.00	00.00
INM-1	3466	H7	3	00.00	22.50	16.10	00.00	00.00	00.00	00.00	00.00
INM-1	3467	H7	3	00.00	22.85	17.39	00.00	00.00	00.00	00.00	00.00
INM-1	3468	H7	3	00.00	24.60	18.85	00.00	00.00	00.00	00.00	00.00
INM-1	3469	H7	3	00.00	24.71	17.70	00.00	00.00	00.00	00.00	00.00
INM-1	3470	H7	3	00.00	21.00	15.40	00.00	00.00	00.00	00.00	00.00
INM-1	3471	H7	3	00.00	25.45	19.70	00.00	00.00	00.00	00.00	00.00
INM-1	3472	H7	3	00.00	23.30	16.80	00.00	00.00	00.00	00.00	00.00
INM-1	3510	H7	3	00.00	20.70	15.10	00.00	00.00	00.00	00.00	00.00
INM-1	3511	H7	3	00.00	25.05	19.30	00.00	00.00	00.00	00.00	00.00
INM-1	2020	D24	3	00.00	24.11	19.61	00.00	00.00	00.00	00.00	00.00
INM-1	1231	D25	3	00.00	00.00	00.00	22.51	17.89	10.33	17.74	13.56
INM-1	1234	D25	3	00.00	00.00	00.00	00.00	00.00	10.31	17.77	13.47
INM-1	1235	D25	3	00.00	00.00	00.00	00.00	00.00	11.06	17.32	12.82
INM-1	1889	E10	3	00.00	00.00	00.00	00.00	00.00	11.35	17.95	13.35
INM-1	6762	E23	3	00.00	23.06	16.20	00.00	00.00	00.00	00.00	00.00
INM-1	6509	E25	3	00.00	23.20	17.55	00.00	00.00	00.00	00.00	00.00
INM-1	189	F24	3	00.00	00.00	00.00	21.49	00.00	10.21	16.59	12.50
INM-1	6528	F25	3	00.00	00.00	00.00	25.16	18.55	11.15	17.65	13.57
INM-1	6530	F25	3	00.00	00.00	00.00	22.30	16.00	10.75	15.85	12.05
INM-1	2442	F25&F26	3	00.00	21.04	16.39	00.00	00.00	00.00	00.00	00.00
INM-1	2443	F25&F26	3	00.00	18.82	14.25	00.00	00.00	00.00	00.00	00.00
INM-1	2209	G10	3	00.00	00.00	00.00	20.78	16.03	9.62	15.94	11.78
INM-1	2905	G10	3	00.00	00.00	00.00	25.74	18.97	11.30	18.43	13.93
INM-1	1269	G11	3	00.00	23.59	19.90	00.00	00.00	00.00	00.00	00.00
INM-1	1270	G11	3	00.00	21.12	16.38	00.00	00.00	00.00	00.00	00.00
INM-1	1272	G11	3	00.00	00.00	00.00	00.00	00.00	9.89	16.95	13.06
INM-1	1274	G11	3	00.00	00.00	00.00	00.00	00.00	9.99	16.65	12.47
INM-1	3747	G11	3	00.00	25.25	18.60	00.00	00.00	00.00	00.00	00.00
INM-1	3748	G11	3	00.00	00.00	00.00	25.20	18.27	11.32	18.00	14.00
INM-1	3749	G11	3	00.00	00.00	00.00	00.00	00.00	10.15	16.65	12.50

INM-1	3750	G11	3	00.00	00.00	00.00	21.45	15.95	9.48	15.81	11.75
INM-1	3751	G11	3	00.00	00.00	00.00	21.00	17.78	9.54	16.80	12.50
INM-1	886	G12	3	00.00	00.00	00.00	21.71	16.75	10.01	16.40	12.86
INM-1	887	G12	3	00.00	00.00	00.00	23.17	17.31	10.54	16.81	13.55
INM-1	888	G12	3	00.00	00.00	00.00	23.01	16.29	10.75	15.83	12.82
INM-1	890	G12	3	00.00	23.69	18.08	00.00	00.00	00.00	00.00	00.00
INM-1	891	G12	3	00.00	23.85	18.85	00.00	00.00	00.00	00.00	00.00
INM-1	893	G12	3	00.00	21.34	16.30	00.00	00.00	00.00	00.00	00.00
INM-1	893	G12	3	00.00	22.42	17.81	00.00	00.00	00.00	00.00	00.00
INM-1	7810	G25	3	00.00	24.15	18.80	00.00	00.00	00.00	00.00	00.00
INM-1	7811	G25	3	00.00	24.75	17.40	00.00	00.00	00.00	00.00	00.00
INM-1	7812	G25	3	00.00	00.00	00.00	23.60	16.70	10.65	16.35	12.60
INM-1	7813	G25	3	00.00	00.00	00.00	24.30	17.50	10.30	17.10	13.00
INM-1	7814	G25	3	00.00	00.00	00.00	23.20	17.25	10.50	17.30	12.95
INM-1	7815	G25	3	00.00	00.00	00.00	00.00	00.00	10.93	17.65	13.15
INM-1	7817	G25	3	00.00	00.00	00.00	00.00	00.00	10.60	16.05	12.40
INM-1	7820	G25	3	00.00	00.00	00.00	00.00	00.00	11.15	17.93	13.45
INM-1	7821	G25	3	00.00	00.00	00.00	00.00	00.00	10.55	17.00	13.45
INM-1	7822	G25	3	00.00	00.00	00.00	00.00	00.00	10.60	16.65	11.95
INM-1	2619	H10	3	00.00	00.00	00.00	23.85	17.52	10.65	17.05	12.50
INM-1	2621	H10	3	00.00	21.19	14.01	00.00	00.00	00.00	00.00	00.00
INM-1	2622	H10	3	00.00	00.00	00.00	00.00	00.00	9.35	16.03	12.76
INM-1	2624	H10	3	00.00	00.00	00.00	00.00	00.00	9.98	16.40	12.28
INM-1	1781	H11	3	00.00	00.00	00.00	21.81	17.10	9.87	17.08	13.10
INM-1	1783	H11	3	00.00	00.00	00.00	00.00	00.00	11.29	18.30	13.23
INM-1	1784	H11	3	00.00	00.00	00.00	00.00	00.00	9.41	15.75	12.60
INM-1	4517	H12	3	00.00	23.82	17.60	00.00	00.00	00.00	00.00	00.00
INM-1	4518	H12	3	00.00	22.65	17.46	00.00	00.00	00.00	00.00	00.00
INM-1	4520	H12	3	00.00	00.00	00.00	24.75	18.23	10.50	17.98	14.45
INM-1	4521	H12	3	00.00	00.00	00.00	24.95	18.62	11.33	18.37	14.24
INM-1	7917	Hno102	3	00.00	22.15	16.40	00.00	00.00	00.00	00.00	00.00
INM-1	6181	Hno106	3	00.00	00.00	00.00	21.57	15.87	10.35	15.45	11.25
INM-1	4873	Hno107	3	00.00	00.00	00.00	27.10	17.27	10.45	17.20	13.30
INM-1	4874	Hno107	3	00.00	00.00	00.00	00.00	00.00	10.90	18.50	13.70
INM-1	4487	Hno128	3	00.00	00.00	00.00	00.00	00.00	12.55	19.05	13.90
INM-1	568	F8	4	00.00	00.00	00.00	22.31	17.37	9.75	16.91	12.49
INM-1	1814	H5	4	00.00	00.00	00.00	24.19	17.64	11.13	17.62	14.04
INM-1	3015	H7	4	00.00	19.09	13.65	00.00	00.00	00.00	00.00	00.00
INM-1	8016	E24	4	00.00	22.70	17.00	00.00	00.00	00.00	00.00	00.00
INM-1	8017	E24	4	00.00	00.00	00.00	23.60	17.30	10.57	17.20	13.50
INM-1	8018	E24	4	00.00	00.00	00.00	22.40	16.02	10.20	16.00	12.65
INM-1	8119	E24	4	00.00	00.00	00.00	25.55	19.15	11.40	18.90	14.10
INM-1	8129	E24	4	00.00	00.00	00.00	00.00	00.00	11.10	17.95	14.15
INM-1	5862	E27	4	13.85	23.30	15.65	00.00	00.00	00.00	00.00	00.00
INM-1	5863	E27	4	00.00	20.25	14.60	00.00	00.00	00.00	00.00	00.00
INM-1	7968	F24	4	00.00	00.00	00.00	25.25	18.60	11.30	18.55	14.71
INM-1	7969	F24	4	00.00	00.00	00.00	23.00	16.20	10.55	15.75	12.20
INM-1	7970	F24	4	00.00	00.00	00.00	24.90	18.55	10.75	17.28	12.35
INM-1	7971	F24	4	00.00	00.00	00.00	22.55	17.20	10.75	17.20	13.55
INM-1	5967	F25	4	00.00	22.89	16.05	00.00	00.00	00.00	00.00	00.00
INM-1	1862	G11	4	00.00	00.00	00.00	23.15	17.72	10.48	17.68	13.48
INM-1	1868	G11	4	00.00	00.00	00.00	00.00	00.00	10.96	00.00	13.39
INM-1	7102	G24	4	00.00	19.85	14.15	00.00	00.00	00.00	00.00	00.00
INM-1	7103	G24	4	00.00	21.90	17.10	00.00	00.00	00.00	00.00	00.00
INM-1	7104	G24	4	00.00	22.20	16.40	00.00	00.00	00.00	00.00	00.00
INM-1	7105	G24	4	00.00	00.00	00.00	24.59	17.20	11.20	16.75	13.12
INM-1	7106	G24	4	00.00	00.00	00.00	21.90	17.65	10.26	17.45	13.30
INM-1	7107	G24	4	00.00	00.00	00.00	24.75	00.00	10.95	19.50	12.30
INM-1	7623	G25	4	00.00	23.70	18.00	00.00	00.00	00.00	00.00	00.00
INM-1	7758	D24&E25	4	00.00	00.00	00.00	21.80	17.05	9.60	17.00	12.88
INM-1	7759	D24&E25	4	00.00	00.00	00.00	23.85	16.70	10.95	16.55	12.45
INM-1	4753	Hno117	4	00.00	24.48	17.92	00.00	00.00	00.00	00.00	00.00
INM-1	4694	Hno121	4	00.00	00.00	00.00	00.00	00.00	10.60	16.55	12.20

INM-1	2975	B3	5	00.00	00.00	00.00	00.00	00.00	11.26	16.98	13.15
INM-1	3809	E8	5	00.00	00.00	00.00	26.16	18.35	12.00	18.16	13.70
INM-1	4326	F8	5	00.00	23.60	16.35	00.00	00.00	00.00	00.00	00.00
INM-1	4327	F8	5	00.00	21.93	16.19	00.00	00.00	00.00	00.00	00.00
INM-1	4328	F8	5	00.00	00.00	00.00	23.05	16.35	10.50	16.00	12.60
INM-1	4566	E5	5	00.00	00.00	00.00	22.60	17.55	10.40	17.25	13.00
INM-1	4422		6	00.00	00.00	00.00	00.00	00.00	11.55	18.30	13.55
INM-1	3767	F8	6	00.00	22.30	18.00	00.00	00.00	00.00	00.00	00.00
INM-1	3768	F8	6	00.00	24.85	18.60	00.00	00.00	00.00	00.00	00.00
INM-1	3769	F8	6	00.00	24.55	17.75	00.00	00.00	00.00	00.00	00.00
INM-1	3770	F8	6	00.00	00.00	00.00	23.50	16.60	10.35	16.05	12.30
INM-1	3771	F8	6	00.00	00.00	00.00	24.85	19.25	10.98	18.55	14.48
INM-1	1504	G4	6	00.00	22.58	16.81	00.00	00.00	00.00	00.00	00.00
INM-1	1836	G4	7	00.00	00.00	00.00	22.62	16.31	10.25	15.83	12.92
INM-1	4846	E4	8	00.00	24.70	17.55	00.00	00.00	00.00	00.00	00.00
INM-1	4494	Hno49	8	00.00	24.10	17.40	00.00	00.00	00.00	00.00	00.00
INM-1	6055	C3	11	00.00	20.09	14.55	00.00	00.00	00.00	00.00	00.00
INM-1	6017	E3	11	00.00	20.25	14.36	00.00	00.00	00.00	00.00	00.00
INM-1	813	C1	12	00.00	00.00	00.00	22.98	16.96	9.97	16.11	13.05
INM-1	4282	C2	12	00.00	00.00	00.00	00.00	00.00	11.83	18.70	13.65
INM-1	4283	C2	12	00.00	00.00	00.00	00.00	00.00	11.60	17.70	13.25
INM-1	5610	C3	12	00.00	22.06	16.50	00.00	00.00	00.00	00.00	00.00
INM-1	5917	B3	14	00.00	00.00	00.00	22.60	16.50	10.13	16.15	12.65
INM-1	5768	D8	14	10.65	00.00	00.00	00.00	00.00	00.00	00.00	00.00
INM-1	6039	Hno87	16	00.00	00.00	00.00	00.00	00.00	10.20	16.75	12.20
INM-1	6040	Hno87	16	00.00	00.00	00.00	22.95	16.90	10.08	16.25	12.55
INM-1	2499	D3	16	00.00	17.97	13.55	00.00	00.00	00.00	00.00	00.00

Table A.16.7. Measurements of Metapodia of ***Antilope cervicapra***

Site	Reg.No.	Trench	Layer	BD	TD	WT	TT	TLT
INM-1	5280	LJP	1-4	00.00	00.00	10.52	16.85	12.70
INM-1	1097	G4	1	00.00	00.00	11.22	17.83	13.37
INM-1	3289	D10	1	00.00	00.00	11.68	18.30	12.85
INM-1	3290	D10	1	00.00	00.00	11.86	17.63	12.30
INM-1	3859	D25	1	00.00	00.00	10.69	18.05	12.15
INM-1	3858	D25	1	00.00	00.00	10.35	17.00	12.60
INM-1	3848	D25	1	00.00	00.00	10.85	17.30	12.65
INM-1	3847	D25	1	00.00	00.00	11.40	18.05	13.63
INM-1	3845	D25	1	00.00	00.00	11.49	17.47	12.35
INM-1	3846	D25	1	00.00	00.00	11.20	18.05	13.20
INM-1	3861	D25	1	00.00	00.00	10.10	18.05	13.40
INM-1	1182	F10	1	00.00	00.00	10.84	16.26	11.81
INM-1	1181	F10	1	23.67	16.42	11.13	16.42	12.24
INM-1	3355	F11	1	00.00	00.00	10.27	16.63	12.85
INM-1	3367	F11	1	00.00	00.00	11.20	18.70	15.05
INM-1	3356	F11	1	24.26	18.26	11.21	17.90	13.63
INM-1	3357	F11	1	00.00	00.00	9.33	15.44	11.64
INM-1	8054	F24	1	00.00	00.00	11.60	18.27	13.25
INM-1	8055	F24	1	24.85	16.75	9.85	16.70	12.30
INM-1	8056	F24	1	00.00	00.00	11.30	18.07	13.75
INM-1	8058	F24	1	00.00	00.00	10.00	16.25	11.80
INM-1	8057	F24	1	00.00	00.00	11.30	18.07	13.75
INM-1	1700	G11	1	00.00	00.00	10.83	17.34	12.63
INM-1	1702	G11	1	00.00	00.00	10.44	17.58	13.00
INM-1	1699	G11	1	00.00	00.00	11.74	18.00	13.22
INM-1	2277	E1	2	00.00	00.00	10.74	17.64	13.48
INM-1	2276	E1	2	00.00	00.00	10.77	17.28	00.00
INM-1	2278	E1	2	00.00	00.00	10.63	19.00	12.67
INM-1	2279	E1	2	00.00	00.00	10.43	17.14	12.88
INM-1	3090	G3	2	00.00	00.00	10.60	17.34	13.19
INM-1	3645	G7	2	00.00	00.00	9.95	17.00	12.90

INM-1	1613	H5	2	00.00	00.00	10.99	17.48	12.64
INM-1	1614	H5	2	00.00	00.00	11.62	18.05	13.46
INM-1	1611	H5	2	00.00	00.00	11.18	17.34	12.36
INM-1	1612	H5	2	00.00	00.00	11.51	17.83	12.61
INM-1	1610	H5	2	00.00	00.00	12.26	18.73	13.76
INM-1	1615	H5	2	00.00	00.00	10.47	17.05	12.81
INM-1	1051	H7	2	00.00	00.00	11.02	17.55	12.34
INM-1	3587	I6	2	00.00	00.00	11.48	17.25	13.38
INM-1	1463	D25	2	00.00	00.00	10.18	00.00	00.00
INM-1	7301	E24	2	00.00	00.00	11.20	17.80	13.70
INM-1	7338	E24	2	00.00	00.00	10.25	16.70	13.20
INM-1	7303	E24	2	00.00	00.00	10.40	16.65	13.65
INM-1	7304	E24	2	00.00	00.00	9.35	16.35	11.56
INM-1	7302	E24	2	00.00	00.00	9.40	15.90	11.65
INM-1	6425	E25	2	00.00	00.00	9.20	15.45	11.40
INM-1	6424	E25	2	00.00	00.00	9.80	16.10	11.90
INM-1	6423	E25	2	00.00	00.00	9.65	15.80	12.26
INM-1	6421	E25	2	00.00	00.00	10.00	16.52	11.60
INM-1	6422	E25	2	00.00	00.00	10.00	16.70	12.15
INM-1	6418	E25	2	00.00	00.00	10.65	17.73	13.79
INM-1	6420	E25	2	00.00	00.00	10.45	16.80	12.87
INM-1	6419	E25	2	00.00	00.00	9.85	16.17	12.45
INM-1	6426	E25	2	00.00	00.00	10.70	17.05	12.55
INM-1	7436	E25	2	00.00	00.00	9.55	15.85	12.17
INM-1	7437	E25	2	00.00	00.00	9.70	16.45	12.25
INM-1	7439	E25	2	00.00	00.00	10.70	17.20	13.52
INM-1	7435	E25	2	00.00	00.00	9.40	16.15	11.73
INM-1	7434	E25	2	00.00	00.00	10.05	16.20	12.15
INM-1	7433	E25	2	00.00	00.00	10.45	16.95	12.45
INM-1	7432	E25	2	00.00	00.00	10.40	16.86	12.70
INM-1	7428	E25	2	00.00	00.00	10.50	16.57	12.50
INM-1	7429	E25	2	00.00	00.00	10.98	17.82	11.96
INM-1	7431	E25	2	00.00	00.00	10.75	17.00	13.20
INM-1	7438	E25	2	00.00	00.00	12.24	19.00	13.90
INM-1	6632	F24	2	00.00	00.00	10.80	17.80	12.65
INM-1	6631	F24	2	00.00	00.00	9.90	16.03	11.30
INM-1	6630	F24	2	00.00	00.00	10.50	16.42	12.25
INM-1	6635	F24	2	00.00	00.00	10.47	16.77	13.05
INM-1	6633	F24	2	00.00	00.00	11.45	17.55	13.00
INM-1	6634	F24	2	00.00	00.00	10.35	16.50	12.35
INM-1	7493	F24	2	00.00	00.00	11.75	17.65	13.62
INM-1	7569	F26	2	00.00	00.00	10.88	1755	12.20
INM-1	7570	F26	2	00.00	00.00	10.95	17.44	12.65
INM-1	7568	F26	2	00.00	00.00	11.00	18.05	13.45
INM-1	7029	G24	2	00.00	00.00	10.65	16.15	11.75
INM-1	7028	G24	2	00.00	00.00	11.15	18.17	13.95
INM-1	7030	G24	2	00.00	00.00	10.70	16.80	12.55
INM-1	7176	G25	2	00.00	00.00	11.45	17.40	13.00
INM-1	7177	G25	2	00.00	00.00	11.00	17.70	13.80
INM-1	7178	G25	2	00.00	00.00	10.30	17.40	13.45
INM-1	7179	G25	2	00.00	00.00	10.55	16.75	11.90
INM-1	7180	G25	2	00.00	00.00	10.55	17.04	13.20
INM-1	749	I12	2	00.00	00.00	11.50	18.32	13.55
INM-1	745	I12	2	00.00	00.00	11.11	17.62	12.77
INM-1	750	I12	2	00.00	00.00	10.67	17.49	13.09
INM-1	748	I12	2	00.00	00.00	11.55	17.71	13.31
INM-1	6110	Hno125	2	23.42	17.40	10.40	16.50	11.90
INM-1	3604	G4	3	00.00	00.00	10.65	16.81	12.40
INM-1	6737	E23	3	00.00	00.00	11.00	18.03	13.07
INM-1	6744	E23	3	00.00	00.00	10.90	17.77	12.55
INM-1	6745	E23	3	00.00	00.00	10.32	16.35	12.42
INM-1	6738	E23	3	00.00	00.00	10.97	17.30	12.90
INM-1	6746	E23	3	00.00	00.00	9.90	16.92	11.75

INM-1	6747	E23	3	00.00	00.00	10.25	17.32	12.95
INM-1	6748	E23	3	00.00	00.00	10.50	16.90	12.49
INM-1	6764	E23	3	00.00	00.00	10.96	16.65	12.60
INM-1	6741	E23	3	00.00	00.00	10.28	16.57	12.73
INM-1	6763	E23	3	00.00	00.00	10.36	16.45	11.60
INM-1	6743	E23	3	00.00	00.00	11.50	18.46	13.52
INM-1	6742	E23	3	00.00	00.00	11.45	17.70	13.20
INM-1	6740	E23	3	00.00	00.00	11.60	16.84	12.92
INM-1	6739	E23	3	00.00	00.00	10.41	16.90	13.20
INM-1	6935	E26	3	00.00	00.00	10.70	16.85	12.70
INM-1	6936	E26	3	00.00	00.00	10.35	17.07	12.85
INM-1	6937	E26	3	00.00	00.00	9.85	16.17	12.30
INM-1	6938	E26	3	00.00	00.00	10.20	16.10	11.80
INM-1	6934	E26	3	00.00	00.00	11.16	17.45	13.00
INM-1	6939	E26	3	00.00	00.00	9.90	17.75	13.00
INM-1	3270	F11	3	00.00	00.00	11.47	17.53	12.15
INM-1	6539	F25	3	00.00	00.00	11.47	18.56	13.24
INM-1	6541	F25	3	00.00	00.00	11.85	17.85	12.70
INM-1	6537	F25	3	00.00	00.00	10.87	17.80	12.30
INM-1	6536	F25	3	00.00	00.00	9.65	14.93	10.40
INM-1	6540	F25	3	00.00	00.00	10.85	17.30	12.55
INM-1	6535	F25	3	00.00	00.00	10.15	16.50	12.20
INM-1	6532	F25	3	00.00	00.00	9.13	14.06	9.65
INM-1	6551	F25	3	00.00	00.00	11.35	18.30	13.20
INM-1	6534	F25	3	00.00	00.00	9.90	15.60	12.15
INM-1	6543	F25	3	00.00	00.00	11.95	19.25	14.30
INM-1	6550	F25	3	00.00	00.00	10.95	17.40	12.35
INM-1	6542	F25	3	00.00	00.00	11.05	17.60	12.75
INM-1	6549	F25	3	00.00	00.00	10.85	17.40	13.20
INM-1	6544	F25	3	00.00	00.00	11.10	18.05	12.75
INM-1	6533	F25	3	00.00	00.00	10.00	16.75	12.76
INM-1	6538	F25	3	00.00	00.00	10.85	17.20	13.20
INM-1	6548	F25	3	00.00	00.00	12.15	18.15	13.85
INM-1	6547	F25	3	00.00	00.00	10.45	16.08	10.50
INM-1	6546	F25	3	00.00	00.00	11.80	17.50	12.53
INM-1	6545	F25	3	00.00	00.00	10.80	18.35	14.00
INM-1	2210	G10	3	00.00	00.00	9.90	16.28	12.60
INM-1	3754	G11	3	00.00	00.00	11.45	16.30	12.66
INM-1	3753	G11	3	00.00	00.00	9.80	15.30	11.32
INM-1	3752	G11	3	00.00	00.00	10.95	17.30	12.95
INM-1	7831	G25	3	00.00	00.00	11.38	16.85	13.30
INM-1	7829	G25	3	00.00	00.00	11.90	19.50	13.00
INM-1	7825	G25	3	00.00	00.00	11.45	17.55	13.85
INM-1	7826	G25	3	00.00	00.00	10.80	18.24	13.45
INM-1	7827	G25	3	00.00	00.00	10.40	16.97	12.87
INM-1	7828	G25	3	00.00	00.00	10.45	16.65	11.70
INM-1	7830	G25	3	00.00	00.00	11.45	17.90	14.12
INM-1	7719	G26	3	00.00	00.00	9.95	16.20	12.10
INM-1	7718	G26	3	00.00	00.00	10.97	18.30	13.85
INM-1	7717	G26	3	00.00	00.00	12.60	19.73	14.83
INM-1	7716	G26	3	00.00	00.00	11.60	18.60	13.90
INM-1	7720	G26	3	00.00	00.00	11.40	17.05	13.32
INM-1	7721	G26	3	00.00	00.00	10.20	16.30	11.08
INM-1	7715	G26	3	21.30	15.58	9.56	15.17	12.08
INM-1	7893	G27	3	00.00	00.00	11.60	18.27	13.65
INM-1	7892	G27	3	00.00	00.00	10.70	17.75	13.00
INM-1	7891	G27	3	00.00	00.00	11.57	18.10	12.65
INM-1	7890	G27	3	00.00	00.00	10.85	17.30	12.30
INM-1	7928	Hno103	3	23.65	16.55	10.40	16.50	12.03
INM-1	7929	Hno103	3	00.00	00.00	10.40	16.74	11.50
INM-1	6183	Hno106	3	00.00	00.00	9.82	15.82	11.25
INM-1	3656	Hno127	3	00.00	00.00	11.80	18.40	13.00
INM-1	974	C1	4	00.00	00.00	10.86	00.00	13.62

INM-1	3056	H7	4	00.00	00.00	10.12	16.84	11.96
INM-1	3057	H7	4	00.00	00.00	10.22	16.03	12.01
INM-1	8127	E24	4	00.00	00.00	10.55	16.90	12.23
INM-1	8125	E24	4	00.00	00.00	10.45	17.25	13.85
INM-1	8126	E24	4	00.00	00.00	10.20	17.10	12.75
INM-1	8124	E24	4	00.00	00.00	10.25	16.95	13.57
INM-1	8020	E24	4	00.00	00.00	9.95	16.65	13.05
INM-1	8019	E24	4	00.00	00.00	10.58	17.15	12.80
INM-1	8123	E24	4	24.35	18.03	11.75	17.95	12.80
INM-1	8031	E24	4	00.00	00.00	9.50	16.05	12.00
INM-1	5887	E25	4	00.00	00.00	11.30	18.00	12.70
INM-1	7978	F24	4	00.00	00.00	10.13	16.25	12.45
INM-1	7979	F24	4	00.00	00.00	10.95	18.05	12.85
INM-1	7981	F24	4	00.00	00.00	9.95	15.65	10.70
INM-1	7973	F24	4	00.00	00.00	10.65	16.95	12.43
INM-1	7972	F24	4	23.70	15.90	10.05	15.85	12.15
INM-1	7980	F24	4	00.00	00.00	9.90	16.05	11.35
INM-1	7975	F24	4	00.00	00.00	11.05	18.10	13.17
INM-1	7977	F24	4	00.00	00.00	11.60	19.40	14.45
INM-1	7976	F24	4	00.00	00.00	11.30	16.95	14.10
INM-1	5974	F25	4	00.00	00.00	11.00	17.55	12.90
INM-1	5971	F25	4	00.00	00.00	11.87	18.70	13.85
INM-1	5975	F25	4	00.00	00.00	11.10	18.10	13.45
INM-1	5973	F25	4	00.00	00.00	10.70	18.25	13.90
INM-1	5972	F25	4	00.00	00.00	10.52	16.70	12.80
INM-1	1864	G11	4	00.00	00.00	10.07	16.64	12.36
INM-1	7121	G24	4	00.00	00.00	9.45	15.85	12.45
INM-1	7120	G24	4	00.00	00.00	9.95	16.35	11.84
INM-1	7119	G24	4	00.00	00.00	9.66	15.30	10.45
INM-1	7118	G24	4	00.00	00.00	10.30	15.90	11.70
INM-1	7117	G24	4	00.00	00.00	10.65	16.50	12.50
INM-1	7116	G24	4	00.00	00.00	10.84	18.23	14.00
INM-1	7115	G24	4	00.00	00.00	10.95	17.46	12.80
INM-1	7114	G24	4	00.00	00.00	10.39	17.25	11.90
INM-1	7142	G24	4	23.35	17.05	10.35	16.65	11.70
INM-1	7112	G24	4	00.00	00.00	11.60	18.25	13.67
INM-1	7111	G24	4	00.00	00.00	11.05	17.70	12.40
INM-1	7110	G24	4	00.00	00.00	11.35	18.09	12.90
INM-1	7109	G24	4	00.00	00.00	10.30	17.80	12.60
INM-1	7113	G24	4	00.00	00.00	11.00	18.52	13.85
INM-1	7633	G25	4	00.00	00.00	12.26	18.25	13.60
INM-1	7632	G25	4	00.00	00.00	10.25	15.73	11.35
INM-1	7634	G25	4	00.00	00.00	11.18	17.95	13.30
INM-1	7635	G25	4	00.00	00.00	10.90	17.57	13.75
INM-1	7638	G25	4	25.30	16.55	11.15	16.55	12.82
INM-1	6821	G26	4	00.00	00.00	11.20	18.02	13.00
INM-1	6823	G26	4	00.00	00.00	10.75	17.05	12.50
INM-1	6822	G26	4	00.00	00.00	10.00	16.40	11.90
INM-1	6824	G26	4	00.00	00.00	9.75	16.27	12.26
INM-1	6825	G26	4	00.00	00.00	11.05	18.10	13.40
INM-1	7240	Hno104	4	00.00	00.00	10.75	17.40	13.80
INM-1	7239	Hno104	4	00.00	00.00	10.70	17.25	13.35
INM-1	7238	Hno104	4	00.00	00.00	10.80	17.10	13.15
INM-1	7241	Hno104	4	00.00	00.00	10.35	16.70	12.46
INM-1	4650	Hno111	4	00.00	00.00	10.55	16.67	12.02
INM-1	4653	Hno111	4	00.00	00.00	10.08	16.18	12.70
INM-1	4651	Hno111	4	00.00	00.00	10.25	17.35	12.27
INM-1	4652	Hno111	4	00.00	00.00	10.33	16.76	11.88
INM-1	4654	Hno111	4	00.00	00.00	11.25	17.45	12.65
INM-1	4063	Hno115	4	00.00	00.00	9.57	15.43	11.25
INM-1	4062	Hno115	4	00.00	00.00	10.40	17.00	11.90
INM-1	4757	Hno117	4	00.00	00.00	11.20	16.65	12.51
INM-1	4626	B2	5	00.00	00.00	10.25	15.44	11.15

INM-1	4335	F8	5	00.00	00.00	11.10	18.75	13.75
INM-1	4334	F8	5	00.00	00.00	10.15	16.45	13.17
INM-1	4331	F8	5	00.00	00.00	10.58	16.73	12.25
INM-1	4332	F8	5	00.00	00.00	11.30	17.64	12.90
INM-1	3772	F8	6	00.00	00.00	10.75	17.70	12.20
INM-1	1837	G4	7	00.00	00.00	10.79	16.65	12.41
INM-1	814	C1	12	00.00	00.00	10.51	17.69	11.97
INM-1	815	C1	12	00.00	00.00	11.24	18.08	12.71
INM-1	529	Hno50	12	00.00	00.00	10.10	16.36	12.60
INM-1	527	Hno50	12	00.00	00.00	10.80	17.09	12.32

Table A.16.8. Measurements of Pelvis of ***Antilope cervicapra***

Site	Reg.No.	Trench	Layer	LA
INM-1	464	LJP	1-4	32.13
INM-1	467	LJP	1-4	33.78
INM-1	469	LJP	1-4	34.54
INM-1	470	LJP	1-4	32.68
INM-1	468	E24	1	32.77
INM-1	7382	F26	1	34.35
INM-1	1585	H5	2	31.71
INM-1	471	C23	2	33.19
INM-1	1474	D25	2	28.08
INM-1	472	E25	2	32.11
INM-1	2586	F11	2	33.55
INM-1	2587	F11	2	31.48
INM-1	465	F20	2	33.67
INM-1	6640	F24	2	32.55
INM-1	3220	G11	2	31.11
INM-1	3221	G11	2	32.75
INM-1	2771	H11	2	31.75
INM-1	6952	E26	3	29.25
INM-1	466	F27	3	33.82
INM-1	7854	G25	3	34.50
INM-1	7855	G25	3	32.26
INM-1	7856	G25	3	34.55
INM-1	1788	H11	3	31.44
INM-1	4526	H12	3	34.10
INM-1	4527	H12	3	33.70
INM-1	463		4	33.09
INM-1	8023	E24	4	29.20
INM-1	5869	E27	4	33.35
INM-1	1870	G11	4	32.29
INM-1	1871	G11	4	33.62
INM-1	7101	G24	4	29.05
INM-1	473		11	34.06

Table A.16.9. Measurements of Femur of ***Antilope cervicapra***

Site	Reg.No.	Trench	Layer	BP	TP	BD	TD	LC	TC
INM-1	291	LJP	1-4	49.19	00.00	00.00	00.00	30.62	21.94
INM-1	394	LJP	1-4	00.00	00.00	00.00	00.00	27.95	20.80
INM-1	395	LJP	1-4	00.00	00.00	39.38	00.00	00.00	00.00
INM-1	5283	LJP	1-4	54.05	22.76	00.00	00.00	00.00	00.00
INM-1	5284	LJP	1-4	00.00	00.00	34.97	00.00	00.00	00.00
INM-1	1010	E10	1	54.75	00.00	00.00	00.00	27.55	00.00
INM-1	1012	E10	1	00.00	00.00	00.00	00.00	00.00	18.74
INM-1	1673	G11	1	50.20	23.47	00.00	00.00	28.59	21.35
INM-1	1943	H11	1	44.06	25.02	00.00	00.00	26.14	19.85
INM-1	2613	G6	1	51.39	24.29	00.00	00.00	28.40	19.45
INM-1	3541	H5	1	00.00	00.00	37.25	56.37	00.00	00.00
INM-1	4914	F13	1	52.52	23.07	00.00	00.00	28.00	21.60

INM-1	4915	F13	1	52.95	22.80	00.00	00.00	28.65	21.55
INM-1	199	C23	2	52.81	00.00	00.00	00.00	30.00	19.57
INM-1	286	F25	2	00.00	00.00	40.69	00.00	00.00	00.00
INM-1	288	G25	2	00.00	00.00	35.95	00.00	00.00	00.00
INM-1	847	F12	2	53.30	25.10	00.00	00.00	00.00	00.00
INM-1	1089	H7	2	37.80	00.00	00.00	00.00	26.99	19.58
INM-1	1575	H5	2	52.54	23.31	00.00	00.00	27.40	20.36
INM-1	2098	G6	2	00.00	00.00	36.63	52.25	00.00	00.00
INM-1	2099	G6	2	00.00	00.00	00.00	00.00	00.00	21.08
INM-1	2341		2	37.52	50.80	00.00	00.00	00.00	00.00
INM-1	2559	F11	2	00.00	00.00	39.68	00.00	00.00	00.00
INM-1	2772	H11	2	00.00	00.00	39.13	50.19	00.00	00.00
INM-1	2773	H11	2	00.00	00.00	41.51	55.69	00.00	00.00
INM-1	2774	H11	2	00.00	00.00	37.29	50.15	00.00	00.00
INM-1	2775	H11	2	00.00	00.00	00.00	00.00	30.81	21.07
INM-1	3003	F10	2	50.00	27.97	00.00	00.00	28.30	19.57
INM-1	3588	I6	2	00.00	00.00	00.00	00.00	27.15	21.12
INM-1	3985	G4	2	52.90	30.00	00.00	00.00	23.35	21.70
INM-1	3986	G4	2	00.00	00.00	33.18	39.45	00.00	00.00
INM-1	6432	E25	2	44.95	25.75	00.00	00.00	25.30	18.30
INM-1	7312	E24	2	00.00	00.00	40.60	53.40	00.00	00.00
INM-1	7313	E24	2	00.00	00.00	00.00	50.85	00.00	00.00
INM-1	922	G12	3	00.00	00.00	40.23	00.00	00.00	00.00
INM-1	2674	G7	3	53.10	26.14	00.00	00.00	29.34	21.61
INM-1	2675	G7	3	00.00	00.00	00.00	00.00	26.91	19.53
INM-1	3465	H7	3	00.00	00.00	00.00	00.00	29.65	20.80
INM-1	3597	I6	3	51.17	00.00	00.00	00.00	28.37	00.00
INM-1	6784	E23	3	00.00	00.00	00.00	00.00	24.05	19.20
INM-1	7643	G25	4	00.00	00.00	00.00	56.08	00.00	00.00
INM-1	8030	E24	4	00.00	00.00	00.00	00.00	29.70	22.45
INM-1	8115	E24	4	00.00	00.00	00.00	00.00	26.65	10.96
INM-1	4118	C3	5	00.00	00.00	00.00	00.00	22.45	18.55
INM-1	4220	A3	6	00.00	00.00	00.00	00.00	18.86	16.95
INM-1	4772	A5	7	00.00	00.00	00.00	00.00	23.60	20.05
INM-1	6124	Hno51	7	50.70	00.00	00.00	00.00	00.00	00.00
INM-1	6227	B6	12	00.00	00.00	00.00	00.00	18.70	18.05
INM-1	6329	B5	12	00.00	00.00	30.80	34.50	00.00	00.00
INM-1	6267	E9	14	00.00	00.00	36.15	00.00	00.00	00.00
INM-1	5481	D3	15	00.00	00.00	36.90	50.35	00.00	00.00
INM-1	6050	Hno87	16	00.00	00.00	00.00	00.00	19.00	15.20

Table A.16.10. Measurements of Tibia of ***Antilope cervicapra***

Site	Reg.No.	Trench	Layer	DD	BP	TP	BD	TD
INM-1	38	LJP	1-4	00.00	00.00	00.00	25.39	10.16
INM-1	53	LJP	1-4	00.00	00.00	00.00	27.61	21.78
INM-1	137	LJP	1-4	00.00	00.00	00.00	22.97	16.93
INM-1	139	LJP	1-4	00.00	00.00	00.00	25.47	19.09
INM-1	140	LJP	1-4	00.00	00.00	00.00	27.32	21.52
INM-1	142	LJP	1-4	00.00	00.00	00.00	26.31	20.75
INM-1	144	LJP	1-4	00.00	00.00	00.00	24.67	18.58
INM-1	145	LJP	1-4	00.00	00.00	00.00	23.36	17.98
INM-1	147	LJP	1-4	00.00	00.00	00.00	23.50	17.57
INM-1	266	LJP	1-4	00.00	00.00	00.00	27.20	20.90
INM-1	268	LJP	1-4	00.00	00.00	00.00	26.88	21.25
INM-1	269	LJP	1-4	25.27	19.29	00.00	00.00	00.00
INM-1	354	LJP	1-4	00.00	44.72	47.44	00.00	00.00
INM-1	356	LJP	1-4	00.00	43.83	48.45	00.00	00.00
INM-1	40	F27	1	17.45	00.00	00.00	27.25	21.90
INM-1	48	F25	1	00.00	00.00	00.00	19.25	22.41
INM-1	1100	G4	1	00.00	00.00	00.00	24.71	20.16
INM-1	1101	G4	1	00.00	00.00	00.00	30.07	22.34

INM-1	1102	G4	1	00.00	40.16	00.00	00.00	00.00
INM-1	1167	G10	1	00.00	00.00	00.00	26.15	21.79
INM-1	1193	F10	1	00.00	00.00	00.00	23.74	28.32
INM-1	1678	G11	1	00.00	00.00	00.00	26.71	20.98
INM-1	1679	G11	1	00.00	00.00	00.00	25.45	20.47
INM-1	1945	H11	1	00.00	00.00	00.00	25.55	20.43
INM-1	2049	G7	1	00.00	00.00	00.00	28.37	28.29
INM-1	2614	G6	1	00.00	00.00	00.00	26.11	20.65
INM-1	2615	G6	1	00.00	00.00	00.00	24.31	20.35
INM-1	2935	H10	1	00.00	00.00	00.00	28.58	21.50
INM-1	3338	F11	1	00.00	00.00	00.00	27.40	22.64
INM-1	3339	F11	1	00.00	00.00	00.00	29.90	25.53
INM-1	3852	D25	1	00.00	00.00	00.00	29.65	22.75
INM-1	7054	E23	1	00.00	00.00	00.00	00.00	18.00
INM-1	7055	E23	1	00.00	37.90	39.30	00.00	00.00
INM-1	8060	F24	1	00.00	00.00	00.00	24.95	20.50
INM-1	8061	F27	1	00.00	00.00	00.00	26.40	20.05
INM-1	8062	F27	1	00.00	00.00	00.00	27.20	20.50
INM-1	2162	HNO132	1	17.98	00.00	00.00	28.31	21.32
INM-1	138	G25	2	00.00	00.00	00.00	26.90	21.80
INM-1	203	G25	2	00.00	00.00	00.00	27.37	21.41
INM-1	264	E25	2	00.00	00.00	00.00	26.44	00.00
INM-1	265	E25	2	00.00	00.00	00.00	26.78	21.86
INM-1	267	E25	2	00.00	00.00	00.00	26.89	2090
INM-1	357	E25	2	00.00	00.00	00.00	26.74	20.56
INM-1	383	E25	2	00.00	00.00	00.00	27.65	22.18
INM-1	1070	H7	2	00.00	00.00	00.00	26.64	20.71
INM-1	1071	H7	2	00.00	00.00	00.00	27.86	21.14
INM-1	1139	E10	2	00.00	39.97	46.04	00.00	00.00
INM-1	1335	G11	2	00.00	00.00	00.00	28.69	23.78
INM-1	1339	G11	2	00.00	00.00	00.00	26.31	19.70
INM-1	1340	G11	2	00.00	00.00	00.00	28.14	22.18
INM-1	1341	G11	2	00.00	00.00	00.00	23.62	18.38
INM-1	1468	D25	2	00.00	00.00	00.00	26.47	20.10
INM-1	1576	H5	2	00.00	00.00	00.00	30.93	23.82
INM-1	1577	H5	2	00.00	00.00	00.00	26.39	22.07
INM-1	1578	H5	2	00.00	00.00	00.00	27.01	13.62
INM-1	2106	G6	2	00.00	00.00	00.00	00.00	21.78
INM-1	2284	E1	2	00.00	00.00	00.00	28.01	21.11
INM-1	2342		2	00.00	00.00	00.00	24.24	21.03
INM-1	2344		2	00.00	00.00	46.90	00.00	00.00
INM-1	2557	F11	2	00.00	00.00	00.00	25.77	19.43
INM-1	2558	F11	2	00.00	00.00	00.00	24.11	18.70
INM-1	2777	H11	2	00.00	00.00	00.00	29.65	22.55
INM-1	3004	F10	2	00.00	42.85	00.00	00.00	00.00
INM-1	3005	F10	2	00.00	00.00	00.00	26.90	21.85
INM-1	3006	F10	2	00.00	00.00	00.00	22.80	17.47
INM-1	3178	G11	2	00.00	40.17	00.00	00.00	00.00
INM-1	3179	G11	2	00.00	43.14	00.00	00.00	00.00
INM-1	3180	G11	2	00.00	00.00	00.00	25.00	20.25
INM-1	3181	G11	2	00.00	00.00	00.00	28.35	22.36
INM-1	3182	G11	2	00.00	00.00	00.00	28.03	22.09
INM-1	3396	H6	2	00.00	43.45	00.00	00.00	00.00
INM-1	3397	H6	2	00.00	42.15	45.72	00.00	00.00
INM-1	3567	I10	2	00.00	42.80	48.86	00.00	00.00
INM-1	3987	G4	2	00.00	00.00	00.00	27.15	22.08
INM-1	3988	G4	2	00.00	00.00	00.00	29.00	22.25
INM-1	3989	G4	2	00.00	00.00	00.00	27.47	20.45
INM-1	3990	G4	2	00.00	00.00	00.00	26.95	21.25
INM-1	4450	B7	2	00.00	00.00	00.00	25.20	17.85
INM-1	6433	E25	2	00.00	00.00	00.00	26.25	21.20
INM-1	6434	E25	2	00.00	00.00	00.00	26.55	20.85
INM-1	6435	E25	2	00.00	00.00	00.00	26.45	17.80

INM-1	6436	E25	2	00.00	00.00	00.00	26.55	21.31
INM-1	7446	E25	2	00.00	00.00	00.00	27.60	22.10
INM-1	6637	F24	2	00.00	00.00	00.00	26.15	20.00
INM-1	6638	F24	2	00.00	00.00	00.00	26.35	20.65
INM-1	6639	F24	2	00.00	00.00	00.00	29.05	22.50
INM-1	7016	G24	2	00.00	00.00	00.00	27.70	22.40
INM-1	7184	G25	2	00.00	00.00	00.00	26.82	20.50
INM-1	7314	E24	2	00.00	40.55	00.00	00.00	00.00
INM-1	7315	E24	2	00.00	00.00	00.00	25.65	20.90
INM-1	7316	E24	2	00.00	00.00	00.00	25.75	19.90
INM-1	7495	F24	2	00.00	00.00	00.00	26.40	21.73
INM-1	7574	F26	2	00.00	00.00	00.00	26.85	21.02
INM-1	916	G12	3	00.00	00.00	00.00	25.74	20.62
INM-1	917	G12	3	00.00	00.00	00.00	28.19	22.29
INM-1	918	G12	3	00.00	00.00	00.00	24.96	19.08
INM-1	919	G12	3	00.00	43.82	45.57	00.00	00.00
INM-1	920	G12	3	00.00	41.02	48.49	00.00	00.00
INM-1	921	G12	3	00.00	44.26	00.00	00.00	00.00
INM-1	1254	D25	3	00.00	00.00	00.00	27.49	21.62
INM-1	1302	G11	3	00.00	00.00	00.00	27.09	21.49
INM-1	1303	G11	3	00.00	00.00	00.00	27.97	21.62
INM-1	1773	H11	3	00.00	42.10	37.46	00.00	00.00
INM-1	1774	H11	3	00.00	00.00	00.00	26.09	20.62
INM-1	2013	D24	3	00.00	00.00	00.00	25.79	19.49
INM-1	2203	G10	3	00.00	00.00	00.00	27.97	21.54
INM-1	2670	G7	3	00.00	00.00	00.00	21.69	16.99
INM-1	2671	G7	3	00.00	00.00	00.00	23.39	20.55
INM-1	2672	G7	3	00.00	00.00	00.00	27.94	21.55
INM-1	2673	G7	3	00.00	41.33	43.33	00.00	00.00
INM-1	3463	H7	3	00.00	00.00	00.00	26.70	21.05
INM-1	3756	G11	3	00.00	00.00	00.00	25.54	19.11
INM-1	3757	G11	3	00.00	00.00	00.00	26.60	22.25
INM-1	3790	H12	3	00.00	45.00	00.00	00.00	00.00
INM-1	4528	H12	3	00.00	42.90	45.50	00.00	00.00
INM-1	4529	H12	3	00.00	00.00	00.00	27.63	22.20
INM-1	4530	H12	3	00.00	00.00	00.00	27.40	21.08
INM-1	4531	H12	3	00.00	00.00	00.00	25.16	20.30
INM-1	6518	F25	3	00.00	00.00	00.00	27.70	21.90
INM-1	6519	F25	3	00.00	00.00	00.00	27.35	21.40
INM-1	6520	F25	3	00.00	00.00	00.00	25.60	21.25
INM-1	6521	F25	3	00.00	00.00	00.00	26.58	21.43
INM-1	6522	F25	3	00.00	00.00	00.00	21.61	16.90
INM-1	6729	E23	3	00.00	00.00	00.00	26.15	20.41
INM-1	6761	E23	3	00.00	00.00	00.00	27.70	21.60
INM-1	6945	E26	3	00.00	00.00	00.00	27.45	21.70
INM-1	6946	E26	3	00.00	00.00	00.00	25.95	20.65
INM-1	6947	E26	3	00.00	00.00	00.00	25.05	21.95
INM-1	6948	E26	3	00.00	00.00	00.00	25.65	19.45
INM-1	7834	G25	3	00.00	00.00	00.00	28.05	21.25
INM-1	7835	G25	3	00.00	00.00	00.00	27.00	21.00
INM-1	7836	G25	3	00.00	00.00	00.00	28.80	22.40
INM-1	7878	G27	3	00.00	00.00	00.00	28.00	21.85
INM-1	7898	G27	3	00.00	38.60	00.00	00.00	00.00
INM-1	6187	Hno106	3	00.00	00.00	00.00	28.65	22.06
INM-1	382	E25	4	00.00	44.14	00.00	00.00	00.00
INM-1	666	Hno110	4	00.00	00.00	00.00	27.40	21.48
INM-1	5890	E25	4	00.00	00.00	00.00	28.60	22.15
INM-1	6827	G26	4	00.00	00.00	00.00	26.25	21.53
INM-1	6828	G26	4	00.00	00.00	00.00	26.60	21.45
INM-1	6829	G26	4	00.00	00.00	00.00	26.20	20.30
INM-1	7093	G24	4	00.00	00.00	00.00	26.75	20.78
INM-1	7094	G24	4	00.00	43.10	00.00	00.00	00.00
INM-1	7095	G24	4	00.00	00.00	00.00	25.55	19.65

INM-1	7096	G24	4	00.00	00.00	00.00	27.80	20.65
INM-1	7097	G24	4	00.00	00.00	00.00	27.67	21.60
INM-1	7098	G24	4	00.00	00.00	00.00	28.45	23.10
INM-1	7099	G24	4	00.00	00.00	00.00	27.95	21.45
INM-1	7644	G25	4	00.00	00.00	00.00	28.95	21.25
INM-1	7645	G25	4	00.00	00.00	00.00	25.70	19.75
INM-1	8004	F24	4	00.00	00.00	00.00	26.85	21.45
INM-1	8021	E24	4	00.00	00.00	00.00	26.50	19.35
INM-1	8113	E24	4	00.00	00.00	00.00	27.85	22.35
INM-1	8114	E24	4	00.00	00.00	00.00	27.80	22.65
INM-1	5021	Hno122	4	00.00	41.40	00.00	00.00	00.00
INM-1	4342	F8	5	00.00	00.00	00.00	27.95	21.65
INM-1	4168	B1	6	00.00	00.00	00.00	27.35	20.70
INM-1	4898	C0	7	00.00	45.15	45.20	00.00	00.00
INM-1	271		8	00.00	00.00	00.00	24.59	18.18
INM-1	355	C24	8	00.00	42.24	46.86	00.00	00.00
INM-1	684	A1	11	00.00	00.00	21.04	00.00	00.00
INM-1	6226	B6	12	00.00	24.76	18.35	00.00	00.00
INM-1	5643	A4	13	12.18	00.00	00.00	24.15	18.18
INM-1	5309	E3	14	00.00	00.00	00.00	27.61	21.03
INM-1	6342	E4	14	00.00	00.00	00.00	24.10	17.85
INM-1	5403	C3	15	00.00	00.00	00.00	27.30	21.45
INM-1	5431	E4	15	00.00	00.00	00.00	28.30	20.70
INM-1	6047	Hno87	16	00.00	41.75	44.50	00.00	00.00

Table A.16.11. Measurements of Metatarsal of ***Antilope cervicapra***

Site	Reg.No.	Trench	Layer	DD	BP	TP	BD	TD	WT	TT	TLT
INM-1	59	LJP	1-4	14.01	00.00	00.00	25.43	00.00	16.52	18.82	13.38
INM-1	80	LJP	1-4	00.00	00.00	00.00	23.72	00.00	10.95	17.65	12.80
INM-1	172	LJP	1-4	00.00	00.00	00.00	00.00	00.00	11.42	17.54	12.75
INM-1	173	LJP	1-4	00.00	00.00	00.00	22.76	00.00	9.79	15.97	11.34
INM-1	175	LJP	1-4	00.00	00.00	00.00	24.26	00.00	10.37	16.64	12.25
INM-1	176	LJP	1-4	00.00	00.00	00.00	24.37	00.00	10.59	17.17	13.55
INM-1	177	LJP	1-4	00.00	00.00	00.00	23.67	00.00	10.56	16.56	11.70
INM-1	319	LJP	1-4	00.00	00.00	00.00	24.87	00.00	11.34	17.55	12.87
INM-1	320	LJP	1-4	00.00	00.00	00.00	23.71	00.00	11.23	17.48	12.54
INM-1	321	LJP	1-4	00.00	00.00	00.00	25.15	00.00	11.67	18.74	13.78
INM-1	322	LJP	1-4	00.00	00.00	00.00	22.73	00.00	10.47	16.60	12.60
INM-1	323	LJP	1-4	00.00	00.00	00.00	22.54	00.00	10.21	16.66	12.37
INM-1	324	LJP	1-4	00.00	00.00	00.00	22.55	00.00	9.16	16.22	11.81
INM-1	325	LJP	1-4	00.00	00.00	00.00	23.05	00.00	10.58	16.88	12.70
INM-1	326	LJP	1-4	00.00	00.00	00.00	00.00	00.00	10.78	17.01	11.70
INM-1	388	LJP	1-4	00.00	00.00	00.00	25.02	00.00	11.30	18.51	13.50
INM-1	5278	LJP	1-4	00.00	22.65	25.00	00.00	00.00	00.00	00.00	00.00
INM-1	5279	LJP	1-4	00.00	22.75	25.45	00.00	00.00	00.00	00.00	00.00
INM-1	2721	G6	1	00.00	22.72	24.31	00.00	00.00	00.00	00.00	00.00
INM-1	2724	G6	1	00.00	00.00	00.00	22.43	17.10	10.50	16.96	11.82
INM-1	2725	G6	1	00.00	00.00	00.00	00.00	00.00	11.29	17.22	11.64
INM-1	3542	H5	1	00.00	22.79	25.90	00.00	00.00	00.00	00.00	00.00
INM-1	3283	D10	1	00.00	22.60	23.35	00.00	00.00	00.00	00.00	00.00
INM-1	3284	D10	1	00.00	22.95	23.53	00.00	00.00	00.00	00.00	00.00
INM-1	3285	D10	1	00.00	00.00	00.00	22.40	16.67	10.30	16.18	11.85
INM-1	3286	D10	1	00.00	00.00	00.00	24.15	18.55	10.90	18.24	13.83
INM-1	3842	D25	1	00.00	22.55	24.40	00.00	00.00	00.00	00.00	00.00
INM-1	3843	D25	1	00.00	00.00	00.00	22.70	16.75	10.00	16.00	11.27
INM-1	3844	D25	1	00.00	00.00	00.00	23.50	16.86	10.25	16.15	11.45
INM-1	3866	D25	1	00.00	20.50	22.75	00.00	00.00	00.00	00.00	00.00
INM-1	1004	E10	1	00.00	22.76	24.07	00.00	00.00	00.00	00.00	00.00
INM-1	1006	E10	1	00.00	00.00	00.00	22.76	17.24	10.59	16.46	12.47
INM-1	7050	E23	1	00.00	21.45	24.35	00.00	00.00	00.00	00.00	00.00
INM-1	1180	F10	1	00.00	00.00	00.00	24.60	17.57	11.44	17.44	13.04

INM-1	1183	F10	1	00.00	00.00	00.00	22.70	16.71	10.66	16.29	12.16
INM-1	184	F10	1	00.00	00.00	00.00	25.19	18.49	10.98	16.60	13.02
INM-1	1185	F10	1	00.00	00.00	00.00	00.00	00.00	10.04	16.42	13.03
INM-1	1186	F10	1	00.00	00.00	00.00	00.00	00.00	11.23	17.42	12.63
INM-1	1196	F10	1	00.00	24.35	25.73	00.00	00.00	00.00	00.00	00.00
INM-1	3343	F11	1	00.00	23.19	00.00	00.00	00.00	00.00	00.00	00.00
INM-1	3344	F11	1	00.00	00.00	00.00	23.65	17.25	10.95	17.00	12.86
INM-1	3345	F11	1	00.00	00.00	00.00	25.10	19.25	11.35	18.55	13.80
INM-1	3346	F11	1	00.00	00.00	00.00	24.02	17.95	10.72	17.45	12.83
INM-1	3347	F11	1	00.00	00.00	00.00	23.45	16.43	10.30	16.33	11.90
INM-1	3348	F11	1	00.00	00.00	00.00	26.36	19.50	11.50	19.05	13.78
INM-1	3349	F11	1	00.00	00.00	00.00	24.22	18.20	10.80	17.69	13.08
INM-1	3354	F11	1	00.00	00.00	00.00	00.00	00.00	10.00	15.30	11.35
INM-1	4910	F13	1	00.00	00.00	00.00	24.34	17.45	11.42	16.80	13.00
INM-1	8051	F24	1	00.00	24.10	17.20	00.00	00.00	00.00	00.00	00.00
INM-1	8052	F24	1	00.00	00.00	00.00	23.60	17.20	10.60	16.85	13.05
INM-1	8053	F24	1	00.00	00.00	00.00	24.75	18.75	11.00	18.10	13.20
INM-1	7276	F26	1	00.00	00.00	00.00	24.93	17.60	11.10	17.45	13.35
INM-1	174	F27	1	00.00	00.00	00.00	22.91	00.00	10.13	15.69	11.32
INM-1	318	F27	1	00.00	00.00	00.00	22.49	00.00	9.38	15.98	11.18
INM-1	1686	G11	1	00.00	21.96	25.20	00.00	00.00	00.00	00.00	00.00
INM-1	1687	G11	1	00.00	00.00	00.00	23.68	17.59	10.79	17.49	12.77
INM-1	1688	G11	1	00.00	00.00	00.00	22.11	16.88	9.90	15.85	11.47
INM-1	1689	G11	1	00.00	00.00	00.00	25.93	18.20	11.76	18.21	13.49
INM-1	2944	H10	1	00.00	00.00	00.00	00.00	00.00	11.66	17.14	12.70
INM-1	2953	H10	1	00.00	21.83	25.96	00.00	00.00	00.00	00.00	00.00
INM-1	1952	H11	1	00.00	22.95	23.62	00.00	00.00	00.00	00.00	00.00
INM-1	1953	H11	1	00.00	21.92	23.16	00.00	00.00	00.00	00.00	00.00
INM-1	1958	H11	1	00.00	00.00	00.00	24.08	17.79	10.41	16.92	12.38
INM-1	1964	H11	1	00.00	00.00	00.00	00.00	00.00	10.25	15.86	12.05
INM-1	1966	H11	1	00.00	00.00	00.00	00.00	00.00	9.74	15.29	11.51
INM-1	2149	Hno132	1	00.00	21.07	22.78	00.00	00.00	00.00	00.00	00.00
INM-1	2150	Hno132	1	00.00	00.00	00.00	25.32	18.63	11.78	18.64	13.41
INM-1	2151	Hno132	1	00.00	00.00	00.00	21.73	16.91	10.18	16.45	12.38
INM-1	2152	Hno132	1	00.00	00.00	00.00	26.12	16.88	11.14	16.44	13.15
INM-1	2153	Hno132	1	00.00	00.00	00.00	26.36	18.05	11.35	17.90	13.25
INM-1	2154	Hno132	1	00.00	00.00	00.00	24.38	18.02	10.70	17.88	13.88
INM-1	2155	Hno132	1	00.00	00.00	00.00	24.12	18.15	11.04	17.78	13.79
INM-1	2156	Hno132	1	00.00	00.00	00.00	00.00	00.00	10.77	17.24	12.09
INM-1	2349		2	00.00	00.00	00.00	24.59	17.33	11.61	17.21	13.81
INM-1	2350		2	00.00	00.00	00.00	24.33	17.93	11.24	17.74	12.77
INM-1	2351		2	00.00	00.00	00.00	23.29	16.54	10.55	16.25	11.69
INM-1	3960	G4	2	00.00	22.30	23.40	00.00	00.00	00.00	00.00	00.00
INM-1	3961	G4	2	00.00	21.90	23.65	00.00	00.00	00.00	00.00	00.00
INM-1	3962	G4	2	00.00	00.00	00.00	23.05	17.90	10.73	17.58	12.55
INM-1	3969	G4	2	00.00	00.00	00.00	00.00	00.00	10.17	16.15	11.30
INM-1	2090	G6	2	00.00	24.03	26.67	00.00	00.00	00.00	00.00	00.00
INM-1	2091	G6	2	00.00	22.71	24.92	00.00	00.00	00.00	00.00	00.00
INM-1	2093	G6	2	00.00	00.00	00.00	25.73	18.59	10.81	18.10	13.17
INM-1	2094	G6	2	00.00	00.00	00.00	24.28	18.71	10.55	17.73	12.47
INM-1	2095	G6	2	00.00	00.00	00.00	26.02	18.48	11.17	17.22	12.01
INM-1	2096	G6	2	00.00	00.00	00.00	23.98	17.01	10.87	17.15	12.75
INM-1	3640	G7	2	00.00	00.00	00.00	25.08	18.70	10.90	17.80	12.15
INM-1	3644	G7	2	00.00	00.00	00.00	23.13	16.54	9.80	15.84	11.10
INM-1	1595	H5	2	00.00	22.02	25.09	00.00	00.00	00.00	00.00	00.00
INM-1	1596	H5	2	00.00	20.54	23.15	00.00	00.00	00.00	00.00	00.00
INM-1	1605	H5	2	00.00	00.00	00.00	23.60	18.69	11.13	18.50	13.41
INM-1	1606	H5	2	00.00	00.00	00.00	23.63	17.53	10.71	17.06	12.99
INM-1	1607	H5	2	00.00	00.00	00.00	23.51	17.57	10.75	17.20	13.12
INM-1	1608	H5	2	00.00	00.00	00.00	24.94	18.89	11.77	18.68	13.65
INM-1	1609	H5	2	00.00	00.00	00.00	00.00	00.00	10.05	16.13	11.31
INM-1	1616	H5	2	00.00	00.00	00.00	00.00	00.00	11.29	17.52	12.87
INM-1	3406	H6	2	00.00	00.00	00.00	22.84	16.65	10.07	16.53	12.33

INM-1	3407	H6	2	00.00	00.00	00.00	22.76	17.69	10.35	17.65	12.41
INM-1	3408	H6	2	00.00	00.00	00.00	22.96	17.53	10.35	17.37	12.54
INM-1	3409	H6	2	00.00	00.00	00.00	23.78	17.51	10.67	17.40	12.90
INM-1	3410	H6	2	00.00	00.00	00.00	00.00	00.00	9.74	15.75	11.67
INM-1	1054	H7	2	00.00	00.00	00.00	24.32	17.47	11.23	17.13	12.75
INM-1	1055	H7	2	00.00	00.00	00.00	26.06	18.44	11.33	17.66	12.75
INM-1	1056	H7	2	00.00	00.00	00.00	25.01	18.71	10.96	17.29	12.60
INM-1	1994	D10	2	00.00	22.94	25.21	00.00	00.00	00.00	00.00	00.00
INM-1	1445	D25	2	00.00	00.00	00.00	22.52	17.12	10.27	16.60	12.32
INM-1	1446	D25	2	00.00	00.00	00.00	23.71	17.33	10.45	17.03	11.32
INM-1	1448	D25	2	00.00	00.00	00.00	22.31	16.00	10.03	15.29	11.81
INM-1	1451	D25	2	00.00	00.00	00.00	00.00	00.00	10.11	15.54	11.88
INM-1	1457	D25	2	00.00	00.00	00.00	00.00	00.00	10.99	17.34	12.41
INM-1	1458	D25	2	00.00	00.00	00.00	00.00	00.00	10.12	16.42	11.49
INM-1	1465	D25	2	00.00	00.00	00.00	00.00	00.00	9.29	16.38	12.14
INM-1	1466	D25	2	00.00	00.00	00.00	00.00	00.00	11.13	17.55	12.28
INM-1	1467	D25	2	00.00	00.00	00.00	00.00	00.00	10.88	16.35	12.39
INM-1	1479	D25	2	00.00	22.46	00.00	00.00	00.00	00.00	00.00	00.00
INM-1	7293	E24	2	00.00	21.61	23.65	00.00	00.00	00.00	00.00	00.00
INM-1	7294	E24	2	00.00	22.40	25.26	00.00	00.00	00.00	00.00	00.00
INM-1	7295	E24	2	00.00	23.10	26.35	00.00	00.00	00.00	00.00	00.00
INM-1	7296	E24	2	00.00	00.00	00.00	25.40	18.66	11.73	18.10	13.55
INM-1	7297	E24	2	00.00	00.00	00.00	00.00	00.00	10.93	17.13	12.19
INM-1	7298	E24	2	00.00	00.00	00.00	00.00	00.00	11.05	16.67	12.14
INM-1	7299	E24	2	00.00	00.00	00.00	00.00	00.00	11.27	17.23	12.80
INM-1	7300	E24	2	00.00	00.00	00.00	00.00	00.00	10.54	17.15	13.05
INM-1	6415	E25	2	00.00	00.00	00.00	23.00	16.70	10.40	16.55	12.40
INM-1	6416	E25	2	00.00	00.00	00.00	25.90	18.40	11.90	18.26	12.95
INM-1	6417	E25	2	00.00	00.00	00.00	27.05	18.95	11.65	18.70	13.90
INM-1	7426	E25	2	00.00	00.00	00.00	23.00	16.80	10.45	16.65	11.75
INM-1	7427	E25	2	00.00	00.00	00.00	00.00	00.00	10.06	15.96	12.10
INM-1	7430	E25	2	00.00	00.00	00.00	00.00	00.00	11.25	17.40	12.75
INM-1	3010	F10	2	00.00	00.00	00.00	23.49	17.99	10.78	17.53	13.11
INM-1	3011	F10	2	00.00	00.00	00.00	23.32	16.62	10.20	15.99	12.29
INM-1	2566	F11	2	00.00	22.73	25.47	00.00	00.00	00.00	00.00	00.00
INM-1	2567	F11	2	00.00	20.04	25.76	00.00	00.00	00.00	00.00	00.00
INM-1	2568	F11	2	00.00	21.80	23.13	00.00	00.00	00.00	00.00	00.00
INM-1	2570	F11	2	00.00	00.00	00.00	24.47	18.60	10.97	18.39	13.41
INM-1	2571	F11	2	00.00	00.00	00.00	25.69	19.34	11.41	18.26	13.45
INM-1	2572	F11	2	00.00	00.00	00.00	25.48	18.23	11.41	18.15	13.27
INM-1	2574	F11	2	00.00	00.00	00.00	00.00	00.00	10.20	16.71	11.46
INM-1	828	F12	2	00.00	00.00	00.00	23.12	17.70	11.28	17.49	12.75
INM-1	830	F12	2	00.00	00.00	00.00	22.79	16.80	10.05	15.87	12.21
INM-1	6621	F24	2	00.00	23.60	24.80	00.00	00.00	00.00	00.00	00.00
INM-1	6627	F24	2	00.00	00.00	00.00	25.05	17.65	10.70	17.20	12.75
INM-1	6628	F24	2	00.00	00.00	00.00	24.35	17.65	10.65	16.55	12.30
INM-1	6629	F24	2	00.00	00.00	00.00	25.60	18.82	11.25	18.05	13.23
INM-1	7491	F24	2	00.00	00.00	00.00	25.73	17.75	10.30	16.90	12.10
INM-1	7492	F24	2	00.00	00.00	00.00	24.55	17.10	11.30	17.10	12.85
INM-1	7565	F26	2	00.00	00.00	00.00	23.40	17.60	10.95	17.20	12.60
INM-1	7566	F26	2	00.00	00.00	00.00	24.60	18.10	10.68	17.93	13.38
INM-1	7567	F26	2	00.00	00.00	00.00	22.70	16.26	10.40	16.07	11.70
INM-1	1350	G11	2	00.00	00.00	00.00	23.25	17.04	10.84	16.95	12.76
INM-1	1351	G11	2	00.00	00.00	00.00	24.47	18.02	10.80	17.40	13.55
INM-1	1352	G11	2	00.00	00.00	00.00	22.76	16.82	10.62	16.15	11.96
INM-1	1355	G11	2	00.00	00.00	00.00	23.17	17.45	10.54	17.12	12.76
INM-1	1356	G11	2	00.00	00.00	00.00	00.00	00.00	10.60	16.81	11.74
INM-1	1357	G11	2	00.00	00.00	00.00	00.00	00.00	10.13	15.54	12.10
INM-1	1358	G11	2	00.00	00.00	00.00	00.00	00.00	10.31	15.53	11.84
INM-1	3196	G11	2	00.00	19.70	22.39	00.00	00.00	00.00	00.00	00.00
INM-1	3197	G11	2	00.00	20.93	23.14	00.00	00.00	00.00	00.00	00.00
INM-1	3198	G11	2	00.00	20.93	23.26	00.00	00.00	00.00	00.00	00.00
INM-1	3199	G11	2	00.00	00.00	00.00	23.46	18.19	10.48	17.06	12.30

INM-1	3200	G11	2	00.00	00.00	00.00	24.46	18.46	10.89	18.33	13.06
INM-1	3201	G11	2	00.00	00.00	00.00	23.87	17.07	11.11	17.04	12.52
INM-1	3202	G11	2	00.00	00.00	00.00	23.15	16.93	10.37	15.33	11.41
INM-1	3203	G11	2	00.00	00.00	00.00	24.04	18.69	10.27	17.38	12.36
INM-1	3207	G11	2	00.00	00.00	00.00	25.95	18.08	11.11	17.95	13.39
INM-1	3208	G11	2	00.00	00.00	00.00	23.57	17.34	10.27	16.19	11.27
INM-1	3209	G11	2	00.00	00.00	00.00	23.26	17.43	10.20	17.40	13.05
INM-1	3210	G11	2	00.00	00.00	00.00	23.40	16.44	10.05	15.81	11.86
INM-1	3212	G11	2	00.00	00.00	00.00	00.00	00.00	11.20	18.09	12.70
INM-1	3215	G11	2	00.00	00.00	00.00	00.00	00.00	10.95	17.45	13.15
INM-1	7025	G24	2	00.00	00.00	00.00	23.60	17.35	11.00	16.90	13.30
INM-1	7026	G24	2	00.00	00.00	00.00	24.30	17.40	10.90	16.98	12.60
INM-1	7027	G24	2	00.00	00.00	00.00	00.00	00.00	10.40	15.65	11.40
INM-1	7175	G25	2	00.00	00.00	00.00	26.17	19.95	12.20	18.75	14.45
INM-1	2784	H11	2	00.00	23.98	26.84	00.00	00.00	00.00	00.00	00.00
INM-1	2785	H11	2	00.00	23.68	26.15	00.00	00.00	00.00	00.00	00.00
INM-1	2786	H11	2	00.00	21.33	24.50	00.00	00.00	00.00	00.00	00.00
INM-1	2791	H11	2	00.00	00.00	00.00	23.96	17.20	11.10	16.67	13.10
INM-1	2792	H11	2	00.00	00.00	00.00	23.81	18.78	11.21	18.46	13.47
INM-1	2793	H11	2	00.00	00.00	00.00	23.43	16.74	10.85	16.83	12.53
INM-1	2794	H11	2	00.00	00.00	00.00	25.95	17.45	11.70	17.34	13.37
INM-1	2795	H11	2	00.00	00.00	00.00	00.00	00.00	10.62	17.80	13.10
INM-1	2796	H11	2	00.00	00.00	00.00	00.00	00.00	10.08	16.16	12.09
INM-1	2797	H11	2	00.00	00.00	00.00	00.00	00.00	9.94	16.30	12.61
INM-1	2812	H11	2	00.00	23.46	25.83	00.00	00.00	00.00	00.00	00.00
INM-1	741	I12	2	00.00	00.00	00.00	23.80	17.62	10.67	17.17	13.51
INM-1	742	I12	2	00.00	00.00	00.00	00.00	00.00	10.93	17.15	12.54
INM-1	743	I12	2	12.53	00.00	00.00	00.00	00.00	00.00	00.00	00.00
INM-1	744	I12	2	00.00	00.00	00.00	00.00	00.00	11.15	17.94	13.15
INM-1	2852	G3	3	00.00	23.80	25.26	00.00	00.00	00.00	00.00	00.00
INM-1	2860	G3	3	00.00	00.00	00.00	22.38	17.47	10.16	17.15	12.59
INM-1	2861	G3	3	00.00	00.00	00.00	23.39	16.99	10.94	16.90	12.33
INM-1	2862	G3	3	00.00	00.00	00.00	00.00	00.00	11.58	17.11	12.86
INM-1	3603	G4	3	00.00	00.00	00.00	00.00	00.00	10.75	17.08	13.05
INM-1	2681	G7	3	00.00	00.00	00.00	00.00	00.00	10.60	17.16	12.09
INM-1	2682	G7	3	00.00	00.00	00.00	00.00	00.00	10.72	17.34	12.27
INM-1	2684	G7	3	00.00	00.00	00.00	00.00	00.00	10.50	16.94	12.39
INM-1	2685	G7	3	00.00	00.00	00.00	00.00	00.00	11.51	18.07	12.70
INM-1	3474	H7	3	00.00	23.94	26.20	00.00	00.00	00.00	00.00	00.00
INM-1	3475	H7	3	00.00	22.13	24.14	00.00	00.00	00.00	00.00	00.00
INM-1	3479	H7	3	00.00	00.00	00.00	23.15	16.69	10.50	16.37	11.75
INM-1	3480	H7	3	00.00	00.00	00.00	22.50	17.20	10.37	17.10	12.60
INM-1	3481	H7	3	00.00	00.00	00.00	24.05	18.30	11.25	17.95	13.15
INM-1	3482	H7	3	00.00	00.00	00.00	22.51	17.50	10.05	17.47	12.40
INM-1	3483	H7	3	00.00	00.00	00.00	22.25	17.44	10.30	17.25	13.00
INM-1	3484	H7	3	00.00	00.00	00.00	23.60	17.82	11.23	17.46	13.15
INM-1	3485	H7	3	00.00	00.00	00.00	00.00	00.00	11.14	17.73	13.28
INM-1	3486	H7	3	00.00	00.00	00.00	00.00	00.00	10.16	16.61	12.28
INM-1	3487	H7	3	00.00	00.00	00.00	00.00	00.00	9.80	16.12	12.62
INM-1	3488	H7	3	00.00	00.00	00.00	00.00	00.00	10.05	15.46	11.77
INM-1	3489	H7	3	00.00	00.00	00.00	00.00	00.00	10.72	17.67	13.59
INM-1	3490	H7	3	00.00	00.00	00.00	00.00	00.00	9.80	15.91	11.75
INM-1	3491	H7	3	00.00	00.00	00.00	00.00	00.00	10.03	15.70	10.66
INM-1	3596	I6	3	00.00	00.00	00.00	24.40	00.00	11.32	17.24	12.74
INM-1	2018	D24	3	00.00	00.00	00.00	00.00	00.00	10.79	16.05	11.60
INM-1	1232	D25	3	00.00	00.00	00.00	24.20	17.26	11.26	17.52	13.08
INM-1	1233	D25	3	00.00	00.00	00.00	00.00	00.00	10.23	16.61	11.55
INM-1	1236	D25	3	00.00	00.00	00.00	00.00	00.00	10.00	15.92	12.25
INM-1	1237	D25	3	00.00	00.00	00.00	00.00	00.00	9.36	15.41	10.33
INM-1	1888	E10	3	11.59	18.04	18.27	00.00	00.00	00.00	00.00	00.00
INM-1	6731	E23	3	00.00	21.25	23.60	00.00	00.00	00.00	00.00	00.00
INM-1	6732	E23	3	00.00	00.00	00.00	24.57	18.23	11.50	17.90	13.50
INM-1	6733	E23	3	00.00	00.00	00.00	23.85	17.20	10.35	16.15	11.75

INM-1	6734	E23	3	00.00	00.00	00.00	24.75	18.90	11.20	18.65	14.05
INM-1	6735	E23	3	00.00	00.00	00.00	23.25	17.90	10.60	17.60	12.65
INM-1	6736	E23	3	00.00	00.00	00.00	23.30	17.55	10.05	17.15	11.90
INM-1	6510	E25	3	00.00	22.75	24.05	00.00	00.00	00.00	00.00	00.00
INM-1	6511	E25	3	00.00	22.80	25.95	00.00	00.00	00.00	00.00	00.00
INM-1	6512	E25	3	00.00	21.11	24.25	00.00	00.00	00.00	00.00	00.00
INM-1	6965	E26	3	00.00	20.75	23.25	00.00	00.00	00.00	00.00	00.00
INM-1	2246	F11	3	00.00	00.00	00.00	00.00	00.00	10.62	17.03	11.83
INM-1	6525	F25	3	00.00	00.00	00.00	22.45	17.00	10.05	16.85	12.50
INM-1	6526	F25	3	00.00	00.00	00.00	22.35	16.65	10.20	16.50	12.45
INM-1	6527	F25	3	00.00	00.00	00.00	26.47	18.50	11.96	18.36	13.90
INM-1	6529	F25	3	00.00	00.00	00.00	24.90	18.30	11.35	18.26	13.70
INM-1	6531	F25	3	00.00	00.00	00.00	24.20	17.35	10.25	16.52	11.65
INM-1	6552	F25	3	00.00	00.00	00.00	00.00	00.00	11.25	17.68	13.25
INM-1	6553	F25	3	00.00	00.00	00.00	00.00	00.00	12.45	19.42	13.97
INM-1	6581	F25	3	00.00	27.45	28.30	00.00	00.00	00.00	00.00	00.00
INM-1	2446	F25&F26	3	00.00	23.01	25.56	00.00	00.00	00.00	00.00	00.00
INM-1	2447	F25&F26	3	00.00	22.34	24.60	00.00	00.00	00.00	00.00	00.00
INM-1	2448	F25&F26	3	00.00	18.99	22.57	00.00	00.00	00.00	00.00	00.00
INM-1	2449	F25&F26	3	00.00	00.00	00.00	00.00	00.00	10.03	16.41	11.83
INM-1	2450	F25&F26	3	00.00	00.00	00.00	00.00	00.00	10.64	16.98	12.27
INM-1	2906	G10	3	00.00	00.00	00.00	23.60	16.06	9.70	15.75	11.56
INM-1	1268	G11	3	00.00	23.23	24.10	00.00	00.00	00.00	00.00	00.00
INM-1	1271	G11	3	00.00	00.00	00.00	23.07	17.16	10.79	17.08	12.50
INM-1	1273	G11	3	00.00	00.00	00.00	00.00	00.00	10.35	16.93	11.56
INM-1	1275	G11	3	00.00	00.00	00.00	25.08	18.86	11.48	18.68	13.47
INM-1	1276	G11	3	00.00	00.00	00.00	00.00	00.00	11.00	17.65	00.00
INM-1	1277	G11	3	00.00	00.00	00.00	00.00	00.00	10.46	16.80	12.94
INM-1	1278	G11	3	00.00	00.00	00.00	00.00	00.00	10.41	17.05	13.30
INM-1	881	G12	3	00.00	00.00	00.00	23.32	17.05	10.87	16.79	12.65
INM-1	882	G12	3	00.00	00.00	00.00	24.86	19.18	11.95	18.65	13.56
INM-1	883	G12	3	00.00	00.00	00.00	22.68	16.35	10.54	16.35	11.66
INM-1	884	G12	3	00.00	00.00	00.00	23.74	17.63	11.15	16.96	13.24
INM-1	885	G12	3	00.00	00.00	00.00	21.94	16.76	10.23	16.76	12.48
INM-1	894	G12	3	00.00	21.29	00.00	00.00	00.00	00.00	00.00	00.00
INM-1	895	G12	3	00.00	22.91	24.64	00.00	00.00	00.00	00.00	00.00
INM-1	896	G12	3	00.00	22.71	25.90	00.00	00.00	00.00	00.00	00.00
INM-1	897	G12	3	00.00	22.65	23.37	00.00	00.00	00.00	00.00	00.00
INM-1	898	G12	3	00.00	21.00	23.35	00.00	00.00	00.00	00.00	00.00
INM-1	7818	G25	3	00.00	00.00	00.00	24.05	17.20	10.90	17.51	11.65
INM-1	7819	G25	3	00.00	00.00	00.00	00.00	00.00	10.95	17.51	11.65
INM-1	7823	G25	3	00.00	00.00	00.00	00.00	00.00	11.55	17.85	12.40
INM-1	7824	G25	3	00.00	00.00	00.00	00.00	00.00	12.30	18.10	13.10
INM-1	7714	G26	3	00.00	00.00	00.00	00.00	00.00	10.23	17.09	11.90
INM-1	7887	G27	3	00.00	22.75	25.50	00.00	00.00	00.00	00.00	00.00
INM-1	7888	G27	3	00.00	00.00	00.00	23.45	17.25	10.70	17.25	12.33
INM-1	7889	G27	3	00.00	00.00	00.00	00.00	00.00	9.90	16.65	12.00
INM-1	2618	H10	3	00.00	00.00	00.00	25.76	18.50	11.48	18.03	13.50
INM-1	2620	H10	3	00.00	24.23	20.35	00.00	00.00	00.00	00.00	00.00
INM-1	2623	H10	3	00.00	00.00	00.00	00.00	00.00	9.50	16.00	11.95
INM-1	1779	H11	3	00.00	00.00	00.00	23.16	16.94	9.68	16.31	11.35
INM-1	1780	H11	3	00.00	00.00	00.00	21.71	15.80	9.80	00.00	10.45
INM-1	1782	H11	3	00.00	00.00	00.00	23.68	17.45	10.80	17.49	12.91
INM-1	1785	H11	3	00.00	00.00	00.00	00.00	00.00	9.74	16.23	12.66
INM-1	2636	H11	3	00.00	00.00	00.00	25.14	18.32	11.89	17.79	13.26
INM-1	2638	H11	3	00.00	00.00	00.00	00.00	00.00	10.58	17.94	12.44
INM-1	2639	H11	3	00.00	00.00	00.00	00.00	00.00	11.38	18.26	13.34
INM-1	3788	H12	3	00.00	00.00	00.00	22.30	15.85	9.85	15.75	11.66
INM-1	3789	H12	3	00.00	00.00	00.00	26.16	20.30	11.90	18.90	14.00
INM-1	4519	H12	3	00.00	22.42	24.40	00.00	00.00	00.00	00.00	00.00
INM-1	6182	Hno106	3	00.00	00.00	00.00	00.00	00.00	10.53	17.00	12.20
INM-1	6184	Hno106	3	00.00	22.30	25.35	00.00	00.00	00.00	00.00	00.00
INM-1	4870	Hno107	3	00.00	22.60	23.60	00.00	00.00	00.00	00.00	00.00

INM-1	4871	Hno107	3	00.00	00.00	00.00	22.05	16.60	9.40	15.34	11.05
INM-1	4872	Hno107	3	00.00	00.00	00.00	26.35	19.05	11.88	18.90	13.95
INM-1	973	C1	4	00.00	00.00	00.00	23.16	17.80	10.60	17.17	12.61
INM-1	1376	E1	4	00.00	00.00	00.00	24.59	16.90	10.14	16.64	11.87
INM-1	569	F8	4	00.00	21.13	24.27	00.00	00.00	00.00	00.00	00.00
INM-1	1812	H5	4	00.00	22.36	24.63	00.00	00.00	00.00	00.00	00.00
INM-1	1813	H5	4	00.00	21.31	24.05	00.00	00.00	00.00	00.00	00.00
INM-1	3052	H7	4	00.00	21.59	23.69	00.00	00.00	00.00	00.00	00.00
INM-1	3053	H7	4	00.00	00.00	00.00	23.09	17.98	10.69	17.09	12.99
INM-1	3054	H7	4	00.00	00.00	00.00	23.38	17.87	10.54	17.70	13.00
INM-1	3055	H7	4	00.00	00.00	00.00	00.00	00.00	11.15	16.50	11.90
INM-1	2960	D10	4	00.00	00.00	00.00	23.21	17.08	10.36	17.09	13.00
INM-1	8121	E24	4	00.00	00.00	00.00	25.25	18.40	11.70	18.00	13.75
INM-1	8122	E24	4	00.00	00.00	00.00	23.50	18.03	11.00	17.93	13.02
INM-1	5879	E25	4	00.00	24.05	25.45	00.00	00.00	00.00	00.00	00.00
INM-1	5880	E25	4	00.00	00.00	00.00	24.80	18.45	11.30	18.58	12.28
INM-1	5881	E25	4	00.00	00.00	00.00	25.40	18.55	11.54	18.27	13.20
INM-1	5882	E25	4	00.00	00.00	00.00	22.95	16.40	10.80	16.25	12.00
INM-1	5883	E25	4	00.00	00.00	00.00	00.00	00.00	12.07	19.05	13.90
INM-1	5884	E25	4	00.00	00.00	00.00	00.00	00.00	10.70	17.75	12.78
INM-1	5886	E25	4	00.00	00.00	00.00	00.00	00.00	10.48	16.60	12.80
INM-1	5858	E27	4	00.00	00.00	00.00	24.85	18.00	11.45	17.90	13.05
INM-1	5860	E27	4	00.00	21.15	23.15	00.00	00.00	00.00	00.00	00.00
INM-1	5861	E27	4	13.15	20.05	23.70	00.00	00.00	00.00	00.00	00.00
INM-1	7964	F24	4	00.00	21.42	23.95	00.00	00.00	00.00	00.00	00.00
INM-1	7965	F24	4	00.00	22.54	25.83	00.00	00.00	00.00	00.00	00.00
INM-1	7966	F24	4	00.00	20.15	23.50	00.00	00.00	00.00	00.00	00.00
INM-1	7967	F24	4	00.00	00.00	00.00	23.85	18.00	10.65	17.57	13.25
INM-1	7974	F24	4	00.00	00.00	00.00	00.00	00.00	10.72	18.46	13.05
INM-1	5970	F25	4	00.00	00.00	00.00	24.40	18.20	10.95	17.95	13.03
INM-1	1860	G11	4	00.00	21.87	24.39	00.00	00.00	00.00	00.00	00.00
INM-1	1861	G11	4	00.00	00.00	00.00	23.62	17.59	10.91	17.62	13.10
INM-1	7108	G24	4	00.00	00.00	00.00	24.95	18.10	11.25	17.90	13.30
INM-1	7624	G25	4	00.00	00.00	00.00	23.66	17.25	10.30	16.55	11.85
INM-1	7625	G25	4	00.00	00.00	00.00	25.50	19.00	12.00	18.95	13.80
INM-1	7626	G25	4	00.00	00.00	00.00	24.82	19.00	11.45	18.90	14.29
INM-1	7627	G25	4	00.00	00.00	00.00	24.25	18.45	11.00	18.25	13.85
INM-1	7628	G25	4	00.00	00.00	00.00	23.50	17.46	11.07	17.40	12.45
INM-1	7629	G25	4	00.00	00.00	00.00	00.00	00.00	10.05	16.35	13.20
INM-1	7630	G25	4	00.00	00.00	00.00	00.00	00.00	10.30	16.55	12.55
INM-1	7631	G25	4	00.00	00.00	00.00	00.00	00.00	9.90	15.92	11.60
INM-1	7694	G25	4	00.00	20.65	21.30	00.00	00.00	00.00	00.00	00.00
INM-1	6816	G26	4	00.00	22.07	25.30	00.00	00.00	00.00	00.00	00.00
INM-1	6817	G26	4	00.00	00.00	00.00	23.05	16.45	10.60	16.34	12.60
INM-1	6818	G26	4	00.00	00.00	00.00	24.45	18.30	11.30	18.00	14.10
INM-1	6819	G26	4	00.00	00.00	00.00	23.16	17.00	10.50	16.60	11.30
INM-1	6820	G26	4	00.00	00.00	00.00	25.42	18.60	11.70	18.75	13.80
INM-1	4786	B0	4	00.00	22.79	25.08	00.00	00.00	00.00	00.00	00.00
INM-1	7757	D24&E25	4	00.00	21.35	24.30	00.00	00.00	00.00	00.00	00.00
INM-1	7760	D24&E25	4	00.00	00.00	00.00	25.25	18.17	11.45	17.93	13.60
INM-1	7235	Hno104	4	00.00	19.98	22.85	00.00	00.00	00.00	00.00	00.00
INM-1	7236	Hno104	4	00.00	00.00	00.00	23.16	17.15	10.75	16.82	12.20
INM-1	7237	Hno104	4	00.00	00.00	00.00	00.00	00.00	10.90	16.10	12.25
INM-1	662	Hno110	4	00.00	00.00	00.00	00.00	00.00	10.06	16.90	12.17
INM-1	663	Hno110	4	00.00	20.45	24.18	00.00	00.00	00.00	00.00	00.00
INM-1	665	Hno110	4	00.00	00.00	00.00	00.00	00.00	10.95	17.32	12.37
INM-1	4754	Hno117	4	00.00	21.25	23.15	00.00	00.00	00.00	00.00	00.00
INM-1	4755	Hno117	4	00.00	00.00	00.00	23.72	16.60	10.25	16.50	12.45
INM-1	4756	Hno117	4	00.00	00.00	00.00	23.96	17.58	10.85	17.07	12.30
INM-1	4329	F8	5	00.00	00.00	00.00	24.70	17.65	11.15	17.37	12.75
INM-1	4330	F8	5	00.00	00.00	00.00	26.87	19.10	11.40	17.95	12.75
INM-1	4333	F8	5	00.00	00.00	00.00	25.00	17.95	10.90	17.60	13.60
INM-1	4346	F8	5	00.00	17.70	21.80	00.00	00.00	00.00	00.00	00.00

INM-1	4565	E5&B0	5	00.00	19.95	25.42	00.00	00.00	00.00	00.00	00.00
INM-1	1503	G4	6	00.00	23.12	26.12	00.00	00.00	00.00	00.00	00.00
INM-1	1506	G4	6	00.00	00.00	00.00	23.95	18.51	11.10	18.13	13.14
INM-1	1834	G4	7	00.00	24.92	21.38	00.00	00.00	00.00	00.00	00.00
INM-1	1835	G4	7	00.00	00.00	00.00	24.04	16.69	10.80	15.96	12.44
INM-1	7932	Hno51	8	00.00	21.90	24.23	00.00	00.00	00.00	00.00	00.00
INM-1	4951	E6	8	00.00	00.00	00.00	00.00	00.00	10.92	17.76	12.50
INM-1	4952	E6	8	00.00	00.00	00.00	24.15	17.32	10.65	16.60	11.70
INM-1	3736	Hno69	8	00.00	00.00	00.00	23.00	16.62	10.58	16.28	11.30
INM-1	738	Hno70&73	8	00.00	22.28	23.61	00.00	00.00	00.00	00.00	00.00
INM-1	3067	C4	9	00.00	23.28	27.35	00.00	00.00	00.00	00.00	00.00
INM-1	6014	B7	11	00.00	20.55	20.85	00.00	00.00	00.00	00.00	00.00
INM-1	4182	B1	12	00.00	00.00	00.00	25.50	19.65	11.95	19.25	14.10
INM-1	530	Hno50	12	00.00	00.00	00.00	25.08	00.00	11.40	17.65	12.37
INM-1	6135	Hno83	12	00.00	00.00	00.00	00.00	00.00	10.05	16.85	12.30
INM-1	6247	E5	15	00.00	00.00	00.00	24.60	18.40	10.85	17.30	12.17

Table A.16.12. Measurements of Astragalus of ***Antilope cervicapra***

Site	Reg.No.	Trench	Layer	GL	GLL	GL	TL	TM	BD
INM-1	54	LJP	1-4	30.82	28.71	18.07	17.76	18.60	00.00
INM-1	55	LJP	1-4	30.13	21.59	10.08	16.78	16.06	17.20
INM-1	56	LJP	1-4	31.56	31.16	28.68	18.64	17.09	19.84
INM-1	57	LJP	1-4	30.15	30.08	28.64	17.83	16.96	18.19
INM-1	78	LJP	1-4	30.18	29.91	28.40	17.90	17.34	17.79
INM-1	83	LJP	1-4	31.40	30.98	28.97	18.23	17.88	18.21
INM-1	86	LJP	1-4	31.86	31.38	29.03	18.73	17.88	18.75
INM-1	90	LJP	1-4	30.31	00.00	29.41	18.40	17.62	18.71
INM-1	122	LJP	1-4	33.35	33.25	31.29	19.43	18.48	19.59
INM-1	123	LJP	1-4	32.30	32.25	30.63	19.37	19.10	19.92
INM-1	124	LJP	1-4	28.60	28.29	26.75	16.79	15.75	17.02
INM-1	133	LJP	1-4	32.33	32.13	30.49	18.99	17.79	19.26
INM-1	215	LJP	1-4	32.37	32.24	30.69	19.18	18.85	19.86
INM-1	216	LJP	1-4	31.77	31.38	29.25	18.43	17.85	17.23
INM-1	217	LJP	1-4	32.45	32.41	30.16	19.06	18.56	19.20
INM-1	218	LP	1-4	31.79	31.24	29.89	19.83	18.26	19.74
INM-1	219	LJP	1-4	30.74	30.70	28.72	17.46	16.65	17.56
INM-1	220	LJP	1-4	32.31	31.93	30.38	18.75	18.23	18.80
INM-1	222	LJP	1-4	31.13	30.96	29.80	17.76	17.32	18.57
INM-1	224	LJP	1-4	00.00	00.00	30.47	1814	18.71	19.21
INM-1	225	LJP	1-4	33.54	33.52	31.91	20.23	18.15	20.02
INM-1	226	LJP	1-4	33.43	32.18	30.11	18.83	16.71	19.48
INM-1	363	LJP	1-4	33.51	33.24	30.96	18.33	18.08	19.54
INM-1	5274	LJP	1-4	31.80	31.75	29.75	17.40	17.55	17.50
INM-1	6355	A3&B3	1-4	30.45	30.15	27.90	16.40	16.90	17.70
INM-1	3797	G6	4	33.95	33.33	31.20	18.85	18.36	19.20
INM-1	3798	G6	4	30.50	30.25	00.00	16.53	00.00	18.30
INM-1	1118	G4	1	30.32	30.22	27.98	17.52	00.00	00.00
INM-1	2035	G7	1	32.77	32.60	30.98	18.84	19.77	20.55
INM-1	3531	H5	1	32.30	32.15	31.55	19.20	18.50	18.65
INM-1	3532	H5	1	31.53	31.35	29.30	17.50	18.78	19.80
INM-1	3533	H5	1	33.15	33.05	31.05	18.88	18.84	19.90
INM-1	3278	D10	1	32.05	31.99	30.20	18.85	00.00	18.90
INM-1	3279	D10	1	31.74	31.37	29.30	17.25	17.45	17.67
INM-1	3280	D10	1	29.13	28.98	27.15	15.45	16.25	15.40
INM-1	3835	D25	1	32.32	31.97	28.90	18.52	17.72	19.40
INM-1	3836	D25	1	27.60	27.25	26.15	00.00	00.00	00.00
INM-1	3837	D25	1	00.00	00.00	30.40	17.97	17.75	20.00
INM-1	3838	D25	1	00.00	00.00	00.00	17.78	00.00	18.57
INM-1	996	E10	1	29.91	29.77	28.15	16.86	16.98	17.28
INM-1	998	E10	1	32.56	32.52	30.30	18.41	17.92	19.04
INM-1	7049	E23	1	28.70	28.50	27.40	14.20	14.95	15.00

INM-1	1205	F10	1	30.94	30.72	29.91	18.82	18.28	19.05
INM-1	1206	F10	1	00.00	00.00	00.00	15.3	17.48	00.00
INM-1	3318	F11	1	31.32	31.28	28.85	17.95	17.38	19.35
INM-1	3319	F11	1	30.40	30.34	28.90	17.60	17.65	18.00
INM-1	3320	F11	1	31.90	31.70	29.30	17.60	17.95	18.65
INM-1	3321	F11	1	31.90	31.56	29.70	00.00	00.00	18.70
INM-1	8043	F24	1	31.28	31.20	28.20	17.45	17.65	18.20
INM-1	8044	F24	1	31.80	31.70	30.30	17.20	17.60	19.80
INM-1	8045	F24	1	00.00	00.00	28.55	00.00	17.45	18.87
INM-1	1176	G10	1	00.00	00.00	28.75	00.00	18.23	00.00
INM-1	1658	G11	1	31.50	30.91	28.73	17.28	16.72	00.00
INM-1	1659	G11	1	32.54	32.28	29.68	17.34	18.09	18.63
INM-1	1660	G11	1	29.69	26.65	28.70	16.49	18.27	17.3
INM-1	1727	G11	1	00.00	00.00	00.00	00.00	16.50	00.00
INM-1	1930	H11	1	32.43	32.35	30.12	18.04	18.58	19.22
INM-1	1931	H11	1	29.13	19.09	17.45	16.62	16.71	17.61
INM-1	1932	H11	1	31.88	31.38	30.18	18.43	17.74	18.70
INM-1	1933	H11	1	32.36	31.90	30.25	17.80	17.84	19.03
INM-1	1929	H11	1	30.93	30.48	29.11	00.00	16.51	17.18
INM-1	2137	Hno132	1	31.43	30.71	29.01	17.02	17.56	17.59
INM-1	2138	Hno132	1	31.59	30.74	28.28	16.58	17.06	17.28
INM-1	2139	Hno132	1	31.13	30.21	28.91	16.70	17.82	17.55
INM-1	2140	Hno132	1	00.00	00.00	00.00	00.00	00.00	21.13
INM-1	2274	E1	2	31.78	31.52	28.93	16.58	17.39	18.96
INM-1	3078	G3	2	32.46	00.00	29.86	17.99	18.29	19.95
INM-1	3079	G3	2	29.73	29.20	27.72	17.13	16.48	17.30
INM-1	2078	G6	2	30.62	30.53	28.51	17.75	16.91	18.58
INM-1	2079	G6	2	32.80	32.67	30.38	18.05	19.39	19.79
INM-1	2080	G6	2	31.11	31.03	28.90	17.45	16.86	18.65
INM-1	3636	G7	2	30.12	30.06	28.0	17.07	16.89	18.26
INM-1	3637	G7	2	30.72	30.54	28.95	17.73	16.94	18.90
INM-1	1571	H5	2	31.65	30.78	28.28	18.10	17.09	19.04
INM-1	3585	I6	2	33.03	32.82	31.01	18.05	19.06	00.00
INM-1	1990	D10	2	34.59	33.57	32.16	18.92	18.48	21.05
INM-1	1426	D25	2	31.22	31.11	30.15	17.96	18.30	18.26
INM-1	1427	D25	2	33.84	33.76	30.97	18.97	19.51	20.20
INM-1	1428	D25	2	30.42	29.92	28.85	17.16	17.93	18.38
INM-1	1429	D25	2	29.74	29.63	00.00	16.54	00.00	00.00
INM-1	1430	D25	2	33.32	32.72	30.20	18.01	18.04	19.02
INM-1	1431	D25	2	29.81	29.40	28.23	16.47	16.14	18.14
INM-1	1432	D25	2	32.34	31.43	29.72	17.10	17.56	19.21
INM-1	1121	E10	2	31.45	31.41	29.08	18.08	17.69	18.48
INM-1	7273	E24	2	30.42	30.11	27.60	16.65	16.95	17.75
INM-1	7274	E24	2	30.85	30.72	28.93	16.30	16.70	17.40
INM-1	7275	E24	2	00.00	00.00	29.17	17.55	18.45	19.15
INM-1	7276	E24	2	32.65	32.47	30.43	18.30	17.70	19.52
INM-1	7277	E24	2	33.15	33.05	31.52	17.90	18.50	19.74
INM-1	223	E25	2	34.49	34.03	31.51	20.20	18.62	20.51
INM-1	6398	E25	2	30.22	30.13	28.50	17.23	00.00	18.55
INM-1	6399	E25	2	31.65	31.60	29.15	18.25	19.05	18.55
INM-1	6400	E25	2	32.00	31.90	00.00	17.60	17.75	19.50
INM-1	6401	E25	2	29.20	29.10	26.88	16.60	16.00	18.07
INM-1	7416	E25	2	29.75	29.50	28.00	16.60	17.50	17.00
INM-1	7417	E25	2	30.95	30.85	29.38	17.30	00.00	18.66
INM-1	2996	F10	2	31.18	31.12	29.62	18.21	18.09	19.40
INM-1	2997	F10	2	31.05	30.97	28.69	18.20	18.50	18.70
INM-1	2542	F11	2	28.23	28.20	26.82	16.15	15.79	17.13
INM-1	2544	F11	2	33.92	32.91	30.54	19.08	18.36	19.54
INM-1	2545	F11	2	32.47	32.38	30.08	17.25	17.89	18.52
INM-1	2546	F11	2	32.75	32.67	30.67	17.59	18.62	19.00
INM-1	2547	F11	2	00.00	00.00	27.75	16.38	15.71	17.37
INM-1	6615	F24	2	32.20	31.95	29.15	18.15	18.35	19.75
INM-1	1336	G11	2	32.92	32.45	30.85	17.93	18.85	20.66

INM-1	1337	G11	2	28.96	28.81	27.29	17.82	16.62	18.31
INM-1	1338	G11	2	31.40	31.04	29.20	17.46	17.48	18.62
INM-1	3153	G11	2	29.78	29.62	27.96	16.52	17.13	19.80
INM-1	3154	G11	2	34.15	33.96	32.62	18.87	00.00	19.80
INM-1	3155	G11	2	33.95	33.49	30.95	19.09	19.85	19.80
INM-1	3156	G11	2	30.68	30.52	28.63	16.71	17.24	18.00
INM-1	7004	G24	2	33.35	33.25	30.70	18.75	18.90	19.95
INM-1	7164	G25	2	31.20	30.45	28.20	17.20	17.75	18.95
INM-1	7165	G25	2	29.70	29.65	28.80	16.50	16.70	16.95
INM-1	7166	G25	2	32.20	31.85	30.00	17.55	17.40	19.75
INM-1	7167	G25	2	31.60	31.55	29.00	17.65	18.35	18.80
INM-1	7168	G25	2	28.70	00.00	27.35	16.35	16.55	17.55
INM-1	7169	G25	2	00.00	00.00	29.50	00.00	16.92	17.50
INM-1	2762	H11	2	31.80	31.52	29.99	17.86	18.18	19.75
INM-1	2763	H11	2	31.20	31.05	28.38	17.70	17.24	18.76
INM-1	751	I12	2	32.23	31.53	29.93	16.33	17.81	18.10
INM-1	752	I12	2	30.88	30.79	29.29	15.51	17.42	17.40
INM-1	7903	Hno108	2	30.80	30.75	29.40	17.26	00.00	19.40
INM-1	2668	G7	3	31.64	31.41	29.19	18.42	17.33	18.44
INM-1	2669	G7	3	33.49	33.39	31.33	18.53	16.06	21.00
INM-1	3452	H7	3	28.92	28.71	28.30	16.16	16.92	17.68
INM-1	3453	H7	3	31.29	30.21	28.60	16.75	16.83	17.75
INM-1	3454	H7	3	32.02	31.93	30.76	18.23	18.47	00.00
INM-1	3455	H7	3	34.20	33.22	30.58	18.79	19.35	00.00
INM-1	1245	D25	3	31.25	31.20	28.43	18.80	00.00	00.00
INM-1	1246	D25	3	30.91	30.82	29.36	17.48	18.46	19.37
INM-1	1247	D25	3	31.81	31.47	29.49	18.10	18.37	19.00
INM-1	4037	C0	3	32.48	32.30	30.25	18.55	17.69	20.05
INM-1	1886	E10	3	30.02	29.75	28.65	17.08	17.41	18.57
INM-1	6724	E23	3	31.85	31.55	29.85	17.40	17.56	18.90
INM-1	6760	E23	3	25.05	24.85	23.41	12.75	13.00	14.65
INM-1	6903	E26	3	30.26	30.05	28.58	17.35	17.60	17.80
INM-1	6904	E26	3	00.00	00.00	28.95	17.10	17.55	17.40
INM-1	6905	E26	3	00.00	00.00	28.95	17.45	17.75	19.55
INM-1	6906	E26	3	31.25	31.20	29.35	16.85	17.20	00.00
INM-1	6907	E26	3	29.45	29.25	27.00	15.65	16.05	16.95
INM-1	6908	E26	3	31.05	30.90	28.90	17.55	18.15	18.20
INM-1	6909	E26	3	30.20	30.00	28.30	17.65	18.00	18.90
INM-1	6910	E26	3	32.80	32.60	31.05	17.65	17.75	20.05
INM-1	2245	F11	3	33.80	33.10	30.95	18.50	19.56	19.26
INM-1	6499	F25	3	28.70	28.65	28.25	16.30	16.90	18.15
INM-1	6500	F25	3	30.95	30.80	29.15	16.95	17.85	18.10
INM-1	6501	F25	3	31.40	31.35	29.10	17.60	17.95	18.40
INM-1	6502	F25	3	30.20	30.07	28.80	18.20	16.95	18.25
INM-1	6503	F25	3	32.30	32.22	00.00	17.86	18.25	00.00
INM-1	6504	F25	3	30.65	30.45	28.50	17.30	17.70	18.40
INM-1	2196	G10	3	31.11	30.99	30.20	17.95	18.69	19.69
INM-1	2197	G10	3	00.00	00.00	28.33	00.00	16.29	17.89
INM-1	2198	G10	3	31.07	30.87	29.01	18.77	17.24	18.03
INM-1	2199	G10	3	32.69	32.56	00.00	19.20	00.00	00.00
INM-1	2898	G10	3	30.59	30.50	29.17	17.11	17.20	17.83
INM-1	2899	G10	3	30.58	30.50	29.41	16.74	00.00	17.50
INM-1	2900	G10	3	00.00	00.00	29.25	17.62	17.70	18.29
INM-1	1296	G11	3	31.66	31.54	29.82	18.30	17.91	18.81
INM-1	1297	G11	3	31.44	31.10	30.36	17.70	18.35	18.42
INM-1	7807	G25	3	32.65	32.05	29.60	00.00	00.00	19.65
INM-1	7808	G25	3	31.10	30.78	29.85	18.07	18.15	19.38
INM-1	7809	G25	3	32.65	32.53	30.30	18.30	18.55	19.25
INM-1	7886	G27	3	32.20	31.75	29.41	18.15	18.30	18.33
INM-1	1769	H11	3	30.87	30.85	29.26	17.37	16.62	18.75
INM-1	1770	H11	3	32.38	32.28	29.56	18.84	17.79	18.75
INM-1	3786	H12	3	30.78	30.62	29.00	17.42	17.26	17.85
INM-1	4511	H12	3	30.06	29.90	28.65	16.70	17.10	17.89

INM-1	4512	H12	3	31.00	30.90	29.00	16.95	17.55	18.62
INM-1	6176	Hno106	3	29.80	29.55	28.29	16.80	17.25	17.55
INM-1	6179	Hno106	3	31.00	30.75	28.80	17.40	17.45	19.35
INM-1	6180	Hno106	3	32.70	32.60	30.35	18.25	18.50	19.12
INM-1	4869	Hno107	3	00.00	00.00	28.65	17.00	17.58	17.85
INM-1	1810	H5	4	29.79	29.33	28.08	16.89	16.77	00.00
INM-1	3045	H7	4	31.51	31.47	28.92	17.53	16.75	18.84
INM-1	3046	H7	4	33.15	33.08	30.85	19.23	19..8	21.98
INM-1	3047	H7	4	30.03	29.78	28.67	16.50	17.18	19.19
INM-1	3048	H7	4	00.00	00.00	30.45	00.00	18.00	18.98
INM-1	2958	D10	4	30.95	30.55	29.79	18.16	16.95	18.18
INM-1	8106	E24	4	31.90	31.70	28.55	17.30	17.67	18.32
INM-1	8107	E24	4	32.65	32.55	30.80	17.80	17.95	00.00
INM-1	8108	E24	4	33.40	33.10	31.70	18.10	18.65	18.95
INM-1	8109	E24	4	00.00	00.00	00.00	18.60	18.75	20.75
INM-1	5874	E25	4	32.88	32.65	00.00	18.90	19.25	20.30
INM-1	5875	E25	4	32.55	32.47	29.28	17.95	18.46	18.88
INM-1	5876	E25	4	31.80	31.66	29.90	17.30	18.10	18.45
INM-1	5865	E27	4	30.00	29.77	29.00	17.45	17.03	17.50
INM-1	5866	E27	4	28.21	28.15	26.60	15.70	16.70	18.30
INM-1	7957	F24	4	29.95	29.90	27.90	16.35	16.55	16.65
INM-1	7958	F24	4	29.05	28.85	27.75	15.75	16.45	17.95
INM-1	5966	F25	4	31.85	31.63	29.90	18.25	18.40	19.44
INM-1	1858	G11	4	33.02	32.85	30.57	17.99	18.41	19.50
INM-1	7081	G24	4	00.00	00.00	28.65	16.50	16.70	18.30
INM-1	7082	G24	4	31.80	31.65	30.30	17.80	18.00	18.35
INM-1	7083	G24	4	32.25	32.05	29.40	18.00	19.00	19.50
INM-1	7084	G24	4	30.16	00.00	29.00	17.30	17.40	00.00
INM-1	7620	G25	4	31.05	30.95	29.20	16.95	17.20	18.31
INM-1	7621	G25	4	29.90	29.80	28.60	17.50	18.20	19.40
INM-1	7622	G25	4	30.78	30.65	28.25	17.23	17.50	18.20
INM-1	7653	G25	4	30.63	30.58	28.25	16.35	16.65	17.75
INM-1	7691	G25	4	30.95	30.90	00.00	17.35	00.00	18.65
INM-1	7692	G25	4	00.00	00.00	00.00	16.72	17.25	18.95
INM-1	6805	G26	4	33.25	33.05	30.10	18.85	18.15	20.00
INM-1	6807	G26	4	32.35	32.00	29.30	17.20	17.95	19.00
INM-1	6808	G26	4	32.75	32.62	31.25	17.80	18.15	19.53
INM-1	6809	G26	4	32.00	31.85	28.95	18.10	18.20	18.70
INM-1	6810	G26	4	30.85	30.75	29.15	16.70	17.80	17.90
INM-1	84	LJP	4	29.24	29.20	27.43	17.13	00.00	16.94
INM-1	4785	B0	4	00.00	30.86	29.72	17.85	18.40	18.20
INM-1	7230	Hno104	4	32.62	32.55	29.50	17.86	00.00	18.70
INM-1	7231	Hno104	4	30.85	30.70	27.56	17.93	17.60	18.90
INM-1	7232	Hno104	4	31.04	30.90	28.92	17.20	17.45	17.95
INM-1	4642	Hno111	4	31.65	31.55	30.05	18.08	19.05	19.40
INM-1	4643	Hno111	4	32.05	32.00	30.18	18.00	18.09	19.80
INM-1	4060	Hno115	4	31.93	31.82	29.00	17.65	17.80	18.70
INM-1	4061	Hno115	4	30.45	00.00	29.17	16.95	17.15	00.00
INM-1	5032	Hno119	4	00.00	00.00	00.00	17.50	17.55	19.05
INM-1	789	Hno124	4	00.00	00.00	00.00	00.00	00.00	19.58
INM-1	4625	B2	5	29.87	29.53	28.90	16.75	16.86	17.38
INM-1	4324	F8	5	31.42	31.06	28.90	16.70	15.80	20.10
INM-1	4325	F8	5	31.55	31.40	29.20	28.05	17.22	19.34
INM-1	4219	A3	6	26.40	26.21	24.47	14.75	14.95	16.60
INM-1	507	F3	6	30.90	30.64	29.44	17.21	16.92	17.76
INM-1	3773	F8	6	00.00	00.00	29.02	00.00	17.85	17.05
INM-1	2008	G3	6	30.75	30.56	27.87	17.80	16.24	17.58
INM-1	4116	C3	7	26.66	26.54	25.70	14.78	14.42	16.70
INM-1	4134	C3	7	31.80	31.41	28.25	18.05	17.62	19.33
INM-1	4135	C3	7	00.00	00.00	00.00	00.00	00.00	19.28
INM-1	4374	E5	7	31.68	31.62	29.80	18.05	18.40	19.70
INM-1	4393	B6	8	33.15	33.10	31.20	18.05	18.68	18.92
INM-1	4948	E6	8	29.54	29.20	27.18	17.07	16.80	17.67

INM-1	4459	E7	8	26.28	26.17	25.28	15.03	15.09	16.67
INM-1	4961	KilnB1	8	00.00	31.70	00.00	18.15	00.00	00.00
INM-1	5157	B2	11	33.07	32.83	00.00	18.16	00.00	19.40
INM-1	5237	C3	11	00.00	00.00	30.15	17.15	17.55	18.40
INM-1	5679	C8	11	30.30	30.20	28.45	17.05	17.40	18.20
INM-1	5344	D5	11	29.68	29.56	28.4	16.34	16.90	17.34
INM-1	5754	C4	12	33.10	32.90	30.60	18.17	18.30	19.35
INM-1	5760	C5	12	29.65	29.60	28.37	16.55	17.20	17.60
INM-1	5850	C6	12	30.86	30.75	29.05	17.95	18.07	19.13
INM-1	5743	E3	12	30.20	30.03	27.30	15.86	15.03	16.45
INM-1	4820	B0	12	30.85	30.67	28.83	17.65	17.86	19.10
INM-1	6053	Hno76	12	29.80	29.45	27.87	17.30	16.65	17.50
INM-1	5332	Hno79	12	00.00	00.00	28.15	17.38	17.90	18.41
INM-1	5819	D3	13	31.85	31.55	28.83	16.90	17.80	18.75
INM-1	6083	E3	13	31.84	31.49	29.45	17.10	17.45	19.35
INM-1	5802	B5	14	31.70	31.30	29.65	17.70	17.80	19.70
INM-1	6338	E4	14	29.05	29.00	27.98	16.98	17.00	18.20
INM-1	5398	C3	15	31.70	31.64	00.00	17.51	17.96	00.00
INM-1	6064	D8	15	30.20	29.95	27.20	17.20	17.00	17.95
INM-1	2501	D3	16	28.89	28.84	27.15	16.83	16.59	17.68
INM-1	4151	D3	16	30.66	30.59	29.59	16.57	17.50	17.80

Table A.16.13. Measurements of Calcaneum of ***Antilope cervicapra***

Site	Reg.No.	Trench	Layer	GL	GB
INM-1	65	LJP	1-4	61.24	18.61
INM-1	131	LJP	1-4	62.29	17.38
INM-1	210	LJP	1-4	62.83	00.00
INM-1	211	LJP	1-4	64.73	19.54
INM-1	212	LJP	1-4	00.00	19.79
INM-1	213	LJP	1-4	00.00	22.67
INM-1	214	LJP	1-4	64.91	18.62
INM-1	372	LJP	1-4	67.47	19.32
INM-1	396	LJP	1-4	64.23	23.30
INM-1	5275	LJP	1-4	60.60	17.45
INM-1	5276	LJP	1-4	73.92	00.00
INM-1	5277	LJP	1-4	67.55	20.15
INM-1	6356	A3&B3	1-4	67.45	19.45
INM-1	3799	G6	4	65.10	19.90
INM-1	1116	G4	1	62.23	19.88
INM-1	2708	G6	1	00.00	20.69
INM-1	2031	G7	1	60.73	20.93
INM-1	2032	G7	1	62.98	19.96
INM-1	2033	G7	1	69.15	21.48
INM-1	3281	D10	1	69.75	23.55
INM-1	3282	D10	1	66.25	22.55
INM-1	3839	D25	1	63.85	20.00
INM-1	992	E10	1	59.21	20.99
INM-1	993	E10	1	66.27	22.18
INM-1	994	E10	1	00.00	16.87
INM-1	995	E10	1	64.39	20.77
INM-1	1188	F10	1	61.78	20.85
INM-1	3324	F11	1	00.00	19.70
INM-1	3325	F11	1	67.25	20.65
INM-1	3326	F11	1	65.60	21.60
INM-1	4909	F13	1	57.40	18.70
INM-1	8041	F24	1	61.70	20.05
INM-1	8042	F24	1	00.00	21.00
INM-1	7373	F26	1	59.35	18.98
INM-1	1662	G11	1	63.09	20.00
INM-1	1663	G11	1	00.00	22.38
INM-1	1664	G11	1	62.85	21.56

INM-1	2930	H10	1	61.91	20.66
INM-1	1927	H11	1	68.07	20.19
INM-1	1928	H11	1	66.06	21.71
INM-1	1929A	H11	1	57.98	22.19
INM-1	100	LJP	1	69.28	19.30
INM-1	2132	Hno132	1	71.69	21.62
INM-1	2133	Hno132	1	00.00	19.39
INM-1	2134	Hno132	1	67.28	21.63
INM-1	2136	Hno132	1	66.31	19.50
INM-1	2334		2	64.06	23.56
INM-1	2335		2	64.41	19.68
INM-1	3119		2	67.58	19.50
INM-1	1749	D1	2	59.85	17.46
INM-1	3082	G3	2	00.00	21.24
INM-1	3083	G3	2	63.13	18.97
INM-1	3953	G4	2	64.55	22.60
INM-1	2081	G6	2	54.61	21.75
INM-1	3631	G7	2	00.00	21.45
INM-1	3632	G7	2	66.70	00.00
INM-1	3633	G7	2	63.60	20.80
INM-1	3634	G7	2	58.85	19.65
INM-1	1563	H5	2	00.00	27.48
INM-1	1564	H5	2	64.80	21.15
INM-1	1565	H5	2	65.58	20.79
INM-1	1566	H5	2	69.53	22.04
INM-1	1567	H5	2	62.95	21.14
INM-1	1568	H5	2	00.00	21.14
INM-1	3385	H6	2	00.00	22.00
INM-1	2819	D10	2	66.44	21.23
INM-1	1434	D25	2	00.00	20.09
INM-1	1435	D25	2	63.40	22.10
INM-1	1436	D25	2	00.00	22.08
INM-1	4073	C0	2	00.00	21.15
INM-1	7278	E24	2	65.00	21.90
INM-1	7279	E24	2	66.10	20.35
INM-1	7280	E24	2	62.15	20.55
INM-1	7281	E24	2	00.00	16.55
INM-1	7282	E24	2	63.45	20.75
INM-1	7283	E24	2	00.00	21.85
INM-1	7339	E24	2	61.55	20.05
INM-1	6403	E25	2	62.85	20.95
INM-1	6404	E25	2	63.15	21.70
INM-1	6448	E25	2	63.15	20.50
INM-1	7420	E25	2	61.45	20.80
INM-1	7421	E25	2	66.46	21.90
INM-1	2998	F10	2	65.96	22.11
INM-1	2999	F10	2	70.32	19.65
INM-1	832	F12	2	69.21	22.36
INM-1	833	F12	2	57.97	17.37
INM-1	834	F12	2	72.31	23.18
INM-1	6612	F24	2	00.00	18.45
INM-1	6613	F24	2	62/30	22.80
INM-1	6614	F24	2	65.95	21.80
INM-1	7485	F24	2	67.15	22.50
INM-1	7560	F26	2	60.70	19.15
INM-1	3157	G11	2	66.97	22.58
INM-1	3158	G11	2	63.07	20.55
INM-1	3159	G11	2	62.39	20.08
INM-1	3160	G11	2	63.75	20.85
INM-1	3161	G11	2	65.96	21.82
INM-1	3162	G11	2	60.77	20.10
INM-1	3163	G11	2	67.04	22.48
INM-1	3164	G11	2	68.68	00.00

INM-1	7005	G24	2	69.45	23.50
INM-1	7006	G24	2	63.70	20.20
INM-1	68	G25	2	65.25	20.81
INM-1	69	G25	2	65.52	17.82
INM-1	105	G25	2	63.90	19.53
INM-1	197	G25	2	68.26	20.31
INM-1	7170	G25	2	66.75	20.10
INM-1	7171	G25	2	63.50	20.10
INM-1	2765	H11	2	00.00	19.37
INM-1	2766	H11	2	63.29	21.18
INM-1	2767	H11	2	00.00	20.90
INM-1	2768	H11	2	00.00	20.00
INM-1	3568	I10	2	61.81	19.40
INM-1	3576	I10	2	00.00	21.85
INM-1	755	I12	2	62.72	20.79
INM-1	756	I12	2	62.86	20.73
INM-1	757	I12	2	59.86	19.00
INM-1	4470	Hno129	2	63.13	21.60
INM-1	2846	G3	3	66.16	21.88
INM-1	3652	G7	3	00.00	20.90
INM-1	3445	H7	3	66.38	22.40
INM-1	3446	H7	3	68.22	22.49
INM-1	3447	H7	3	65.08	21.91
INM-1	3448	H7	3	00.00	22.32
INM-1	3449	H7	3	64.94	22.47
INM-1	3450	H7	3	00.00	23.30
INM-1	1240	D25	3	68.28	21.10
INM-1	1241	D25	3	00.00	21.51
INM-1	1242	D25	3	64.43	20.06
INM-1	1252	D25	3	00.00	18.15
INM-1	6721	E23	3	58.00	22.25
INM-1	6722	E23	3	68.65	22.65
INM-1	6723	E26	3	64/15	20.70
INM-1	6917	E26	3	00.00	20.40
INM-1	6918	E26	3	63.20	20.30
INM-1	6919	E26	3	63.20	20.50
INM-1	6920	E26	3	65.85	20.35
INM-1	6921	E26	3	63.95	19.30
INM-1	6922	E26	3	62.45	21.05
INM-1	6964	E26	3	00.00	18.15
INM-1	6498	F25	3	64.25	21.75
INM-1	2200	G10	3	66.57	21.98
INM-1	2201	G10	3	61.61	20.57
INM-1	2901	G10	3	00.00	26.26
INM-1	1298	G11	3	63.80	21.40
INM-1	1299	G11	3	00.00	20.81
INM-1	901	G12	3	66.62	21.46
INM-1	902	G12	3	68.26	22.28
INM-1	903	G12	3	64.05	21.04
INM-1	904	G12	3	63.73	18.10
INM-1	905	G12	3	60.10	18.63
INM-1	906	G12	3	68.23	20.39
INM-1	70	G25	3	47.74	00.00
INM-1	7803	G25	3	00.00	19.00
INM-1	7804	G25	3	62.55	21.00
INM-1	7805	G25	3	00.00	20.40
INM-1	7734	G26	3	66.85	22.00
INM-1	7884	G27	3	62.65	21.30
INM-1	7885	G27	3	59.76	18.80
INM-1	1771	H11	3	00.00	18.92
INM-1	3785	H12	3	65.90	21.65
INM-1	4513	H12	3	65.08	22.55
INM-1	4514	H12	3	67.85	23.62

INM-1	4515	H12	3	62.48	19.55
INM-1	4516	H12	3	71.85	24.85
INM-1	7918	Hno102	3	54.70	19.85
INM-1	6183	Hno106	3	00.00	19.95
INM-1	4875	Hno107	3	00.00	21.35
INM-1	1811	H5	4	65.57	00.00
INM-1	3050	H7	4	63.86	18.95
INM-1	8111	E24	4	63.85	21.75
INM-1	5868	E27	4	64.85	20.90
INM-1	7963	F24	4	00.00	17.90
INM-1	5962	F25	4	61.82	19.10
INM-1	1859	G11	4	67.35	00.00
INM-1	7077	G24	4	66.40	23.90
INM-1	7618	G25	4	64.20	21.05
INM-1	7619	G25	4	00.00	20.25
INM-1	7693	G25	4	60.60	21.60
INM-1	6812	G26	4	61.90	20.50
INM-1	6813	G26	4	66.32	21.80
INM-1	6814	G26	4	60.70	20.60
INM-1	6815	G26	4	00.00	20.25
INM-1	4784	B0	4	54.35	19.87
INM-1	7233	Hno104	4	00.00	20.35
INM-1	7234	Hno104	4	63.95	21.55
INM-1	6109	Hno105	4	62.48	17.75
INM-1	4645	Hno111	4	63.10	18.65
INM-1	4059	Hno115	4	70.25	22.00
INM-1	5022	Hno122	4	00.00	20.30
INM-1	2974	B3	5	00.00	22.85
INM-1	3807	E8	5	70.05	24.65
INM-1	4323	F8	5	64.76	21.10
INM-1	3774	F8	6	00.00	20.80
INM-1	1501	G4	6	69.36	20.63
INM-1	1502	G4	6	00.00	19.39
INM-1	2263	G5	6	66.81	22.59
INM-1	6121	Hno51	7	66.78	21.55
INM-1	4960	Kiln B1	8	59.95	17.70
INM-1	4601	Hno65	8	00.00	21.05
INM-1	4147	Hno73	8	55.43	18.30
INM-1	4353	C1	9	00.00	18.80
INM-1	5939	B3	12	59.85	20.25
INM-1	5783	Hno75	12	61.45	20.30
INM-1	5331	Hno79	12	00.00	20.08
INM-1	5443	B4	13	61.62	19.35
INM-1	5624	A4	15	69.90	23.15
INM-1	6369	B2	15	00.00	20.05
INM-1	5399	C3	15	00.00	20.90
INM-1	5687	D7	15	00.00	20.55
INM-1	6041	Hno87	16	61.80	20.65
INM-1	6042	Hno87	16	64.80	20.40

Table A.16.14. Measurements of Centrotarsal of ***Antilope cervicapra***

Site	Reg.No.	Trench	Layer	GB
INM-1	592	LJP	1-4	25.41
INM-1	1117	G4	1	26.66
INM-1	3832	D25	1	25.15
INM-1	3833	D25	1	24.40
INM-1	8046	F24	1	26.70
INM-1	1661	G11	1	23.21
INM-1	2932	H10	1	22.98
INM-1	1935	H11	1	25.24
INM-1	91	LJP	1	25.55

INM-1	92	LJP	1	23.07
INM-1	4252	Hno132	1	23.53
INM-1	3080	G3	2	24.75
INM-1	3081	G3	2	26.34
INM-1	3635	G7	2	22.94
INM-1	1628	H5	2	22.85
INM-1	3389	H6	2	22.60
INM-1	3390	H6	2	24.95
INM-1	7271	E24	2	23.55
INM-1	7272	E24	2	24.00
INM-1	6402	E25	2	23.20
INM-1	7418	E25	2	25.25
INM-1	2769	H11	2	26.77
INM-1	4985	C2	3	26.05
INM-1	2700	G7	3	20.88
INM-1	3456	H7	3	24.98
INM-1	1249	D25	3	24.87
INM-1	1250	D25	3	22.34
INM-1	1251	D25	3	21.75
INM-1	6725	E23	3	26.30
INM-1	6726	E23	3	22.50
INM-1	6727	E23	3	23.50
INM-1	6783	E23	3	22.75
INM-1	6786	E23	3	24.70
INM-1	6911	E26	3	23.90
INM-1	6912	E26	3	24.00
INM-1	6913	E26	3	25.55
INM-1	6914	E26	3	24.30
INM-1	6915	E26	3	24.70
INM-1	6916	E26	3	23.80
INM-1	6505	F25	3	22.30
INM-1	6506	F25	3	21.30
INM-1	2195	G10	3	24.66
INM-1	1293	G11	3	22.06
INM-1	2640	H11	3	24.04
INM-1	6181	Hno106	3	24.30
INM-1	6182	Hno106	3	23.46
INM-1	8110	E24	4	25.50
INM-1	5877	E25	4	25.00
INM-1	5878	E25	4	23.45
INM-1	5867	E27	4	23.86
INM-1	7959	F24	4	25.65
INM-1	7960	F24	4	22.25
INM-1	7961	F24	4	22.50
INM-1	7962	F24	4	23.35
INM-1	7079	G24	4	24.80
INM-1	7080	G24	4	24.70
INM-1	7125	G24	4	24.10
INM-1	7690	G25	4	23.17
INM-1	6811	G26	4	22.37
INM-1	672	Hno110	4	25.31
INM-1	673	Hno110	4	24.88
INM-1	3808	E8	5	26.69
INM-1	2007	G3	6	23.02
INM-1	1500	G4	6	23.10
INM-1	6122	Hno51	7	24.67
INM-1	7933	Hno51	7	23.47
INM-1	4949	E6	8	22.05
INM-1	3693	B1	12	28.38
INM-1	6328	B5	12	23.63
INM-1	6225	B6	12	25.30
INM-1	520	Hno50	12	24.89

Table A.16.15. Measurements of First Phalanx of ***Antilope cervicapra***

Site	Reg.No.	Trench	Layer	GL	BP	TP	BD	TD
INM-1	14	LJP	1-4	00.00	8.16	00.00	00.00	00.00
INM-1	73	LJP	1-4	42.73	12.58	16.82	9.63	10.43
INM-1	74	LJP	1-4	43.42	12.13	16.30	10.03	11.02
INM-1	75	LJP	1-4	48.72	11.69	16.52	9.79	11.12
INM-1	76	LJP	1-4	47.51	10.79	16.30	9.71	10.82
INM-1	97	LJP	1-4	48.16	10.70	15.91	9.69	10.25
INM-1	98	LJP	1-4	52.56	11.53	18.57	10.24	11.52
INM-1	99	LJP	1-4	55.20	11.78	18.08	10.40	11.57
INM-1	114	LJP	1-4	49.68	10.64	16.53	9.43	11.45
INM-1	115	LJP	1-4	44.60	11.86	17.20	9.52	10.64
INM-1	162	LJP	1-4	49.37	10.74	16.43	9.60	11.70
INM-1	163	LJP	1-4	58.29	11.96	17.62	10.46	11.45
INM-1	196	LJP	1-4	48.18	10.86	16.25	9.51	11.63
INM-1	246	LJP	1-4	46.97	12.90	16.50	10.58	10.27
INM-1	247	LJP	1-4	46.08	11.44	00.00	10.22	00.00
INM-1	250	LJP	1-4	54.24	11.69	17.41	10.60	11.59
INM-1	251	LJP	1-4	51.66	11.54	17.09	10.48	10.60
INM-1	252	LJP	1-4	50.08	11.27	16.76	9.58	10.77
INM-1	253	LJP	1-4	44.75	12.38	17.14	10.57	11.00
INM-1	254	LJP	1-4	50.56	11.41	17.28	10.43	11.34
INM-1	255	LJP	1-4	39.40	10.77	15.71	9.10	9.82
INM-1	260	LJP	1-4	44.52	10.77	15.87	9.63	10.33
INM-1	261	LJP	1-4	43.43	11.58	15.83	8.84	9.43
INM-1	262	LJP	1-4	49.33	11.16	15.08	9.86	10.99
INM-1	263	LJP	1-4	46.40	11.82	16.74	10.73	11.11
INM-1	340	LJP	1-4	48.61	12.01	16.25	9.85	10.58
INM-1	341	LJP	1-4	44.02	12.17	16.53	10.40	11.17
INM-1	358	LJP	1-4	46.21	12.77	17.28	10.54	11.04
INM-1	5268	LJP	1-4	52.50	11.95	17.36	11.00	12.05
INM-1	5269	LJP	1-4	55.75	11.95	16.65	10.69	11.65
INM-1	3794	G6	4	53.07	11.85	17.65	11.35	11.68
INM-1	3795	G6	4	51.90	11.50	16.95	10.30	11.60
INM-1	1111	G4	1	00.00	12.06	17.93	00.00	00.00
INM-1	1112	G4	1	43.85	12.02	16.92	10.59	10.76
INM-1	1113	G4	1	42.69	12.35	16.00	9.24	9.99
INM-1	2703	G6	1	49.66	11.83	16.59	9.88	11.26
INM-1	2704	G6	1	49.21	10.50	15.73	9.38	11.44
INM-1	2705	G6	1	00.00	10.46	16.10	00.00	00.00
INM-1	2024	G7	1	44.60	12.31	16.15	10.07	10.77
INM-1	2026	G7	1	00.00	00.00	00.00	10.72	12.07
INM-1	3515	H5	1	44.10	12.92	17.05	10.37	10.54
INM-1	3516	H5	1	00.00	11.45	17.24	00.00	00.00
INM-1	3517	H5	1	00.00	12.92	17.90	00.00	00.00
INM-1	3518	H5	1	00.00	12.32	16.45	00.00	00.00
INM-1	3519	H5	1	00.00	12.62	16.85	00.00	00.00
INM-1	3520	H5	1	00.00	00.00	00.00	9.90	12.30
INM-1	3521	H5	1	00.00	00.00	00.00	10.60	12.12
INM-1	3522	H5	1	00.00	00.00	00.00	10.40	11.98
INM-1	3820	D25	1	52.40	11.00	16.54	10.00	11.40
INM-1	3821	D25	1	00.00	00.00	00.00	10.45	12.20
INM-1	3822	D25	1	00.00	11.88	17.16	00.00	00.00
INM-1	3823	D25	1	41.90	12.45	15.85	9.90	10.40
INM-1	3824	D25	1	49.51	10.65	15.75	9.32	10.80
INM-1	3825	D25	1	00.00	00.00	00.00	10.90	12.20
INM-1	3854	D25	1	00.00	11.70	16.25	00.00	00.00
INM-1	3855	D25	1	00.00	00.00	00.00	10.60	12.05
INM-1	1000	E10	1	47.77	12.93	18.22	10.85	11.92
INM-1	1001	E10	1	51.67	12.44	17.35	10.87	12.09

INM-1	7046	E23	1	43.15	11.55	15.70	9.85	9.55
INM-1	7047	E23	1	00.00	11.15	19.30	00.00	00.00
INM-1	7048	E23	1	00.00	00.00	00.00	9.75	9.90
INM-1	1179	F10	1	41.40	10.77	15.32	9.37	9.78
INM-1	1210	F10	1	00.00	00.00	00.00	10.33	12.69
INM-1	3302	F11	1	40.10	12.85	16.05	9.85	10.20
INM-1	3303	F11	1	44.00	12.75	16.95	10.70	10.45
INM-1	3304	F11	1	47.95	13.50	18.40	11.10	12.30
INM-1	3305	F11	1	48.45	11.25	16.10	10.20	11.10
INM-1	3306	F11	1	53.90	12.30	17.40	10.95	11.95
INM-1	3307	F11	1	48.45	10.18	15.24	9.95	10.85
INM-1	4904	F13	1	49.75	10.30	16.30	9.95	10.70
INM-1	4905	F13	1	50.37	11.75	17.65	10.55	12.05
INM-1	4906	F13	1	00.00	13.15	19.15	00.00	00.00
INM-1	8036	F24	1	48.00	10.45	15.85	10.10	11.25
INM-1	8037	F24	1	48.55	10.30	15.95	10.05	11.15
INM-1	8038	F24	1	43.50	12.15	16.90	10.30	10.75
INM-1	8039	F24	1	41.95	11.82	16.30	9.90	10.15
INM-1	8040	F24	1	00.00	00.00	00.00	9.60	10.60
INM-1	7372	F26	1	52.92	10.95	16.42	9.76	10.85
INM-1	7393	F26	1	00.00	10.55	15.80	00.00	00.00
INM-1	1172	G10	1	44.34	12.34	16.63	10.93	10.79
INM-1	1637	G11	1	52.30	11.97	17.11	10.56	11.56
INM-1	1638	G11	1	52.52	12.11	17.76	10.91	11.49
INM-1	1639	G11	1	51.9	11.01	15.91	9.66	11.49
INM-1	1640	G11	1	51.29	11.15	17.74	10.08	11.29
INM-1	1641	G11	1	48.95	11.16	15.78	10.40	11.23
INM-1	1642	G11	1	45.41	12.51	16.52	10.69	11.11
INM-1	1643	G11	1	43.45	11.69	16.32	9.73	9.97
INM-1	1644	G11	1	40.82	11.95	16.26	10.44	10.84
INM-1	1645	G11	1	00.00	11.17	16.19	00.00	00.00
INM-1	1646	G11	1	00.00	00.00	00.00	10.48	10.97
INM-1	1647	G11	1	00.00	00.00	00.00	9.55	9.41
INM-1	1648	G11	1	00.00	00.00	00.00	10.04	11.50
INM-1	7525	G24	1	48.90	11.70	17.15	10.25	11.55
INM-1	7526	G24	1	53.55	11.45	16.15	00.00	12.20
INM-1	7527	G24	1	00.00	10.17	15.35	00.00	00.00
INM-1	7528	G24	1	00.00	11.95	16.10	00.00	00.00
INM-1	7529	G24	1	00.00	00.00	00.00	9.55	11.70
INM-1	2919	H10	1	44.88	11.08	15.32	9.27	9.60
INM-1	2920	H10	1	48.45	12.60	17.21	10.84	10.96
INM-1	2921	H10	1	42.22	11.54	16.05	9.86	10.28
INM-1	2922	H10	1	00.00	00.00	00.00	9.97	10.92
INM-1	2923	H10	1	00.00	00.00	00.00	9.54	11.15
INM-1	2924	H10	1	00.00	00.00	00.00	9.88	11.33
INM-1	1908	H11	1	00.00	21.88	30.73	00.00	00.00
INM-1	1909	H11	1	53.24	11.55	17.40	10.74	12.76
INM-1	1910	H11	1	49.11	10.62	16.10	10.14	11.73
INM-1	1911	H11	1	50.27	12.34	18.04	11.68	11.80
INM-1	1912	H11	1	50.61	11.44	17.23	10.51	12.20
INM-1	1913	H11	1	44.98	12.92	17.11	00.00	11.35
INM-1	1914	H11	1	46.14	11.58	16.86	10.28	10.56
INM-1	1915	H11	1	44.42	12.77	17.38	10.59	11.22
INM-1	1916	H11	1	39.78	11.52	16.25	9.40	10.14
INM-1	1917	H11	1	40.97	12.27	16.61	10.48	00.00
INM-1	1918	H11	1	00.00	00.00	00.00	10.11	10.91
INM-1	1919	H11	1	00.00	00.00	00.00	10.03	11.39
INM-1	1920	H11	1	00.00	00.00	00.00	10.26	10.18
INM-1	1921	H11	1	00.00	00.00	00.00	9.61	9.41
INM-1	3761	I11	1	43.85	11.90	16.81	10.20	11.35
INM-1	2116	Hno132	1	53.21	11.09	16.72	10.16	11.17
INM-1	2117	Hno132	1	51.58	11.65	17.34	10.37	12.01
INM-1	2118	Hno132	1	51.47	10.97	16.46	10.61	11.04

INM-1	2119	Hno132	1	47.58	10.41	15.74	8.97	10.42
INM-1	2120	Hno132	1	45.41	12.33	16.59	9.96	10.14
INM-1	2121	Hno132	1	45.02	12.18	17.14	10.15	10.43
INM-1	2122	Hno132	1	44.62	12.35	16.82	10.49	11.10
INM-1	2123	Hno132	1	37.12	11.04	14.59	9.22	9.69
INM-1	2124	Hno132	1	00.00	11.04	17.97	00.00	00.00
INM-1	2125	Hno132	1	00.00	11.03	15.54	00.00	00.00
INM-1	2126	Hno132	1	00.00	00.00	00.00	9.61	11.00
INM-1	2174	Hno132	1	45.05	11.33	16.78	9.95	10.08
INM-1	2330		2	52.72	11.00	16.44	9.53	11.57
INM-1	2331		2	51.89	11.61	17.19	10.76	11.50
INM-1	2770	E1	2	50.06	11.18	17.24	10.19	10.88
INM-1	2306	E1	2	00.00	00.00	00.00	8.97	10.02
INM-1	3108	E2	2	00.00	10.65	14.80	00.00	00.00
INM-1	3109	E2	2	00.00	11.60	17.15	00.00	00.00
INM-1	3110	E2	2	00.00	00.00	00.00	11.66	10.54
INM-1	3930	G4	2	44.75	12.55	17.85	10.65	10.60
INM-1	3931	G4	2	50.80	11.65	16.95	10.45	10.75
INM-1	3932	G4	2	47.00	13.35	17.85	11.20	12.05
INM-1	3933	G4	2	41.60	11.40	16.00	9.90	10.00
INM-1	3934	G4	2	51.12	11.75	18.00	10.80	11.95
INM-1	3935	G4	2	46.85	12.55	17.70	10.75	12.40
INM-1	3936	G4	2	40.30	11.75	16.65	9.95	10.65
INM-1	3937	G4	2	00.00	00.00	00.00	9.85	10.75
INM-1	3938	G4	2	00.00	00.00	00.00	9.47	10.70
INM-1	3939	G4	2	00.00	13.20	17.45	00.00	00.00
INM-1	3940	G4	2	00.00	12.55	17.45	00.00	00.00
INM-1	3941	G4	2	00.00	00.00	00.00	8.55	10.25
INM-1	3942	G4	2	00.00	00.00	00.00	9.70	11.00
INM-1	2068	G6	2	49.64	10.45	16.31	10.18	11.55
INM-1	2069	G6	2	00.00	12.42	17.55	00.00	00.00
INM-1	2070	G6	2	00.00	11.87	17.43	00.00	00.00
INM-1	2071	G6	2	00.00	10.50	14.62	00.00	00.00
INM-1	2072	G6	2	00.00	10.75	17.34	00.00	00.00
INM-1	2073	G6	2	00.00	11.96	17.64	00.00	00.00
INM-1	2074	G6	2	00.00	00.00	00.00	8.23	10.34
INM-1	2075	G6	2	00.00	00.00	00.00	10.03	11.74
INM-1	3617	G7	2	46.45	13.00	17.80	10.95	11.25
INM-1	3618	G7	2	47.80	10.45	15.65	9.30	11.00
INM-1	3619	G7	2	48.47	10.90	15.60	9.45	11.10
INM-1	3620	G7	2	45.20	12.20	16.30	10.50	10.85
INM-1	3621	G7	2	50.00	11.80	16.85	9.70	10.85
INM-1	3622	G7	2	00.00	11.60	17.25	00.00	00.00
INM-1	3623	G7	2	00.00	10.70	16.65	00.00	00.00
INM-1	3624	G7	2	00.00	00.00	00.00	10.36	10.85
INM-1	3625	G7	2	00.00	00.00	00.00	11.60	12.17
INM-1	1531	H5	2	50.60	10.65	16.52	9.95	11.34
INM-1	1532	H5	2	43.72	12.47	16.55	10.75	10.62
INM-1	1533	H5	2	45.72	10.94	15.25	9.48	10.25
INM-1	1534	H5	2	45.15	12.82	17.90	10.58	10.69
INM-1	1535	H5	2	42.71	10.93	15.77	9.58	9.99
INM-1	1536	H5	2	40.17	12.28	16.17	9.88	9.96
INM-1	1537	H5	2	00.00	11.65	17.57	00.00	00.00
INM-1	1538	H5	2	00.00	11.08	15.73	00.00	00.00
INM-1	1539	H5	2	00.00	11.75	16.62	00.00	00.00
INM-1	1541	H5	2	00.00	00.00	00.00	10.40	11.33
INM-1	1542	H5	2	00.00	00.00	00.00	9.98	10.47
INM-1	1543	H5	2	00.00	00.00	00.00	10.87	11.16
INM-1	1544	H5	2	00.00	00.00	00.00	10.34	11.34
INM-1	1545	H5	2	00.00	00.00	00.00	10.15	11.41
INM-1	1546	H5	2	00.00	00.00	00.00	9.50	9.74
INM-1	1547	H5	2	00.00	00.00	00.00	11.12	11.75
INM-1	1548	H5	2	00.00	00.00	00.00	11.05	10.71

INM-1	3369	H6	2	44.21	10.62	15.30	8.65	9.50
INM-1	3370	H6	2	43.41	11.95	16.35	10.30	10.88
INM-1	3371	H6	2	48.15	12.60	17.80	10.85	10.96
INM-1	3372	H6	2	50.25	10.90	15.77	9.60	10.94
INM-1	3373	H6	2	48.91	11.20	15.45	9.85	10.36
INM-1	3375	H6	2	00.00	10.65	15.13	00.00	00.00
INM-1	3376	H6	2	00.00	12.30	17.30	00.00	00.00
INM-1	3377	H6	2	00.00	12.00	17.55	00.00	00.00
INM-1	3378	H6	2	00.00	00.00	00.00	9.10	10.68
INM-1	1023	H7	2	54.76	11.96	17.47	10.93	12.91
INM-1	1024	H7	2	47.79	12.33	17.69	00.00	11.63
INM-1	1025	H7	2	52.06	10.99	17.56	9.89	11.32
INM-1	1026	H7	2	42.92	12.11	16.77	10.29	10.65
INM-1	1027	H7	2	00.00	11.08	16.74	00.00	00.00
INM-1	1028	H7	2	42.67	11.87	14.81	10.00	9.79
INM-1	1029	H7	2	48.93	10.68	15.67	9.57	11.54
INM-1	1030	H7	2	47.48	12.74	17.85	12.74	10.72
INM-1	1031	H7	2	49.12	10.57	16.06	9.66	11.03
INM-1	1032	H7	2	00.00	00.00	00.00	11.52	9.81
INM-1	1033	H7	2	00.00	00.00	00.00	12.19	10.65
INM-1	1034	H7	2	00.00	00.00	00.00	12.13	10.40
INM-1	1035	H7	2	00.00	00.00	00.00	8.52	8.44
INM-1	1036	H7	2	00.00	00.00	00.00	00.00	9.73
INM-1	1037	H7	2	00.00	00.00	00.00	10.12	11.29
INM-1	1038	H7	2	00.00	00.00	00.00	10.06	00.00
INM-1	1039	H7	2	00.00	00.00	00.00	00.00	11.51
INM-1	1040	H7	2	00.00	00.00	00.00	10.11	11.89
INM-1	3581	I6	2	41.62	11.83	16.74	10.09	10.00
INM-1	1991	D10	2	42.63	11.65	16.17	10.43	10.34
INM-1	1998	D10	2	00.00	00.00	00.00	9.74	11.52
INM-1	1381	D25	2	00.00	00.00	00.00	10.35	12.22
INM-1	1382	D25	2	00.00	00.00	00.00	9.79	19.57
INM-1	1383	D25	2	00.00	00.00	00.00	10.09	11.75
INM-1	1384	D25	2	00.00	00.00	00.00	10.82	11.28
INM-1	1385	D25	2	00.00	00.00	00.00	11.50	12.09
INM-1	1386	D25	2	00.00	00.00	00.00	10.10	11.05
INM-1	1387	D25	2	00.00	00.00	00.00	9.52	10.04
INM-1	1388	D25	2	00.00	00.00	00.00	10.19	11.07
INM-1	1389	D25	2	00.00	00.00	00.00	10.43	10.57
INM-1	1390	D25	2	00.00	11.37	16.27	00.00	00.00
INM-1	1391	D25	2	00.00	11.67	16.96	00.00	00.00
INM-1	1392	D25	2	00.00	10.99	15.89	00.00	00.00
INM-1	1394	D25	2	00.00	11.97	19.14	00.00	00.00
INM-1	1395	D25	2	00.00	12.03	17.98	00.00	00.00
INM-1	1396	D25	2	36.73	00.00	00.00	8.98	9.87
INM-1	1397	D25	2	38.56	11.01	15.61	8.37	8.34
INM-1	1398	D25	2	42.07	12.33	16.32	10.26	9.93
INM-1	1399	D25	2	41.88	11.20	16.15	10.12	10.20
INM-1	1400	D25	2	49.37	10.28	15.52	9.66	11.37
INM-1	1401	D25	2	50.14	11.47	16.95	10.36	11.45
INM-1	1402	D25	2	53.50	11.21	17.12	10.39	11.05
INM-1	1403	D25	2	54.79	12.44	18.38	11.44	13.43
INM-1	1404	D25	2	50.27	11.14	16.95	9.99	11.10
INM-1	1405	D25	2	51.04	10.34	15.95	10.07	11.10
INM-1	1406	D25	2	50.30	11.94	16.42	9.91	11.79
INM-1	1407	D25	2	00.00	00.00	00.00	9.89	10.96
INM-1	1408	D25	2	50.48	11.35	15.95	10.63	11.49
INM-1	1409	D25	2	54.24	11.26	17.41	10.95	11.46
INM-1	1478	D25	2	00.00	00.00	00.00	10.71	10.85
INM-1	7253	D25	2	46.87	11.10	15.90	10.10	10.78
INM-1	7254	E24	2	44.30	12.25	17.30	10.88	10.40
INM-1	7255	E24	2	47.40	10.45	15.10	9.35	10.60
INM-1	7256	E24	2	41.66	10.50	15.30	9.60	9.70

INM-1	7257	E24	2	38.55	10.65	14.45	8.40	9.55
INM-1	7258	E24	2	00.00	12.10	17.25	00.00	00.00
INM-1	7259	E24	2	00.00	10.85	16.30	00.00	00.00
INM-1	7260	E24	2	00.00	10.15	14.90	00.00	00.00
INM-1	7261	E24	2	00.00	10.20	14.90	00.00	00.00
INM-1	7262	E24	2	00.00	00.00	00.00	10.50	12.00
INM-1	7337	E24	2	40.75	10.55	14.75	9.40	9.10
INM-1	6376	E24	2	46.70	10.85	16.40	9.80	10.50
INM-1	6377	E25	2	46.55	10.25	15.76	9.70	00.00
INM-1	6378	E25	2	00.00	00.00	14.70	9.19	9.90
INM-1	6379	E25	2	44.30	11.10	14.85	10.20	9.90
INM-1	6380	E25	2	49.55	11.65	15.45	9.60	10.80
INM-1	6381	E25	2	00.00	11.75	16.30	00.00	00.00
INM-1	6382	E25	2	00.00	11.70	16.95	00.00	00.00
INM-1	6383	E25	2	00.00	11.60	16.60	00.00	00.00
INM-1	6384	E25	2	00.00	00.00	00.00	9.45	10.00
INM-1	6385	E25	2	00.00	00.00	00.00	9.75	10.30
INM-1	7394	E25	2	42.70	10.70	15.10	9.15	9.45
INM-1	7395	E25	2	46.35	10.70	15.65	9.95	11.45
INM-1	7396	E25	2	45.15	11.95	17.90	10.55	10.65
INM-1	7397	E25	2	47.25	10.35	15.94	9.55	11.00
INM-1	7398	E25	2	43.80	11.50	16.00	10.05	10.75
INM-1	7399	E25	2	40.20	12.00	15.20	9.45	9.80
INM-1	7400	E25	2	00.00	11.40	15.70	00.00	00.00
INM-1	7401	E25	2	00.00	11.10	16.35	00.00	00.00
INM-1	7402	E25	2	00.00	11.85	17.15	00.00	00.00
INM-1	7403	E25	2	00.00	11.30	16.05	00.00	00.00
INM-1	7404	E25	2	00.00	00.00	00.00	9.57	10.60
INM-1	7405	E25	2	00.00	00.00	00.00	8.25	9.85
INM-1	7406	E25	2	00.00	00.00	00.00	9.65	10.80
INM-1	2515	F11	2	50.15	11.24	17.02	10.18	12.06
INM-1	2516	F11	2	51.12	12.32	15.95	10.30	1187
INM-1	2517	F11	2	51.51	13.39	20.20	10.19	11.42
INM-1	2518	F11	2	50.29	10.69	17.14	9.92	11.37
INM-1	2519	F11	2	52.17	11.23	16.48	10.20	11.72
INM-1	2520	F11	2	50.77	11.91	16.96	9.98	11.81
INM-1	2521	F11	2	43.38	10.86	15.83	9.05	9.25
INM-1	2522	F11	2	00.00	11.97	18.24	00.00	00.00
INM-1	2523	F11	2	11.65	17.22	00.00	00.00	00.00
INM-1	2524	F11	2	00.00	12.22	17.07	00.00	00.00
INM-1	2525	F11	2	00.00	00.00	00.00	10.48	10.90
INM-1	2526	F11	2	00.00	00.00	00.00	10.56	11.16
INM-1	2527	F11	2	00.00	00.00	00.00	10.37	10.28
INM-1	6599	F24	2	41.95	12.10	17.25	10.62	11.20
INM-1	6600	F24	2	46.90	11.95	15.95	11.10	11.25
INM-1	6601	F24	2	00.00	11.80	16.05	00.00	00.00
INM-1	6602	F24	2	00.00	12.30	15.85	00.00	00.00
INM-1	6603	F24	2	00.00	00.00	00.00	10.40	11.95
INM-1	6604	F24	2	00.00	00.00	00.00	11.25	10.80
INM-1	6605	F24	2	00.00	00.00	00.00	10.65	12.45
INM-1	6606	F24	2	00.00	00.00	00.00	8.90	10.05
INM-1	7474	F24	2	51.65	12.00	17.00	10.50	11.70
INM-1	7475	F24	2	50.85	11.90	16.80	9.90	11.30
INM-1	7476	F24	2	46.60	9.65	14.55	9.35	10.48
INM-1	7477	F24	2	43.35	10.75	16.35	9.83	9.80
INM-1	7478	F24	2	43.05	11.85	15.90	00.00	9.80
INM-1	7479	F24	2	00.00	00.00	00.00	9.40	11.40
INM-1	7480	F24	2	00.00	00.00	00.00	10.65	11.90
INM-1	7546	F26	2	48.25	9.85	16.25	9.55	11.20
INM-1	7547	F26	2	41.27	11.75	16.84	10.40	11.05
INM-1	7548	F26	2	44.36	11.95	16.15	10.30	10.45
INM-1	7549	F26	2	43.75	11.10	16.12	10.55	9.75
INM-1	7550	F26	2	43.36	12.25	17.06	10.80	10.70

INM-1	7551	F26	2	44.25	11.45	16.50	10.95	10.17
INM-1	7552	F26	2	00.00	00.00	00.00	8.96	10.10
INM-1	7553	F26	2	00.00	11.15	16.52	00.00	00.00
INM-1	7554	F26	2	00.00	11.20	16.15	00.00	00.00
INM-1	7555	F26	2	00.00	00.00	00.00	10.85	11.95
INM-1	7556	F26	2	00.00	00.00	00.00	10.70	10.90
INM-1	7557	F26	2	00.00	00.00	00.00	8.00	9.85
INM-1	7558	F26	2	00.00	00.00	00.00	9.65	9.70
INM-1	1312	G11	2	55.10	12.07	17.83	10.99	11.76
INM-1	1313	G11	2	52.80	12.08	17.56	10.56	11.51
INM-1	1314	G11	2	00.00	00.00	00.00	11.22	13.25
INM-1	1315	G11	2	00.00	12.37	17.45	00.00	00.00
INM-1	1316	G11	2	41.82	12.47	16.67	19.54	11.24
INM-1	1317	G11	2	42.07	12.76	16.61	9.63	10.10
INM-1	1318	G11	2	46.46	13.12	17.84	10.83	10.96
INM-1	1319	G11	2	51.87	11.24	16.86	9.56	10.92
INM-1	1320	G11	2	49.41	11.25	15.43	10.17	10.64
INM-1	1321	G11	2	50.66	12.42	16.66	10.79	11.27
INM-1	1322	G11	2	46.97	12.74	17.50	10.35	11.24
INM-1	1323	G11	2	00.00	00.00	00.00	9.90	9.68
INM-1	3125	G11	2	52.17	11.55	17.87	11.55	11.61
INM-1	3126	G11	2	52.72	11.37	16.51	10.48	11.90
INM-1	3127	G11	2	49.82	11.35	15.65	10.33	11.72
INM-1	3128	G11	2	49.71	10.77	15.81	8.90	11.10
INM-1	3129	G11	2	44.90	12.67	16.86	10.92	11.40
INM-1	3130	G11	2	48.54	11.40	16.68	10.90	11.50
INM-1	3131	G11	2	53.62	11.35	16.97	9.95	10.88
INM-1	3132	G11	2	52.82	11.70	16.29	10.79	11.73
INM-1	3133	G11	2	43.54	11.70	16.11	9.76	10.90
INM-1	3134	G11	2	47.15	12.42	17.62	10.56	10.93
INM-1	3135	G11	2	52.24	12.26	18.03	12.45	10.97
INM-1	3136	G11	2	48.41	9.93	15.43	9.21	11.10
INM-1	3137	G11	2	52.64	11.02	16.35	10.00	11.85
INM-1	3138	G11	2	51.70	11.32	17.28	10.44	11.00
INM-1	3139	G11	2	44.63	11.63	15.90	10.16	10.90
INM-1	3140	G11	2	49.12	10.73	15.57	9.97	11.91
INM-1	3141	G11	2	57.20	11.47	18.60	10.62	12.36
INM-1	3142	G11	2	00.00	12.22	17.37	00.00	00.00
INM-1	3143	G11	2	00.00	11.99	16.81	00.00	00.00
INM-1	3144	G11	2	00.00	10.98	17.03	00.00	00.00
INM-1	3145	G11	2	00.00	00.00	00.00	10.76	11.87
INM-1	3146	G11	2	00.00	00.00	00.00	9.35	10.16
INM-1	6995	G24	2	48.00	10.15	15.55	9.95	11.10
INM-1	6996	G24	2	42.60	11.55	16.30	9.70	9.85
INM-1	6997	G24	2	41.90	11.75	16.20	10.05	10.15
INM-1	6998	G24	2	00.00	10.70	16.95	00.00	00.00
INM-1	6999	G24	2	00.00	11.70	16.45	00.00	00.00
INM-1	7000	G24	2	00.00	00.00	00.00	9.85	10.75
INM-1	7001	G24	2	00.00	00.00	00.00	11.05	10.96
INM-1	7152	G25	2	47.30	13.05	17.10	10.80	12.10
INM-1	7153	G25	2	49.70	10.80	15.90	10.30	11.90
INM-1	7154	G25	2	54.95	12.75	18.45	11.55	12.30
INM-1	7155	G25	2	50.35	11.46	16.65	10.45	11.45
INM-1	7156	G25	2	50.05	11.25	16.35	9.70	11.48
INM-1	7157	G25	2	00.00	11.25	15.50	00.00	00.00
INM-1	2749	H11	2	00.00	00.00	00.00	9.51	9.65
INM-1	2750	H11	2	00.00	00.00	00.00	00.00	11.05
INM-1	2751	H11	2	00.00	00.00	00.00	9.77	10.07
INM-1	2752	H11	2	00.00	11.41	15.95	00.00	00.00
INM-1	2753	H11	2	00.00	11.67	15.86	00.00	00.00
INM-1	3565	I10	2	44.85	11.55	16.40	10.15	10.80
INM-1	3566	I10	2	47.45	12.20	17.30	10.55	10.92
INM-1	753	I12	2	48.46	11.01	16.05	10.32	11.22

INM-1	754	I12	2	44.87	11.58	17.35	10.22	11.21
INM-1	364	LJP	2	52.23	10.59	16.39	9.32	10.96
INM-1	7902	Hno108	2	49.30	00.00	17.20	9.82	11.08
INM-1	7903	Hno108	2	41.85	11.90	15.30	10.65	10.15
INM-1	4462	Hno129	2	41.73	11.79	15.35	9.90	10.25
INM-1	4463	Hno129	2	00.00	11.33	15.52	00.00	00.00
INM-1	4464	Hno129	2	00.00	10.92	15.15	00.00	00.00
INM-1	4249	Hno132	2	46.09	00.00	17.33	9.80	10.80
INM-1	2837	G3	3	53.16	11.65	17.57	10.75	11.89
INM-1	2838	G3	3	44.84	11.91	16.77	10.31	11.00
INM-1	2839	G3	3	51.52	11.13	16.99	10.50	11.74
INM-1	2840	G3	3	00.00	11.96	16.79	00.00	00.00
INM-1	2841	G3	3	00.00	11.26	16.57	00.00	00.00
INM-1	2842	G3	3	00.00	00.00	00.00	9.50	9.51
INM-1	3602	G4	3	44.15	11.55	15.80	00.00	10.35
INM-1	2648	G7	3	50.14	11.14	17.12	10.43	11.14
INM-1	2649	G7	3	47.02	11.23	16.12	9.81	10.51
INM-1	2650	G7	3	51.14	11.12	16.60	10.10	11.80
INM-1	2651	G7	3	43.26	11.35	15.75	10.01	10.64
INM-1	2652	G7	3	48.84	12.84	17.93	10.83	11.21
INM-1	2653	G7	3	41.21	11.64	15.57	8.99	9.52
INM-1	2654	G7	3	00.00	00.00	00.00	10.96	12.36
INM-1	2655	G7	3	00.00	00.00	00.00	9.54	11.45
INM-1	2656	G7	3	00.00	00.00	00.00	10.14	11.35
INM-1	2657	G7	3	00.00	10.52	15.93	00.00	00.00
INM-1	2658	G7	3	00.00	12.34	16.72	00.00	00.00
INM-1	2699	G7	3	00.00	10.63	15.11	00.00	00.00
INM-1	3416	H7	3	00.00	10.66	16.15	10.23	11.15
INM-1	3417	H7	3	42.85	12.60	17.03	9.96	10.63
INM-1	3418	H7	3	46.25	10.25	15.40	8.90	10.30
INM-1	3419	H7	3	44.25	11.63	16.05	9.71	10.30
INM-1	3420	H7	3	00.00	00.00	00.00	8.35	9.30
INM-1	3421	H7	3	00.00	00.00	00.00	10.10	11.40
INM-1	3422	H7	3	43.00	12.45	16.31	9.65	10.70
INM-1	3423	H7	3	00.00	10.75	15.45	00.00	00.00
INM-1	3424	H7	3	00.00	10.60	15.95	00.00	00.00
INM-1	3425	H7	3	00.00	10.71	16.05	00.00	00.00
INM-1	3426	H7	3	00.00	11.54	15.45	00.00	00.00
INM-1	3427	H7	3	00.00	12.45	17.90	00.00	00.00
INM-1	3428	H7	3	00.00	11.45	17.40	00.00	00.00
INM-1	3429	H7	3	00.00	00.00	00.00	10.00	11.25
INM-1	3430	H7	3	00.00	00.00	00.00	10.25	10.45
INM-1	3593	I6	3	47.34	11.15	15.45	10.05	10.75
INM-1	4095	D10	3	00.00	00.00	00.00	9.95	12.15
INM-1	1211	D25	3	44.72	12.04	17.56	10.69	10.21
INM-1	1212	D25	3	00.00	00.00	00.00	9.64	9.85
INM-1	1213	D25	3	00.00	11.57	16.82	00.00	00.00
INM-1	1214	D25	3	44.09	11.51	16.31	10.24	10.26
INM-1	1215	D25	3	44.75	11.80	15.92	10.27	10.43
INM-1	1216	D25	3	42.09	9.98	14.20	9.10	9.01
INM-1	1878	E10	3	00.00	11.98	16.68	00.00	00.00
INM-1	1879	E10	3	00.00	00.00	00.00	8.53	10.84
INM-1	6690	E23	3	51.75	11.05	16.60	10.00	10.35
INM-1	6691	E23	3	48.95	10.80	16.50	9.85	11.25
INM-1	6692	E23	3	49.00	10.71	16.45	9.85	11.10
INM-1	6693	E23	3	48.95	10.70	16.10	9.70	10.86
INM-1	6694	E23	3	44.40	11.70	16.35	9.65	10.25
INM-1	6695	E23	3	41.45	11.90	16.70	9.80	10.25
INM-1	6696	E23	3	00.00	11.65	16.65	00.00	00.00
INM-1	6697	E23	3	00.00	10.65	16.00	00.00	00.00
INM-1	6698	E23	3	00.00	11.45	15.80	00.00	00.00
INM-1	6699	E23	3	00.00	12.10	16.45	00.00	00.00
INM-1	6700	E23	3	00.00	00.00	00.00	10.05	11.15

INM-1	6701	E23	3	00.00	00.00	00.00	10.15	10.70
INM-1	6702	E23	3	00.00	00.00	00.00	10.45	11.55
INM-1	6752	E23	3	47.45	10.40	14.96	9.85	10.70
INM-1	6753	E23	3	42.70	11.20	15.50	9.55	10.00
INM-1	6754	E23	3	41.30	12.55	16.10	9.95	10.06
INM-1	6868	E26	3	52.35	11.40	16.94	10.90	11.65
INM-1	6869	E26	3	46.40	10.45	15.30	9.75	10.30
INM-1	6870	E26	3	53.20	11.85	16.90	10.25	12.25
INM-1	6871	E26	3	43.40	11.90	16.20	9.85	9.70
INM-1	6872	E26	3	47.25	00.00	00.00	9.20	10.45
INM-1	6873	E26	3	00.00	10.45	15.90	00.00	00.00
INM-1	6874	E26	3	00.00	11.64	16.50	00.00	00.00
INM-1	6875	E26	3	00.00	11.10	16.90	00.00	00.00
INM-1	6876	E26	3	00.00	11.70	17.64	00.00	00.00
INM-1	6877	E26	3	00.00	00.00	00.00	8.86	9.45
INM-1	6878	E26	3	00.00	00.00	00.00	10.30	10.68
INM-1	6879	E26	3	00.00	00.00	00.00	10.15	10.95
INM-1	6880	E26	3	00.00	00.00	00.00	10.26	11.70
INM-1	6881	E26	3	00.00	00.00	00.00	10.55	11.50
INM-1	6882	E26	3	00.00	00.00	00.00	9.40	9.75
INM-1	6883	E26	3	00.00	00.00	00.00	9.50	10.15
INM-1	6884	E26	3	00.00	00.00	00.00	9.57	11.40
INM-1	1817	F10	3	49.41	10.74	16.30	9.93	11.14
INM-1	2242	F11	3	47.66	10.13	16.18	9.34	10.94
INM-1	2243	F11	3	44.99	11.84	17.83	10.58	11.34
INM-1	3268	F11	3	00.00	10.35	15.00	00.00	00.00
INM-1	6474	F25	3	44.75	11.88	16.10	10.15	10.40
INM-1	6475	F25	3	50.66	11.15	16.65	10.00	11.25
INM-1	6476	F25	3	00.00	12.05	17.25	00.00	00.00
INM-1	6477	F25	3	00.00	10.70	16.05	00.00	00.00
INM-1	6478	F25	3	00.00	11.34	15.60	00.00	00.00
INM-1	6479	F25	3	00.00	12.25	16.65	00.00	00.00
INM-1	6480	F25	3	00.00	00.00	00.00	9.65	10.69
INM-1	6481	F25	3	00.00	00.00	00.00	10.90	11.45
INM-1	6482	F25	3	00.00	00.00	00.00	10.37	11.40
INM-1	6483	F25	3	00.00	00.00	00.00	9.65	10.75
INM-1	6484	F25	3	00.00	00.00	00.00	9.73	11.25
INM-1	6485	F25	3	00.00	00.00	00.00	8.45	9.70
INM-1	2182	G10	3	40.81	12.09	16.68	9.49	10.13
INM-1	2183	G10	3	48.59	10.95	15.73	9.67	11.39
INM-1	2184	G10	3	51.30	11.57	16.59	10.47	12.24
INM-1	2185	G10	3	45.47	12.53	16.43	10.45	11.02
INM-1	2186	G10	3	00.00	00.00	00.00	10.60	10.80
INM-1	2187	G10	3	00.00	11.24	17.31	00.00	00.00
INM-1	2188	G10	3	00.00	12.04	17.89	00.00	00.00
INM-1	2889	G10	3	47.88	10.64	15.26	9.79	10.73
INM-1	2890	G10	3	49.48	11.43	15.96	9.45	11.03
INM-1	2891	G10	3	46.87	10.18	14.91	9.16	10.23
INM-1	2892	G10	3	50.06	12.18	16.55	10.02	12.12
INM-1	1290	G11	3	52.76	12.72	16.53	10.87	12.32
INM-1	1291	G11	3	44.57	10.93	16.18	10.02	10.38
INM-1	3741	G11	3	41.17	10.40	15.15	8.65	9.29
INM-1	3742	G11	3	44.44	12.40	16.50	10.05	11.15
INM-1	3743	G11	3	00.00	11.30	16.84	00.00	00.00
INM-1	3744	G11	3	00.00	10.85	11.75	00.00	00.00
INM-1	874	G12	3	00.00	12.22	16.70	9.90	10.78
INM-1	875	G12	3	50.93	10.85	15.64	9.70	10.95
INM-1	876	G12	3	50.71	11.52	16.16	10.14	10.95
INM-1	877	G12	3	53.47	12.12	17.02	10.35	11.39
INM-1	7780	G25	3	46.85	10.35	15.05	9.37	10.15
INM-1	7781	G25	3	50.90	11.30	16.40	10.35	11.60
INM-1	7782	G25	3	50.20	10.05	15.85	9.85	10.85
INM-1	7783	G25	3	54.75	11.55	17.65	10.40	11.90

INM-1	7784	G25	3	52.96	11.60	16.40	10.15	10.95
INM-1	7785	G25	3	47.57	10.70	15.90	10.40	10.55
INM-1	7786	G25	3	42.80	12.20	17.25	10.30	11.05
INM-1	7787	G25	3	00.00	12.00	17.55	00.00	00.00
INM-1	7788	G25	3	00.00	11.55	16.07	00.00	00.00
INM-1	7789	G25	3	00.00	12.30	17.95	00.00	00.00
INM-1	7790	G25	3	00.00	11.80	17.15	00.00	00.00
INM-1	7791	G25	3	00.00	11.70	16.25	00.00	00.00
INM-1	7792	G25	3	00.00	11.50	16.80	00.00	00.00
INM-1	7793	G25	3	00.00	00.00	00.00	10.15	12.15
INM-1	7794	G25	3	00.00	00.00	00.00	10.85	11.20
INM-1	7795	G25	3	00.00	00.00	00.00	10.70	12.00
INM-1	7859	G25	3	00.00	00.00	00.00	9.76	11.15
INM-1	7700	G26	3	46.80	12.10	17.30	10..46	10.75
INM-1	7701	G26	3	45.55	12.43	17.55	10.35	10.77
INM-1	7702	G26	3	43.45	11.20	15.50	9.75	9.85
INM-1	7703	G26	3	00.00	00.00	15.30	9.50	10.35
INM-1	7704	G26	3	00.00	10.10	14.95	00.00	00.00
INM-1	7705	G26	3	00.00	11.60	17.10	00.00	00.00
INM-1	7706	G26	3	00.00	11.55	15.45	00.00	00.00
INM-1	7707	G26	3	00.00	00.00	00.00	9.72	10.78
INM-1	7733	G26	3	00.00	10.95	15.40	00.00	00.00
INM-1	7870	G27	3	44.20	11.00	16.35	9.80	10.30
INM-1	7871	G27	3	42.60	12.20	16.75	10.48	11.10
INM-1	7872	G27	3	42.70	11.65	15.43	9.55	10.40
INM-1	7873	G27	3	41.70	11.37	15.60	9.25	9.55
INM-1	7874	G27	3	00.00	10.20	15.75	00.00	00.00
INM-1	7875	G27	3	40.75	11.45	15.95	9.75	10.20
INM-1	7876	G27	3	00.00	00.00	00.00	10.05	11.70
INM-1	7877	G27	3	00.00	00.00	00.00	10.00	10.95
INM-1	2611	H10	3	00.00	00.00	00.00	10.84	10.75
INM-1	2612	H10	3	00.00	00.00	00.00	00.00	11.65
INM-1	1754	H11	3	00.00	12.59	17.12	00.00	00.00
INM-1	1755	H11	3	00.00	11.18	17.49	00.00	00.00
INM-1	1756	H11	3	00.00	12.86	17.89	00.00	00.00
INM-1	1757	H11	3	00.00	11.64	15.58	00.00	00.00
INM-1	1758	H11	3	47.38	00.00	00.00	9.52	10.70
INM-1	1759	H11	3	44.84	11.12	15.50	9.47	9.65
INM-1	1760	H11	3	00.00	00.00	00.00	8.30	9.91
INM-1	1761	H11	3	00.00	00.00	00.00	9.51	10.31
INM-1	2634	H12	3	53.27	11.42	17.42	10.84	11.62
INM-1	3782	H12	3	46.90	12.28	16.20	10.38	11.00
INM-1	3783	H12	3	46.75	10.39	15.65	9.75	11.45
INM-1	4505	H12	3	44.57	12.30	16.80	10.75	10.85
INM-1	4506	H12	3	50.78	11.65	17.40	10.45	11.56
INM-1	4507	H12	3	51.30	11.50	16.50	10.05	11.23
INM-1	4508	H12	3	51.70	12.40	17.40	10.45	11.70
INM-1	4509	H12	3	44.52	12.80	17.15	10.15	10.80
INM-1	4510	H12	3	52.18	10.79	16.90	10.21	11.50
INM-1	7915	Hno102	3	50.38	11.26	15.75	10.01	11.15
INM-1	7916	Hno102	3	00.00	11.65	15.95	00.00	00.00
INM-1	6172	Hno106	3	44.35	11.35	16.55	10.10	10.70
INM-1	4861	Hno107	3	42.80	12.30	15.95	10.00	10.58
INM-1	4862	Hno107	3	00.00	1060	16.30	00.00	00.00
INM-1	4863	Hno107	3	00.00	00.00	00.00	10.05	11.70
INM-1	4864	Hno107	3	00.00	00.00	00.00	9.30	10.52
INM-1	4483	Hno128	3	46.80	10.53	15.95	9.95	11.12
INM-1	4484	Hno128	3	46.75	10.40	15.30	9.40	00.00
INM-1	577	F8	4	00.00	11.51	15.96	00.00	00.00
INM-1	1798	H5	4	50.68	10.36	16.17	9.24	10.68
INM-1	1799	H5	4	48.60	10.66	16.32	10.14	10.85
INM-1	1800	H5	4	41.01	11.83	16.14	9.34	9.93
INM-1	1801	H5	4	41.03	00.00	00.00	9.34	10.30

INM-1	3033	H7	4	49.69	10.94	16.75	10.20	11.52
INM-1	3034	H7	4	00.00	11.73	15.73	00.00	00.00
INM-1	3035	H7	4	00.00	00.00	00.00	9.82	10.31
INM-1	3036	H7	4	00.00	00.00	00.00	10.71	11.90
INM-1	3037	H7	4	00.00	00.00	00.00	9.64	9.48
INM-1	3038	H7	4	00.00	00.00	00.00	9.81	10.12
INM-1	2954	D10	4	00.00	12.01	14.75	00.00	00.00
INM-1	2955	D10	4	43.87	11.68	16.35	10.12	10.26
INM-1	8092	E24	4	42.45	11.50	16.10	9.77	10.20
INM-1	8093	E24	4	51.85	11.40	16.85	10.57	12.05
INM-1	8094	E24	4	53.06	11.25	16.35	10.18	11.55
INM-1	8095	E24	4	51.75	11.70	16.40	11.20	12.35
INM-1	8096	E24	4	00.00	10.95	15.00	00.00	00.00
INM-1	8097	E24	4	00.00	11.70	16.65	00.00	00.00
INM-1	8098	E24	4	00.00	00.00	00.00	9.95	11.53
INM-1	8099	E24	4	00.00	00.00	00.00	9.60	10.60
INM-1	8100	E24	4	00.00	00.00	00.00	9.75	11.35
INM-1	8101	E24	4	00.00	00.00	00.00	8.86	10.85
INM-1	5863A	E25	4	50.46	12.35	16.50	10.15	11.20
INM-1	5864A	E25	4	40.65	11.14	15.17	9.17	9.85
INM-1	5865A	E25	4	00.00	00.00	00.00	9.85	11.15
INM-1	5866A	E25	4	00.00	10.70	16.15	00.00	00.00
INM-1	5848	E27	4	45.21	11.83	16.62	10.10	10.50
INM-1	5849	E27	4	00.00	00.00	00.00	9.60	10.50
INM-1	5850	E27	4	00.00	00.00	00.00	9.60	10.50
INM-1	7934	F24	4	52.45	11.55	17.30	10.40	11.76
INM-1	7935	F24	4	00.00	11.07	15.13	00.00	00.00
INM-1	7936	F24	4	00.00	10.85	16.65	00.00	00.00
INM-1	7937	F24	4	00.00	00.00	00.00	9.75	10.50
INM-1	7938	F24	4	00.00	00.00	00.00	9.05	10.25
INM-1	7939	F24	4	00.00	00.00	00.00	10.50	12.10
INM-1	7940	F24	4	00.00	00.00	00.00	10.80	12.40
INM-1	7941	F24	4	00.00	00.00	00.00	8.75	8.00
INM-1	7942	F24	4	00.00	00.00	00.00	9.95	10.35
INM-1	5948	F25	4	00.00	11.45	16.55	00.00	00.00
INM-1	5949	F25	4	51.72	11.36	17.06	10.50	11.49
INM-1	1853	G11	4	48.85	10.92	15.36	9.79	11.18
INM-1	1854	G11	4	41.37	10.69	15.17	8.75	10.26
INM-1	1857	G11	4	00.00	12.00	18.01	00.00	00.00
INM-1	7064	G24	4	40.30	11.15	15.65	9.25	9.90
INM-1	7065	G24	4	00.00	00.00	00.00	10.55	11.85
INM-1	7066	G24	4	00.00	11.85	16.00	00.00	00.00
INM-1	7067	G24	4	00.00	10.50	14.67	00.00	00.00
INM-1	7068	G24	4	00.00	10.85	16.35	00.00	00.00
INM-1	7069	G24	4	00.00	00.00	00.00	9.86	11.85
INM-1	7070	G24	4	00.00	00.00	00.00	10.30	11.15
INM-1	7071	G24	4	00.00	00.00	00.00	10.50	11.55
INM-1	7158	G24	4	42.50	10.70	15.40	9.75	9.90
INM-1	7159	G24	4	00.00	10.35	15.55	9.60	9.93
INM-1	7160	G24	4	00.00	10.75	15.80	00.00	00.00
INM-1	7161	G24	4	00.00	11.75	16.30	00.00	00.00
INM-1	7162	G24	4	00.00	8.85	10.45	00.00	00.00
INM-1	7602	G25	4	44.25	13.45	18.70	10.60	11.00
INM-1	7603	G25	4	45.17	12.60	17.15	10.65	10.95
INM-1	7604	G25	4	50.60	11.47	16.50	10.40	11.40
INM-1	7605	G25	4	46.30	10.55	15.00	9.57	10.00
INM-1	7606	G25	4	00.00	00.00	00.00	8.57	8.70
INM-1	7607	G25	4	00.00	11.55	17.05	00.00	00.00
INM-1	7608	G25	4	00.00	10.70	15.78	00.00	00.00
INM-1	7609	G25	4	00.00	11.45	16.15	00.00	00.00
INM-1	7610	G25	4	00.00	00.00	00.00	9.65	11.21
INM-1	7611	G25	4	00.00	00.00	00.00	9.40	10.95
INM-1	7686	G25	4	48.71	11.05	15.65	9.50	10.65

INM-1	7687	G25	4	00.00	11.10	17.00	00.00	00.00
INM-1	6790	G26	4	48.75	9.93	15.30	8.87	10.50
INM-1	6791	G26	4	41.80	12.50	16.05	9.80	10.15
INM-1	6792	G26	4	50.70	10.40	15.70	9.70	11.40
INM-1	6793	G26	4	47.05	11.35	16.40	9.50	9.55
INM-1	6795	G26	4	00.00	12.20	16.95	00.00	00.00
INM-1	6796	G26	4	00.00	11.20	15.40	00.00	00.00
INM-1	6797	G26	4	00.00	10.80	15.65	00.00	00.00
INM-1	6798	G26	4	00.00	00.00	00.00	9.65	10.80
INM-1	6799	G26	4	00.00	00.00	00.00	9.55	10.71
INM-1	7750	G26	4	49.40	10.40	15.90	9.80	11.35
INM-1	7751	D24&E25	4	51.65	10.70	15.55	9.80	10.95
INM-1	7752	D24&E25	4	00.00	10.50	16.50	00.00	00.00
INM-1	7753	D24&E25	4	00.00	00.00	00.00	10.15	10.60
INM-1	7226	Hno104	4	00.00	10.60	16.30	00.00	00.00
INM-1	7227	Hno104	4	00.00	10.05	15.68	00.00	00.00
INM-1	6104	Hno105	4	41.38	10.45	14.65	9.80	9.90
INM-1	6105	Hno105	4	52.90	11.45	17.08	9.90	11.10
INM-1	653	Hno110	4	54.87	11.59	17.65	10.8	12.36
INM-1	654	Hno110	4	44.82	11.55	16.27	9.64	10.44
INM-1	655	Hno110	4	00.00	00.00	00.00	10.93	11.53
INM-1	4635	Hno111	4	44.80	12.40	16.75	10.85	10.30
INM-1	4636	Hno111	4	00.00	10.65	15.31	00.00	00.00
INM-1	4042	Hno115	4	47.30	10.60	15.70	10.20	11.05
INM-1	4043	Hno115	4	49.15	10.85	15.95	9.75	10.78
INM-1	4044	Hno115	4	00.00	10.65	15.96	00.00	00.00
INM-1	4045	Hno115	4	00.00	10.45	15.30	00.00	00.00
INM-1	4046	Hno115	4	00.00	10.65	15.55	00.00	00.00
INM-1	4047	Hno115	4	00.00	10.05	15.75	00.00	00.00
INM-1	4048	Hno115	4	00.00	00.00	00.00	10.07	11.90
INM-1	4049	Hno115	4	00.00	00.00	00.00	9.65	10.50
INM-1	4050	Hno115	4	00.00	00.00	00.00	9.45	11.20
INM-1	4051	Hno115	4	00.00	00.00	00.00	12.55	13.40
INM-1	4748	Hno117	4	00.00	00.00	00.00	10.25	10.55
INM-1	4749	Hno117	4	00.00	00.00	00.00	9.35	10.60
INM-1	4750	Hno117	4	00.00	00.00	00.00	10.05	10.50
INM-1	4751	Hno117	4	00.00	10.42	15.37	00.00	00.00
INM-1	4689	Hno121	4	00.00	00.00	00.00	9.85	10.60
INM-1	4690	Hno121	4	00.00	00.00	00.00	9.60	10.05
INM-1	4100	C3	5	00.00	00.00	00.00	10.05	10.10
INM-1	4309	F8	5	50.95	10.63	16.09	9.30	10.65
INM-1	4310	F8	5	50.95	11.65	16.75	9.80	10.75
INM-1	4311	F8	5	48.90	10.55	15.50	9.60	10.80
INM-1	4312	F8	5	44.15	12.60	17.25	11.00	11.16
INM-1	4313	F8	5	48.20	10.33	15.40	9.67	11.30
INM-1	4314	F8	5	00.00	00.00	00.00	10.60	12.30
INM-1	3764	F8	6	46.71	10.80	16.90	10.60	10.95
INM-1	3765	F8	6	00.00	12.25	17.95	00.00	00.00
INM-1	2005	G3	6	00.00	11.60	16.97	00.00	00.00
INM-1	1488	G4	6	41.17	10.31	14.71	8.54	8.53
INM-1	1489	G4	6	00.00	00.00	00.00	9.97	10.97
INM-1	2259	G5	6	48.72	10.58	16.09	9.38	10.36
INM-1	2260	G5	6	44.26	10.98	15.30	9.57	10.20
INM-1	4192	E8	7	44.05	10.93	15.68	10.15	9.70
INM-1	1828	G4	7	00.00	00.00	00.00	10.63	10.19
INM-1	4850	C0	7	57.62	12.03	17.25	10.83	11.90
INM-1	4728	C1	8	40.80	11.00	15.87	9.75	9.67
INM-1	4971	C2	8	44.80	11.30	16.60	10.45	10.65
INM-1	1741	A1	9	00.00	12.07	17.07	00.00	00.00
INM-1	4572	E3	9	00.00	00.00	00.00	9.22	8.30
INM-1	697	A1	11	48.70	10.76	15.96	9.33	10.87
INM-1	5240	B1	11	41.20	11.00	15.80	9.69	10.10
INM-1	6099	B4	11	00.00	00.00	00.00	9.05	9.80

INM-1	6011	B7	11	00.00	9.95	12.20	00.00	00.00
INM-1	6012	B7	11	00.00	00.00	00.00	10.45	11.80
INM-1	5100	C1	11	00.00	00.00	00.00	10.48	11.48
INM-1	6091	C7	11	00.00	11.76	16.05	00.00	00.00
INM-1	3687	B1	12	00.00	10.80	15.83	00.00	00.00
INM-1	818	C1	12	00.00	12.07	17.61	00.00	00.00
INM-1	5609	C3	12	00.00	10.60	15.09	00.00	00.00
INM-1	5839A	C6	12	42.70	10.20	00.00	9.95	10.38
INM-1	6290	C7	12	43.60	11.85	15.85	9.85	10.85
INM-1	6052	Hno76	12	47.52	12.80	18.30	10.80	11.85
INM-1	5595	Hno84	12	43.80	11.80	16.05	9.95	10.20
INM-1	526	Hno50	12	00.00	00.00	00.00	9.28	11.02
INM-1	6133	D3	13	38.45	10.70	14.86	9.45	10.15
INM-1	6144	D3	13	33.70	9.52	12.15	00.00	8.55
INM-1	5811	E4	13	00.00	00.00	00.00	10.25	10.90
INM-1	5916	B3	14	00.00	11.95	16.84	00.00	00.00
INM-1	5396	C3	15	44.48	12.50	16.75	10.05	10.10
INM-1	5397	C3	15	00.00	12.13	17.22	00.00	00.00
INM-1	5639	C5	15	39.45	10.19	15.07	9.20	8.90
INM-1	5559	D4	15	00.00	00.00	00.00	10.09	10.53
INM-1	5999	A5	16	51.65	11.00	16.60	10.60	11.35

Table A.16.16. Measurements of Second Phalanx of ***Antilope cervicapra***

Site	Reg.No.	Trench	Layer	GL	BP	TP	BD	TD
INM-1	66	LJP	1-4	25.10	9.86	13.96	9.02	11.53
INM-1	108	LJP	1-4	24.35	10.15	14.22	9.01	11.57
INM-1	109	LJP	1-4	25.91	10.09	14.19	8.92	11.44
INM-1	110	LJP	1-4	24.77	9.80	00.00	8.51	00.00
INM-1	234	LJP	1-4	24.96	10.54	12.89	8.72	10.21
INM-1	235	LJP	1-4	24.62	9.88	14.56	8.85	11.39
INM-1	236	LJP	1-4	25.19	9.40	13.24	8.23	10.44
INM-1	360	LJP	1-4	27.53	10.09	14.59	8.33	00.00
INM-1	366	LJP	1-4	25.85	9.97	13.65	8.78	11.42
INM-1	2706	G6	1	25.28	10.28	14.84	8.37	11.42
INM-1	2707	G6	1	26.57	11.15	15.27	10.40	12.46
INM-1	3526	H5	1	25.95	11.17	15.00	10.05	12.45
INM-1	3527	H5	1	26.73	11.25	15.60	9.60	12.30
INM-1	3826	D25	1	23.70	10.90	14.15	8.93	10.95
INM-1	3827	D25	1	23.00	10.55	14.75	9.32	10.92
INM-1	3828	D25	1	25.60	10.20	13.65	8.90	12.05
INM-1	3829	D25	1	26.45	9.90	14.15	9.10	12.25
INM-1	3856	D25	1	24.50	10.45	14.35	9.20	11.50
INM-1	998	E10	1	25.63	9.92	13.90	7.41	10.71
INM-1	999	E10	1	00.00	9.92	14.78	7.86	11.34
INM-1	3314	F11	1	23.70	11.30	14.75	8.95	10.70
INM-1	3315	F11	1	23.90	10.25	12.65	8.45	11.15
INM-1	4908	F13	1	25.10	10.70	14.85	8.90	12.21
INM-1	8086	F27	1	23.65	9.82	13.75	8.80	11.35
INM-1	1650	G11	1	27.67	10.77	15.27	9.94	12.60
INM-1	1726	G11	1	00.00	8.40	11.90	00.00	00.00
INM-1	2926	H10	1	23.17	9.40	13.35	9.25	11.32
INM-1	2927	H10	1	24.3	10.27	14.65	9.17	11.55
INM-1	2928	H10	1	00.00	10.72	13.88	00.00	00.00
INM-1	1923	H11	1	26.92	10.85	14.24	9.56	12.54
INM-1	1924	H11	1	26.82	10.81	14.70	9.67	11.60
INM-1	2129	Hno132	1	26.10	10.86	15.93	00.00	12.65
INM-1	2130	Hno132	1	23.18	9.68	13.41	8.43	11.14
INM-1	2131	Hno132	1	25.85	10.99	15.41	10.46	12.81
INM-1	3075	G3	2	24.45	10.65	15.25	9.72	11.98
INM-1	3943	G4	2	25.35	11.15	14.90	9.68	11.83
INM-1	3944	G4	2	23.65	9.50	13.75	7.96	11.00

INM-1	3945	G4	2	25.08	10.36	13.45	9.11	11.80
INM-1	3946	G4	2	22.10	8.40	12.60	7.15	10.05
INM-1	4031	G4	2	00.00	7.65	10.50	00.00	00.00
INM-1	2063	G6	2	28.87	11.76	15.78	10.64	13.22
INM-1	2103	G6	2	00.00	00.00	00.00	9.81	12.80
INM-1	3628	G7	2	00.00	00.00	00.00	9.08	10.58
INM-1	3629	G7	2	00.00	00.00	00.00	8.75	10.86
INM-1	1551	H5	2	24.87	9.43	12.72	8.11	11.83
INM-1	1552	H5	2	25.98	10.53	14.53	9.59	11.80
INM-1	1553	H5	2	27.03	10.75	14.60	9.89	12.15
INM-1	1554	H5	2	23.98	10.13	12.75	00.00	11.04
INM-1	3381	H6	2	27.20	10.70	14.36	9.68	12.40
INM-1	1043	H7	2	25.08	10.10	14.34	9.28	11.83
INM-1	1045	H7	2	23.79	9.09	13.53	8.10	11.27
INM-1	3582	I6	2	25.40	11.70	14.50	10.50	11.70
INM-1	1410	D25	2	26.78	10.82	15.49	9.78	12.54
INM-1	1411	D25	2	25.26	10.80	13.84	9.47	11.25
INM-1	1412	D25	2	24.51	10.22	13.42	8.71	11.66
INM-1	1413	D25	2	22.79	9.31	12.56	8.31	10.67
INM-1	1414	D25	2	25.10	10.42	14.31	9.54	11.96
INM-1	1415	D25	2	25.61	10.52	14.12	8.88	12.16
INM-1	1416	D25	2	23.54	9.47	13.78	8.78	11.39
INM-1	1417	D25	2	24.67	10.85	13.64	9.33	11.17
INM-1	1418	D25	2	23.98	11.43	15.30	9.41	11.79
INM-1	7263	E24	2	26.35	8.93	10.60	8.90	11.25
INM-1	7264	E24	2	26.75	10.50	15.60	9.13	12.35
INM-1	7265	E24	2	23.50	9.55	12.60	8.30	10.35
INM-1	7266	E24	2	23.75	10.39	13.27	9.05	11.15
INM-1	7267	E24	2	23.25	9.45	12.90	8.60	11.13
INM-1	7268	E24	2	22.65	9.05	13.45	8.55	11.00
INM-1	7269	E24	2	23.97	10.55	13.40	9.10	11.45
INM-1	7270	E24	2	25.15	9.92	14.30	9.53	11.55
INM-1	7481	E24	2	22.65	9.40	12.27	8.63	10.30
INM-1	7482	E24	2	26.80	9.50	13.10	8.88	11.95
INM-1	6388	E25	2	24.50	10.95	14.15	9.50	11.65
INM-1	6389	E25	2	00.00	10.70	13.81	00.00	00.00
INM-1	6390	E25	2	27.49	11.40	15.80	10.47	13.00
INM-1	6391	E25	2	25.20	10.60	14.45	9.65	11.55
INM-1	6392	E25	2	25.25	10.30	14.46	9.20	11.40
INM-1	6393	E25	2	27.06	9.85	14.20	8.75	12.05
INM-1	6394	E25	2	22.85	9.88	13.10	8.60	10.20
INM-1	6395	E25	2	00.00	9.75	12.50	00.00	00.00
INM-1	7407	E25	2	24.65	9.25	13.65	9.05	11.55
INM-1	7408	E25	2	22.50	9.47	12.40	8.20	10.00
INM-1	7409	E25	2	21.95	9.40	12.70	7.95	10.20
INM-1	7410	E25	2	26.15	10.15	13.60	9.35	11.85
INM-1	7411	E25	2	24.45	10.85	14.30	9.73	11.50
INM-1	7412	E25	2	21.50	9.65	12.42	7.90	10.40
INM-1	7413	E25	2	20.20	8.05	10.45	6.45	8.20
INM-1	7414	E25	2	00.00	9.90	13.67	00.00	00.00
INM-1	2993	F10	2	22.45	8.46	12.34	7.72	10.56
INM-1	2994	F10	2	23.71	8.99	13.76	7.47	11.42
INM-1	2995	F10	2	23.64	9.14	13.87	7.75	10.42
INM-1	2532	F11	2	25.88	00.00	14.87	9.49	12.28
INM-1	2533	F11	2	26.33	10.22	14.16	9.69	11.85
INM-1	2534	F11	2	25.74	11.28	14.36	9.61	11.01
INM-1	2535	F11	2	24.77	9.78	13.64	8.84	11.13
INM-1	2536	F11	2	24.71	10.56	13.13	9.22	11.28
INM-1	2537	F11	2	23.55	10.12	13.53	8.95	10.69
INM-1	6607	F24	2	23.35	9.75	12.50	9.00	11.05
INM-1	6608	F24	2	24.23	10.35	14.20	9.60	12.20
INM-1	6609	F24	2	00.00	00.00	00.00	9.30	12.10
INM-1	7559	F26	2	22.92	9.75	13.55	9.15	11.80

INM-1	1324	G11	2	24.87	10.15	00.00	9.02	10.96
INM-1	1325	G11	2	22.46	9.71	13.57	7.79	10.58
INM-1	1326	G11	2	24.44	10.59	14.37	9.45	11.12
INM-1	1328	G11	2	23.13	9.38	13.11	00.00	9.68
INM-1	1329	G11	2	25.61	10.36	15.86	9.54	11.72
INM-1	3151	G11	2	21.99	9.73	13.35	8.82	10.77
INM-1	3152	G11	2	25.00	10.99	14.40	9.23	13.25
INM-1	7002	G24	2	22.75	9.75	12.26	8.65	10.20
INM-1	7003	G24	2	24.65	9.45	14.10	8.70	11.80
INM-1	2754	H11	2	26.78	10.46	14.28	9.48	12.85
INM-1	2755	H11	2	24.75	9.10	13.63	8.02	11.69
INM-1	2756	H11	2	25.50	9.62	12.94	8.89	11.25
INM-1	2757	H11	2	24.37	10.03	14.03	8.82	10.87
INM-1	4251	Hno130	2	25.83	9.95	13.65	9.20	11.35
INM-1	2661	G7	3	22.97	10.54	14.96	8.18	00.00
INM-1	2701	G7	3	00.00	10.35	14.62	00.00	00.00
INM-1	3437	H7	3	25.60	10.71	14.44	9.64	10.90
INM-1	3438	H7	3	24.82	10.58	14.46	9.60	11.10
INM-1	3439	H7	3	23.35	10.27	13.75	8.70	10.40
INM-1	3440	H7	3	24.75	10.30	14.55	9.10	11.75
INM-1	3441	H7	3	24.35	9.75	12.65	8.70	11.20
INM-1	3595	I6	3	24.46	10.23	13.46	9.11	11.88
INM-1	4082	D10	3	25.28	10.30	13.60	9.45	12.20
INM-1	2017	D24	3	22.53	9.47	13.13	8.11	10.26
INM-1	1221	D25	3	27.19	10.39	15.04	8.77	12.02
INM-1	1222	D25	3	25.49	9.81	14.72	8.73	11.49
INM-1	1223	D25	3	23.19	10.39	14.29	8.31	10.98
INM-1	1224	D25	3	26.78	10.57	15.00	10.31	12.96
INM-1	1225	D25	3	26.26	10.09	14.35	7.97	11.05
INM-1	1226	D25	3	23.28	9.52	12.16	8.53	10.45
INM-1	1227	D25	3	26.12	10.70	14.86	9.48	11.94
INM-1	1228	D25	3	25.60	11.00	14.20	9.72	11.60
INM-1	1229	D25	3	23.17	9.93	13.00	8.66	11.35
INM-1	1258	D25	3	27.27	10.43	14.64	9.67	12.12
INM-1	1260	D25	3	00.00	9.36	10.79	7.98	12.12
INM-1	6703	E23	3	27.60	10.15	15.15	10.05	12.85
INM-1	6704	E23	3	24.50	11.00	13.85	9.27	11.50
INM-1	6705	E23	3	23.05	10.00	13.50	8.30	10.66
INM-1	6706	E23	3	23.75	10.20	13.75	8.35	10.78
INM-1	6707	E23	3	23.60	9.05	13.00	8.65	10.80
INM-1	6708	E23	3	23.40	10.40	14.20	9.00	10.65
INM-1	6709	E23	3	24.25	10.10	13.90	8.95	11.70
INM-1	6710	E23	3	23.05	10.00	13.75	9.26	11.05
INM-1	6711	E23	3	23.85	9.85	13.95	8.55	10.60
INM-1	6712	E23	3	24.25	11.18	14.48	9.60	11.80
INM-1	6713	E23	3	00.00	10.25	13.65	00.00	00.00
INM-1	6714	E23	3	00.00	00.00	00.00	9.00	11.25
INM-1	6755	E23	3	23.15	9.47	14.35	8.90	11.35
INM-1	6885	E26	3	24.70	9.75	13.75	8.55	10.56
INM-1	6886	E26	3	26.65	10.45	14.95	00.00	11.70
INM-1	6887	E26	3	23.75	9.02	12.75	7.65	10.90
INM-1	6888	E26	3	24.80	10.00	14.35	9.20	11.35
INM-1	6889	E26	3	23.90	10.10	14.80	9.15	11.90
INM-1	6890	E26	3	26.07	10.95	15.09	9.62	12.35
INM-1	6891	E26	3	25.40	10.45	14.30	9.47	11.90
INM-1	6892	E26	3	24.20	8.95	12.40	8.05	10.90
INM-1	6893	E26	3	24.25	10.00	13.35	8.40	11.36
INM-1	6894	E26	3	22.90	9.90	12.85	8.67	10.60
INM-1	6895	E26	3	25.35	9.58	13.30	9.60	12.35
INM-1	6896	E26	3	25.20	10.55	13.80	9.00	11.75
INM-1	1820	F10	3	27.54	10.22	16.01	8.36	12.23
INM-1	6486	F25	3	23.41	10.00	13.80	9.55	11.80
INM-1	6487	F25	3	24.55	11.00	14.15	9.35	11.25

INM-1	6488	F25	3	23.90	9.70	14.35	9.15	10.92
INM-1	6489	F25	3	23.55	9.65	11.93	8.95	10.76
INM-1	6490	F25	3	23.80	10.05	13.20	9.25	11.80
INM-1	6491	F25	3	22.90	9.35	13.40	8.60	10.70
INM-1	6492	F25	3	24.50	9.80	13.40	8.90	11.70
INM-1	6493	F25	3	23.35	9.45	12.60	8.60	10.96
INM-1	6494	F25	3	22.85	10.30	14.05	8.85	11.15
INM-1	6495	F25	3	00.00	9.55	13.25	00.00	00.00
INM-1	6496	F25	3	00.00	9.30	11.18	00.00	00.00
INM-1	6580	F25	3	22.90	10.10	13.60	9.15	10.86
INM-1	2190	G10	3	25.61	10.29	14.80	8.69	11.91
INM-1	2191	G10	3	23.92	10.13	14.22	9.28	11.30
INM-1	2192	G10	3	24.37	9.91	14.99	9.33	13.58
INM-1	2895	G10	3	24.54	10.14	13.63	8.19	9.86
INM-1	2896	G10	3	21.93	8.73	11.68	00.00	00.00
INM-1	1283	G11	3	23.91	10.21	13.18	8.77	11.93
INM-1	1284	G11	3	23.15	9.45	13.66	7.70	10.70
INM-1	1285	G11	3	23.73	10.24	13.86	8.80	11.21
INM-1	3745	G11	3	22.98	9.96	13.60	8.25	10.45
INM-1	7796	G25	3	23.35	10.00	13.35	9.25	11.75
INM-1	7797	G25	3	25.20	10.97	13.85	9.95	00.00
INM-1	7798	G25	3	23.90	9.75	13.55	8.55	11.07
INM-1	7799	G25	3	20.67	9.20	11.75	7.12	9.10
INM-1	7800	G25	3	25.20	9.90	00.00	9.30	11.40
INM-1	7801	G25	3	25.80	9.20	13.50	7.80	11.00
INM-1	7708	G26	3	22.05	8.97	12.10	7.95	10.30
INM-1	7709	G26	3	24.30	10.45	13.75	8.98	11.30
INM-1	7710	G26	3	22.55	10.18	14.10	8.45	10.70
INM-1	7881	G27	3	25.50	10.40	14.20	9.25	12.50
INM-1	7882	G27	3	23.45	9.95	14.65	8.60	11.25
INM-1	2613	H10	3	25.82	9.66	13.43	7.85	10.98
INM-1	1762	H11	3	25.45	9.83	14.71	8.34	11.59
INM-1	1763	H11	3	24.98	10.68	13.40	9.49	11.66
INM-1	1764	H11	3	25.04	10.99	15.39	9.34	11.95
INM-1	6106	Hno106	3	22.45	11.10	00.00	8.80	10.35
INM-1	6107	Hno106	3	25.70	10.68	14.72	10.20	12.65
INM-1	6108	Hno106	3	22.46	9.61	13.40	8.45	10.45
INM-1	6175	Hno106	3	25.25	10.35	00.00	9.15	10.70
INM-1	4865	Hno107	3	24.85	10.95	14.80	9.70	11.60
INM-1	4866	Hno107	3	24.25	9.65	13.80	8.80	12.15
INM-1	4867	Hno107	3	00.00	9.35	12.35	00.00	00.00
INM-1	4868	Hno107	3	24.90	10.20	13.80	8.85	12.00
INM-1	4485	Hno128	3	22.05	8.57	11.76	8.15	10.30
INM-1	3030	B3	4	26.28	10.65	14.98	9.27	11.78
INM-1	1804	H5	4	24.63	9.57	13.39	8.56	11.68
INM-1	1805	H5	4	24.60	10.67	14.87	8.85	12.35
INM-1	1806	H5	4	22.55	9.15	12.22	8.59	10.89
INM-1	1807	H5	4	23.01	9.67	12.91	8.77	10.87
INM-1	3042	H7	4	24.43	10.03	13.25	8.42	10.03
INM-1	3043	H7	4	25.99	10.30	13.79	9.52	11.76
INM-1	8010	E24	4	24.35	10.30	13.40	9.59	11.70
INM-1	8011	E24	4	24.50	10.45	13.40	9.07	11.02
INM-1	8012	E24	4	00.00	9.70	12.90	8.35	00.00
INM-1	8102	E24	4	22.75	9.30	12.35	9.00	10.55
INM-1	5868A	E25	4	23.80	9.70	13.10	8.50	10.45
INM-1	5869A	E25	4	23.85	10.15	13.05	8.40	11.55
INM-1	5870A	E25	4	24.35	9.88	13.45	8.96	11.28
INM-1	5871A	E25	4	22.18	8.67	12.17	6.97	9.60
INM-1	5850	E27	4	23.24	10.00	13.75	8.90	10.73
INM-1	5851	E27	4	23.73	10.15	14.55	8.75	10.73
INM-1	5852	E27	4	23.12	10.35	14.00	9.30	11.40
INM-1	7943	F24	4	24.60	9.35	13.05	8.15	10.90
INM-1	7944	F24	4	24.58	9.70	12.60	8.65	10.40

INM-1	7945	F24	4	21.40	9.40	13.02	8.20	9.40
INM-1	7946	F24	4	23.85	9.65	13.00	8.75	11.15
INM-1	7947	F24	4	00.00	10.65	13.45	00.00	00.00
INM-1	7948	F24	4	22.97	9.35	11.50	7.15	9.40
INM-1	7949	F24	4	25.85	9.28	12.70	8.45	11.50
INM-1	7950	F24	4	24.25	9.20	12.45	7.45	10.15
INM-1	7951	F24	4	23.40	8.95	12.25	7.95	10.15
INM-1	7952	F24	4	24.05	9.90	12.25	9.15	11.75
INM-1	5951	F25	4	22.55	10.00	14.36	8.35	10.50
INM-1	5952	F25	4	23.20	9.20	12.70	8.30	10.55
INM-1	5953	F25	4	25.40	9.67	14.00	8.35	11.65
INM-1	5954	F25	4	23.67	10.15	12.96	8.75	10.24
INM-1	7072	G24	4	25.35	11.10	14.00	00.00	12.15
INM-1	7073	G24	4	23.75	10.68	14.25	8.95	11.02
INM-1	7074	G24	4	22.71	9.90	13.85	9.17	10.93
INM-1	7163	G24	4	23.20	9.40	13.35	8.06	10.70
INM-1	7612	G25	4	25.90	10.00	13.10	8.57	11.15
INM-1	7613	G25	4	24.95	10.00	14.25	9.00	11.55
INM-1	7614	G25	4	23.30	9.30	12.45	8.85	10.29
INM-1	7615	G25	4	24.85	9.90	13.50	8.35	11.25
INM-1	7652	G25	4	24.10	10.85	13.94	9.45	11.20
INM-1	7688	G25	4	22.90	9.25	13.30	8.10	10.85
INM-1	7697	G25	4	00.00	10.49	13.05	00.00	00.00
INM-1	6801	G26	4	23.25	9.70	13.64	8.35	10.68
INM-1	6802	G26	4	25.95	9.95	13.70	9.40	12.00
INM-1	6803	G26	4	23.30	9.32	11.95	00.00	00.00
INM-1	7228	Hno104	4	00.00	8.60	11.70	00.00	00.00
INM-1	7229	Hno104	4	22.35	9.26	11.65	7.63	9.55
INM-1	788	Hno113	4	24.71	10.05	13.23	8.26	10.40
INM-1	4054	Hno115	4	23.98	10.30	14.40	9.80	12.75
INM-1	4055	Hno115	4	23.20	9.65	13.55	8.95	11.33
INM-1	4056	Hno115	4	25.00	9.50	13.30	8.85	11.70
INM-1	4057	Hno115	4	23.33	9.10	12.75	8.87	10.90
INM-1	4058	Hno115	4	00.00	10.00	14.20	00.00	00.00
INM-1	4691	Hno121	4	25.60	10.65	15.05	10.08	11.60
INM-1	4692	Hno121	4	21.65	9.25	13.47	8.07	10.35
INM-1	4695	Hno121	4	00.00	00.00	00.00	9.15	11.57
INM-1	4111	C3	5	23.00	9.50	12.10	8.23	00.00
INM-1	4318	F8	5	25.15	10.42	14.65	9.35	11.45
INM-1	4319	F8	5	23.30	9.60	14.32	8.98	10.50
INM-1	4320	F8	5	24.85	9.65	14.80	9.20	10.93
INM-1	4321	F8	5	25.73	10.20	13.55	8.85	11.40
INM-1	3766	F8	6	26.38	9.18	13.05	8.20	11.25
INM-1	2261	G5	6	25.67	9.55	13.75	9.16	11.71
INM-1	3864		7	00.00	00.00	00.00	7.03	8.71
INM-1	1830	G4	7	26.70	11.20	15.42	9.95	12.54
INM-1	1831	G4	7	23.33	9.10	12.12	7.64	11.15
INM-1	1832	G4	7	23.26	10.61	13.42	8.97	11.36
INM-1	4232	C1	10	21.16	8.40	12.35	7.15	9.40
INM-1	3690	B1	12	00.00	00.00	00.00	9.63	11.67
INM-1	819	C1	12	22.75	9.57	15.51	7.02	9.22
INM-1	4274	C2	12	00.00	10.88	14.75	00.00	00.00
INM-1	5753	C4	12	22.95	8.75	11.95	7.35	9.35
INM-1	523	Hno50	12	24.65	11.09	14.93	9.63	12.05
INM-1	6296	Hno81	12	24.75	11.66	15.10	10.50	12.05
INM-1	5813	E4	13	23.10	9.25	12.70	7.75	10.30
INM-1	8162	A1	14	24.35	9.57	12.35	7.85	9.35
INM-1	8154	A2	14	26.80	9.80	12.75	8.30	11.30
INM-1	5560	D4	15	25.40	10.65	14.40	9.20	12.20

Table A.16.17 Measurements of Third Phalanx of ***Antilope cervicapra***

Site	Reg.No.	Trench	Layer	GL	DL	BP	TP
INM-1	15	LJP	1-4	25.37	20.28	8.94	00.00
INM-1	71	LJP	1-4	25.53	20.31	8.05	00.00
INM-1	245	LJP	1-4	26.55	22.03	8.64	00.00
INM-1	368	LJP	1-4	27.44	22.15	8.96	00.00
INM-1	369	LJP	1-4	29.19	24.46	9.14	00.00
INM-1	3530	H5	1	27.00	21.85	8.25	19.00
INM-1	1195	F10	1	25.44	20.63	8.49	18.32
INM-1	1207	F10	1	26.05	00.00	7.78	16.78
INM-1	1655	G11	1	26.94	21.88	8.59	18.75
INM-1	1656	G11	1	23.92	18.83	8.69	18.18
INM-1	2929	H10	1	27.79	10.41	9.63	16.91
INM-1	1925	H11	1	25.01	20.92	8.42	16.33
INM-1	3076	G3	2	31.49	26.82	10.17	20.89
INM-1	3951	G4	2	29.55	25.15	9.00	19.40
INM-1	2057	G6	2	31.76	24.29	9.59	23.05
INM-1	2058	G6	2	27.27	23.90	9.48	20.11
INM-1	2059	G6	2	27.54	21.58	8.83	19.98
INM-1	2060	G6	2	24.50	19.40	8.65	19.22
INM-1	1630	H5	2	26.36	00.00	8.03	00.00
INM-1	3382	H6	2	32.37	25.29	9.95	21.15
INM-1	3383	H6	2	00.00	00.00	9.02	18.08
INM-1	1048	H7	2	28.52	24.58	9.25	20.56
INM-1	1049	H7	2	31.15	26.75	10.23	21.98
INM-1	1050	H7	2	28.66	24.28	9.46	20.01
INM-1	3583	I6	2	27.21	21.25	8.08	19.34
INM-1	3584	I6	2	28.20	23.95	9.65	19.74
INM-1	1422	D25	2	25.14	21.11	8.58	19.10
INM-1	7483	E24	2	25.70	19.35	8.15	18.20
INM-1	7484	E24	2	30.35	24.81	9.60	21.05
INM-1	7415	E25	2	28.60	23.37	8.75	19.90
INM-1	2540	F11	2	27.39	22.91	9.93	16.54
INM-1	2541	F11	2	25.35	00.00	8.35	15.14
INM-1	6610	F24	2	26.85	21.60	8.10	19.55
INM-1	6611	F24	2	29.65	24.82	9.25	19.90
INM-1	1332	G11	2	29.40	24.53	10.81	20.78
INM-1	2759	H11	2	27.11	00.00	8.26	15.89
INM-1	2760	H11	2	25.87	20.44	8.44	19.33
INM-1	2663	G7	3	29.22	23.96	10.12	23.75
INM-1	3442	H7	3	27.09	20.70	8.75	17.55
INM-1	1264	D25	3	25.33	19.17	8.84	19.45
INM-1	6715	E23	3	28.05	23.30	9.00	19.85
INM-1	6716	E23	3	28.95	24.20	9.45	18.80
INM-1	6717	E23	3	25.70	21.25	7.90	17.65
INM-1	6718	E23	3	28.90	23.10	9.30	20.18
INM-1	6719	E23	3	27.05	22.75	8.15	17.82
INM-1	6720	E23	3	25.70	22.00	9.00	19.15
INM-1	6759	E23	3	29.15	23.85	9.10	19.65
INM-1	6757	E23	3	28.10	22.85	8.80	22.90
INM-1	6897	E26	3	28.50	23.60	7.80	18.05
INM-1	6898	E26	3	31.32	23.70	9.25	19.05
INM-1	6899	E26	3	26.20	20.26	6.65	15.95
INM-1	6900	E26	3	31.25	23.10	9.60	00.00
INM-1	6901	E26	3	28.70	23.15	9.10	16.00
INM-1	6902	F11	3	25.55	19.30	8.36	16.00
INM-1	3265	F25	3	28.20	22.02	9.27	20.70
INM-1	6497	F25	3	27.40	23.35	8.73	17.81
INM-1	2193	G10	3	25.55	21.58	8.09	13.58
INM-1	2897	G10	3	29.49	23.22	9.45	15.65
INM-1	7802	G25	3	25.40	20.00	8.45	15.65

INM-1	7712	G26	3	23.25	19.35	8.30	16.85
INM-1	7883	G27	3	27.75	22.85	8.90	19.70
INM-1	1765	H11	3	30.67	23.31	8.81	20.70
INM-1	2637	H11	3	26.46	20.79	9.22	20.88
INM-1	541	Hno126	3	24.25	19.80	8.04	00.00
INM-1	8013	E24	4	25.69	21.20	8.85	17.35
INM-1	8014	E24	4	27.65	21.10	9.60	21.40
INM-1	8015	E24	4	29.35	23.74	8.45	19.70
INM-1	8103	E24	4	30.40	23.90	8.85	20.96
INM-1	8104	E24	4	25.75	20.55	7.63	18.85
INM-1	8105	E24	4	31.50	23.95	9.85	21.20
INM-1	5872A	E25	4	26.25	21.95	8.85	18.90
INM-1	5873A	E25	4	25.90	00.00	8.95	00.00
INM-1	5856	E27	4	29.65	24.20	8.95	18.25
INM-1	7953	F24	4	28.20	22.65	9.30	19.25
INM-1	7954	F24	4	29.62	23.30	9.35	19.10
INM-1	7955	F24	4	00.00	00.00	9.20	18.40
INM-1	7956	F24	4	28.70	00.00	8.80	00.00
INM-1	5959	F25	4	29.85	23.50	9.20	19.60
INM-1	5960	F25	4	25.35	19.30	8.25	16.20
INM-1	5961	F25	4	31.45	24.75	9.10	20.70
INM-1	5986	F25	4	29.22	00.00	9.97	00.00
INM-1	7075	G24	4	26.05	20.50	8.15	19.15
INM-1	7076	G24	4	25.90	23.05	9.15	00.00
INM-1	7616	G25	4	27.30	21.15	8.40	18.00
INM-1	7617	G25	4	28.60	24.20	9.35	20.25
INM-1	7689	G25	4	25.25	20.00	8.30	18.00
INM-1	6804	G26	4	29.20	23.35	8.40	19.55
INM-1	4693	Hno121	4	27.20	22.70	9.35	18.15
INM-1	1515	G4	6	30.56	25.09	9.24	21.56
INM-1	4108	A1&A2	7	28.10	22.35	9.60	21.15
INM-1	688	A1	11	27.59	22.33	8.48	00.00
INM-1	6013	B7	11	27.50	23.65	9.46	19.60
INM-1	348	C1	11	29.98	25.91	10.33	00.00
INM-1	349	C1	11	28.10	23.37	8.62	00.00
INM-1	5596	Hno84	12	28.00	23.60	8.60	17.70

Table A.16.18. Measurements of isolated Maxillary second premolar of ***Antilope cervicapra***

Site	Reg.No.	Trench	Layer	GL	GB
INM-1	3513	H7	3	7.75	5.40

Table A.16.19. Measurements of isolated Maxillary first/second molar of ***Antilope cervicapra***

Site	Reg.No.	Trench	Layer	GL	GB
INM-1	1128	G4	1	15.63	11.06
INM-1	2735	G6	1	16.68	11.75
INM-1	2737	G6	1	15.81	12.27
INM-1	2740	G6	1	16.59	11.82
INM-1	2734	G6	1	15.72	10.14
INM-1	2738	G6	1	16.15	11.45
INM-1	2741	G6	1	13.00	11.29
INM-1	3559	H5	1	13.30	9.40
INM-1	7056	E23	1	14.17	8.90
INM-1	1197	F10	1	15.05	10.48
INM-1	3360	F11	1	15.78	11.44
INM-1	4924	F13	1	17.07	12.38
INM-1	7384	F26	1	15.02	10.50
INM-1	7386	F26	1	15.90	11.15
INM-1	7545	G24	1	15.32	11.95
INM-1	3103	G3	2	14.75	10.58
INM-1	4020	G4	2	14.70	9.50

INM-1	4019	G4	2	16.10	10.75
INM-1	1624	H5	2	15.27	12.00
INM-1	3412	H6	2	14.05	11.46
INM-1	7319	E24	2	16.60	10.95
INM-1	7320	E24	2	13.80	10.20
INM-1	7318	E24	2	13.90	11.00
INM-1	6444	E25	2	14.10	11.00
INM-1	2594	F11	2	16.23	12.59
INM-1	2596	F11	2	15.00	11.05
INM-1	6668	F24	2	17.90	12.40
INM-1	6667	F24	2	14.98	10.10
INM-1	6661	F24	2	13.83	9.80
INM-1	7576	F26	2	15.70	10.75
INM-1	7577	F26	2	15.77	11.25
INM-1	3232	G11	2	15.20	9.60
INM-1	7034	G24	2	15.00	11.00
INM-1	7189	G25	2	15.40	11.65
INM-1	3504	H7	3	14.21	9.50
INM-1	2217	G10	3	15.53	10.89
INM-1	7863	G25	3	17.05	11.70
INM-1	7864	G25	3	15.50	9.95
INM-1	7745	G26	3	10.80	9.35
INM-1	5879	E27	4	14.15	11.45
INM-1	1866	G11	4	13.63	9.52
INM-1	7123	G24	4	14.05	9.40
INM-1	6191	Hno106	3	14.87	11.80
INM-1	5108	C1	11	14.15	8.15
INM-1	6330	B5	12	11.91	6.30
INM-1	5858A	C6	12	19.55	14.33
INM-1	5688	D3	12	15.90	10.30
INM-1	533	Hno50	12	12.33	9.58
INM-1	532	Hno50	12	11.07	8.55
INM-1	5829	C4	13	11.40	10.15
INM-1	528	E4	13	14.85	9.40
INM-1	6103	B3	15	14.60	9.65
INM-1	6248	E5	15	15.23	10.50
INM-1	6249	E5	15	14.98	10.45
INM-1	6250	E5	15	17.28	11.00
INM-1	6049	Hno87	16	16.25	10.65
INM-1	7440	E25	2	15.00	11.04
INM-1	2430	F25&F26	3	13.72	9.12

Table A.16.20. Measurements of Maxillary second molar of ***Antilope cervicapra***

Site	Reg.No.	Trench	Layer	GL	GB
INM-1	1364	G11	2	15.40	11.86

Table A.16.21. Measurements of Maxillary third molar of ***Antilope cervicapra***

Site	Reg.No.	Trench	Layer	GL	GB
INM-1	21.69	Hno132	1	16.00	9.86
INM-1	2739	G6	1	15.22	9.99
INM-1	2736	G6	1	15.98	10.25
INM-1	2055	G7	1	15.94	11.29
INM-1	2054	G7	1	16.68	10.33
INM-1	1979	H11	1	15.79	10.39
INM-1	4018	G4	2	16.07	11.22
INM-1	7321	E24	2	16.45	13.00
INM-1	3021	F10	2	14.93	10.34
INM-1	3019	F10	2	16.14	9.64
INM-1	2595	F11	2	16.32	11.47
INM-1	6669	F24	2	17.30	13.10

INM-1	6670	F24	2	18.55	12.50
INM-1	7035	G24	2	13.90	10.30
INM-1	2803	H11	2	15.60	9.19
INM-1	3502	H7	3	15.56	10.60
INM-1	3503	H7	3	14.70	9.40
INM-1	7862	G25	3	14.55	11.30
INM-1	7739	G26	3	16.65	11.28
INM-1	7740	G26	3	13.75	8.35
INM-1	4573	E3	9	18.00	15.00
INM-1	4821	B0	12	15.02	9.52

Table A.16.22. Measurements of Mandibular deciduous premolar of ***Antilope cervicapra***

Site	Reg.No.	Trench	Layer	GL	GB
INM-1	5993	B3	14	15.60	5.95

Table A.16.23. Measurements of Manibular first/second premolar of ***Antilope cervicapra***

Site	Reg.No.	Trench	Layer	GL	GB
INM-1	1634	H5	2	8.51	5.41

Table A.16.24. Measurements of Mandibular third premolar of ***Antilope cervicapra***

Site	Reg.No.	Trench	Layer	GL	GB
INM-1	3513	H7	3	7.75	5.40

Table A.16.25. Measurements of Mandibular fourth premolar of ***Antilope cervicapra***

Site	Reg.No.	Trench	Layer	GL	GB
INM-1	1307	G111	3	15.08	5.41
INM-1	5539	D8	11	17.10	5.90

Table A.16.26. Measurements of Mandibular first/second molars of ***Antilope cervicapra***

Site	Reg.No.	Trench	Layer	GL	GB
INM-1	3556	H5	1	14.75	8.40
INM-1	7385	F26	1	15.80	9.60
INM-1	7392	F26	1	12.90	7.60
INM-1	1729	G11	1	15.13	7.61
INM-1	1715	G11	1	14.91	6.86
INM-1	1976	H11	1	15.94	6.58
INM-1	1977	H11	1	15.30	6.80
INM-1	4452	B7	2	15.75	7.52
INM-1	4030	G4	2	15.10	7.93
INM-1	4022	G4	2	15.80	7.73
INM-1	1083	H7	2	15.02	8.15
INM-1	1085	H7	2	13.46	6.47
INM-1	1442	D25	2	15.43	7.89
INM-1	7340	E24	2	14.65	7.65
INM-1	6445	E25	2	15.90	8.00
INM-1	3020	F10	2	15.77	7.16
INM-1	6659	F24	2	14.05	6.55
INM-1	6666	F24	2	15.35	7.10
INM-1	7575	F26	2	14.00	7.65
INM-1	3235	G11	2	14.90	7.90
INM-1	3233	G11	2	16.15	7.20
INM-1	1365	G11	2	14.66	7.50
INM-1	2802	H11	2	14.63	7.15
INM-1	772	I12	2	15.35	6.84
INM-1	2694	G7	3	13.81	6.81

INM-1	3500	H7	3	15.95	6.80
INM-1	6993	E26	3	16.25	7.90
INM-1	1825	F10	3	16.28	7.06
INM-1	6588	F25	3	16.10	6.90
INM-1	6587	F25	3	12.86	6.75
INM-1	2429	F25&F26	3	15.28	8.18
INM-1	2428	F25&F26	3	15.34	8.25
INM-1	2218	G10	3	14.92	7.93
INM-1	2219	G10	3	16.01	6.13
INM-1	1308	G11	3	17.44	7.15
INM-1	6192	Hno106	3	16.09	6.55
INM-1	6193	Hno106	3	16.05	7.50
INM-1	8024	E24	4	14.45	6.07
INM-1	7988	F24	4	13.75	7.90
INM-1	7141	G24	4	15.35	7.55
INM-1	4663	Hno111	4	15.20	7.95
INM-1	786	Hno124	4	14.91	10.62
INM-1	4123	C3	5	14.15	6.45
INM-1	3271	F11	5	16.32	7.25
INM-1	3775	F8	6	17.40	7.50
INM-1	1520	G4	6	12.37	7.17
INM-1	1519	G4	6	9.04	6.20
INM-1	1518	G4	6	14.63	7.54
INM-1	4701	E2	7	15.90	7.45
INM-1	1840	G4	7	16.19	7.49
INM-1	4714	Hno61	8	14.22	6.40
INM-1	4603	Hno65	8	14.20	7.10
INM-1	4149	Hno73	8	16.85	7.20
INM-1	5538	D8	11	15.26	6.50
INM-1	6228	B6	12	14.34	8.60
INM-1	821	C1	12	16.68	7.49
INM-1	5932	B3	14	14.90	6.65
INM-1	5407	C3	15	14.83	7.93

Table A.16.27. Measurements of Mandibular second molars of ***Antilope cervicapra***

Site	Reg.No.	Trench	Layer	GL	GB
INM-1	427	LJP	1-4	20.10	7.91
INM-1	3557	H5	1	16.55	7.32
INM-1	3558	H5	1	15.37	7.60
INM-1	4021	G4	2	20.30	6.88
INM-1	942	G12	3	14.80	7.85

Table A.16.28. Measurements of Mandibular third molars of ***Antilope cervicapra***

Site	Reg.No.	Trench	Layer	GL	GB
INM-1	7057	E23	1	21.30	7.35
INM-1	1714	G1	1	18.95	7.40
INM-1	1713	G11	1	19.69	7.74
INM-1	1723	G11	1	18.39	6.75
INM-1	1970	H11	1	20.95	7.85
INM-1	2170	Hno132	1	20.35	6.75
INM-1	1082	H7	2	20.95	8.03
INM-1	1443	D25	2	19.70	7.10
INM-1	7322	E24	2	20.00	6.82
INM-1	2593	F11	2	20.05	9.05
INM-1	6660	F24	2	19.50	7.15
INM-1	7033	G24	2	21.80	8.50
INM-1	7188	G25	2	18.13	7.20
INM-1	2693	G7	3	19.56	6.76
INM-1	2692	G7	3	20.60	6.90
INM-1	3499	H7	3	20.99	6.95

INM-1	3498	H7	3	20.60	7.90
INM-1	6994	E26	3	19.73	8.17
INM-1	943	G12	3	20.35	7.65
INM-1	7840	G25	3	20.20	6.78
INM-1	7865	G25	3	20.20	6.78
INM-1	7987	F24	4	20.50	8.21
INM-1	2425	F25&F26	3	19.96	7.15
INM-1	2424	F25&F26	3	20.24	7.06
INM-1	2420	F25&F26	3	18.30	6.67
INM-1	2422	F25&F26	3	21.66	7.55
INM-1	2421	F25&F26	3	19.88	6.79
INM-1	2423	F25&F26	3	20.17	7.96
INM-1	7122	G24	4	21.60	8.15
INM-1	7646	G25	4	19.53	7.40
INM-1	7647	G25	4	19.15	7.60
INM-1	5408	C3	15	21.00	7.69
INM-1	5573	D4	15	19.58	7.57

Table A.16.29. Measurements of Mandibles with teeth of ***Antilope cervicapra***

dp2, dp3, dp4- deciduous mandibular second, third and fourth premolar
p2 –second premolar, p3- third premolar, p4- fourth premolar, m1- first molar, m2- second molar, m3- third molar, MC-Mandibular condyle, MS- Mandibular symphysis.

Site	Reg.No.	Trench	Layer	Mandible	Length	Width	Height
INM-1	434	LJP	1-4	m3	19.29	7.09	00.00
INM-1	443	LJP	1-4	p2	00.00	00.00	13.91
INM-1				p3	00.00	00.00	19.61
INM-1				p4	00.00	00.00	23.29
INM-1				m1	13.10	8.45	23.04
INM-1				m2	00.00	00.00	24.18
INM-1	446	LJP	1-4	m2	14.44	7.60	00.00
INM-1				m3	18.70	6.69	25.76
INM-1	437	LJP	1-4	m3	17.76	6.31	32.38
INM-1	444	LJP	1-4	dp3	7.99	4.69	00.00
INM-1				dp4	16.64	6.54	00.00
INM-1				m1	15.77	6.63	00.00
INM-1	432	LJP	1-4	dp3	6.64	4.29	16.31
INM-1				dp4	13.98	5.67	20.25
INM-1				m1	12.10	6.64	21.26
INM-1				m2	00.00	00.00	25.49
INM-1	448	LJP	1-4	dp2	00.00	00.00	14.09
INM-1				dp3	00.00	00.00	17.04
INM-1				dp4	15.10	6.13	19.45
INM-1				m1	14.59	6.66	22.65
INM-1				m2	00.00	00.00	28.79
INM-1	451	LJP	1-4	m2	17.00	6.46	00.00
INM-1	452	LJP	1-4	p3	00.00	00.00	19.58
INM-1				p4	00.00	00.00	21.66
INM-1				m1	00.00	00.00	23.17
INM-1				m2	00.00	00.00	24.07
INM-1				m3	00.00	00.00	26.12
INM-1	5287	LJP	1-4	MC	00.00	19.35	00.00
INM-1	5292	LJP	1-4	m3	18.60	7.75	00.00
INM-1	447	E7	1	p2	00.00	00.00	15.19
INM-1				p3	5.43	4.91	18.06
INM-1				p4	9.93	5.91	19.45
INM-1				m1	11.24	7.14	20.35
INM-1				m2	00.00	00.00	23.04

INM-1	439	F26	1	dp2	00.00	00.00	12.39
INM-1				dp3	00.00	00.00	16.61
INM-1				dp4	15.70	6.21	17.78
INM-1				m1	00.00	00.00	21.96
INM-1	441	E27	1	m2	14.09	7.71	00.00
INM-1				m3	2-.56	7.98	25.86
INM-1	1122	G4	1	m3	18.95	7.08	00.00
INM-1	1123	G4	1	dp2	00.00	00.00	10.84
INM-1				dp3	00.00	00.00	13.30
INM-1				dp4	15.28	5.65	16.31
INM-1				m1	00.00	00.00	20.19
INM-1				m2	14.36	6.51	00.00
INM-1	1124	G4	1	m1	11.11	7.74	00.00
INM-1				m2	15.65	8.82	00.00
INM-1	1125	G4	1	m2	17.84	7.49	00.00
INM-1	1127	G4	1	MC	00.00	17.63	00.00
INM-1	1177	G10	1	p2	00.00	00.00	15.26
INM-1	1198	F10	1	m3	21.20	8.23	38.38
INM-1	1703	G11	1	MC	00.00	16.41	00.00
INM-1	1704	G11	1	MC	00.00	16.42	00.00
INM-1	1705	G11	1	MC	00.00	18.56	00.00
INM-1	1706	G11	1	MC	00.00	16.85	00.00
INM-1	1707	G11	1	MC	00.00	18.46	00.00
INM-1	1708	G11	1	MC	00.00	19.40	00.00
INM-1	1709	G11	1	p2	00.00	00.00	17.69
INM-1	1710	G11	1	p2	00.00	00.00	17.19
INM-1				p3	9.75	5.58	20.54
INM-1	1711	G11	1	dp2	00.00	00.00	11.73
INM-1				dp3	9.63	4.59	13.71
INM-1				dp4	16.64	5.95	15.64
INM-1	1712	G11	1	m1	00.00	00.00	19.46
INM-1	1725	G11	1	MS	00.00	00.00	16.94
INM-1	1740	G11	1	MC	00.00	20.90	00.00
INM-1	1971	H11	1	dp4	17.44	6.78	00.00
INM-1	1972	H11	1	p4	8.87	4.66	00.00
INM-1				m1	10.38	6.12	00.00
INM-1	1973	H11	1	dp2	3.13	2.19	00.00
INM-1				dp3	8.12	4.27	13.87
INM-1				dp4	17.07	5.96	15.57
INM-1	1975	H11	1	p2	00.00	00.00	19.08
INM-1				p3	7.57	5.21	21.71
INM-1	2726	G6	1	MC	00.00	19.51	00.00
INM-1	2727	G6	1	MC	00.00	27.25	00.00
INM-1	2728	G6	1	MC	00.00	18.45	00.00
INM-1	2729	G6	1	MS	00.00	00.00	15.10
INM-1	2730	G6	1	MS	00.00	00.00	17.35
INM-1	2731	G6	1	m1	11.71	6.96	00.00
INM-1				m2	15.72	8.26	00.00
INM-1				m3	21.44	7.92	00.00
INM-1	2732	G6	1	p3	6.61	4.68	00.00
INM-1				p4	7.93	5.45	00.00
INM-1				m1	8.46	6.01	00.00
INM-1				m2	12.85	7.25	00.00
INM-1	2733	G6	1	dp4	15.74	6.24	00.00
INM-1				m1	13.60	6.04	00.00
INM-1	2945	H10	1	m1	13.59	7.02	21.66
INM-1				m2	16.04	7.13	24.98
INM-1	2946	H10	1	MS	00.00	00.00	19.84

INM-1	3358	F11	1	m1	13.99	6.49	23.58
INM-1				m2	15.55	6.50	27.65
INM-1				m3	00.00	00.00	33.20
INM-1	3359	F11	1	p3	00.00	00.00	16.79
INM-1				p4	8.87	5.67	18.40
INM-1				m1	9.60	6.75	20.35
INM-1	3552	H5	1	p4	00.00	00.00	20.10
INM-1				m1	13.25	6.90	21.92
INM-1				m2	15.58	6.98	26.06
INM-1				m3	20.05	6.80	30.65
INM-1	3553	H5	1	p4	8.17	5.05	19.11
INM-1				m1	10.88	6.88	21.75
INM-1				m2	14.75	7.38	25.20
INM-1	3554	H5	1	m1	15.00	7.25	00.00
INM-1	3555	H5	1	MC	00.00	17.10	00.00
INM-1	3862	D25	1	MC	00.00	16.75	00.00
INM-1	8064	F27	1	p4	9.05	5.20	00.00
INM-1				m1	10.50	6.70	00.00
INM-1				m2	13.60	9.45	00.00
INM-1	8083	F27	1	m2	16.95	6.18	00.00
INM-1	487	J9	1	p4	9.09	4.50	20.28
INM-1				m1	12.33	7.03	20.98
INM-1				m2	14.49	6.88	23.61
INM-1				m3	00.00	00.00	28.09
INM-1	2166	Hno132	1	MS	00.00	00.00	15.84
INM-1				p3	00.00	00.00	17.27
INM-1				p4	00.00	00.00	19.51
INM-1				m1	11.13	6.83	20.44
INM-1				m2	14.07	7.80	21.62
INM-1				m3	00.00	00.00	25.14
INM-1	4925	F13	1	MS	00.00	00.00	17.25
INM-1				p2	00.00	00.00	18.00
INM-1				p3	9.90	5.10	21.50
INM-1				p4	00.00	00.00	23.05
INM-1	4926	F13	1	dp2	5.30	3.50	00.00
INM-1				dp3	10.25	5.80	00.00
INM-1				dp4	17.65	6.20	00.00
INM-1	436	G24	2	p4	8.82	00.00	00.00
INM-1				m1	12.02	6.67	21.95
INM-1				m2	15.37	7.46	24.97
INM-1				m3	00.00	00.00	28.82
INM-1	438	G24	2	p3	00.00	00.00	17.26
INM-1				p4	8.84	4.37	00.00
INM-1				m1	13.25	7.19	00.00
INM-1				m2	15.04	7.26	00.00
INM-1	449	E24	2	p2	00.00	00.00	20.68
INM-1				p3	00.00	00.00	23.26
INM-1				p4	00.00	00.00	23.57
INM-1				m1	13.32	7.05	23.59
INM-1	767	I12	2	p2	00.00	00.00	14.83
INM-1				p3	00.00	00.00	18.43
INM-1				m1	00.00	00.00	19.73
INM-1	768	I12	2	m2	15.69	6.62	00.00
INM-1	1076	H7	2	m1	12.13	7.45	21.85
INM-1				m2	14.84	7.62	25.28
INM-1				m3	00.00	00.00	29.13
INM-1	1077	H7	2	MC	00.00	18.95	00.00
INM-1	1079	H7	2	MC	00.00	19.00	00.00

INM-1	1080	H7	2	dp4	14.80	00.00	00.00
INM-1				m1	14.74	00.00	21.78
INM-1				m2	00.00	00.00	26.86
INM-1	1084	H7	2	MC	00.00	19.31	00.00
INM-1	1346	G11	2	p3	00.00	00.00	14.00
INM-1				p4	00.00	00.00	15.88
INM-1				m1	00.00	00.00	19.02
INM-1	1347	G11	2	MC	00.00	17.38	00.00
INM-1	1348	G11	2	MC	00.00	16.71	00.00
INM-1	1368	G11	2	MC	00.00	16.47	00.00
INM-1	1438	D25	2	MC	00.00	18.26	00.00
INM-1	1439	D25	2	MC	00.00	16.77	00.00
INM-1	1441	D25	2	m1	13.89	6.65	00.00
INM-1				m2	15.65	7.55	00.00
INM-1	1477	D25	2	MS	00.00	00.00	14.16
INM-1	840	F12	2	MC	00.00	19.80	00.00
INM-1	1618	H5	2	dp4	14.61	5.88	00.00
INM-1				m1	14.47	5.82	00.00
INM-1	1619	H5	2	m3	18.48	7.20	00.00
INM-1	1620	H5	2	dp4	00.00	6.42	19.60
INM-1				m1	13.30	7.13	21.57
INM-1				m2	16.31	6.88	25.92
INM-1				m3	00.00	00.00	31.88
INM-1	1623	H5	2	MC	00.00	18.70	00.00
INM-1	1632	H5	2	MS	00.00	00.00	16.63
INM-1	2102	G6	2	p2	00.00	00.00	13.86
INM-1				p3	00.00	00.00	15.70
INM-1				p4	00.00	00.00	18.84
INM-1				m1	10.87	7.24	21.34
INM-1				m2	14.25	8.36	24.39
INM-1				m3	00.00	7.59	27.60
INM-1	2104	G6	2	MC	00.00	18.04	00.00
INM-1	2285	E1	2	MC	00.00	15.68	00.00
INM-1	2286	E1	2	MC	00.00	16.01	00.00
INM-1	2288	E1	2	MS	00.00	00.00	17.71
INM-1				p3	00.00	00.00	17.10
INM-1				p4	00.00	00.00	20.27
INM-1				m1	13.87	6.81	20.79
INM-1				m2	16.11	6.57	25.69
INM-1				m3	16.62	5.04	30.34
INM-1	2290	E1	2	p3	7.08	3.08	17.68
INM-1				p4	9.41	4.25	20.72
INM-1				m1	00.00	00.00	22.62
INM-1	2291	E1	2	MS	00.00	00.00	14.93
INM-1				p2	00.00	00.00	17.37
INM-1				p3	00.00	00.00	17.38
INM-1				p4	00.00	00.00	16.45
INM-1				m1	00.00	00.00	15.38
INM-1				m2	00.00	00.00	15.85
INM-1				m3	00.00	00.00	20.00
INM-1	2463	F24	2	MS	00.00	00.00	15.20
INM-1				p3	00.00	00.00	18.41
INM-1				p4	00.00	00.00	21.59
INM-1				m1	00.00	00.00	23.34
INM-1	2464	F24	2	p2	00.00	00.00	14.69
INM-1				p3	00.00	00.00	17.23
INM-1				p4	00.00	00.00	19.91
INM-1				m1	00.00	00.00	21.59

INM-1	2465	F26	2	m1	10.76	7.06	21.23
INM-1				m2	00.00	00.00	23.71
INM-1	2466	F24	2	m1	11.79	7.27	00.00
INM-1				m2	13.02	7.69	00.00
INM-1	2592	F11	2	MC	00.00	20.53	00.00
INM-1	2597	F11	2	m1	12.38	6.48	23.76
INM-1				m2	14.61	6.55	27.13
INM-1				MC	00.00	18.06	00.00
INM-1	3113	E2	2	MS	00.00	00.00	18.20
INM-1	3222	G11	2	MC	00.00	19.58	00.00
INM-1	3223	G11	2	MC	00.00	22.51	00.00
INM-1	3224	G11	2	MC	00.00	18.65	00.00
INM-1	3225	G11	2	MC	00.00	16.73	00.00
INM-1	3226	G11	2	p3	00.00	00.00	19.69
INM-1				p4	9.67	4.53	22.56
INM-1				m1	12.40	8.25	23.80
INM-1				m2	14.80	6.84	26.35
INM-1	3227	G11	2	p2	00.00	00.00	16.10
INM-1				p3	00.00	00.00	16.80
INM-1				p4	00.00	00.00	18.36
INM-1	3228	G11	2	p3	00.00	00.00	15.55
INM-1				p4	00.00	00.00	18.35
INM-1				m1	00.00	00.00	20.05
INM-1	3229	G11	2	p2	3.45	00.00	00.00
INM-1				p3	6.05	4.45	00.00
INM-1				p4	7.85	5.75	00.00
INM-1				m1	9.50	6.08	00.00
INM-1	3411	H6	2	p2	00.00	00.00	11.30
INM-1				p3	8.20	4.10	13.76
INM-1	3998	G4	2	MS	00.00	00.00	21.45
INM-1	3999	G4	2	MS	00.00	00.00	19.39
INM-1	4000	G4	2	m1	10.67	6.90	19.75
INM-1				m2	14.00	7.50	21.83
INM-1				m3	19.05	7.15	26.10
INM-1	4001	G4	2	m3	19.00	7.55	30.90
INM-1	4003	G4	2	dp4	18.10	8.95	00.00
INM-1				m1	15.60	10.05	00.00
INM-1	4073	C0	2	m3	19.50	7.20	00.00
INM-1	3586	I6	2	m3	21.92	8.27	00.00
INM-1	3600	I6	2	MS	00.00	00.00	17.75
INM-1	3647	G7	2	dp2	00.00	00.00	12.05
INM-1				dp3	7.95	3.82	14.30
INM-1				dp4	16.50	5.78	16.05
INM-1				m1	00.00	00.00	20.15
INM-1	3648	G7	2	dp2	00.00	00.00	11.65
INM-1				dp3	7.98	3.82	13.85
INM-1				dp4	00.00	00.00	15.20
INM-1	3649	G7	2	MC	00.00	13.64	00.00
INM-1	3650	G7	2	MC	00.00	19.65	00.00
INM-1	6438	E25	2	MC	00.00	16,32	00.00
INM-1	6439	E25	2	MC	00.00	15.75	00.00
INM-1	6440	E25	2	MC	00.00	15.90	00.00
INM-1	6441	E25	2	p4	8.55	5.30	00.00
INM-1	6442	E25	2	m1	8.90	6.82	21.15
INM-1				m2	11.80	7.75	22.60
INM-1				m3	20.50	8.00	25.70
INM-1	6657	F24	2	MC	00.00	16.15	00.00
INM-1	6658	F24	2	m1	13.00	6.57	00.00

INM-1				m2	15.70	7.55	00.00
INM-1	6665	F24	2	m2	13.45	8.25	00.00
INM-1				m3	22.30	9.50	00.00
INM-1	7032	G24	2	MC	00.00	17.40	00.00
INM-1	7185	G25	2	m1	14.90	8.00	00.00
INM-1	7186	G25	2	dp3	7.25	3.95	00.00
INM-1				dp4	15.85	5.45	16.80
INM-1				m1	12.85	5.80	21.10
INM-1	7317	E24	2	MC	00.00	16.18	00.00
INM-1	7441	E25	2	m2	16.95	7.70	00.00
INM-1	7442	E25	2	MC	00.00	16.95	00.00
INM-1	7443	E25	2	MC	00.00	17.25	00.00
INM-1	7456	E25	2	MC	00.00	16.70	00.00
INM-1	7497	F24	2	MC	00.00	17.15	00.00
INM-1	442	G25	3	p2	00.00	00.00	12.89
INM-1				p3	00.00	00.00	17.55
INM-1				p4	10.05	5.67	20.50
INM-1				m1	13.94	7.37	21.80
INM-1	440	F25	3	m1	13.50	9.31	00.00
INM-1				m2	13.29	7.94	00.00
INM-1				m3	14.84	9.46	00.00
INM-1	433	F24	3	dp4	13.95	6.51	18.42
INM-1				m1	14.62	6.53	21.84
INM-1	429	G26	3	dp4	12.75	8.48	00.00
INM-1				m1	13.40	10.49	00.00
INM-1	538	Hno126	3	MC	00.00	19.39	00.00
INM-1	936	G12	3	p4	00.00	00.00	19.96
INM-1				m1	13.18	6.79	21.64
INM-1				m2	15.42	7.28	26.64
INM-1				m3	17.75	6.42	30.02
INM-1	937	G12	3	p3	5.98	4.93	16.71
INM-1				p4	7.14	5.57	19.16
INM-1				m1	9.25	00.00	20.78
INM-1				m2	13.32	00.00	23.51
INM-1				m3	21.72	8.37	26.70
INM-1	938	G12	3	m1	9.01	6.11	00.00
INM-1				m2	12.87	7.39	00.00
INM-1				m3	21.23	7.92	24.35
INM-1	939	G12	3	m2	12.24	7.76	00.00
INM-1				m3	19.33	8.11	26.65
INM-1	940	G12	3	p4	8.80	5.21	00.00
INM-1				m1	12.15	7.06	00.00
INM-1				m2	14.12	7.54	00.00
INM-1	944	G12	3	m2	15.55	00.00	00.00
INM-1				m3	00.00	00.00	33.05
INM-1	945	G12	3	m1	00.00	00.00	20.78
INM-1				m2	00.00	00.00	23.80
INM-1	1824	F10	3	p2	00.00	00.00	16.03
INM-1	2212	G10	3	m1	12.24	6.22	20.94
INM-1				m2	14.23	6.32	24.42
INM-1				m3	17.66	5.94	29.97
INM-1	2232	G10	3	MS	00.00	00.00	13.90
INM-1	2254	F11	3	MC	00.00	17.84	00.00
INM-1	2255	F11	3	MC	00.00	18.15	00.00
INM-1	2320	E10	3	MS	00.00	00.00	32.20
INM-1	2358	E10	3	m1	14.22	6.94	00.00
INM-1	2690	G7	3	p2	00.00	00.00	15.27
INM-1				p3	00.00	00.00	18.64

INM-1	2691	G7	3	MC	00.00	16.83	00.00
INM-1	2865	G3	3	MC	00.00	20.77	00.00
INM-1	2866	G3	3	MS	00.00	00.00	16.57
INM-1	2868	G3	3	dp2	3.95	8.46	00.00
INM-1				dp3	8.46	4.35	00.00
INM-1				dp4	16.57	4.86	00.00
INM-1	2869	G3	3	p2	00.00	00.00	15.19
INM-1				p3	00.00	00.00	19.75
INM-1				p4	00.00	00.00	21.58
INM-1				m1	14.16	6.83	23.38
INM-1				m2	15.21	6.73	27.83
INM-1				m3	20.80	7.21	00.00
INM-1	2870	G3	3	p4	10.05	6.68	00.00
INM-1				m1	11.40	7.40	00.00
INM-1	3272	F11	3	m2	13.02	7.45	00.00
INM-1	3273	F11	3	m1	10.65	8.04	00.00
INM-1	3492	H7	3	m1	10.30	6.90	00.00
INM-1				m2	13.67	7.80	00.00
INM-1				m3	19.61	7.05	00.00
INM-1	3493	H7	3	p4	10.06	5.03	00.00
INM-1				m1	12.65	7.07	00.00
INM-1				m2	15.62	7.48	00.00
INM-1				m3	18.90	6.95	00.00
INM-1	3494	H7	3	p2	00.00	00.00	16.25
INM-1				p3	7.40	3.80	20.45
INM-1				p4	8.97	5.05	21.10
INM-1	3495	H7	3	MC	00.00	18.65	00.00
INM-1	3651	H7	3	MC	00.00	13.68	00.00
INM-1	3792	H12	3	p3	00.00	00.00	16.23
INM-1				p4	7.30	5.08	19.00
INM-1				m1	10.27	6.60	19.35
INM-1				m2	13.27	7.75	21.30
INM-1	4486	Hno128	3	p2	00.00	00.00	13.10
INM-1				p3	00.00	00.00	14.24
INM-1				p4	00.00	00.00	16.00
INM-1	4532	H12	3	p3	00.00	00.00	15.37
INM-1				p4	00.00	00.00	18.47
INM-1				m1	12.95	6.52	19.85
INM-1				m2	15.39	6.65	23.15
INM-1				m3	00.00	00.00	28.39
INM-1	4533	H12	3	m1	12.85	7.00	00.00
INM-1				m2	15.08	7.40	24.36
INM-1				m3	19.20	7.50	29.60
INM-1	4534	H12	3	m1	12.80	6.67	21.13
INM-1				m2	00.00	00.00	25.80
INM-1	4537	H12	3	m2	15.48	10.20	00.00
INM-1				m3	13.90	8.45	00.00
INM-1	6188	Hno106	3	MC	00.00	14.95	00.00
INM-1	6189	Hno106	3	p4	9.25	5.20	00.00
INM-1	6523	F25	3	MC	00.00	16.05	00.00
INM-1	6524	F25	3	MC	00.00	16.90	00.00
INM-1	6781	E23	3	MC	00.00	18.25	00.00
INM-1	6953	E26	3	MC	00.00	17.50	00.00
INM-1	6954	E26	3	MC	00.00	18.55	00.00
INM-1	6966	E26	3	MC	00.00	17.90	00.00
INM-1	7724	G26	3	MC	00.00	17.26	00.00
INM-1	7839	G25	3	MC	00.00	20.80	00.00
INM-1	676	Hno110	4	dp2	9.95	4.87	17.83

INM-1				dp3	00.00	00.00	20.66
INM-1				dp4	00.00	00.00	21.05
INM-1	972	C1	4	p3	00.00	00.00	20.87
INM-1				p4	8.57	4.96	21.83
INM-1				ml	10.78	6.78	23.26
INM-1				m2	14.68	7.43	26.66
INM-1				m3	22.00	7.53	30.85
INM-1	1867	G11	4	p2	00.00	00.00	17.81
INM-1				p3	00.00	00.00	18.96
INM-1	1872	G11	4	MC	00.00	16.18	00.00
INM-1	4761	Hno117	4	MC	00.00	15.40	00.00
INM-1	4762	Hno117	4	MC	00.00	15.65	00.00
INM-1	4763	Hno117	4	MC	00.00	18.60	00.00
INM-1	5977	F25	4	MC	00.00	17.05	00.00
INM-1	5978	F25	4	MC	00.00	15.40	00.00
INM-1	5979	F25	4	MC	00.00	16.70	00.00
INM-1	6111	Hno105	4	MC	00.00	19.35	00.00
INM-1	6857	G26	4	MC	00.00	20.10	00.00
INM-1	6858	G26	4	MC	00.00	18.50	00.00
INM-1	7648	G25	4	MC	00.00	19.65	00.00
INM-1	7762	D24&E25	4	MC	00.00	16.75	00.00
INM-1	7986	F24	4	MS	00.00	00.00	15.25
INM-1	8153	E24	4	MC	00.00	17.30	00.00
INM-1	1516	G4	6	MC	00.00	19.10	00.00
INM-1	1839	G4	7	p2	00.00	00.00	18.35
INM-1				p3	7.68	4.32	20.88
INM-1				p4	10.93	5.03	22.49
INM-1				ml	13.58	7.02	23.21
INM-1				m2	15.05	7.05	27.66
INM-1				m3	00.00	00.00	29.27
INM-1	3902	B3&C3	8	MS	00.00	00.00	34.25
INM-1	4602	Hno65	8	p3	00.00	00.00	10.90
INM-1				p4	00.00	00.00	12.40
INM-1				ml	00.00	00.00	14.20
INM-1	5067	E5	8	MC	00.00	19.55	00.00
INM-1	5105	C1	11	MC	00.00	18.45	00.00
INM-1	820	C1	12	p2	00.00	00.00	11.05
INM-1				p3	00.00	00.00	12.97
INM-1	5925	B3	14	MC	00.00	14.95	00.00
INM-1	5400	C3	15	MC	00.00	20.25	00.00
INM-1	5401	C3	15	MC	00.00	20.29	00.00

Table A.16.30. Measurements of Maxilla with teeth of ***Antilope cervicapra***

P2 –second premolar, P3- third premolar, P4- fourth premolar,
M1- first molar, M2- second molar, M3- third molar, M0- 1st/2nd/3rd molar

Site	Reg.No.	Trench	Layer	Maxilla	Length	Width	Height
INM-1	2168	Hno132	1	M2	14.13	10.06	00.00
INM-1				M3	13.97	9.22	00.00
INM-1	1967	H11	1	M1	9.45	9.05	00.00
INM-1				M2	13.73	9.46	00.00
INM-1	1081	H7	2	P4	8.15	10.05	00.00
INM-1				M1	13.55	11.40	00.00
INM-1	7036	G24	2	M0	16.63	10.95	00.00
INM-1	7037	G24	2	M0	15.47	10.95	00.00
INM-1	7124	G24	4	P3	6.75	6.45	00.00

INM-1				M1	7.90	7.55	00.00
INM-1				M2	12.88	10.45	00.00
INM-1	4536	H12	3	P4	6.37	00.00	00.00
INM-1				M1	10.45	10.25	00.00
INM-1				M2	12.95	10.75	00.00
INM-1				M3	16.50	10.25	00.00
INM-1	4006	G4	2	M3	16.01	11.50	00.00
INM-1	4007	G4	2	P3	6.15	5.90	00.00
INM-1				P4	8.30	8.55	00.00
INM-1	3590	I6	2	P2	5.20	4.18	00.00
INM-1				P3	6.63	6.55	00.00
INM-1				P4	8.34	8.45	00.00
INM-1	3496	H7	3	M1	9.04	9.90	00.00
INM-1				M2	12.37	11.90	00.00
INM-1				M3	15.10	10.90	00.00
INM-1	3497	H7	3	M0	16.67	12.06	00.00

Table A.17.Measurements of Bones of *Gazella bennetti*

Table A.17.1. Measurements of Scapula of *Gazella bennetti*

Site	Reg.No.	Trench	Layer	GLP	LG	BG	SLC
INM-1	304	C23	2	31.90	24.40	20.67	18.03
INM-1	308	E24	2	00.00	00.00	18.56	19.46

Table A.17.2. Measurements of Humerus of *Gazella bennetti*

Site	Reg.No.	Trench	Layer	BD	TD
INM-1	282	LJP	1-4	26.50	13.76

Table A.17.3. Measurements of Radius of *Gazella bennetti*

Site	Reg.No.	Trench	Layer	BP	TP
INM-1	3393	H6	2	00.00	13.30

Table A.17.4. Measurements of Tibia of *Gazella bennetti*

Site	Reg.No.	Trench	Layer	BD	TD
INM-1	3340	F11	1	20.33	14.85
INM-1	1255	D25	3	23.71	18.25
INM-1	3913	B3&C3	8	21.05	17.10

Table A.17.5. Measurements of Metatarsal of *Gazella bennetti*

Site	Reg.No.	Trench	Layer	BP	TP
INM-1	3401	H6	2	17.55	17.90

Table A.17.6. Measurements of Astragalus of *Gazella bennetti*

Site	Reg.No.	Trench	Layer	GL	GLL	GLM	TL	TM	BD
INM-1	5781	Hno75	12	25.00	24.27	24.00	13.30	13.65	16.53
INM-1	6016	E3	11	24.10	23.98	23.60	13.85	15.15	15.74

Table A.17.7. Measurements of First Phalanx of *Gazella bennetti*

Site	Reg.No.	Trench	Layer	GL	BP	TP	BD	TD
INM-1	2025	G7	1	42.73	11.05	14.82	9.22	9.80
INM-1	6758	E23	3	00.00	00.00	00.00	7.05	9.50
INM-1	1802	H5	4	00.00	00.00	00.00	6.56	7.30
INM-1	6794	G26	4	00.00	9.05	14.75	00.00	00.00
INM-1	5810	C4	13	35.55	10.00	12.60	9.00	8.35

Table A.17.8. Measurements of Second Phalanx of *Gazella bennetti*

Site	Reg.No.	Trench	Layer	GL	BP	TP	BD	TD
INM-1	3316	F11	1	21.80	7.85	10.75	7.75	9.50
INM-1	1651	G11	1	24.44	9.37	13.07	7.44	9.80
INM-1	2538	F11	2	21.47	8.17	12.09	7.49	9.38
INM-1	6759	E23	3	22.00	7.35	10.00	6.90	10.00
INM-1	6800	G26	4	26.85	9.05	13.70	8.30	00.00

Table A.17.9. Measurements of Maxillary third molar of *Gazella bennetti*

Site	Reg.No.	Trench	Layer	GL	GB
INM-1	773	I12	2	15.67	9.92
INM-1	774	I12	2	17.27	9.46

A.18. Measurements of Bones of *Tetracerus quadricornis*

Table A.18.1. Measurements of Scapula of *Tetracerus quadricornis*

Site	Reg.No.	Trench	Layer	GLP	LG	BG	SLC
INM-1	186	E25	4	00.00	17.69	15.11	13.73
INM-1	6045	B4	15	24.85	20.03	16.43	00.00

Table A.18.2. Measurements of Humerus of *Tetracerus quadricornis*

Site	Reg.No.	Trench	Layer	BP	TP
INM-1	3646	G7	2	25.95	13.55

Table A.18.3. Measurements of Radius of *Tetracerus quadricornis*

Site	Reg.No.	Trench	Layer	BD	TD
INM-1	393	F25	3	22.65	16.74

Table A.18.4. Measurements of Metacarpal of *Tetracerus quadricornis*

Site	Reg.No.	Trench	Layer	DD	BD	TD	WT	TT	TLT
INM-1	191	LJP	1-4	00.00	22.12	00.00	10.13	16.67	12.72
INM-1	330	LJP	1-4	00.00	00.00	00.00	6.76	00.00	00.00
INM-1	6072	B5	15	11.00	17.95	14.65	8.20	14.35	11.05

Table A.18.5. Measurements of Femur of *Tetracerus quadricornis*

Site	Reg.No.	Trench	Layer	DD	BD
INM-1	1105	G4	1	35.38	43.55

Table A.18.6. Measurements of Tibia of *Tetracerus quadricornis*

Site	Reg.No.	Trench	Layer	DD	BD
INM-1	3007	F10	2	19.60	15.36
INM-1	4725	B1	5	21.37	15.90
INM-1	3912	B3&C3	8	21.11	15.45
INM-1	3818	C0	12	22.56	16.88

Table A.18.7. Measurements of Metacarpal of *Tetracerus quadricornis*

Site	Reg.No.	Trench	Layer	BP	TP	BD	TD	WT	TT	TLT
INM-1	1447	D25	2	00.00	00.00	18.69	13.95	8.33	13.71	10.32
INM-1	1453	D25	2	00.00	00.00	00.00	00.00	9.07	14.62	10.15
INM-1	2445	F25&F26	3	18.68	17.80	00.00	00.00	00.00	00.00	00.00

Table A.18.8. Measurements of Astragalus of *Tetracerus quadricornis*

Site	Reg.No.	Trench	Layer	GL	GLL	GLM	TL	TM	BD
INM-1	89	LJP	1-4	25.37	24.73	22.72	13.80	13.55	15.18
INM-1	4038	C0	3	00.00	00.00	23.20	13.15	13.40	00.00
INM-1	7085	G24	4	27.68	27.58	25.95	15.60	15.45	16.50
INM-1	6070	B5	1	27.50	27.30	27.03	15.40	15.53	17.05
INM-1	6246	E5	1	24.73	24.70	22.90	13.35	13.50	15.57

Table A.18.9. Measurements of First Phalanx of *Tetracerus quadricornis*

Site	Reg.No.	Trench	Layer	BP	TP
INM-1	1393	D25	2	9.28	14.81

Table A.18.10. Measurements of Second Phalanx of *Tetracerus quadricornis*

Site	Reg.No.	Trench	Layer	GL	BP	TP	BD	TD
INM-1	361	LJP	1-4	22.39	8.32	12.30	7.12	10.10

Table A.18.11. Measurements of Maxillary first/second molar of *Tetracerus quadricornis*

Site	Reg.No.	Trench	Layer	BP	TP
INM-1	2266	G5	6	11.76	8.62
INM-1	6104	B3	15	13.23	8.90

A.19. Measurements of Bones of *Boselaphus tragocamelus*

Table A.19.1. Measurements of Humerus of ***Boselaphus tragocamelus***

Site	Reg.No.	Trench	Layer	BD	TD
INM-1	776	I12	2	57.47	30.02

Table A.19.2. Measurements of Radius of ***Boselaphus tragocamelus***

Site	Reg.No.	Trench	Layer	BP	TP
INM-1	333	LJP	1-4	65.74	30.98
INM-1	4290	C2	12	53.80	27.62

Table A.19.3. Measurements of Ulna of ***Boselaphus tragocamelus***

Site	Reg.No.	Trench	Layer	BP
INM-1	7509	F24	2	31.15
INM-1	4474	Hno129	2	35.32
INM-1	5737	E3	12	30.40
INM-1	5664	A5	15	28.45
INM-1	5665	A5	15	31.45

Table A.19.4. Measurements of Metacarpal of ***Boselaphus tragocamelus***

Site	Reg.No.	Trench	Layer	BP	TP	BD	TD	WT	TT	TLT
INM-1	1681	G11	1	41.59	26.77	00.00	00.00	00.00	00.00	00.00
INM-1	1627	H5	2	39.85	24.84	00.00	00.00	00.00	00.00	00.00
INM-1	3473	H7	3	44.15	28.15	00.00	00.00	00.00	00.00	00.00
INM-1	900	G12	3	00.00	00.00	44.75	31.82	20.55	28.37	23.28

Table A.19.5. Measurements of Metacarpal of ***Boselaphus tragocamelus***

Site	Reg.No.	Trench	Layer	WT	TT	TLT
INM-1	6577	D25	3	19.30	29.00	22.70
INM-1	7666	G25	4	18.65	30.90	23.45
INM-1	4338	F8	5	19.80	30.00	24.05
INM-1	4339	F8	5	19.54	30.00	23.39

INM-1	5731	C3	13	20.70	31.15	23.70
INM-1	5914	B3	14	18.75	28.80	22.55

Table A.19.6. Measurements of Tibia of ***Boselaphus tragocamelus***

Site	Reg.No.	Trench	Layer	BD	TD
INM-1	45	LJP	1-4	46.41	39.28
INM-1	1099	G4	1	48.31	39.12

Table A.19.7. Measurements of Metatarsal of ***Boselaphus tragocamelus***

Site	Reg.No.	Trench	Layer	BP	TP	BD	TD	WT	TT	TLT
INM-1	1095	G4	1	47.38	44.23	00.00	00.00	00.00	00.00	00.00
INM-1	3543	H5	1	46.90	42.27	00.00	00.00	00.00	00.00	00.00
INM-1	1697	G11	1	00.00	00.00	48.36	28.05	21.52	26.64	19.80
INM-1	2092	G6	2	00.00	00.00	43.10	32.40	20.40	32.53	25.38
INM-1	1597	H5	2	39.67	41.30	00.00	00.00	00.00	00.00	00.00
INM-1	3402	H6	2	41.92	43.40	00.00	00.00	00.00	00.00	00.00
INM-1	3403	H6	2	36.40	37.80	00.00	00.00	00.00	00.00	00.00
INM-1	2853	G3	3	00.00	00.00	39.95	29.35	17.57	29.24	22.48
INM-1	899	G12	3	00.00	00.00	44.14	31.55	20.32	31.57	23.70
INM-1	2451	F25-F26	3	00.00	00.00	38.81	29.26	18.57	28.29	22.08
INM-1	5895	E25	4	00.00	00.00	00.00	00.00	21.30	30.50	22.70
INM-1	3679	C6	8	39.64	40.18	00.00	00.00	00.00	00.00	00.00
INM-1	5090	B2	13	42.58	42.23	00.00	00.00	00.00	00.00	00.00
INM-1	6284	B3	14	44.70	42.10	00.00	00.00	00.00	00.00	00.00
INM-1	6285	B3	14	38.26	39.45	00.00	00.00	00.00	00.00	00.00
INM-1	5378	C3	15	00.00	00.00	47.45	32.26	20.35	32.09	25.18
INM-1	5415	C3	15	40.80	41.39	00.00	00.00	00.00	00.00	00.00

Table A.19.8. Measurements of First Phalanx of ***Boselaphus tragocamelus***

Site	Reg.No.	Trench	Layer	GL	BP	TP	BD	TD
INM-1	93	LJP	1-4	68.41	21.60	29.19	21.58	18.25
INM-1	164	LJP	1-4	67.67	21.10	29.40	22.45	18.81
INM-1	256	LJP	1-4	73.91	22.92	30.04	22.09	18.03
INM-1	2021	G7	1	00.00	00.00	00.00	22.22	18.11
INM-1	2023	G7	1	00.00	00.00	00.00	20.84	17.50
INM-1	3514	H5	1	67.40	20.50	28.65	20.55	18.90
INM-1	3819	D25	1	00.00	21.15	31.80	00.00	00.00
INM-1	4903	F13	1	00.00	21.00	28.15	00.00	00.00
INM-1	2114	Hno132	1	00.00	00.00	00.00	21.08	17.86
INM-1	2115	Hno132	1	00.00	00.00	00.00	20.58	17.67
INM-1	3925	G4	2	00.00	20.55	27.10	00.00	00.00
INM-1	3926	G4	2	00.00	00.00	00.00	21.80	17.25
INM-1	3927	G4	2	00.00	22.60	29.65	00.00	00.00
INM-1	3928	G4	2	64.65	20.95	27.25	19.05	17.20
INM-1	2064	G6	2	00.00	20.84	27.33	00.00	00.00
INM-1	2065	G6	2	00.00	00.00	00.00	19.79	16.46
INM-1	1528	H5	2	68.38	21.70	30.10	21.26	16.75
INM-1	1529	H5	2	00.00	24.89	33.14	00.00	00.00
INM-1	1629	H5	2	00.00	00.00	00.00	18.89	16.86
INM-1	3367	H6	2	00.00	20.50	28.75	00.00	00.00
INM-1	3368	H6	2	00.00	00.00	00.00	21.00	17.90
INM-1	6375	E25	2	60.45	20.20	26.60	19.60	16.25
INM-1	846	F12	2	00.00	24.04	31.21	00.00	00.00
INM-1	3122	G11	2	61.29	19.60	25.83	18.74	17.07
INM-1	3123	G11	2	66.68	20.88	29.74	20.80	17.62
INM-1	3124	G11	2	00.00	21.32	28.70	00.00	00.00
INM-1	4461	Hno129	2	00.00	17.65	23.25	00.00	00.00
INM-1	2660	G7	3	00.00	00.00	00.00	18.35	00.00
INM-1	3415	H7	3	00.00	00.00	00.00	23.68	19.45
INM-1	2014	D24	3	00.00	24.28	29.12	00.00	00.00

INM-1	6960	E26	3	66.25	20.15	29.45	22.05	18.60
INM-1	6554	F25	3	66.40	20.95	27.25	20.20	17.12
INM-1	6555	F25	3	63.20	19.95	26.38	20.70	17.36
INM-1	6556	F25	3	00.00	00.00	00.00	19.75	16.45
INM-1	2181	G10	3	00.00	00.00	00.00	20.10	18.04
INM-1	879	G12	3	61.65	20.73	25.78	19.32	17.70
INM-1	880	G12	3	00.00	00.00	00.00	22.02	17.19
INM-1	7727	G26	3	61.60	19.52	24.05	18.80	15.76
INM-1	7728	G26	3	00.00	19.23	26.28	00.00	00.00
INM-1	2610	H10	3	00.00	27.02	33.92	00.00	00.00
INM-1	2633	H10	3	00.00	00.00	00.00	17.56	16.18
INM-1	4539	H12	3	64.89	18.70	23.55	18.41	15.75
INM-1	1375	E1	4	62.26	20.04	26.89	19.76	16.94
INM-1	1797	H5	4	00.00	19.78	25.52	00.00	00.00
INM-1	7126	G24	4	00.00	00.00	00.00	19.05	16.75
INM-1	7127	G24	4	00.00	00.00	00.00	18.30	15.98
INM-1	7128	G24	4	00.00	00.00	00.00	20.05	16.90
INM-1	7129	G24	4	00.00	00.00	00.00	21.10	18.25
INM-1	7656	G25	4	69.65	20.32	25.34	20.20	16.80
INM-1	7657	G25	4	00.00	20.25	26.20	00.00	00.00
INM-1	7670	G25	4	00.00	20.58	25.68	00.00	00.00
INM-1	6847	G26	4	64.18	20.45	29.15	20.36	16.35
INM-1	6855	G26	4	00.00	13.10	17.20	00.00	00.00
INM-1	6856	G26	4	00.00	10.45	15.65	00.00	00.00
INM-1	645	Hno110	4	69.29	21.81	30.09	21.30	18.29
INM-1	646	Hno110	4	69.56	23.30	30.17	23.74	18.63
INM-1	647	Hno110	4	68.89	22.11	27.32	21.93	19.08
INM-1	4747	Hno117	4	64.15	19.95	24.80	18.70	17.30
INM-1	4306	F8	5	00.00	00.00	00.00	20.55	16.70
INM-1	4307	F8	5	00.00	21.65	28.63	00.00	00.00
INM-1	4390	B6	8	65.27	20.45	28.16	19.56	18.30
INM-1	4453	E7	8	00.00	00.00	00.00	24.05	18.95
INM-1	3609	C0	8	00.00	20.78	28.75	00.00	00.00
INM-1	4883	Hno71	8	00.00	00.00	00.00	21.95	18.05
INM-1	4764	C0	10	00.00	00.00	00.00	19.60	15.93
INM-1	5098	C1	11	00.00	21.36	30.40	00.00	00.00
INM-1	3688	B1	12	00.00	00.00	00.00	21.50	18.88
INM-1	5807		13	67.93	20.75	29.55	22.10	17.90
INM-1	5448	B1	14	00.00	00.00	00.00	23.30	17.53
INM-1	5449	B1	14	00.00	25.00	29.10	00.00	00.00
INM-1	5898	B3	14	00.00	21.05	27.55	00.00	00.00
INM-1	5901	B3	14	00.00	00.00	00.00	21.97	16.40
INM-1	5791	B5	14	73.95	00.00	31.78	23.50	19.05
INM-1	5366	C3	15	00.00	20.62	28.90	00.00	00.00
INM-1	6232	C4	15	00.00	21.90	29.70	00.00	00.00
INM-1	6233	C4	15	00.00	20.95	29.00	00.00	00.00
INM-1	5466	D3	15	67.05	20.45	28.45	20.57	17.80
INM-1	5556	D4	15	69.05	22.45	31.00	22.32	17.77
INM-1	5557	D4	15	63.35	27.00	28.50	19.60	16.76
INM-1	5558	D4	15	67.20	00.00	27.36	20.80	17.00
INM-1	5421	E4	15	00.00	00.00	31.37	00.00	00.00
INM-1	4813	B0	15	71.23	21.35	28.80	21.65	17.85
INM-1	6087	B3	16	00.00	20.55	26.55	00.00	00.00

Table A.19.9. Measurements of Second Phalanx of ***Boselaphus tragocamelus***

Site	Reg.No.	Trench	Layer	GL	BP	TP	BD	TD
INM-1	227	LJP	1-4	36.21	18.87	00.00	15.98	21.12
INM-1	237	LJP	1-4	38.57	23.17	00.00	18.89	26.08
INM-1	5270	LJP	1-4	45.10	23.55	29.05	19.00	22.90
INM-1	1108	G4	1	44.79	24.68	28.79	18.67	25.14
INM-1	1109	G4	1	40.84	22.88	27.02	18.04	22.98
INM-1	3525	H5	1	42.65	00.00	27.48	18.60	26.46

INM-1	3830	D25	1	43.00	21.60	28.45	18.85	21.40
INM-1	2925	H10	1	36.30	19.47	24.78	17.88	19.76
INM-1	1922	H11	1	00.00	22.39	27.20	00.00	00.00
INM-1	2127	Hno132	1	00.00	00.00	00.00	16.57	22.06
INM-1	2128	Hno132	1	42.62	24.13	30.40	18.37	23.82
INM-1	3947	G4	2	00.00	00.00	00.00	18.40	22.95
INM-1	2062	G6	2	30.47	21.59	26.33	17.52	21.77
INM-1	3627	G7	2	41.05	21.35	26.35	15.43	21.50
INM-1	1420	D25	2	41.85	20.95	25.99	16.90	23.55
INM-1	1421	D25	2	43.30	22.35	26.96	17.96	25.95
INM-1	2530	F11	2	39.64	23.37	27.94	18.49	23.05
INM-1	2531	F11	2	37.62	20.50	24.65	16.54	21.94
INM-1	6642	F24	2	42.05	23.07	28.65	18.85	24.05
INM-1	1331	G11	2	39.70	23.52	26.96	16.90	20.23
INM-1	3150	G11	2	40.67	20.83	26.55	16.43	21.23
INM-1	4465	Hno129	2	40.85	20.68	26.51	16.41	23.35
INM-1	4466	Hno129	2	00.00	22.65	26.37	00.00	00.00
INM-1	3434	H7	3	39.75	20.60	25.45	17.20	23.55
INM-1	3435	H7	3	41.15	22.79	29.05	18.05	23.15
INM-1	3436	H7	3	41.15	21.65	25.05	18.15	21.60
INM-1	1217	D25	3	00.00	00.00	00.00	17.02	14.97
INM-1	1218	D25	3	40.74	22.00	26.42	17.11	24.54
INM-1	1219	D25	3	42.14	21.83	27.47	18.4	23.50
INM-1	6770	E23	3	45.25	25.35	28.40	19.95	25.05
INM-1	6785	E23	3	00.00	00.00	00.00	15.80	20.45
INM-1	2244	F11	3	40.24	21.94	28.02	18.68	23.09
INM-1	6557	F25	3	39.65	21.00	26.65	15.60	21.55
INM-1	2371	F25&F26	3	45.72	23.96	29.84	19.56	26.36
INM-1	2372	F25&F26	3	43.32	22.03	26.58	19.16	23.35
INM-1	2373	F25&F26	3	41.91	22.36	28.45	17.96	22.38
INM-1	2374	F25&F26	3	41.82	21.78	00.00	00.00	00.00
INM-1	2375	F25&F26	3	40.58	20.88	24.08	16.95	22.82
INM-1	2389	F25&F26	3	00.00	00.00	00.00	16.04	20.04
INM-1	7841	G25	3	45.70	24.70	27.65	18.70	24.55
INM-1	7842	G25	3	44.90	00.00	27.95	19.95	24.85
INM-1	7729	G26	3	43.81	23.03	28.35	18.58	24.57
INM-1	7894	G27	3	39.65	20.80	27.08	16.13	21.70
INM-1	7921	Hno102	3	44.65	23.65	28.00	20.20	25.80
INM-1	3040	H7	4	39.49	20.27	24.64	16.94	23.03
INM-1	3041	H7	4	00.00	19.85	23.95	00.00	00.00
INM-1	7989	F24	4	00.00	00.00	00.00	10.95	15.80
INM-1	7990	F24	4	00.00	22.00	00.00	19.05	23.40
INM-1	7991	F24	4	37.00	20.53	25.15	15.05	20.23
INM-1	7992	F24	4	00.00	19.45	23.25	00.00	00.00
INM-1	7658	G25	4	38.80	20.63	24.85	16.40	21.90
INM-1	6848	G26	4	00.00	22.50	30.00	00.00	00.00
INM-1	6849	G26	4	40.90	22.15	27.40	17.15	22.30
INM-1	6850	G26	4	39.30	20.70	25.05	16.70	00.00
INM-1	658	Hno110	4	43.18	00.00	29.46	17.96	23.55
INM-1	4315	F8	5	00.00	20.40	25.10	00.00	00.00
INM-1	4316	F8	5	44.03	23.90	27.35	19.46	25.20
INM-1	5162	D6	8	44.35	21.80	28.47	19.75	24.20
INM-1	4766	C0	10	36.78	23.90	29.70	22.70	24.90
INM-1	5243	B1	11	37.15	20.70	26.25	16.35	20.90
INM-1	5780	Hno75	12	37.80	20.07	24.95	16.12	20.23
INM-1	5814	E4	13	39.95	22.30	26.30	17.35	21.25
INM-1	4676	B0	13	38.72	22.30	25.85	17.70	21.97
INM-1	8155	A2	14	41.65	20.50	25.40	15.85	20.83
INM-1	5452	B1	14	36.31	20.10	00.00	18.10	21.15
INM-1	5453	B1	14	36.45	20.18	22.60	18.10	21.15
INM-1	5908	B3	14	00.00	22.43	00.00	18.02	22.05
INM-1	5993	D6	14	43.40	24.35	28.80	18.17	21.90
INM-1	5655	A5	15	00.00	21.56	00.00	17.80	22.40

INM-1	5656	A5	15	43.06	00.00	29.05	21.10	25.12
INM-1	5657	A5	15	38.41	23.35	23.60	18.41	22.88
INM-1	5369	C3	15	40.36	21.10	25.10	17.27	22.25
INM-1	5561	D4	15	36.48	20.55	23.35	18.40	20.34
INM-1	6234	E5	15	39.05	20.25	25.55	16.65	21.30
INM-1	6235	E5	15	00.00	22.30	28.80	00.00	00.00
INM-1	4814	B0	15	46.72	23.21	30.08	18.42	24.12
INM-1	5355	Hno95	15	42.40	24.08	28.75	19.67	24.55

Table A.19.10. Measurements of Third Phalanx of ***Boselaphus tragocamelus***

Site	Reg.No.	Trench	Layer	GL	DL	BP	TP
INM-1	242	LJP	1-4	54.74	45.87	17.98	00.00
INM-1	2028	G7	1	54.56	44.17	20.24	34.02
INM-1	3529	H5	1	49.10	42.30	18.75	32.90
INM-1	3831	D25	1	50.80	42.25	17.00	32.55
INM-1	2056	G6	2	48.42	40.85	16.29	30.79
INM-1	1556	H5	2	53.79	47.97	18.71	34.69
INM-1	1557	H5	2	53.08	46.20	20.65	35.81
INM-1	1558	H5	2	53.84	46.10	19.06	32.82
INM-1	1559	H5	2	49.97	00.00	18.36	00.00
INM-1	1560	H5	2	53.41	45.71	18.62	32.98
INM-1	3384	H6	2	43.95	30.72	14.96	25.75
INM-1	6643	F24	2	51.50	44.60	17.75	31.15
INM-1	6644	F24	2	53.85	00.00	19.50	00.00
INM-1	1333	G11	2	00.00	00.00	16.90	31.09
INM-1	7009	G24	2	46.89	00.00	17.60	30.90
INM-1	2666	G7	3	00.00	00.00	17.12	28.39
INM-1	3443	H7	3	57.80	47.70	19.93	37.28
INM-1	1243	D25	3	51.92	42.59	16.86	34.27
INM-1	6558	F25	3	58.75	48.30	19.85	37.55
INM-1	6559	F25	3	55.00	50.21	21.00	38.00
INM-1	6560	F25	3	51.00	44.36	17.15	32.85
INM-1	6561	F25	3	48.10	41.30	14.90	29.55
INM-1	7843	G25	3	53.57	47.20	17.75	35.15
INM-1	7844	G25	3	55.65	45.05	18.45	36.00
INM-1	7845	G25	3	53.78	43.45	16.50	34.00
INM-1	7896	G27	3	00.00	00.00	21.55	40.67
INM-1	7922	Hno102	3	54.88	47.60	18.40	39.40
INM-1	7993	F24	4	49.65	37.25	16.20	30.15
INM-1	5955	F25	4	53.22	46.35	19.95	35.08
INM-1	5957	F25	4	48.55	41.11	17.05	32.15
INM-1	7130	G24	4	57.55	48.70	18.40	36.35
INM-1	7659	G25	4	54.50	45.60	18.15	36.25
INM-1	7660	G25	4	57.25	47.92	16.85	33.10
INM-1	6801	G26	4	51.65	00.00	16.50	00.00
INM-1	6840	G26	4	53.24	46.75	17.70	33.65
INM-1	6841	G26	4	55.75	47.95	17.50	32.00
INM-1	6844	G26	4	45.74	43.12	16.80	29.30
INM-1	1495	G4	6	59.44	00.00	19.52	00.00
INM-1	1497	G4	6	48.87	41.44	17.61	32.00
INM-1	5244	B1	11	55.65	50.30	21.37	37.80
INM-1	816	C1	12	60.34	49.28	21.83	00.00
INM-1	817	C1	12	53.60	49.14	21.70	00.00
INM-1	5370	C3	15	56.64	49.25	20.05	38.35
INM-1	5371	C3	15	55.90	46.95	18.30	34.00
INM-1	5372	C3	15	00.00	49.20	19.55	00.00
INM-1	6238	E5	15	58.60	48.80	18.70	40.26
INM-1	6027	Hno87	16	52.42	45.40	18.05	34.66

Table A.19.11. Measurements of Maxillary first/second molar of ***Boselaphus tragocamelus***

Site	Reg.No.	Trench	Layer	GL	GB
INM-1	3101	G3	2	25.07	18.47
INM-1	1517	G4	6	20.08	22.25

Table A.19.12. Measurements of Mandibular first/second molar of ***Boselaphus tragocamelus***

Site	Reg.No.	Trench	Layer	GL	GB
INM-1	2417	F25&F26	3	24.54	16.19

Table A.19.13. Measurements of Mandibular second molar of ***Boselaphus tragocamelus***

Site	Reg.No.	Trench	Layer	GL	GB
INM-1	769	I12	2	24.42	14.33

A.20. Measurements of Bones of *Axis axis*

Table A.20.1. Measurements of Humerus of *Axis*

Site	Reg.No.	Trench	Layer	BD	TD
INM-1	1190	F10	1	00.00	16.63
INM-1	1573	H5	2	36.37	16.17
INM-1	1253	D25	3	33.59	16.59
INM-1	5923	B3	14	37.50	19.50

Table A.20.2. Measurements of Radius of *Axis axis*

Site	Reg.No.	Trench	Layer	BP	TP
INM-1	5265	D4	9	37.25	19.20

Table A.20.3. Measurements of Metacarpal of *Axis axis*

Site	Reg.No.	Trench	Layer	BD	TD	WT	TT	TLT
INM-1	328	LJP	1-4	24.12	00.00	10.88	17.09	13.28

Table A.20.4. Measurements of Tibia of *Axis axis*

Site	Reg.No.	Trench	Layer	BD	TD
INM-1	171	G10	1	26.78	00.00
INM-1	351		6	37.25	27.66

Table A.20.5. Measurements of Astragalus of *Axis axis*

Site	Reg.No.	Trench	Layer	GL	GLL	GLM	TL	TM	BD
INM-1	1248	D25	3	00.00	00.00	33.31	18.84	19.28	22.51
INM-1	5864	E27	4	38.91	38.85	35.50	22.85	23.10	25.10
INM-1	5246	B1	11	37.80	37.25	34.65	20.40	20.10	22.85
INM-1	5919	B3	14	36.20	35.76	32.40	19.80	19.90	23.25
INM-1	5995	D6	14	39.40	39.27	36.22	21.68	21.90	24.60
INM-1	2502	D3	16	40.21	38.80	35.49	00.00	21.26	00.00

Table A.20.6. Measurements of Calcaneum of *Axis axis*

Site	Reg.No.	Trench	Layer	GL	GB
INM-1	4644	Hno111	4	00.00	24.92
INM-1	7078	G24	4	9.70	30.50

Table A.20.7. Measurements of Centrotarsal of *Axis axis*

Site	Reg.No.	Trench	Layer	GB
INM-1	339	LJP	1-4	32.78
INM-1	3746	G11	3	30.33

Table A.20.8. Measurements of First Phalanx of *Axis axis*

Site	Reg.No.	Trench	Layer	GL	BP	TP	BD	TD
INM-1	165	LJP	1-4	48.17	15.11	21.02	00.00	00.00
INM-1	342	LJP	1-4	43.03	13.54	17.54	12.23	11.78
INM-1	343	LJP	1-4	48.90	17.28	22.30	13.95	13.55
INM-1	367	LJP	1-4	00.00	15.12	16.93	00.00	00.00
INM-1	5862A	E25	4	40.65	12.50	15.55	11.20	10.55

Table A.20.9. Measurements of Second Phalanx of *Axis axis*

Site	Reg.No.	Trench	Layer	GL	BP	TP	BD	TD
INM-1	111	LJP	1-4	25.75	14.16	20.63	12.23	18.00
INM-1	365	LJP	1-4	29.00	11.76	17.19	9.72	14.58
INM-1	4640	Hno111	4	27.60	11.50	00.00	9.60	13.75
INM-1	1808	H5	4	00.00	13.17	18.15	00.00	00.00

Table A.20.10. Measurements of Mandibular third molar of *Axis axis*

Site	Reg.No.	Trench	Layer	GL	GB
INM-1	3616	C0	14	23.29	9.30

Table A.20.11. Measurements of Maxillary first/second molar of *Axis axis*

Site	Reg.No.	Trench	Layer	GL	GB
INM-1	3683	C6	8	24.40	13.80

Table A.20.12. Measurements of Maxillary second molar of *Axis axis*

Site	Reg.No.	Trench	Layer	GL	GB
INM-1	2171	Hno132	1	00.00	14.96

Table A.20.13. Measurements of Maxillary third molar of *Axis axis*

Site	Reg.No.	Trench	Layer	GL	GB
INM-1	2172	Hno132	1	21.52	14.96
INM-1	1733	G11	1	22.57	15.87

A.21. Measurements of Bones of *Cervus unicolor*

Table A.21.1. Measurements of Humerus of ***Cervus unicolor***

Site	Reg.No.	Trench	Layer	DD
INM-1	926	G12	3	24.17

Table A.21.2. Measurements of Ulna of ***Cervus unicolor***

Site	Reg.No.	Trench	Layer	BP
INM-1	931	G12	3	30.26

Table A.21.3. Measurements of Tibia of ***Cervus unicolor***

Site	Reg.No.	Trench	Layer	BD	TD
INM-1	2202	G10	3	46.44	36.56

Table A.21.4. Measurements of Metatarsal of ***Cervus unicolor***

Site	Reg.No.	Trench	Layer	BP	TP
INM-1	583	E6	7	42.04	40.16

Table A.21.5. Measurements of First Phalanx of ***Cervus unicolor***

Site	Reg.No.	Trench	Layer	BD	TD
INM-1	596	B1	7	14.65	00.00

Table A.21.6. Measurements of Second Phalanx of ***Cervus unicolor***

Site	Reg.No.	Trench	Layer	GL	BP	TP	BD	TD
INM-1	2027	G7	1	39.21	18.43	25.26	14.28	20.22

Table A.21.7. Measurements of Second Phalanx of ***Cervus unicolor***

Site	Reg.No.	Trench	Layer	GL	DL	BP	TP
INM-1	239	LJP	1-4	51.03	45.56	18.08	00.00
INM-1	2029	G7	1	53.86	44.84	17.31	32.94
INM-1	2273	E1	2	00.00	00.00	17.27	31.34
INM-1	3044	H7	4	43.28	35.44	13.66	25.14
INM-1	2498	D3	16	58.42	47.44	19.43	38.88

A.22. Measurements of Bones of *Muntiacus muntjak*

Table A.22.1. Measurements of Tibia of ***Muntiacus muntjak***

Site	Reg.No.	Trench	Layer	BD	TD
INM-1	5404	C3	15	22.50	16.48

Table A.22.2. Measurements of Metacarpal of ***Muntiacus muntjak***

Site	Reg.No.	Trench	Layer	DD	BP	TP
INM-1	3914	B3&C3	8	12.05	17.25	12.32

Table A.22.3. Measurements of Metatarsal of ***Muntiacus muntjak***

Site	Reg.No.	Trench	Layer	BD	TD	WT	TT	TLT
INM-1	2043	G7	1	19.11	14.55	8.54	14.17	11.14

A.23. Measurements of Bones of *Herpestes edwardsi*

Table A.23.1. Measurements of Scapula of ***Herpestes edwardsi***

Site	Reg.No.	Trench	Layer	LG	BG
INM-1	5785	Hno75	12	14.10	8.85

Table A.23.2. Measurements of Femur of ***Herpestes edwardsi***

Site	Reg.No.	Trench	Layer	GL	DD	BP	TP	BD	TD
INM-1	2365	G27	3	74.07	7.54	16.26	8.21	13.97	14.35

A.24. Measurements of Bones of *Lepus nigricoliis*

Table A.24.1. Measurements of Humerus of ***Lepus nigricoliis***

Site	Reg.No.	Trench	Layer	BD	TD
INM-1	4476	Hno129	2	10.65	5.60

Table A.24.2. Measurements of Femur of ***Lepus nigricoliis***

Site	Reg.No.	Trench	Layer	LC	TC
INM-1	1257	D25	3	10.52	9.31

Table A.24.3. Measurements of Tibia of ***Lepus nigricoliis***

Site	Reg.No.	Trench	Layer	BP	TP
INM-1	565	LJP	1-4	25.89	21.15
INM-1	566	LJP	1-4	25.89	21.15
INM-1	564		10	16.81	18.40

A.25. Measurements of Bones of *Gallus gallus*

Table A.25.1. Measurements of Tibia of ***Gallus gallus***

Site	Reg.No.	Trench	Layer	BP	TP	BD	TD
INM-1	590	LJP	1-4	17.19	12.50	10.49	00.00

A.26. Measurements of Molluscan Shell

Table A. 26.1. Measurements of *Lamellidens marginalis*

Site	Reg.No.	Trench	Layer	GL	GB
INM-1	548	LJP	1-4	57.29	39.41
INM-1	556	LJP	1-4	00.00	36.53
INM-1	555	LJP	1-4	50.59	35.54
INM-1	554	LJP	1-4	81.57	42.65
INM-1	553	LJP	1-4	40.01	00.00
INM-1	557	LJP	1-4	49.93	36.18
INM-1	558	LJP	1-4	00.00	35.75
INM-1	551	LJP	1-4	59.70	35.49
INM-1	549	LJP	1-4	00.00	36.49
INM-1	562	LJP	1-4	25.57	00.00
INM-1	561	LJP	1-4	30.09	00.00
INM-1	560	LJP	1-4	40.34	00.00
INM-1	550	LJP	1-4	63.14	42.41
INM-1	552	LJP	1-4	45.36	31.38
INM-1	3301	D10	1	00.00	28.85
INM-1	1987	H11	1	00.00	32.40
INM-1	1985	H11	1	61.22	39.52
INM-1	1983	H11	1	50.17	34.05
INM-1	1988	H11	1	00.00	23.42
INM-1	1984	H11	1	50.28	36.41
INM-1	1986	H11	1	00.00	37.21
INM-1	7544	G24	1	43.05	20.80
INM-1	2302	E1	2	00.00	38.19
INM-1	3114	E2	2	30.30	23.44
INM-1	3104	G3	2	47.89	24.19
INM-1	2112	G6	2	00.00	36.70
INM-1	2111	G6	2	00.00	34.21
INM-1	2113	G6	2	00.00	31.48
INM-1	2001	D10	2	34.35	26.67
INM-1	2003	D10	2	00.00	34.70
INM-1	2002	D10	2	45.24	31.31
INM-1	1487	D25	2	35.83	17.82
INM-1	1482	D25	2	00.00	36.57
INM-1	1483	D25	2	40.10	27.69
INM-1	1484	D25	2	35.72	26.95
INM-1	1485	D25	2	36.84	27.56
INM-1	1486	D25	2	00.00	32.13
INM-1	559	E24	2	46.83	33.80
INM-1	7370	E24	2	41.80	21.55
INM-1	6472	E25	2	36.65	18.70
INM-1	7473	E25	2	46.60	23.41
INM-1	7472	E25	2	34.45	17.27
INM-1	7523	F24	2	81.00	00.00
INM-1	6688	F24	2	41.65	19.25

INM-1	7524	F24	2	00.00	13.50
INM-1	6689	F24	2	31.96	15.30
INM-1	6687	F24	2	63.25	37.35
INM-1	7522	F24	2	00.00	50.80
INM-1	7222	G25	2	81.40	45.71
INM-1	7223	G25	2	44.50	21.65
INM-1	3022	F10	2	24.28	13.46
INM-1	3026	F10	2	78.81	38.45
INM-1	2809	H11	2	00.00	18.25
INM-1	2808	H11	2	42.13	22.71
INM-1	7911	Hno108	2	00.00	41.60
INM-1	2695	G7	3	00.00	30.04
INM-1	3512	H7	3	84.00	40.55
INM-1	1905	E10	3	00.00	35.42
INM-1	1904	E10	3	00.00	32.40
INM-1	1902	E10	3	00.00	43.18
INM-1	1901	E10	3	00.00	31.90
INM-1	1900	E10	3	00.00	40.11
INM-1	1903	E10	3	00.00	30.00
INM-1	2239	G10	3	00.00	28.96
INM-1	2236	G10	3	45.08	32.46
INM-1	2237	G10	3	46.16	32.54
INM-1	2916	G10	3	37.31	19.27
INM-1	2917	H11	3	00.00	19.34
INM-1	1796	H11	3	77.14	41.37
INM-1	1794	H11	3	50.86	35.31
INM-1	1793	H11	3	57.48	36.50
INM-1	1795	E23	3	51.51	35.17
INM-1	6780	E26	3	80.05	41.15
INM-1	6992	E26	3	34.20	17.10
INM-1	6991	F25	3	00.00	45.65
INM-1	6596	F25	3	00.00	18.75
INM-1	6597	G26	3	59.15	00.00
INM-1	7747	G26	3	32.80	16.60
INM-1	547	Hno126	3	81.65	42.60
INM-1	1874	G11	4	36.84	00.00
INM-1	8149	E24	4	00.00	30.22
INM-1	5886	E27	4	76.00	40.95
INM-1	7151	G24	4	00.00	34.05
INM-1	7150	G24	4	77.25	43.30
INM-1	5045	Hno119	4	00.00	42.85
INM-1	5044	Hno119	4	00.00	33.30
INM-1	2492	B1	5	00.00	16.79
INM-1	2493	B1	5	00.00	38.79
INM-1	2485	B1	5	00.00	39.91
INM-1	2983	B3	5	00.00	35.86
INM-1	2984	B3	5	65.87	34.93
INM-1	4125	C3	5	87.29	41.00
INM-1	1522	G4	6	62.66	39.05
INM-1	1521	G4	6	48.17	35.06
INM-1	4154	D3	7	00.00	44.80
INM-1	4153	D3	7	86.50	44.25
INM-1	1852	G4	7	28.25	20.37
INM-1	1851	G4	7	73.05	37.00
INM-1	1850	G4	7	00.00	40.23
INM-1	1849	G4	7	00.00	38.21
INM-1	4495	Hno49	8	00.00	39.20
INM-1	4496	Hno49	8	00.00	30.55
INM-1	5159	B2	11	48.24	37.20
INM-1	5110	C1	11	73.75	38.80
INM-1	5112	C1	11	36.30	19.98
INM-1	6371	Hno80	12	53.37	24.65
INM-1	5314	E3	14	00.00	51.75

INM-1	8161	A1	14	00.00	49.50
INM-1	6347	E4	14	80.10	40.00
INM-1	6346	E4	14	86.60	46.20
INM-1	6345	E4	14	00.00	47.65

Table A. 26.2. Measurements of *Digoniostoma pulchella*

Site	Reg.No.	Trench	Layer	GL	GB
INM-1	566	LJP	1-4	23.53	00.00
INM-1	5602	Hno84	16	16.00	24.60
INM-1	4155	D3	7	00.00	17.85

Table A. 26.3. Measurements of *Parreyssia corrugata*

Site	Reg.No.	Trench	Layer	GL	GB
INM-1	5294	LJP	1-4	36.72	00.00
INM-1	3298	D10	1	00.00	33.48
INM-1	3297	D10	1	52.72	36.10
INM-1	3300	D10	1	27.17	20.90
INM-1	3299	D10	1	42.45	30.05
INM-1	3296	D10	1	43.08	31.63
INM-1	2952	H10	1	28.25	22.70
INM-1	2951	H10	1	40.56	31.12
INM-1	2950	H10	1	59.78	39.00
INM-1	2949	H10	1	51.37	33.88
INM-1	7063	E23	1	00.00	38.80
INM-1	7061	E23	1	52.30	35.10
INM-1	7058	E23	1	43.25	32.57
INM-1	7062	E23	1	55.35	36.35
INM-1	7059	E23	1	43.40	32.10
INM-1	7060	E23	1	58.95	39.40
INM-1	7388	F26	1	66.26	44.85
INM-1	7389	F26	1	68.50	42.20
INM-1	7390	F26	1	43.72	29.70
INM-1	7391	F26	1	00.00	22.60
INM-1	7542	G24	1	44.05	31.95
INM-1	7540	G24	1	53.75	36.83
INM-1	7535	G24	1	52.75	36.65
INM-1	7536	G24	1	61.25	40.41
INM-1	7537	G24	1	52.85	39.07
INM-1	7541	G24	1	00.00	33.75
INM-1	7534	G24	1	56.45	37.45
INM-1	7538	G24	1	43.83	31.63
INM-1	7539	G24	1	48.45	31.98
INM-1	7543	G24	1	50.00	34.15
INM-1	2301	E1	2	00.00	35.50
INM-1	4024	G4	2	54.45	35.95
INM-1	2829	D10	2	48.56	32.22
INM-1	2828	D10	2	00.00	44.40
INM-1	2830	D10	2	00.00	38.98
INM-1	3023	F10	2	51.63	35.10
INM-1	3024	F10	2	50.20	32.92
INM-1	3025	F10	2	42.09	31.46
INM-1	2804	H11	2	00.00	41.40
INM-1	2805	H11	2	00.00	47.84
INM-1	2806	H11	2	40.83	28.03
INM-1	2807	H11	2	61.60	40.04
INM-1	2811	H11	2	52.22	34.60
INM-1	7365	E24	2	00.00	34.30
INM-1	7347	E24	2	00.00	30.65
INM-1	7344	E24	2	39.40	29.20
INM-1	7343	E24	2	51.12	33.47
INM-1	7349	E24	2	32.55	25.40

INM-1	7352	E24	2	37.47	27.85
INM-1	7353	E24	2	51.25	33.38
INM-1	7363	E24	2	37.75	27.57
INM-1	7354	E24	2	40.30	29.67
INM-1	7355	E24	2	49.00	35.40
INM-1	7356	E24	2	46.65	32.90
INM-1	7357	E24	2	43.40	30.27
INM-1	7358	E24	2	52.76	35.50
INM-1	7350	E24	2	25.97	21.55
INM-1	7351	E24	2	19.88	14.60
INM-1	7359	E24	2	50.10	33.82
INM-1	7368	E24	2	00.00	29.40
INM-1	7361	E24	2	47.20	30.55
INM-1	7366	E24	2	00.00	38.30
INM-1	7362	E24	2	39.90	27.75
INM-1	7360	E24	2	48.24	31.86
INM-1	7367	E24	2	00.00	35.90
INM-1	7369	E24	2	00.00	25.05
INM-1	7345	E24	2	51.60	36.10
INM-1	7346	E24	2	00.00	42.15
INM-1	7364	E24	2	29.95	21.53
INM-1	7348	E24	2	42.70	30.00
INM-1	6452	E25	2	46.10	32.05
INM-1	6451	E25	2	00.00	38.55
INM-1	6453	E25	2	51.50	34.40
INM-1	6461	E25	2	00.00	32.90
INM-1	6464	E25	2	00.00	40.95
INM-1	6465	E25	2	51.05	00.00
INM-1	6450	E25	2	40.10	28.85
INM-1	6473	E25	2	00.00	39.50
INM-1	6449	E25	2	00.00	30.65
INM-1	6471	E25	2	28.26	20.50
INM-1	6470	E25	2	26.56	19.08
INM-1	6454	E25	2	46.90	33.85
INM-1	6455	E25	2	53.00	38.40
INM-1	6469	E25	2	33.77	24.15
INM-1	6456	E25	2	00.00	37.10
INM-1	6457	E25	2	51.55	35.35
INM-1	6458	E25	2	62.26	41.35
INM-1	6459	E25	2	00.00	36.36
INM-1	6460	E25	2	00.00	32.65
INM-1	6466	E25	2	00.00	42.25
INM-1	6462	E25	2	54.80	39.03
INM-1	6463	E25	2	00.00	38.25
INM-1	6467	E25	2	00.00	43.80
INM-1	6468	E25	2	33.40	24.10
INM-1	7461	R25	2	44.25	33.50
INM-1	7460	E25	2	62.00	39.56
INM-1	7464	E25	2	43.26	30.20
INM-1	7467	E25	2	41.33	30.00
INM-1	7468	E25	2	45.80	30.76
INM-1	7458	E25	2	46.55	32.00
INM-1	7463	E25	2	41.35	29.90
INM-1	7462	E25	2	38.90	29.45
INM-1	7466	E25	2	29.90	21.77
INM-1	7465	E25	2	57.30	37.35
INM-1	7469	E25	2	00.00	35.45
INM-1	7471	E25	2	32.05	25.10
INM-1	7459	E25	2	46.60	32.27
INM-1	7470	E25	2	00.00	31.40
INM-1	6679	F24	2	45.80	32.50
INM-1	7515	F24	2	41.40	28.55
INM-1	7514	F24	2	28.90	21.13

INM-1	7513	F24	2	00.00	33.15
INM-1	6680	F24	2	42.85	32.80
INM-1	6676	F24	2	55.00	38.85
INM-1	6675	F24	2	54.95	39.90
INM-1	6674	F24	2	54.60	35.05
INM-1	6673	F24	2	68.80	45.45
INM-1	7517	F24	2	00.00	35.10
INM-1	6678	F24	2	00.00	36.35
INM-1	7520	F24	2	00.00	40.55
INM-1	6681	F24	2	41.55	29.20
INM-1	6682	F24	2	00.00	32.45
INM-1	7518	F24	2	41.00	29.40
INM-1	7512	F24	2	52.40	39.51
INM-1	7519	F24	2	00.00	36.10
INM-1	6684	F24	2	00.00	28.15
INM-1	7521	F24	2	00.00	36.05
INM-1	6686	F24	2	32.55	23.30
INM-1	6685	F24	2	00.00	26.08
INM-1	6683	F24	2	38.00	29.40
INM-1	7516	F24	2	48.50	33.90
INM-1	7596	F26	2	39.56	29.78
INM-1	7591	F26	2	00.00	39.75
INM-1	7592	F26	2	53.85	36.56
INM-1	7593	F26	2	45.26	31.60
INM-1	7595	F26	2	41.80	30.30
INM-1	7590	F26	2	00.00	38.22
INM-1	7597	F26	2	51.25	36.70
INM-1	7598	F26	2	30.55	23.65
INM-1	7599	F26	2	00.00	32.80
INM-1	7600	F26	2	34.10	00.00
INM-1	7601	F26	2	00.00	27.85
INM-1	7594	F26	2	49.55	35.85
INM-1	7039	G24	2	51.55	35.20
INM-1	7038	G24	2	56.82	37.50
INM-1	7202	G25	2	55.85	36.94
INM-1	7203	G25	2	40.93	28.62
INM-1	7204	G25	2	39.75	28.40
INM-1	7219	G25	2	00.00	37.00
INM-1	7220	G25	2	00.00	26.85
INM-1	7206	G25	2	00.00	22.65
INM-1	7207	G25	2	00.00	35.65
INM-1	7205	G25	2	33.47	24.35
INM-1	7221	G25	2	27.95	00.00
INM-1	7209	G25	2	00.00	33.00
INM-1	7201	G25	2	57.65	38.40
INM-1	7211	G25	2	00.00	36.05
INM-1	7212	G25	2	00.00	35.20
INM-1	7213	G25	2	55.30	41.65
INM-1	7214	G25	2	44.90	32.40
INM-1	7215	G25	2	00.00	34.49
INM-1	7200	G25	2	49.75	34.60
INM-1	7208	G25	2	00.00	35.07
INM-1	7216	G25	2	00.00	43.49
INM-1	7217	G25	2	00.00	37.55
INM-1	7210	G25	2	45.96	31.32
INM-1	7218	G25	2	00.00	35.20
INM-1	7913	Hno108	2	54.80	40.20
INM-1	2473	F25	3	41.90	29.25
INM-1	2472	F26	3	41.88	30.57
INM-1	2238	G10	3	00.00	30.96
INM-1	2915	G10	3	41.83	29.91
INM-1	6990	E26	3	28.80	23.80
INM-1	6985	E26	3	45.80	33.80

INM-1	6970	E26	3	42.70	30.55
INM-1	6987	E26	3	00.00	37.00
INM-1	6975	E26	3	00.00	38.00
INM-1	6989	E26	3	40.20	29.65
INM-1	6974	E26	3	00.00	44.45
INM-1	6984	E26	3	56.90	38.00
INM-1	6980	E26	3	27.30	20.80
INM-1	6983	E26	3	57.35	38.30
INM-1	6979	E26	3	33.05	25.50
INM-1	6977	E26	3	42.05	29.55
INM-1	6971	E26	3	43.30	29.90
INM-1	6982	E26	3	56.15	37.35
INM-1	6969	E26	3	00.00	36.35
INM-1	6978	E26	3	32.83	25.10
INM-1	6988	E26	3	00.00	31.35
INM-1	6976	E26	3	43.10	31.75
INM-1	6973	E26	3	51.30	35.75
INM-1	6981	E26	3	50.60	36.05
INM-1	6986	E26	3	37.80	28.80
INM-1	6972	E26	3	53.15	38.00
INM-1	6594	F25	3	44.65	31.30
INM-1	6593	F25	3	33.40	25.85
INM-1	6595	F25	3	00.00	34.85
INM-1	7869	G25	3	57.00	40.20
INM-1	7866	G25	3	45.35	34.15
INM-1	7867	G25	3	50.35	35.75
INM-1	7868	G25	3	00.00	24.95
INM-1	7748	G26	3	00.00	32.40
INM-1	7749	G26	3	00.00	27.88
INM-1	7899	G27	3	00.00	37.50
INM-1	7924	Hno102	3	55.75	38.85
INM-1	7925	Hno102	3	50.00	37.40
INM-1	6197	Hno106	3	42.00	30.40
INM-1	6199	Hno106	3	00.00	29.75
INM-1	6198	Hno106	3	41.70	28.85
INM-1	6195	Hno106	3	50.10	34.53
INM-1	6196	Hno106	3	41.80	31.45
INM-1	6200	Hno106	3	43.85	32.35
INM-1	4881	Hno107	3	00.00	27.55
INM-1	4880	Hno107	3	59.05	39.60
INM-1	4879	Hno107	3	35.95	26.69
INM-1	3657	Hno127	3	39.79	29.75
INM-1	8150	E24	4	42.35	33.15
INM-1	8033	E24	4	00.00	32.60
INM-1	8034	E24	4	28.85	00.00
INM-1	5882	E27	4	59.00	39.85
INM-1	5885	E27	4	00.00	23.05
INM-1	5884	E27	4	36.65	28.85
INM-1	5883	E27	4	46.15	32.10
INM-1	5989	F25	4	00.00	29.75
INM-1	5990	F25	4	00.00	33.40
INM-1	5988	F25	4	42.30	30.80
INM-1	7149	G24	4	44.70	32.65
INM-1	7148	G24	4	48.80	35.05
INM-1	7147	G24	4	50.40	33.40
INM-1	7146	G24	4	44.50	31.50
INM-1	6863	G26	4	70.80	45.90
INM-1	6866	G26	4	00.00	25.45
INM-1	6865	G26	4	51.60	36.60
INM-1	6864	G26	4	55.22	34.90
INM-1	7901	G27	4	40.30	30.95
INM-1	7777	D4&E26	4	38.95	30.19
INM-1	7778	D4&E26	4	00.00	38.30

INM-1	7779	D4&E26	4	00.00	37.15
INM-1	7776	D4&E26	4	23.25	17.88
INM-1	6116	Hno105	4	38.86	27.25
INM-1	5039	Hno119	4	00.00	32.10
INM-1	5043	Hno119	4	00.00	31.62
INM-1	5040	Hno119	4	00.00	42.85
INM-1	5041	Hno119	4	00.00	41.20
INM-1	5042	Hno119	4	35.20	27.20
INM-1	2488	B1	5	00.00	44.42
INM-1	2487	B1	5	56.73	35.14
INM-1	2486	B1	5	00.00	33.11
INM-1	2490	B1	5	54.72	38.69
INM-1	2494	B1	5	00.00	30.60
INM-1	2491	B1	5	00.00	38.98
INM-1	2489	B1	5	54.96	38.52
INM-1	4124	C3	5	38.05	29.30
INM-1	4561		7	00.00	39.18
INM-1	2479	B1	7	34.79	26.06
INM-1	2478	B1	7	00.00	27.90
INM-1	2477	B1	7	44.96	32.48
INM-1	2476	B1	7	41.45	29.47
INM-1	2475	B1	7	00.00	35.42
INM-1	2474	B1	7	44.45	28.37
INM-1	5051	B1	7	00.00	41.85
INM-1	4223	B4	7	00.00	29.50
INM-1	4222	B4	7	00.00	31.00
INM-1	4224	B4	7	00.00	34.20
INM-1	4106	A1&A2	7	29.90	16.77
INM-1	4105	A1&A2	7	22.98	17.10
INM-1	4103	A1&A2	7	57.55	30.23
INM-1	4104	A1&A2	7	00.00	25.70
INM-1	4155	D3	7	30.30	25.40
INM-1	4963	B1	8	00.00	45.00
INM-1	2471	B1	8	49.05	32.82
INM-1	2483	B1	8	32.04	00.00
INM-1	2470	B1	8	50.94	36.55
INM-1	2468	B1	8	55.63	38.49
INM-1	2469	B1	8	00.00	43.17
INM-1	2480	B1	8	28.96	21.27
INM-1	2482	B1	8	36.45	25.95
INM-1	2484	B1	8	31.37	22.50
INM-1	2481	B1	8	00.00	26.27
INM-1	5260	E2	9	00.00	39.55
INM-1	4843	E1	10	53.25	36.80
INM-1	5111	C1	11	53.35	35.35
INM-1	6307	Hno81	12	00.00	44.30
INM-1	6306	Hno81	12	00.00	37.85
INM-1	5576	D4	15	51.05	35.55
INM-1	5577	D4	15	00.00	42.90

Table A. 26.4. Measurements of *Viviparus bengalensis*

Site	Reg.No.	Trench	Layer	GL	GB
INM-1	2810	H11	12	25.55	18.25
INM-1	3698	B1	12	18.75	14.50

Table A. 26.5. Measurements of *Turbo coronatus*

Site	Reg.No.	Trench	Layer	GL	GB
INM-1	3924	B3&C3	8	41.15	38.65

Table A. 26.6. Measurements of *Meretrix meretrix*

Site	Reg.No.	Trench	Layer	GL	GB
INM-1	3918	B3&C3	8	29.75	25.60
INM-1	3920	B3&C3	8	34.85	30.25
INM-1	3919	B3&C3	8	33.95	30.35
INM-1	3921	B3&C3	8	28.00	24.35

Table A. 26.7. Measurements of *Marcia recens*

Site	Reg.No.	Trench	Layer	GL	GB
INM-1	3922	B3&C3	8	31.65	25.55

Table A. 26.8. Measurements of *Paphia gallus*

Site	Reg.No.	Trench	Layer	GL	GB
INM-1	3923	B3&C3	8	31.60	23.80

SOUTH ASIAN ARCHAEOLOGY SERIES

EDITED BY ALOK K. KANUNGO

SAA No 1. Kanungo, Alok Kumar 2004 *Glass Beads in Ancient India: An Ethnoarchaeological Approach* (*British Archaeological Reports, International Series* S1242) Oxford. ISBN 1 84171 364 3.

SAA No 2. Kanungo, Alok Kumar (Ed) 2005 *Gurudakshina: Facets of Indian Archaeology, Essays presented to Prof. V.N. Misra* (*British Archaeological Reports, International Series* S1433) Oxford. ISBN 1 84171 723 1.

SAA No 3. Swayam, S. 2006 *Invisible People: Pastoral life in Proto-Historic Gujurat* (*British Archaeological Reports, International Series* S1464) Oxford. ISBN 1 84171 732 0.

SAA No 4. Mushrif-Tripathy, Veena & Walimbe S.R. 2006 *Human Skeletal Remains from Chalcolithic Nevasa: Osteobiographic Analysis (British Archaeological Reports, International Series* S1476) Oxford. ISBN 1 84171 737 1.

SAA No 5. Jahan, Shahnaj Husne 2006 *Excavating Waves and Winds of (Ex)change: A Study of Maritime Trade in Early Bengal (British Archaeological Reports, International Series* S1533) Oxford. ISBN 1 84171 753 3.

SAA No 6. Pawankar, Seema J. 2007 *Man and Animal Relationship in Early Farming Communities of Western India, with Special Reference to Inamgaon (British Archaeological Reports, International Series* S1639) Oxford. ISBN 978 1 4073 0062 7.

9 781407 300627

www.ingramcontent.com/pod-product-compliance
Lightning Source LLC
LaVergne TN
LVHW071204210926
848482LV00015B/721